To one of the wisest visionaries I know

Bob

From Arthur
in appreciation

15 February 2021

Ludwig Wittgenstein: Dictating Philosophy

Arthur Gibson • Niamh O'Mahony

Ludwig Wittgenstein: Dictating Philosophy

To Francis Skinner – The Wittgenstein-Skinner Manuscripts

Transcribed and Edited with Introductory Chapters and Notes

Arthur Gibson
Department of Pure Mathematics
and Mathematical Statistics, Centre
for Mathematical Sciences
University of Cambridge
Cambridge, UK

Niamh O'Mahony
Cambridge, UK

This book is published as a special edition of the Nordic Wittgenstein Studies with the website of the series.

The cover uses photos of Wittgenstein's amanuensis, Francis Skinner, are virtually unknown. This hitherto unknown one from 1936, kindly retrieved by his family, needed extensive restoration for it to be publishable. We are grateful to the mathematician Dr Simone Parisotto, for achieving this outcome – so faithful to the original – amid his research at the Cambridge Mathematics Faculty and Fitzwilliam Museum.

ISBN 978-3-030-36086-3 ISBN 978-3-030-36087-0 (eBook)
https://doi.org/10.1007/978-3-030-36087-0

This Springer imprint is published by the registered company Springer Nature Switzerland AG
The registered company address is: Gewerbestrasse 11, 6330 Cham, Switzerland

Plate 1 Wittgenstein's amanuensis and friend, Francis Skinner, summer 1936, at the wedding of Francis' sister. (The effects of the metal leg brace due to polio show on Francis' outer right trouser leg. See also Appendix [Q] for Francis' sister's notes on him.) We are grateful to the Skinner family for retrieving this photograph. (Published images of Wittgenstein's amanuensis, Francis Skinner, are virtually unknown. The condition of this rare 1936 photograph of Francis Skinner needed extensive restoration for it to be publishable. We are grateful to Mathematician Dr. Simone Parisotto for achieving this so faithfully to the original, amid his research at the Cambridge Mathematics Faculty and Fitzwilliam Museum.)

For Nadia and Mark

Foreword

Arthur Gibson was the first to recognise the importance of the new discovery of the Wittgenstein-Skinner hitherto unpublished Archive. He deserves many congratulations on the publication. Its contents enable us (or rather primarily him) to reconstruct a whole phase in Wittgenstein's philosophy. His Introduction is exemplary and provides new ways forward to generate future insights. Professor Gibson affords us fresh insights into Wittgenstein's methods of composition, of thought, and particularly reveals what Wittgenstein's difficulties with mathematics were; but that this could all only (or best) be done by someone with Gibson's breadth of skills and ability to pull in a relevant insight from remote areas. This book throws so much light on a crucial period of the development of the work and the publishing plans of Wittgenstein. It is very revelatory of his attitude towards processes of creative collaboration and even friendship.

The book abounds with good qualities: this Edition's transcriptions of the original manuscripts; its profound, original yet accessible Introduction; the informative importance of its footnotes, also for their evidence signalling Wittgenstein's thought processes; the wide range of valuable, hitherto inaccessible as well as new ranges of information and insights in its Appendices – all amount to a treasury for readers, and for further understanding of the subject.

In Cambridge, Professor Gibson has, in effect, crafted his own informal research phrontisterion. His expertise and researches, which have conquered areas of the learned world sometimes in a dramatic manner, are spread across subjects and faculties. This focus reflects aspects of the importance of his own original researches in a range of subjects, which also can be brought to bear on Wittgenstein, as witnessed in the present book. Various authorities and his colleagues in diverse subjects have celebrated and highlighted aspects of this picture. Bernard Williams stated, “Arthur Gibson is a philosopher of significant international distinction – at the very least”. Considering Wittgenstein’s engagement (e.g. in Chap. 4 below) with philosophy of language, an Oxford Professor of French, Malcolm Bowie’s observation is pertinent: “Arthur Gibson is a writer of phenomenal intellectual prowess, and everything he writes demands to be taken seriously. Readers who stay with him over long distances will reap rich rewards: he is an enlarger of horizons”. His approach to Wittgenstein’s concerns with language games is informed by linguistic expertise: a Cambridge Regius Professor of Hebrew, Robert Gordon reported: “Gibson’s book *Text and Tablet* brings a logician’s insights to bear on Biblical and Near Eastern traditions and on their modern interpreters. . . this book proposes original solutions to a wide range of problems”. Other examples abound; Martin Rees stated, “Arthur Gibson’s *God and the Universe* is an erudite and impressive achievement” – relevant for the present book, with Wittgenstein engaging with Newton (Chap. 8) and James Jeans (Chap. 5), and others. In assessing how Wittgenstein’s writings in the present book inform twenty-first-century debates, it is worth reading another of the Editor’s books, which engages with such concerns as *Metaphysics and Transcendence*. Notice the view of its publisher’s (Routledge) anonymous Continental referee, stating that “This book is outstanding and of major significance. It is a distinguished and important contribution to the subject, as well as original writing of high order...a product of a mature and deep mind. It falls into that class of both surprising potential and quality that could outstrip Foucault”. Frank Kermode, who over years researched with the Editor on numerous subjects, concluded: “Arthur Gibson combines a temperament as cheerful as a good open coal fire with a mind like a pressure cooker. . . He is a Renaissance man. . .and a genius”.

This Edition is also fortunate to benefit from its other Editor, Niamh O’Mahony, not least with her expertise as a sometime Trinity College Cambridge Russianist researcher and a former writer, researcher and reviewer on the staff of the BBC Russian Service, which place her in an expert position to guide us. This is evidenced by her impressive new translations of the Pushkin poems (see Appendix [J]) – poems that Wittgenstein utilized and copied out by hand in Russian; they appear immediately after the time of Skinner’s death. Valuable is the way, for example, in which the Edition incorporates O’Mahony’s presentation of the newly discovered (Trinity) Pat Sloan Moscow letter about Wittgenstein (Appendix [R]). These draw on her published research and lecturing on the subject (cf. O’Mahony 2013). Even O’Mahony’s and Gibson’s original gathering and interpretation of apparently secondary resources – for example, the first publication of the ancestry of Francis Skinner (Appendix B), which launches and facilitates new initiatives – are of considerable value for future research.

I personally – and readers should – offer profuse thanks for this Edition, which presents Wittgenstein’s hitherto unpublished output, depicting his thoughts and cognitive processes. Congratulations to Arthur Gibson and Niamh O'Mahony on the quality of their achievement and for what impressive resources they are, in their insights and analyses, as well as to what time-consuming lengths they have gone, to provide us with this volume.

The Queen's College
Oxford, UK

University of Siena
Siena, Italy

Professor Brian McGuinness

Reference

O'Mahony, N. (2013). Russian matters for Wittgenstein. In J. Padilla Galvez & M. Gaffal (Eds.), *Doubtful certainties: Language-Games, forms of life, relativism* (pp. 149–180). Heusenstamm: Ontos, & Walter de Gruyter, Series: Aporia 7.

Preface

Purpose

The aim of this Edition is to present the contents of the hitherto unpublished Wittgenstein-Skinner Archive[1], which is singular in its embodiment of Wittgenstein's creative thought processes.

How Much Do Readers Prefer to Read?

If you wish only to read the Wittgenstein-Skinner transcriptions from the Archive, go straight to Chap. 3, onwards.

- If readers prefer only a brief outline before engaging with Chap. 3, read this Preface.
- If you have some appetite – but not a large one – to read about Wittgenstein and his context, read Sect. 1 in Chap. 1, which is the Introduction.
- If you would like to read a description of what is in the Archive, read Sect. 2 of the Introduction in Chap. 1.
- In case you are interested in the manuscripts themselves and their physical characteristics, go to Chap. 2, Sect. 1.

[1]The word "Archive" here does not imply that all the papers, which comprise it all, share uniform features. Although "Archive" is used for convenience to indicate all the papers, two points help to explain the scope of it. First, Wittgenstein posted by mail all these papers to be considered together for publication. Secondly, while the papers sometimes have different styles, they are variously interrelated by subject and theme. Wittgenstein often deployed various styles/genres to develop the same subjects in order to portray their different aspects and varying perspectives.

- It may be that some readers wish to see how this newly published Archive fits in with, and adds to, our picture of Wittgenstein's previously published writings; if so, read Chap. 2, Sect. 2.[2]
- It might be that you also wish to read about Wittgenstein's partner and amanuensis, Francis Skinner. This can be achieved by going to Chap. 2, Sect. 3.

A Partially Neglected Partnership Gives Birth to the Archive

This Archive is available, not least, due to the respective labours of and interactions between Wittgenstein and his amanuensis Francis Skinner – who was also his partner. Within days of Skinner's death in 1941, Wittgenstein selected these manuscripts and posted them off to be considered for publication. They were never published until now. So, this Edition responds to Wittgenstein's act and to his request. Since we do not know that he ever expressed himself thus with any other amanuensis, and because for some years Skinner was his partner, often living together while composing the contents of these manuscript here published, in this Archive, we have a unique means to read of aspects that echo Wittgenstein's creative thought processes.

Wittgenstein berated himself for what he considered his lack of writerly elucidation – lucid though he was in many ways, with his multiple crisscross analytical journeying and his seeming inability to complete (to his satisfaction) book manuscripts for publication. In consequence, editing an unpublished collection of Wittgenstein's expressions of thought transmitted to an amanuensis is not without hazard; one could liken it to throwing dice aboard ship in order to calm a storm. Despite Wittgenstein's self-criticism, in this book's reproduction of the Archive, we can read, for the first time, significant original insights, explorations and some creative growth in his philosophy.

The "Introduction" (Chap. 1) presents the personal and historical contexts, which include details of how the book's manuscripts came to be composed. Chapter 2's review of the manuscripts might be read as a tour of a museum, in which there are puzzles, secrets and routes to solutions. As Christopher de Hamel observed, we should "engage with manuscripts as individuals to discover what they can tell us, which can be learned from nowhere else".[3] In straightforward and complex ways, manuscripts personify their writers – as with the author Wittgenstein and his amanuensis, working to create new philosophy.

---

[2]Readers wishing to examine further research on Wittgenstein about the 1930s' background, outside the scope of the present book, will very much profit from David's Stern's (2018) Introduction and this book's wide variety of distinguished contributions by other researchers that comprise it, some of which are referred to in the present volume.

[3]de Hamel (2016: 569).

Wittgenstein on Maths and Science in the Archive

Wittgenstein's two leading biographers have positioned the importance of mathematics in and for his philosophy.

According to Brian McGuinness, "It was not for nothing that Hardy and Littlewood would discuss [Wittgenstein's] views on mathematics before the First War or that many who were to be leaders of their profession went to his classes 'for mathematicians' in the early thirties".[4] Some 20 years later in his published Cambridge Rede Lecture, Hardy also engaged with Wittgenstein's later views.[5] It is very likely that Hardy saw in Wittgenstein the prospect that Wittgenstein's research would stimulate original mathematical thinking. Hardy stated that someone "who could give a convincing account of mathematical reality would have solved very many of the most difficult problems of metaphysics. If he [Wittgenstein] could include physical reality in his account, he would have solved them all". Hardy continues, "I believe... that our function is to discover or *observe*" [mathematical reality]. So, as David Stern notes, Littlewood and Hardy wanted him to teach their mathematics students.[6]

Ray Monk epitomises Wittgenstein's picture of himself by citing his friend and collaborator Rush Rhees: "Wittgenstein's chief contribution has been in the philosophy of mathematics. This, it seems, was Wittgenstein's own judgement on himself".[7]

It assists us to position this view which Rhees reports, by attending to Wittgenstein's own context in the Archive's only use of the phrase "philosophy of mathematics". This phrase is introduced in the Archive by Wittgenstein counselling that "when we talked about the philosophy of mathematics" – "Here the use of a phrase seems to justify certain philosophic theories. I.e., are there existential theorems which construct the entities whose existence they prove? We must say that there are two things we call existential theorems: We can never predict in philosophy. We can't say 'there can't be existential propositions of a certain sort in mathematics'."[8]

May it seem odd to discover that Wittgenstein – whose overall interest is directed at other general problems of philosophy – should also preoccupy himself with mathematics? Not so. Wittgenstein aims to lift the clarity of mathematical thinking

---

[4]For example, in the Michaelmas Term 1933, the University Reporter announced that Wittgenstein would give classes on "Philosophy for Mathematicians"; this announcement appears for the Mathematics Faculty. In this issue of the Reporter, he offers to meet students; times to be arranged to suit the convenience of students, who are requested to call upon Dr. Wittgenstein at Trinity College on Friday, 13 October, at 5 pm.

[5]Hardy (1929).

[6]See Hardy (1967: 123); and Stern (2018: 14).

[7]See Ray Monk's (2007: 273). Cf. Monk's whole chapter from which the quotation came; Monk (1990: 466); Monk (1995).

[8]Quoted from Chap. 3 below (Pink Book, PART I), dated 27 November [1933].

into ordinary language in order to do original philosophy, as well as to probe the ways mathematicians slip, or need to refine their mathematical language in discussion. Wittgenstein used patterns, forms of questioning and new connections in maths as analogies for his philosophy; and even for mathematics – consider his proposals about games. This was not an attempt to push maths into philosophy. It is about new ways of thinking.

His redeployment of questioning in maths – applied to other subjects as well as maths – was original; and it was sometimes prescient, as well as mirroring to some degree how gifted original mathematicians did and do explore. How so? The Cambridge mathematician Holly Krieger responded to the following question about mathematics – "What's the cutting-edge of the unknown now?": "Often more difficult than finding mathematical answers is asking the right mathematical questions! Here is... something...that is very exciting in pure mathematics right now, called the principle of unlikely intersections".[9] Such unexpected connections are at the heart of Wittgenstein's explorations, documented in this Archive, which sometimes can contribute to solving intractable problems.[10] Wittgenstein explains how life, maths and language can be depicted as games and ones that use family resemblances as well as other models in his philosophy. He shows that such insights enable us to link hitherto unconnected games, which expose further new aspects and identities.[11] In this perspective – not least in Chap. 8 below – Wittgenstein denies the *self-evidence* of truths in certain fundamental key areas of logic and mathematics. He maintained Russell claimed that they are self-evident; Russell's statements support this as his main position, while others about memory are sensitive to degrees of self-evidence.[12] Wittgenstein argues in this book that non-self-evidence occurs in these areas of logic and maths. He also maintains that, in different ways, our capacity to have knowledge is often bewitched by our perceptions, even in some mathematics and science. Wittgenstein had an eye to the following point when arguing such matters: some of the deepest areas of maths and physics are extraordinarily weird and not self-evident, even to the specialists in those subjects.

---

[9]Cf. Krieger (2017). For further details, see Chap. 1, Sect. 1, subsection "Connecting: Wittgenstein and Langlands".

[10]See Chap. 3 below, which comprises Wittgenstein's hitherto unknown manuscript entitled "Pink Book".

[11]For example, in Chap. 3, Pink Book, PART I, Wittgenstein uses an example from a symmetry group and compares it to some mental phenomena. Behind such explorations, we find that similarities also exist between Poincaré's and Wittgenstein's views that reflect the latter's knowledge of Poincaré (cf. Rodych 2018); cp. footnote 16 below re. the use of Poincaré in Hawking (Haco et al. 2017).

[12]Russell (1912), Chap 11: "It seems, however, highly probable that two different notions are combined in 'self-evidence' as above explained; that one of them, which corresponds to the highest degree of self-evidence, is really an infallible guarantee of truth, while the other, which corresponds to all the other degrees, does not give an infallible guarantee, but only a greater or less presumption. This, however, is only a suggestion, which we cannot as yet develop further". For clarification of self-evidence here see Schroeder (2020: §3.6).

In the Archive, as elsewhere, Wittgenstein periodically engages with such scientific issues. These include not only Freud's relation to science but also James Jeans' physics;[13] and Newton's laws, which he contrasts with Russell's use of logic.[14] Wittgenstein's references to Newton bear on Martin Rees'[15] assessment of Newton and Stephen Hawking: whereas Newton discovered solutions to fundamental questions that themselves were a new scientific worldview, Hawking's achievement centres on his creating an original worldview by posing and grounding a range of original questions – ones that provoke new ways of thinking, and discovery of unrecognised connections.[16] Certainly, many scientists raise worthy questions as well as contribute to original research. Lord Rees' focus presupposes this while duly elevating Hawking's questioning, which substantially contributes to a new way of thinking about the science of the universe and its relation to humanity. (Hawking's probing of the active surfaces of black holes furnishes a route that may lead to solving his information paradox – the paradox discovered by Hawking in his earlier research in 1974.[17])

In this perspective, it is instructive to investigate a parallel between Rees' views of Hawking, here posed with Wittgenstein's way of crafting philosophy, while appreciating their differences. Wittgenstein's philosophy often amounted to breaking through our limitations by means of launching fundamentally new styles of questioning, as devices to expose new forms of grasping and excavating knowledge and understanding. Not least, by grappling to overcome our failure to recognise what is hidden in familiar knowledge. He composed or discovered fresh ranges of intertwining patterns, often non-self-evident ways of questioning. These were often centred in the varieties of original games, family resemblances and other distinctions in Wittgenstein's explorations.[18] These were not merely ornamental strategies. For example, he constructed original ways to revise logic and how to see anew its relations to some maths.[19] We find some great mathematicians admire

[13]'Wednesday, Jan. 24th'. This is not for mere novelty. It is because Wittgenstein's various alternative styles are often codes to signal different purposes, and therefore since he did, we, must preserve this form. This is partly because research on these codes is still underway, and if they are not preserved it will remove evidence for readers to learn from our book. For a discussion to expand the above outline concerning how Wittgenstein's philosophy relates to mathematical sciences, see Gibson (2013).

[14]See Chap. 8, Part VIII.

[15]Lord Rees' address delivered at Hawking's Interment Ceremony, Westminster Abbey, Friday, 15 June 2018.

[16]For example, Sachs (1962) did not see the link between his symmetry group and relevant black hole mathematics. This was exposed by Hawking, Malcolm Perry and co-authors (Haco et al. 2017, 2018). Using Poincaré, they developed a solution for black hole entropy. For Wittgenstein's view of Poincaré, see Chap. 1 (Introduction) below, Section "The Archive, Wittgenstein and Russell"; note Schroeder (2020: §4.2) on Poincaré.

[17]Strominger (2018), Sec.7:30.

[18]For this, see the Introduction below, subsection "Game to Wittgenstein: Conway and Atiyah".

[19]Wittgenstein interprets sets as games in some logic and maths. Contemporary complementary research breakthroughs have independently realised and developed profound potential in such connections; for example, Hylands and Ong (2000) and a relatable project by Hyland (2015); also Conway (1976).

Wittgenstein's polymathic abilities and his philosophy. A reason for this can be seen by the way his philosophy complements aspects of their research thinking, as Chap. 1 below explains.[20]

Signs of Wittgenstein's Thought Processes

Some manuscripts written by great writers betray almost nothing about their emergence or about the inner strivings that take place as they craft their compositions. Wittgenstein sits at the extreme opposite to this, especially regarding the nature of the Archive represented in the present book. There is evidence of his reworking of drafts. Wittgenstein's engagement with his amanuensis and partner appears to have been intense – whether by means of dictation or under direction to take notes. These were sometimes then revised under the close, obsessive attention of Wittgenstein, who sometimes inserted his handwritten refinements on the same pages where he was also instructing Skinner to rephrase expressions. Even where a manuscript seems to have no revision, there is often evidence that it is the product of such previous drafts. Wittgenstein was engaged in unrelenting battles with his own mind and his manuscripts in his attempts to achieve progress, as well as simultaneously challenging his and other notions of "progress".

The footnotes of this Edition reproduce key aspects of this engagement. They report alternative phrasing, rewriting and other indications of Wittgenstein's thought processes that appear *in* the manuscripts. Therefore, footnotes in this book are not pedantic distractions. They are literally internal to the manuscripts, in flagging authorial decisions about changes of content or tone. Skinner must have been an extraordinarily attentive and patient amanuensis. Accordingly, these manuscripts can be perceived as signalling devices to alert the reader to a range of Wittgenstein's agonised inner processes.[21] Appropriately examined, such interplay is a means to excavate or infer elements in the active mind of Wittgenstein. In the Archive, we can derive evidence of some of these mental obstetrics in select aspects of the very birth pangs of Wittgenstein's philosophy.[22] It is a fierce yet delicate process. As Wittgenstein stated when attempting to explain his own struggles to compose philosophy:

> As one can sometimes reproduce music... So sometimes the voice of a philosophical thought is so soft that the noise of spoken words is enough to drown it. (*Zettel* §453)

---

[20]M. Atiyah argued for this assessment in discussions with Gibson at the Cambridge Faculty of Mathematics (2006–2017).

[21]This use of "processes" derives from, and draws attention to, the pioneering research by Wittgenstein Archivist Jonathan Smith (2013). The subtitle of his chapter is "Reading Between the Lines", which aptly indicates that Wittgenstein's writing is more than the sum of the parts, not least when we scrutinize the conjunction between the wording of the main text and the manuscript revision work documented in the footnotes of the present book.

[22]A way of adding to our grasp of this aspect is by reading parts of his student's (Maurice Drury) writing (Drury 2017; see review by Luke Gibbons 2018).

The qualitative texture of the writing is vulnerable, and it needs sensitive handling. It is as if, at one tonal level, his dictations and remarks to his amanuensis are aimed at being the musical score that he struggles to compose in his mind, and Skinner is writing the score, as Wittgenstein performs it. It assists us to recall here Maurice Drury's (2017) report of Wittgenstein's remark, and his emphasis, that music was something very central to and deep in his life: "[Wittgenstein] told me that this he could not express [this] in his writings, and yet it was so important to him that he felt without it he was sure to be misunderstood. I will never forget the emphasis with which he quoted Schopenhauer's dictum: 'Music is a world in itself'." Occasionally, a few select members[23] in his master class were allowed to respond with a detail that Wittgenstein re-crafted to emerge in or as a note in his attempt to compose a new game – of a world in itself.

Wittgenstein lived the ancient dilemma, depicted in Plato's *Phaedrus*:[24] How can mental composing, expressed in oral performance, be expressed, without loss of quality, in written form? In the perspective of Wittgenstein's struggles, we might be forgiven for thinking that he was doomed before he began. Attempting to thwart this dilemma, he exercised deep obsessive concern for detail, with polymathic interplay implementing an inexorable quest for refinement of presentation. It was exacerbated by his uncertainty about philosophical progress, against which he battled. This struggle looks like an attempt to compose a bridge over the dilemma, a bridge he thought may not exist. Consider his glumly quoting Vischer's "a speech is no writ" and Wittgenstein adding "and a think all the less so".[25]

Certain successes battling these problems, and meeting the dilemma on his own terms, were to a considerable degree afforded by the unity and quality that appear to have been internal to his relationship with Francis Skinner – in his role as amanuensis. It would demean Skinner's role to limit him to the amanuensis' duties of scribe and secretary. Both personally and intellectually Skinner's dedication went well beyond those roles. It seems clear, even from his handwritten copies of the *Blue and Brown Book* manuscripts including typed copies[26] – with their handwriting crafting new variants in the text – that Skinner's writerly and intricate editing was deployed to mould the narrative under Wittgenstein's command. Certainly, in this situation,

---

[23]For example, John Wisdom, who along with others, is named in the Archive.

[24]Brian McGuinness (2002: 197) presents this *Phaedrus* dilemma applied to Wittgenstein, suggesting that perhaps Wittgenstein's solution to it was to write many not-quite completed alternative books. This might be borne in mind when we recognize that some of the manuscripts presented in this edition do not seem quite to finish off.

[25]We are grateful to J. C. Klagge and A. Nordmann (2003) for presenting the diary and this translation of Vischer's aphorism. For further details, see Klagge and Nordmann (2003: 35), and footnotes.

[26]Cf. J. Smith (2013).

Skinner was a partner under combat conditions, ruled by Wittgenstein's possessed and vexed creative pathways.[27] The Archive is a treasure trove, with its jewels set within new or familiar patterns. Even in the latter, due to their subtle changes, new light dawns.

Cambridge, UK — Arthur Gibson

[27] David Loner's (2018b) research opens up new lines for further enquiry in this area.

Acknowledgements

"Wittgenstein: Philosophising is an illness"

As our book enters Springer's proofs and production processes, the world is struck by the Covid-19 crisis. The Editors record their appreciation of the Springer staff's great efforts to produce this book, and maintaining the highest standards, during this fraught situation, in particular Ties Nijssen and Werner Hermens, as well as Nagarajan Paramasivam. Such work involves complex coordination and technical cooperation over a number of countries.

The work was also exacerbated by having to recognise the tricky styles and 'therapeutic solutions' in Wittgenstein's manuscripts. In transcribing and preparing these manuscripts for publication, traps lay even for the wary, because Wittgenstein excavates unfamiliar senses of the familiar. For example, he drives readers to recognise that what he terms "a criterion of death" can be pushed aside and that language has no limits – sometimes.

As the world war against Covid-19 continues, pause to ponder Wittgenstein's parallel worlds: World War I – when he, an Austrian prisoner of war in Italy, writes his first book, and emerges to the 1918 flu epidemic, and the death of his unrequited love. World War II – now as new British citizen, he is denied his wish to join the Allied Forces; his amanuensis for this book, his partner Francis Skinner, assisted by the chaos of a hospital flooded with bomb victims – dies of polio at the age of 29. Traumatised, Wittgenstein posts their manuscripts (comprising the present book) off with a prospect of publication, only for them to disappear.

The Editors are grateful for permission to publish the Wittgenstein-Skinner manuscript collection from its owner – The Mathematical Association. Here, we duly acknowledge the Association's – its Council's and its various Members' – roles in drawing the Archive to our attention, also its subsequent advice. Much appreciated is The Mathematical Association's decision in 2001 to loan the collection to

Trinity College, as the best way to bring the Archive to serious scholarly attention.[28] Accordingly, we are grateful for the assistance to its Presidents and successive Chairs of its Council, with appreciation to the Association's Librarian, Dr. Michael Price, and its Archivist, Dr. Mary Walmsley.

Permission to use Wittgenstein's intellectual property in the manuscripts comes from the Master and Fellows of Trinity College Cambridge, with appreciation to the former Wren Librarian, David McKitterick – and not least to his successor, Nicolas Bell.

We much appreciate the enthusiastic support and personal contact with the Wittgenstein family, including especially Pierre and Francoise Stoneborough, Stephan Stockert and Walter Springer.

Our gratitude extends to the descendants of Francis Skinner's extensive family – Dr. Bill Truscott, Professor Ruth Lynden-Bell, Diana Brown, Dr. Harriet Harper, David Woodbridge, Jane Woodbridge, Kate Denning, Alyson Skinner (Coney) and Marion Eastmond. They have, with other of their relations, happily, readily and variously given assistance and aided our research into Francis while we investigated aspects of Francis' biography and his work as amanuensis and mathematician. This includes helpful and patient responses to our questions and extensive, prolonged correspondence from various parts of the world. They have also furnished us with hitherto unpublished photographs.

Professor Brian McGuinness holds a central place with us, the Editors – Arthur Gibson and Niamh O'Mahony – for his advice, discussion and insights over many years and especially during this book project. He composed the Foreword for this book in the extended period of researching and preparing our book manuscript, prior to his later illness. As the book was entering its proofs' stage, we received the sad news of his death on 22 December 2019.

This project has vastly benefited from the scholarship of Archivist Jonathan Smith of Trinity College Cambridge Library. Not infrequently, this has amounted to research collaboration over our different yet complementary projects. For many years, a wide range of Fellows from various subjects in Trinity has readily given their time to engage with us in research discussion that profits this book and beyond. Imre Leader happily plays a comparable role for us in mathematical culture, over many hundreds of hours of lecturing and discussion. Research with Malcolm Perry has helped stir deep matters in applied mathematics of the universe.

Many thanks to Alois Pichler for deep insights, and for his direction of the Wittgenstein Archives at the University of Bergen; and for making the Wittgenstein Nachlass available open access online in both transcription and facsimile. For all our citations from and references to the Nachlass, the reader will be able to trace the larger context on the Wittgenstein Archives' "Nachlass transcriptions" (Wittgenstein, Pichler, 2016) and sites of Bergen Nachlass Edition (Wittgenstein, Pichler, 2015). Of invaluable help is equally the Bergen Archives and Munich Centrum für

[28]Dr. Price and Dr. Walmsley (2013: 406) report the relevant history and have been helpful in facilitating this process, further details of which are cited below in Chap. 1.

Informationsund Sprachverarbeitung cooperation on WiTTFind (http://wittfind.cis.uni-muenchen.de/) that offers free online text search of the Archives' transcriptions. We also thank Alois Pichler for the book's author index and Max Hadersbeck for assisting Alois and the Editors, very sadly noting Max's early decease in July 2020. Also for the support of Niklas Forsberg.

Joachim Schulte gave fundamentally valuable strategic and technical counsel during this enterprise, while Juliet Floyd and others have offered stimulating discussion and insight. We much appreciate the good offices of various people in Vienna, not least Ambassador Emil Brix, Radmila Schweitzer and Alfred Schmidt.

Of those who are no longer with us, and also others who are still research active, there is a special wish to mention research discussions over 30 years with some of my former teachers and later colleagues, including Elizabeth Anscombe,[29] Peter Geach,[30] Bernard Williams, John Conway, Ernst Gombrich, Karl Popper, Timothy Smiley, Arthur Danto and Jimmy Altham. In Trinity, I had cryptic helpful discussions with Piero Sraffa (while an undergraduate). This was also developed in discussions much later by one of his last students, Robert Neild. Theodore Redpath's personal engagement with Wittgenstein was helpful in discussions with me. New avenues for thinking were opened up by teaching for a few years a history Tripos Paper for Boyd Hilton, who has always been an encouraging wise source. Stimulation from Martin Rees began in my undergraduate days and continues apace. Researching over years with Frank Kermode at King's College stimulated fresh approaches to Wittgenstein and other subjects. There were significant periodic discussions with Michael Atiyah in his appreciation of Wittgenstein's research and its yet future potential; Freeman Dyson advised on a variety of technical matters. Coincident with completing this book, Stephen Hawking died, which ended some years of intermittent dialoguing, often including Malcom Perry. It is apt to recall Stephen's challenge to me to "lecture him" on Wittgenstein in the context of Stephen's pronouncement that philosophy is dead.[31] The unusual potential of French Philosopher Jérôme Letourneur was cut short by his early tragic death, though we had the opportunity to research with him and invaluably tease out obscured depths.

We are grateful to Sophia Singer for valuable discussion about her father Reuben Goodstein. In particular to the Hon. Mrs. Anne Keynes and Simon Keynes for discussions about and access to David Pinsent's unpublished manuscripts. Not least, we are also grateful to the library staff at the Trinity College Wren Library, including Bernadette Scully and James Kirwan; also to Jacky Cox for expert guidance concerning Cambridge University's official records in its library. The Editor for Philosophy at Springer, Ties Nijssen, is to be congratulated for his sustained support and efforts, often in difficult circumstances.

[29]Cf. Gibson (2019).

[30]Cf. Gibson (2015).

[31]On 23 July 2012; fulfilled 24 July 2012, in the Potter Room, Centre for Mathematical Sciences, Cambridge University.

We owe special thanks to Kelsey Gibson for helping us to check the manuscripts against the transcriptions, for critical analysis, and for a range of insights that we have incorporated into the book.

Varieties of philosophers and colleagues in other subjects have been of particular help in discussion: James Klagge has been enormously helpful by examining the typescript at an earlier stage, and over years, offering acute, significant observations. Appreciation as well to Ian Hacking, Brook Pearson, Yasemin Erden, David Stern, Nuno Venturhina, Volker Munz, Heinz Kurz, Amartya Sen, John Eatwell, Mathieu Marion, Gabriel Citron, Mark Gianni Chinca, Nicholas Boyle, Myles Burnyeat, Bernt Österman, Ajit Sinha, Volker Schlue, Thomas Wallgren, Josef Rothhapt, André Maury, Jesus Padilla-Galvez, Margit Gaffal, Mauro Luiz Engelmann, Désirée Weber and Rory Ross. It was a pleasure to be able to discuss aspects of Maurice O'Connor Drury's time with, and writings about, Wittgenstein, with his son Luke Drury; also, thanks to him and to Werner Nahm for research discussions and hospitality at The Dublin Institute for Advanced Studies.

A variety of colleagues in the Cambridge Mathematics Faculty have spared time to discuss various subjects, not least, Gary Gibbons, Tim Gowers, Martin Hyland, Mihalis Dafermos, Adrian Matthias, David Tong, Michael Green, Paul Townsend, David Skinner, John Toland, Thomas Forster, John Thompson, Jack Thorne, Tom Körner; and elsewhere, Enrico Bombieri, Roger Penrose, Endre Szemerédi, Danny Calegari, Bhargav Narayanan and Harold Andrés Helfgott, amongst others.

Thanks especially to Bob Hall; also Lily O'Mahony-Gibson for her help in reading parts of the manuscript. We appreciate the assistance of John Roger Gibson and Rod Ramsden and also the support at a later stage by a Stanhill Foundation Fellowship Award, and Ilyas Khan. We have had valued technical computer advice from Debbie Finucane and John Sutton in the Cambridge Faculty of Mathematics. Moreover, we appreciate the assistance of Kate Mole at the British Academy Archive and that of Cornell University Library.

References

Gibson, A. (2015). Peter Geach: a Few Personal Remarks. *Philosophical Investigations, 38*(1/2), 25–33.

Gibson, A. (2019). Anscombe, Cambridge, and the challenge of Wittgenstein. In J. Haldane (Ed.), *The life and philosophy of Elizabeth Anscombe (St Andrews studies in philosophy and public affairs)* (pp. 23–41).

O'Mahony, N. (2013). Russian matters for Wittgenstein. In J. Padilla Galvez & M. Gaffal (Eds.), Doubtful certainties: Language-games, forms of life, relativism (pp *Heusenstamm: Ontos, & Walter de Gruyter* (Vol. 149-180, p. 7). Aporia: Series.

Contents

List of Plates

Part I
The Emergence of the Archive

Plate 1 This is the only known, and hitherto unpublished picture of Francis' Skinner's face in close-up, from summer 1936. We are grateful to the Skinner family for retrieving the photograph. We are also grateful to Dr Simone Parisotto of Cambridge University for restoring it.

Chapter 1
Introduction

Section 1: Death, Disappearance and Discovery

Wittgenstein's Scribe

The present book publishes the manuscripts that, for good reasons, can be entitled the Wittgenstein-Skinner Archive, for reasons that will become obvious below.[1]

The Archive is written in the handwriting of Wittgenstein's amanuensis Francis Skinner.[2] The first-person pronoun throughout is presented as Wittgenstein's.[3] In support of this, various pages are also interwoven with paragraphs, sentences and revisions, in Wittgenstein's own hand.[4] These manuscripts express different ranges of style in the language of Wittgenstein's thought processes. The manuscripts appear to be the result of systematic work to achieve compositions in a variety of styles. They are neither ad hoc notes nor *aides-memoire*.

[1]The term 'Archive' will be used throughout this book to represent the manuscripts that comprise the Wittgenstein-Skinner collection.

[2]To read about Francis Skinner and his setting, see Monk's pioneering work (1990), Chapters 15–18.

[3]For reasons given in this Edition, a presupposition used throughout it is that the authorial voice is Wittgenstein's. So on the basis of this inductive situation, Wittgenstein will be assigned, and referred to, as the ultimate author of the material, while acknowledging the creative role of Francis Skinner as amanuensis.

[4]Presupposed here is the scope of Wittgenstein's uses of 'revision' (i.e., *Umarbeitung*) as investigated in Schulte's (2013); what it is to be a 'revision' is not transparent, and is complexly intertwined, requiring further research. (The present book also uses 'intertwine' to mark a notion whose sense is cumulatively built up in the following pages. For, intertwining as a mathematical concept, see this Introduction's 'Wittgenstein & Langlands'.)

A. Gibson, N. O'Mahony, *Ludwig Wittgenstein: Dictating Philosophy*,
https://doi.org/10.1007/978-3-030-36087-0_1

This creativity was partly the outcome of Wittgenstein's and Skinner's personal relationship, during a number of years when they were inseparable – as their Russian teacher, Fania Pascal, attests.[5] Wittgenstein's own handwritten revisions and additions in the Archive's manuscripts exemplify creative editorial interplay in and with their engagement.[6] This Edition publishes the first discovery of Skinner's papers, which reveal him to be more important than other 1930s' students who took down dictation or made notes of Wittgenstein's lectures. With respect to accuracy *as an amanuensis*, Skinner's work in this sphere places him alongside Wittgenstein's earlier amanuenses – such as Moore and Ramsey. We locate Skinner as unique in personal respects, amongst the network of note-takers, dictatees, and amanuenses with which Wittgenstein engaged in varying ways.

Wittgenstein's assignment of a shelf life to his assistants and aides is worthy of separate investigation. Wittgenstein seemed to need a changing entourage to sustain his processing of oral to written expression. Understanding of this impacts upon how we may measure the unique role of Skinner amongst the variety of Wittgenstein's aides in relation to their functions as causal ingredients in Wittgenstein's creative thought process, and its transmission to manuscripts. How such circumstances influence and facilitate composition is becoming clearer as researchers investigate the various actors in this network, not least regarding Waismann, who in Vienna engaged with Wittgenstein at the same period as Skinner in ways that embody some but not all in parallel with Skinner.[7]

Although Wittgenstein was inclined to hold Skinner at a distance in periods over the last few years of the latter's life, interspersed with sporadic phases of warmth, yet in the period in which the present Archive was written, the two were "inseparable".[8] Sometime after 1937 Wittgenstein appears to have accorded only decreasing episodes of close solace to Skinner, which consigned Francis at times into distant lonely orbits. In the event, tragedy was to play an endgame against Skinner and Wittgenstein.

[5]Pascal in Rhees (1984: 26, 27, 36); see McGuinness (2002: 45, 46).

[6]There are in Skinner's Archive notes on various measurements used in Russia and Spain. These may reflect Skinner's and Wittgenstein's various plans – learning Russian in preparation for emigration to Russia and to go to Spain re. the Spanish Civil War.

[7]Cf. Schulte (2011a); Waismann (1979: 11–32); Stern et al. (2013); Oakes and Pichler (2013); Venturinha (2013); Smith (2013).

[8]Pascal in Rhees (1984: 26).

The Scribe Dies

On the 3rd and 4th of October 1941, German bombers attacked multiple Royal Air Force bases near Cambridge. RAF and civilian victims of the bombings were rushed *en masse* to the hospital for infectious diseases in Cambridge.[9] Amid this traumatic insurgence, the slightly later admission of a recurrent polio patient – taken in by "Dr. Wittgenstein", was by-passed. For too many hours, Francis was left in a corridor, unseen and untreated – to the stage where retrieval of his otherwise controllable polio condition was impossible.[10] Katherine Skinner – Francis' elder sister – reported that his family members "would never forgive Dr. Wittgenstein for not telling them"[11] of Francis' impending death. Perhaps Francis himself asked Wittgenstein to hold off from contacting his family. Only shortly before Francis died, did the senior nurse come upon Francis and Wittgenstein in a corridor. She immediately phoned his family – urging them to come quickly. Francis died before they arrived. In this manner, Wittgenstein's principal amanuensis, closest friend, and also an emerging Trinity College Cambridge Mathematician – Sidney George Francis Guy Skinner – came to die, on 11 October 1941 at the age of 29.[12]

As Francis lay mortally ill on that Saturday, Wittgenstein temporarily left his bedside – hurrying to meet his friend Sraffa in Trinity College, as his Diary notes, 8.45 pm.[13] This may have been at Francis' request to put some of his affairs in order or for support. To some unknown degree, they both shared Skinner on occasions. (One of Francis' roles was that of a go-between between them, to exchange parts of manuscripts – including some pages of the Archive's version of the *Brown Book*).[14] Wittgenstein quickly returned to Francis' side, remaining there until he died the following day. As Sraffa's University Diary records, Wittgenstein immediately

[9]See Michael Bowyer (1990).

[10]Communicated to me by Professor Ruth Lynden-Bell (in a meeting at St Edmund's, Cambridge, March 2008), with information from Francis' sister Katharine (in her notes, reproduced in Appendix Q below).

[11]This new information is published for the first time in the letter reproduced in Appendix Q below; the quotation is on page 4 of Katherine Skinner's letter.

[12]*The Times* Oct. 13th 1941, Issue 49,055, p. 1, col. A, announced his death: "Gallant in a long struggle with ill-health. Funeral. The Parish church, Letchworth, 11.30 a.m. to-morrow (Tuesday)". His death certificate, for which see Appendix [M], is dated 13 October 1941, though he died on 11th – Saturday. (This is probably due to 13th being the day on which the Death Certificate was written, i.e. the first day the Registrar's office was open – Monday. The Death Certificate seems to spell 'Sydney' incorrectly, since his birth certificate and *The Times*' announcement is 'Sidney'. Francis' relation William Truscott observes that the variant spelling derives from changes of preference between Francis' father and others, rather than a registrar's error.)

[13]Sraffa, a close friend of Wittgenstein's – both having been made Trinity Fellows on the same day in 1929; Cf. Smith' (2011) study on Sraffa's life in Trinity, which sets aspects of the scene for their lengthy disputes.

[14]For example, the revised page of the *Brown Book*, identified and reproduced by McGuinness (2008: 227–28).

returned again to be with Sraffa, at 8.45 pm.[15] Such behaviour may intimate that more than appreciative courtesy is behind Wittgenstein's gesture, of giving to Sraffa the copy of the *Blue Book* that Wittgenstein dedicated to Francis.[16] This deathbed visiting pattern was familiar: Wittgenstein had also been at Frank Ramsey's deathbed, sometimes with Sraffa, a decade before.[17]

The heavy emotions presupposed in Wittgenstein's and Sraffa's meetings, would have added to the fierce intensity that usually charged their interactions. Already by early 1934 Wittgenstein had written to Sraffa: "I don't exaggerate that it gives me a tragic feeling when I see how impossible it is to make myself understood to you... I express my great respect for the Strength... of your thinking... I believe that what... makes it impossible for you to follow the way I think is a certain kind of crookedness of my thought... I feel extremely anxious that I should not lose the [great] benefit of your influence on my mind through some sort of obstinacy on my part".[18]

As the anniversary of Francis's death loomed – in October, a year later, we find Wittgenstein writing down his dream, perhaps as a memorial moment, where he was still in thrall to Francis' mental company.[19] In Wittgenstein's dream Francis was residing with him in a hostelry, alongside Michael Drobil. The sculptor Drobil had been a WWI prisoner with Wittgenstein. Probably this dream was written down at Guy's Hospital while Wittgenstein was working there, implementing his and Francis' thwarted joint medical plan to do health work in WW2.

Prior to Wittgenstein's engagement with Francis's demise, WWI had furnished him with a precedent. Shortly before the Great War, Wittgenstein also had a close personal relationship with a young man who died young – David Pinsent.[20] Both Skinner and Pinsent – a generation apart – were undergraduates in Trinity in Cambridge, both read mathematics, probed Wittgenstein with philosophical questions, and both variously helped him to produce compositions in English.[21] David

[15]This information is in Sraffa's personal diary (in Trinity College Library, Sraffa: E13), for the days Oct. 10th and Oct 11th, 1941, which entries mention the time. Wittgenstein had not, at that time, been in the habit of meeting Sraffa as frequently as this. The previous diary entry to these mentions that Wittgenstein came to meet him on Sept. 30 at 8 pm; there is no entry soon after Wittgenstein's 11th October meeting with Sraffa.

[16]The inscription in Sraffa's copy reads, followed by his initials, and Wittgenstein's with those of Francis: "Given to me by Wittgenstein after the death of Skinner. To whom it had belonged. P.S 1/5/ [year unclear] / L.to F." Sraffa added critical notes to this copy, for which see Venturinha (2012). I came upon this original handwritten copy in the Wittgenstein Collection held by Michael Nedo.

[17]Frank P. Ramsey died at the age of 26 in early 1930.

[18]This letter is dated 21.2.34. Appreciation to M. De Iaco (2019: 101) for this transcription.

[19]This dream text was discovered, and the background was retrieved by Alois Pichler (2018a).

[20]Read the pioneering research on Pinsent, related to Wittgenstein, in Brian McGuinness (2005b).

[21]See Chap. 2 below (last Section "Skinner Matters"), which resumes the "in English" theme.

Pinsent substantially assisted Ludwig to draft his only book review in English.[22] Pinsent, a descendant of David Hume, emulating yet already disputing with Wittgenstein, began composing a philosophical work in a post-*Tractatus* style, more akin to a later Wittgenstein.[23]

In contrast, Francis Skinner more completely submerged himself in Wittgenstein's identity via composition. Francis' commitment took over his life. At different periods both Francis (in the 1930s) and Pinsent (prior to WWI) lived in Whewell's Court in Trinity, in rooms near Wittgenstein's. David and Francis each separately spent time with Ludwig in Norway, variously writing under his influence, and travelled elsewhere together. Both were encouraged by Wittgenstein to work in engineering related to airplanes.[24] Pinsent died testing an experimental aircraft over Sussex just at the end of World War I.[25] This parallel with David Pinsent surely deepened Wittgenstein's distress at the death of Francis.

Wittgenstein's Literary Dedications

Wittgenstein had dedicated the *Tractatus* to Pinsent. In the 1930s Wittgenstein dedicated it, in his own hand: "To Francis from Ludwig" (Plate 1.1):[26]

Some major 1930s German manuscripts, written entirely in Wittgenstein's hand,[27] which were composed while he and Skinner were living together, contain Wittgenstein's coded dedication of the volumes to Francis – entrusting him as editor *cum* literary heir, in the event of Wittgenstein's death (Plate 1.2).

> **In the event of my death before the completion or publication of this book my notes should be published in fragmentary form under the title: *Philosophical Remarks*, with the dedication: 'To FRANCIS SKINNER' If this note is read after my death, he is to be informed of my intention at the Address: Trinity College.**

---

[22]See Pinsent's Diary (Book XVI (Feb. 4th 1913-June 27th 1913, p.17; Feb.11th 1913). Pinsent knew Hardy, Russell, Keynes, and Ogden – future co-translator of the *Tractatus* (cf. Pinsent letter Oct. 30th, 1913, p.2; Pinsent letter Oct. 18th, 1913, pp.1–2).

[23]We thank the Hon. Ann Keynes for allowing us to study this unpublished manuscript.

[24]On 30th March 1916 Pinsent arrived at the Royal Aircraft Factory [Establishment] South Farnborough to begin working on turning machines and setting tools – very similar work to that of Skinner 20 years later. In this, both were variously following Wittgenstein's pre-WW I interest in aeronautics. Pinsent's diaries attest to their discussions on the subject (as does the letter from the Superintendent of RAE Farnborough, to Pinsent's mother, announcing his death, May 14th 1918).

[25]8th May 1918.

[26]This previously unpublished detail is due to William Truscott, descended from the Skinner family, who owns Francis' copy of the *Tractatus*.

[27]Volumes X and XI „Philosophische Grammar"and „Philosophische Bemerkungen". These titles do not denote the manuscripts that have been translated into English headed by parallel English titles; see Hilmy (1987: 2, 10–11, 28–29, etc.).

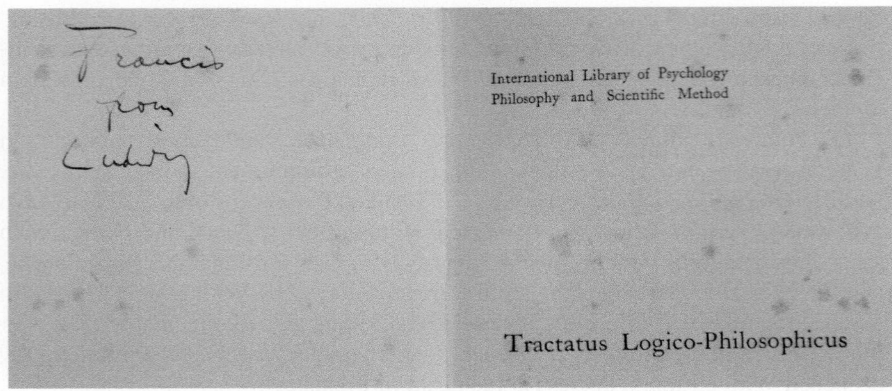

Plate 1.1 Wittgenstein's English Tractatus dedication to Skinner

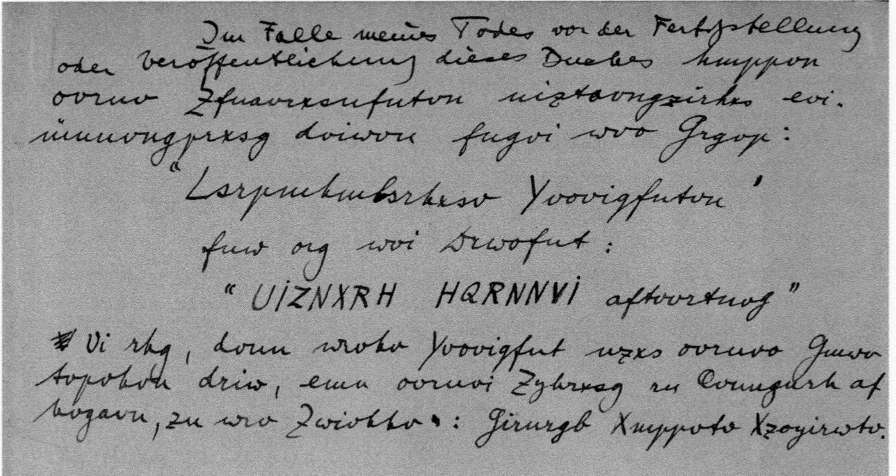
Im Falle meines Todes vor der Fertigstellung oder Veröffentlichung dieses Buches

Plate 1.2 Wittgenstein's German Philosophische Grammatik dedication to Skinner. (This inscription is from the inside cover in the Wittgensein MS.114.f001r. The caption is a translation of the inscription)

These inscriptions about Skinner were not included in any of the English Editions by Wittgenstein's literary heirs, in view of the scope of their publications, which differ from the later German Edition that published the inscriptions.[28] This absence effectively contributed to relegating Skinner to a shadow – written out of the place in Wittgenstein's archival history accorded by Wittgenstein. It is poignant, then, that when Wittgenstein posted off this Archive for safe-keeping, it also contained a brass rubbing of the "unknown medieval scribe" (illustrated below) at Amwell church in

[28]Cf. Wittgenstein: Nedo (Ed.) (1993–2000). See McGuinness (2008), Letter 294 and footnotes.

Plate 1.3 Amwell brass rubbing, 1400 AD, in the Archive. (Trinity Wren Library designation: Add. MS.a.407.rubbing)

Skinner's home vicinity – perhaps produced by both of them. Aptly, the original brass also achieved anonymity: later, it was stolen (Plate 1.3).[29]

Wittgenstein ponders death in the Archive published below, and of what a shadow would be for a tragic context. He states that, "the shadow is a picture".[30] He offers the ironic example: "I wish that Mrs. Braithwaite is dead".[31] He presents an explanation of this, which includes the question: "What relation has the shadow to what it is a shadow of?" answering it with: "The shadow is a portrait." At Francis's bedside as he died, Wittgenstein's desperate state must have been exacerbated, were he to have recalled his thought: *"it would be fair & good if he died and was thereby liberated from my 'foolishness'. But I [Wittgenstein] only half mean it"*.[32]

On 14th October at 11.30, Wittgenstein attended Skinner's funeral, traumatized by his partner's death, even though during 1936 Wittgenstein had written that he found any funeral "repellent".[33]

---

[29]The original brass artwork is that of an unknown friar, originally laid down in the chancel of the Church of St John the Bapist, Great Amwell *circa* 1400; it was reported as stolen in 1968, though one source supposes 1945. See an article on this topic that Skinner might have read: Andrews (1935); also, Pevsner (1999).

[30]Chapter 5, entry for 7th Feb. [1934]. For other relevant background see Monk (1990: 426–28).

[31]Mrs. Braithwaite *née* Masterman. See Chapter 5 below, entry for 7th Jan. [1934], MS p.9.

[32]Wittgenstein, L. (1998d). MS 120 (Entry: 4.1.1938). The translation here is due to advice from Mark Gianni Chinca. See his book Chinca (2020), which provides a new framework for the Medieval allusions and death.

[33]Wittgenstein (1935b) here embeds his aversion to funerals in remarks on mathematical analysis in the Large C Notebook MS 151 (1935b: 19). In this text Wittgenstein is considering Euler's proof, which translates some infinite series into finite ones. Note his observation concerning the reaction that "the infinite…might appeal to some of us and might, on the other hand, repel some of us. e.g. me." Wittgenstein inserts this over the line after 'me' "Funeral", which aligns his revulsion at (mathematical) infinity with attending a funeral. He researched Euler's aim to derive finite series from infinity, which may be why he encouraged Francis to study Euler.

Within 11 days of Francis' death, Wittgenstein had fulfilled an impressive compressed timetable:[34] on 14th October – the day of Skinner's funeral – Wittgenstein both attended it and wrote to the Cambridge Vice-Chancellor, informing him that he would take up work the following week as a "dispensary porter" at Guys Hospital London. His flight was an attenuated trace of their plans for joint medical careers. Within this short period, he also managed to contact Francis' father for the return of the Archive, which was sent *post-haste* from his home in Letchworth to Wittgenstein in Cambridge, then posted to Wittgenstein's former student, Reuben Goodstein[35] – also a long-time friend of Skinner's.[36]

Wittgenstein cancelled his undertaking to deliver the British Academy 'Philosophical Lecture'[37] for March 1942, notes for the beginning of which were written in his Notebook MS 166.[38] Remarkably, the date of the letter is in the week of Skinner's funeral – 20th October. Here is Wittgenstein's hitherto unpublished letter to the British Academy:

It reads:[39]

Trinity College
Cambridge
20.10.41

Dear Sir,

I very much regret to inform you that I shall be unable to deliver the Philosophical Lecture of the British Academy which I promised to deliver next March. I am leaving the University to take up some war-work which will leave me no time to write the lecture. I am sorry to have to trouble and to let you down in this way. I could, if you think that this is all right, deliver a purely oral lecture and should, of course, in the case not expect to be paid for it.

Yours sincerely
L. Wittgenstein

---

34 Cf. McGuinness 2008: Letter 294.

35 See the valuable account about Goodstein by M. Price & M. Walmsley (2013, on-line 2015).

36 Professor Ruth Lynden-Bell reports a family memory is that of these manuscripts being posted by Francis' father from Letchworth to Wittgenstein in Cambridge. (It may be that only some of them were posted, with the remainder already in Wittgenstein's possession.)

37 This title is written in Wittgenstein's pencil hand: "Notes for the 'Philosophical Lecture'." (MS 166, f.1r).

38 Note David Stern's valued analysis in Stern 1993: 445, 458, footnote 5 (in Klagge & Nordmann. Eds.). The draft of the opening of Wittgenstein's lecture is in the same ref.: pp.446–58. This together with Klagge's (2003: 363) research, in contact with A. Kenny, shows that the date of Notebook 166 centres on 1941, not 1935. (We thank Kate Mole at the British Academy, who helped us retrieve and identify this letter, as well as its date. It is in the BA's Archives, ref. Box BA 361.)

39 The original is reproduced in Appendix [C].

Despite this, Wittgenstein had attempted to draft the opening of the cancelled Lecture that was the subject of the above letter. This draft is in his notebook (MS 166). Perhaps with guilt, and cathartic intent, Wittgenstein writes in it of his experience, as if he were a solipsist in asserting:

"Nobody but I can see it. . .nobody except myself knows what it's like"[40]

Wittgenstein continues, as if inside the pain, perhaps mirroring his grief over Francis: "You can't *describe* the phenomenon of describing people feeling pain by describing their pain behaviour."[41]

In the lecture draft Wittgenstein forces his above point by wielding the role of pretence – that of acting in a Shakespearean tragedy. His aim is to show us that it is a pretence to claim understanding of another's pain in the circumstance of experiencing dying or grieving for someone dying. He affirms that there are "Different criteria for pretending in a play and out of a play". Death in a play is so different from someone "dying in reality". He asserts that "utterly dissimilar" criteria obtain, because these words are artificial ***in ordinary life***".[42] Presumably this boldly marked phrase includes really dying. Empathy lures, as if to mislead us into experiencing death in a tragedy: "are we *justified* to say that Lear dies at the end of the play?" We cannot protest: "Oh no, they don't all pretend"; "Edgar pretends to lead Gloucester to the edge of a cliff". Do we *not* experience that: "Gloucester is 'really' blind". He abruptly summarizes, without ironic use of a mathematical term:

> "The difference between:
> 'Now I know the formula'[43] and
> 'Now I can go on'.

This expression 'Now I can go on' is folded into his *Philosophical Investigations* (§151). Wittgenstein uses this sentence, "Now I can go on", of someone who suddenly, in a moment, has understood – is startled, by finding the rule to master a number series. It is not merely knowing a formula but understanding the pattern within the series.[44]

Wittgenstein's draft of the undelivered British Academy Lecture is movingly complemented by the only other writing in Notebook MS 166, in its final pages. Leaving the body of the notebook blank, Wittgenstein had turned to the back-end of the notebook and turned it upside down. He commenced writing at the end of the notebook, in Russian – the language that he and Francis, the now-deceased amanuensis, had learnt together as part of their plan to emigrate to Moscow. It seems as if he turns the understanding of pretence he has just sketched to castigate himself, as an agent in some way contributing to the death of Francis.

---

[40]In his Notebook MS 166, 31r-32v.

[41]Notebook MS 166, 26r.

[42]In bold, because Wittgenstein has added his heavy wavy underline to this phrase.

[43]Here logic is being extended: "the formula of logic" (See below Chapter X, 'Self-evidence', VIII.12).

[44]Cf. R. Williams (2014: 68–69).

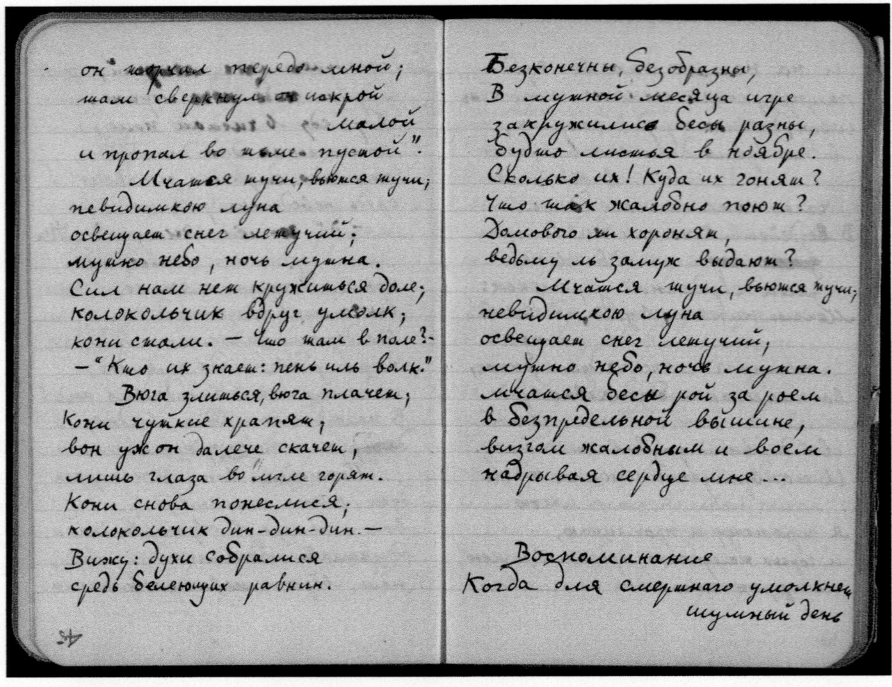

он торчал передо мной;
там сверкнул он искрой
малой
и пропал во тьме пустой".
Мчатся тучи, вьются тучи;
невидимкою луна
освещает снег летучий;
мутно небо, ночь мутна.
Сил нам нет кружиться доле;
колокольчик вдруг умолк;
кони стали. — Что там в поле?-
— "Кто их знает: пень иль волк."
Вюга злится, вюга плачет;
Кони чуткие храпят,
вон уж он далече скачет,
лишь глаза во мгле горят.
Кони снова понеслися;
колокольчик дин-дин-дин.—
Вижу: духи собралися
средь белеющих равнин.

Безконечны, безобразны,
В мутной месяца игре
закружилися бесы разны,
будто листья в ноябре.
Сколько их! Куда их гонят?
Что так жалобно поют?
Домового ли хоронят,
ведьму ль замуж выдают?
Мчатся тучи, вьются тучи;
невидимкою луна
освещает снег летучий,
мутно небо, ночь мутна.
Мчатся бесы рой за роем
в безпредельной вышине,
визгом жалобным и воем
надрывая сердце мне...

Воспоминание
Когда для смертнаго умолкнет
шумный день

Plate 1.4 A sample from the five poems Wittgenstein wrote in Notebook MS166: the last part of Pushkin's Бесы ("The Devils"), the opening of Воспоминание ("Remembrance"). All five are in Appendix [J] translated by N. O'Mahony. (Trinity Wren Library designation: Wittgenstein MS 166. f.044v-f.045r. For a German translation of the five Russian poems, see Biesenbach (2014: 413–417))

Here Wittgenstein deployed his beautiful, cursive Cyrillic handwriting to reproduce doom-filled samples of Pushkin's poetry, rather than Pushkin's more typical lyrical form. It is worth readers viewing samples of Wittgenstein's writerly craft here.[45] It is not unlikely that the smudging on the first two lines is caused by teardrops (see the poem's ending, "Fierce acrid tears stream down my face. Yet, the words I cannot wash away", of his amanuensis?), rather than from a slip of his very precise, confident Russian handwriting:

In the above Plate 1.4, second column, the poem Воспоминание (Remembrance 1828) commences. The second line continues:

> when night shadows descend on mute city squares
> . . .
> That's when the torture starts –
> . . .

[45]Plate 4, and see Appendix [J] for translations of all Pushkin poems that Wittgenstein writes in his Notebook, MS 166. See Pushkin (1954: 223–288) for poems in original Russian.

and during the sluggish night
jagged pangs of remorse rear up in me
like serpents.
My imagination goes berserk
and my head hurts from the incessant onslaught
of agonising thoughts.
Memory silently unfurls before me
its long winding scroll.
Fierce acrid tears stream down my face.
Yet, the words I cannot wash away.[46]

The congruence of this poem with the memorable Francis Skinner as amanuensis, together with Wittgenstein's later self-confessed neglect of him, has its "shadows" in this poem. Perhaps Wittgenstein recalled that "shadow" was used by himself to modulate as a tone the notion of wishing a person dead,[47] as he had wished Francis so to be – not long before the latter's demise.

Clearly these are far from Pushkin's typical Romantic love poems. Wittgenstein embraces five of Pushkin's darkest poems. The others also read as projections of himself: 'reading his life with revulsion'. In the poem entitled Бесы (The Demons), they 'howl plaintively, tearing my heart in twain'.[48] In Анчар (The Poison Tree), Wittgenstein seems to be the master who has knowingly sent his servant to collect its special fruit that kills the servant. Both poems contain the snake motif. Wittgenstein concludes with Елегия (Elegy), which is hardly positively elegiac – with such sentiments as:

My future is set, and looks bleak.
The turbulent sea of the future forebodes
only work and grief.

Here, in grief, Wittgenstein seems to present himself as a deathly language game, drawn from Pushkin's poetry to personify himself as the demoniac, poisoning, tortured prophet, who has betrayed his own vision.[49] This fits in with Wittgenstein's writing concerned as it is with themes of surprise, non-self-evidence, miscalculation, the unpredicted, the unexpected.

This set of features is also what Pushkin is variously concerned with in these and some other poems. These patterns are internal to Pushkin's poetical form: – what J. Thomas Shaw has entitled, *Pushkin's Poetics of the Unexpected*.[50] Wittgenstein no doubt recognised this vein of the unexpected in Pushkin's poetry, which to some extent mirrors aspects of Wittgenstein's own philosophical writings.

---

[46]Translated by N. O'Mahony.

[47]As noted in earlier in this Introduction Chapter 5 below, entry for 7th Jan. [1934], MS p. 9 Chapter 5 below, entry for 7th Jan. [1934], MS p. 9.

[48]See Appendix [J] for translations of all Pushkin poems that Wittgenstein wrote in his Notebook, MS 166. See Pushkin (1954: 223–288) for poems in original Russian.

[49]For further details researching Wittgenstein's relation to Russian and Skinner, as well as his visit to Russia, see N. O'Mahony (2013).

[50]Cf. Shaw (1993).

The themes well suit much of his engagement with quotations from Augustine's *Confessions*, part of which is woven into a manuscript in the Archive, which appears in Chap. 4 below.[51] While living with Skinner, Wittgenstein pondered connections between his philosophy and death, tersely penning the assertion in one of his large format lecture notebooks, of the so-called C-series: "Philosophising is an illness".[52] This seems to hold to a theoretical or autobiographical role, conjoining with Skinner's periodic attacks of polio, one of which prevented him from travelling with Wittgenstein to Russia in September 1935.

The above elements readily support the view that Wittgenstein, unable to bear the presence of the Archive that Francis as faithful servant had produced, pushed it away from himself. It is a not altogether rare response: rejecting the documents associated with the deceased person, which remind us of the absent presence of a loved one. The Archive was posted to the care of their faithful friend, Goodstein. Taken in the round, we would do well to view the manuscripts reproduced in the present Edition as a lost and found labour of Wittgenstein's love. Perhaps the pain of this intimate knowledge was so intensely held in confidence by Goodstein, which may assist us to comprehend some of the reason why Goodstein held the manuscripts in private, at no small cost to his own grief.

The Archive Disappears

Reuben Goodstein was a student at Magdalene College, and previously was a close friend of Skinner's at St Paul's School as well as at Cambridge; both studied mathematics and attended some of Wittgenstein's lectures. Goodstein graduated in the same year as Skinner, both appearing in the same Finals' Tripos list, with identical grading as Wranglers.[53] Ray Monk reports that Professor Goodstein thought Francis might have had a promising career as a professional mathematician, which he gave up due to what Goodstein judged to be Wittgenstein's unfortunate influence.[54] It was a convenient opportunity for Wittgenstein's own pursuit of

---

[51]In the period running up to his death, Skinner writes letters to Wittgenstein in which Skinner asks if he has offended Wittgenstein.

[52]Wittgenstein, L. (1935a) C-Series Notebook, MS 148, p, 57.

[53]Namely, Wrangler *b* – then, almost the highest distinction. Here '*b*' is technically *not* a grade – there is no *a*. It is used in the Tripos lists to denote those who have satisfied the moderators and examiners in schedule b of Part II of the Mathematical Tripos. *The Historical Register of the University of Cambridge Supplements* (p.83) notes that $b*$ is reserved as the highest special distinction; P. H. Wickramasinghe obtained $b*$ in the same year. (One of Skinner's three Trinity examiners for Part I was Besicovitch, on whom Sraffa and Keynes sometimes relied. Hardy was a Moderator for Skinner's Part II.)

[54]See Monk (1990: 359). In contrast, Goodstein was happy to benefit from Wittgenstein's influence on his own research, for example the uniqueness role for primitive recursive arithmetics; cf. Marion and Okada (2018).

mathematics that early on during his Manchester days he met the great Trinity mathematician J. E. Littlewood, who also taught Skinner.[55]

Goodstein had been at Magdalene reading mathematics; he also attended some of Wittgenstein's lectures.[56] Francis Skinner's niece – Professor Ruth Lynden-Bell – mentioned her family memories recalling Goodstein sometimes visiting Francis at his parents' home in Letchworth. Wittgenstein was also one of Goodstein's referees, by which he obtained his post of senior lecturer at Reading University, after working in a drapery shop in London.[57]

Some six years later, in the week following Francis' funeral, Wittgenstein wrote more than one letter to Goodstein, concerning posting off the Archive to him in Reading; and obtained permission from him to do this. To Wittgenstein's letter, arriving on 22nd October 1941, Goodstein replied:

> Dear Dr. Wittgenstein,
>
> Very many thanks for your letters. Three parcels of Francis' papers reached me on the 22nd. I am very grateful to you for sending them to me, as I very much wanted to have them. I have so far sorted out two of the parcels. These consist of
>
> Work done at School
>
> University lecture notes and worked examples
>
> One volume of rough notes on your lectures taken by Francis Skinner himself and a fair copy of these notes dated Michaelmas 1934.
>
> I need not tell you how beautifully neat, careful and thorough all the work is – a true piece of Francis himself.
>
> I suppose the dictated notes to which you refer are in the third parcel.
>
> If I find anything of Francis work sufficiently complete for publication, I shall get in touch with you about it.
>
> What Francis's family perhaps don't realise is that his chief work was his *life* and now that we have lost him the most precious thing that is left is the *memory* of that life, not something that can be dressed in words for a philosophical article. You say that they don't seem to realise what it is they have lost, but perhaps it is that *they* lost him already years ago. If I do hear from his family, which I think very unlikely, though of course I wrote to them, I shall try to explain to them that what they are attempting is the very last thing Francis would have wished.
>
> I shall let you know the next time I am going to London in the hope that we shall be able to meet.
>
> Ever yours
>
> Louis Goodstein.[58]

---

[55]Cf. Littlewood's (1986) *Miscellany*. Note Littlewood's report recommending him for a Senior Research Fellowship at Trinity (see Littlewood 1986: 138; and McGuinness 2008: Letter 133 and notes). Wittgenstein and Littlewood had the same doctor – Bevan, who himself had been a Trinity student and Olympic Gold Medalist in rowing in 1928.

[56]See Chap. 5 (i.e., Lecture for Jan. 22nd, MS p.5), where Wittgenstein entertainingly employs the expression 'I hate Goodstein'.

[57]Frank Ramsey's father, Arthur Stanley Ramsey (mathematician and President of Magdalene, Cambridge) was also Goodstein's referee for this post.

[58]I thank Brian McGuinness for showing me this letter in advance of its publication, as in McGuinness (2008), Letter 295.

There are puzzles about this letter, though not of Professor Goodstein's making. Not least is the isolation of Francis' "*life*" in contrast with (written) work. We would do well here to recall Brian McGuinness's[59] exemplary interpretation, in discussing the identity of such publishable composition: "brings us back to Wittgenstein's life, which we have been treating as key to his work." Even from a distance Wittgenstein's relationship with Skinner is an especially rich case of this interplay. Recognising Skinner's role as amanuensis in the Archive is a key to Wittgenstein's expression of a range of his compositional thought processes in the 1930s. It is a matter for future investigation, to discover how this impacts on Goodstein's contrasting Skinner's *life* as if it were a function bereft of compositional creativity or contribution to Wittgenstein's philosophical 'work'.

Our incomplete knowledge of how complete these literary remains are restricts our capability to reach understanding of some matters. Although he mentions that he does not expect to hear from Francis's family, Goodstein's own engagement with Francis's family over years was good. Professor Goodstein had periodically visited them, due to their enthusiastic invitation, to see Skinner.

We have no extant record of Wittgenstein writing to and visiting Skinner's family after his death, which contrasts with his intense letters to David Pinsent's mother and visit to her after her son David's death. This may of course have been due to the attitude of Skinner's parents to Wittgenstein, or his wish not to add to their, or his own, distress.

The Skinner family enjoyed closely-knit happy family relations, also with their extended family. (In Plate 1.5 see Francis in the foreground between his two sisters, with two USA cousins behind.) In contrast with Pinsent's family, Skinner's family recollections are that Francis's parents, especially his mother, felt that Wittgenstein purposely neglected to consider Francis's delicate vulnerable constitution, substantially and already badly compromised as it was by polio. It is said by members of his family that they knew Francis always had gay inclinations. Since Francis' parents also had academic distinction – his father, from early years had research publications with the Royal Society Proceedings,[60] it was natural for them to expect of Francis that he would publish his research done during the time Trinity College had given him a research award. And his devotion to the much older philosopher had taken him off this trajectory. At the side of this it is worth noting that in his letters Francis Skinner wrote of his appreciation of Wittgenstein encouraging him in his research and research reading, for example in studying Euler.[61]

In his letter quoted above, Goodstein was taking what he supposed, no doubt with good reason, to be Wittgenstein's side, which contrasts with a position held by Skinner's family, partly charged by what appears to have been an expectation of some publication (outlined above) such as "a philosophical article". Goodstein

[59]McGuinness (2002: 7).

[60]See some of the father's publications listed in Appendix [G], Part 4.

[61]For example, Skinner's letters of 16th August [1933], and 22nd July 1933.

Plate 1.5 The Skinner children, circa 1920. At the front from left to right: Katharine, Francis, Priscilla, with two USA cousins behind

reports: "If I find anything of Francis's work sufficiently complete for publication, I shall get in touch with you about it". The Archive contains a unique, unprecedented form of the manuscript of the *Brown Book*. It is clear that it is a fair copy deriving from Wittgenstein, with some revision in the latter's hand. It is unclear how much Goodstein knew in detail about or recollected the full identity and status of some aspects of this manuscript, not least since, for one of the years of the later *Brown Book* dictation, he was away from Cambridge, working in London. Nevertheless, let us dwell on Goodstein's beautiful compliment on the quality of the amanuensis' work by Francis: "I need not tell you how beautifully neat, careful and thorough all the work is – a true piece of Francis himself." Given Goodstein's own distinguished professional standards and adherence to the highest requirements of precision, this amounts to some strong inductive evidence to assign reliability and authenticity to Skinner's reproduction of Wittgenstein's words.[62]

There is no record of Wittgenstein's reply. Perhaps Goodstein's prose correctly presupposed that there would not be one. It seems unlikely that Wittgenstein would regard Francis as the author of the content of this Archive. The references in Goodstein's letter to Skinner's 'work' causes us to pause to investigate what this 'work' might consist of. It seems that Wittgenstein wishes to accord a sense of 'work' to Skinner, which may be publishable. It is the move from this point to how this could be done without naming Skinner as a contributory role in authorship, which is an open question.

Or, is Goodstein's letter a knowing, sincere, albeit rhetorical reply to a Wittgenstein – aware that Skinner's parents demurred from his contact with Francis. They blamed Wittgenstein for being the cause of their son's failure to publish some fruit of his short-lived life, due to Francis's intensive 'postgraduate work' for Wittgenstein?[63] It seems clear that Skinner would have made no claim to authorship, judging by his letters to Wittgenstein, even where he employed expressions such as

[62]See notes on Professor Goodstein's letter in Appendix [F].

[63]I.e., not unlike a senior mathematician who formulates a problem and the framing of a route to a conclusion, while engaging a research student who has devised a clever technique that aids or facilitates the solution of a problem to become a junior second author.

"our work".[64] It is a familiar risk: both author and amanuensis run the gauntlet of the amanuensis' perceived or actual contribution merging to become an aspect of the author. Or does the author exhaustively use the amanuensis to such a degree that the author risks slipping facets of the amanuensis' creativity to become content in the composition?

Professor Goodstein states in his Preface to his 1951 book *Constructive Formalism*:[65]

> My last word is for my dear friend Francis Skinner, who died at Cambridge in 1941, and left no other record of his work and of his great good gifts of heart and mind than lies in the recollections of those who had the good fortune to know him.

In different ways, the cores of Goodstein's letter and his dedication stand in puzzling, no doubt intentionally caring, tension in their views:

> "the most precious thing that we have left is the *memory* of that life, not something that can be dressed in words for a philosophical article"

as well as in the dedication of Goodstein's book:

> left no other record of his work and of his great good gifts of heart and mind.

Of course, the Archive is not a 'philosophical article'; and Goodstein's view "*left no other record* of his work" might require that we should read Professor Goodstein justifiably presupposing that Skinner's handwriting is evidence of his work as amanuensis, not as author.

The Archive appears to signify, by appropriate interpretation, that Wittgenstein's thought processes reveal some of Wittgenstein's compositional creativity, guided by the written records of Skinner's handwriting and revisionary work in it. Professor Goodstein was a very honourable scholar. So much of the above may be resolved or harmonised by concluding that he did not have any opportunity to pore over the manuscripts in the sort of requisite and exquisite detail as did Wittgenstein with Skinner; or, that the trusted relationship Reuben Goodstein had with Wittgenstein placed confident constraints on Goodstein about which we do not know. Given that Goodstein was working away from Cambridge for part of the time of the Archive's dictation,[66] he may not have been in a position to recognise what either the full scale

---

[64]See this letter quoted at the opening of Chap. 2, Sect. 2.3, below, which includes relevant material to propose fresh research areas for some elements pertinent to this topic.

[65]R. K. Goodstein, published in the year of Wittgenstein's death, in the Preface of his *Constructive Formalism* (Leicester: University College Leicester, 1951: 10); this dedication remains unchanged in the 1976 revision of the book). I was able to locate another similar dedication in R. L. Goodstein & E. J. F. Primrose's (1953) *Axiomatic Projective Geometry*, with the help of M. Price and M. Walmsley's (2013: 403) study.

[66]See Gibson (2010a: 351–54). We also need to be cautious about assuming Professor Goodstein was always accurate about his account of Francis. He explicitly reported to Moran's (1972) enquiry that Skinner was in good health at the time of the planned trip of Wittgenstein to Moscow. Goodstein was incorrect in reporting that Skinner was in good health and able to travel with Wittgenstein. Contrariwise, there is decisive evidence that Francis was ill in bed at the time, and

of the identity of the Archive amounts to, nor have appreciated detailed nuances – which to Professor Goodstein's professional logistic eyes might not pass muster. Furthermore, in respectful protective silence, he seems not to have shared the Archive with Wittgenstein's literary heirs; and would perhaps have been ignorant of the heirs' knowledge, or incomplete in his knowledge of the incomplete carbon copy from which the *Brown Book* was published 1958; and Reuben Goodstein might not even have agreed with the literary heirs' decision and kept the Skinner *Brown Book* to himself, since Wittgenstein did not have the *Brown Book* published, and had moved on in his view of how he should write.

Nevertheless, the Skinner copy of the *Brown Book* in the Archive is more extensive, and to our editorial eyes, more completely a finished work – as if it were ready for publication – in terms of Wittgenstein's then-current senses of progress in, say, 1935–1936. The handwritten *Brown Book* text in the Skinner Archive has revisions in Skinner's and in Wittgenstein's hands. Almost all of those sentences or expressions in Wittgenstein's handwriting in Skinner's *Brown Book* are absent from any other manuscript or edition. The physical structure of the Skinner *Brown Book* – into a series of different numbered volumes – is totally different from all other extant version of the *Brown Book*, including Rhees's. Basically, the two-part format of Rhees's edition, and the carbon copies like it, contrast with the Skinner *Brown Book's* multi-volume style. There is no extant evidence that Rhees either knew of, or read, or drew on this Wittgenstein-Skinner version of the *Brown Book*, or was aware of its lengthy extension.

Rush Rhees's 1958 publication appears to be an Edition of the *Brown Book* derived from a carbon copy that seems to be an incomplete draft – such as is Von Wright's copy, preserved in the Trinity Wren Library. For example, some margins of the carbon copy Von Wright owned have parentheses inserted for projected numeration, yet without any numbers. In contrast, the whole Skinner *Brown Book* manuscript is replete with numbering, including cross-referencing of numbered sections – within the Skinner version's hitherto unknown lengthy extension. Even so, most of those corrections in Skinner's hands in the Skinner *Brown Book*, are reproduced in the carbon copy and in Rhees's edition.

Miss Alice Ambrose was with Francis Skinner when Wittgenstein dictated much of the *Brown Book* to them, though probably not all – due to the disputes between Wittgenstein and her.[67] We have no known original extant manuscript of her work to this end, though the Von Wright carbon copy may derive from her typing industry.[68]

under medical care – expressing regret in letters that he was no able to be with his partner. We can estimate from other evidence that Goodstein judged that they were too close, for some reasons.

[67]See Appendices [B, K, N], and McGuinness's (2008) analysis of Miss Ambrose's letters.

[68]See Appendix [B]: the previously unpublished 1936 letter from Ambrose to Mrs. G. E. Moore mentions that she has just finished typing the *Brown Book* from work in Cambridge. Parts of the letter display an extreme state of condemnation of aspects of Wittgenstein, which perhaps bears on his opinion of her talent. In a previous year, she borrowed some manuscript or notes from Francis and returned them before she left Cambridge. The carbon copy in the Trinity Library is a later one made by von Wright, which may derive from the typed version made by Ambrose mentioned in the

It is possible that the presence of Skinner's handwriting corrections reproduced for the most part in the carbon copy, may derive from Miss Ambrose borrowing parts of the only partially-finished handwritten copy, not least since almost all of Wittgenstein's handwriting additions, and some of Skinner's, do not appear in the carbon copy, and so may have been added after Miss Ambrose ceased to have contact with the *Brown Book* manuscript in its written form in Skinner's and Wittgenstein's handwriting, without any of Ambrose's. As De Pellegrin (2019) shows, there is some editorial interplay with Ambrose. However, it is Skinner's full version of the *Brown Book* that Wittgenstein sends to Goodstein to consider for publication; not Ambrose's notes of it, which include her own adumbrations.

It seems evident that not all of the *Brown Book*'s dictation, or revisions, were concluded by the time Miss Ambrose left for the USA. Furthermore, Miss Ambrose seems to have broken with Wittgenstein in the latter part of dictation. Wittgenstein and Skinner continued without her. Perhaps this may help to explain why the Rhees version does not contain the different and lengthier version that is in the Skinner Archive. Given that we have examples of Wittgenstein's own handwriting in the Skinner version, yet not in the manuscript on which the Rhees version is based, we should acknowledge the priority of the Skinner Archive version. As Brian McGuinness (2012: 265) keenly observes, this extra revision work throws things back into the melting pot, or is an unraveling rather than a tying-up, since Ambrose brought her own brand of tergiversation to compete with Wittgenstein's.[69] She was certainly not, to say the least, much present at private times when Ludwig and Francis lived and worked together on revision.

So, amongst other researches, there are unfinished tasks: one, that of assessing if the *Brown Book* here is indeed a finished 'work' of Wittgenstein's (at least in shifting terms of criteria of incompletion manifested by Wittgenstein[70]). We may yet have to re-assess the *Brown Book* manuscript in the Wittgenstein-Skinner Archive. It is worth considering a facet of Wittgenstein's letter (Autumn 1935) when he sent Russell a copy of the *Blue Book*; here is the relevant part:[71]

1936 letter in Appendix [B]. This carbon copy does not contain the Skinner manuscript's extension, nor does it possess a number of new sentences written in Wittgenstein's hand, which are placed on the first three pages of the Skinner manuscript – and have no precedent in any version; neither does it have the six-volume structure that is present in the Skinner manuscript, which is displaced in this carbon copy, and the published *Blue and Brown Books*, by dividing the work into two-Parts. Even some numeration for the Remarks is absent from the carbon copy, where in their places are pairs of empty parentheses.

[69]Certainly, we should sympathise with the difficulty a woman would experience having to submit to Wittgenstein's moribund mores about women. The account by J. D. Loner (2018a) richly adds to our grasp of Miss Ambrose's subjection to him.

[70]Guided by Schulte's analysis (2005b).

[71]In this first reference to Russell in the present book, the reader may wish to read further about him, for which see Monk (1997, 2000).

Dear Russell,
Two years ago, or so, I promised to send you a manuscript of mine. Now the one I am sending you today isn't *that* manuscript. I'm still pottering about with it, and God knows whether I will ever publish it, or any of it. . . .So I'm sending you [another] one. . . .
Yours ever,
Ludwig Wittgenstein

Wittgenstein may well be alluding to either the Big Typescript[72] or the *Brown Book.* Certainly, the letter fits well with his and Skinner's continuing revision of the *Brown Book* and their correspondence about their work. (A topic I return to in Chap. 2 below).

Goodstein's subsequent 1976 "*revised*" Edition of *Constructive Formalism*, leaves the Preface about Skinner unchanged. It is possible to read his letter to Wittgenstein (above), by noting that it refers to a missing third parcel, which had yet to arrive, as containing Skinner's own work. Maybe Goodstein had received and read the contents of this parcel, and decided its contents were not suitable for publication? There is the question as to why Goodstein never published an account of the Archive nor – apparently – ever communicated them to Wittgenstein's literary heirs and editors. One way of tackling this difficulty is to pose the conclusion that Goodstein could not produce an Edition or account of the Archive that he found acceptable. This still leaves us with the problem, discussed above, of why he stated that Skinner "left no other record of his work and of his great good gifts of heart and mind". This possibly connects with the answer to the question of why Wittgenstein came to post the Archive to Goodstein. Brian McGuinness remarked to me that "it's funny that Wittgenstein was so anxious to rid himself of this material so soon after Skinner's death".[73] Perhaps Goodstein believed that the Archive did not meet the high standard set by Wittgenstein. Or, it may be that the practicalities of Goodstein's life and its demands left him unable to realise either the decision to act or the resources to fulfil what he hoped to do.

Discovering the Wittgenstein-Skinner Archive

It was mentioned above in the Acknowledgements that the Archive was identified by officers in The Mathematical Association amongst gifts that its former President Professor Goodstein had donated to the Association. Its Council decided in 2001 that a loan to Trinity College was the best way to bring the Archive's manuscripts to serious scholarly attention.[74] As its officers Michael Price and Mary Walmsley helpfully report: "curiously, [Professor Goodstein] appears to have done nothing with them.[75]

[72]See also, McGuinness (2008), p. 25's footnote.

[73]Email dated 27.1.2013.

[74]The first box is recorded as received by Trinity in April 2002, and the second on 24 May 2002.

[75]Drs. Price and Walmsley (2013: 406) continue their report with helpful details: "However, in the mid-1970s, when [Professor `Goodstein] started to clear his office in preparation for his retirement in 1977, he chose to donate to the MA Library not only some of his library books but also his own typescript copy of *The Blue Book* and Skinner's manuscript notes of Wittgenstein's lectures and dictations, 1931–1935. For personal reasons, Goodstein kept these papers throughout his time both

Later, Brian McGuinness saw parts of the Archive very briefly when he was attending to another manuscript project. His view was that they were of some importance. Subsequently the present editor was invited to examine the Archive.

Understanding the disappearance of the Archive is to a certain degree bound up with the presuppositions, sometimes invisibly tied to Goodstein's silence. The Archive appears largely to have been kept as a private matter by Goodstein throughout his life. According to his daughter, he would never have taken the Archive home in case his wife saw it, since it would risk her destroying anything associated with Wittgenstein. Goodstein's silence might amount to a dilemma. He might have genuinely thought the MSS were not worthy of publication. *If* this were the case, and since the Archive contains the *Brown Book*, Goodstein might have disagreed with Rhees's and others' decision to publish it, which may then have further entrenched Goodstein to remain silent about the Archive. A number of people who knew him, said, at least where Wittgenstein was concerned, he always modestly played his cards close to his chest. That be as it may: we can be grateful that Goodstein carefully preserved the Archive.

The foregoing can give pause to reflect on some of Goodstein's assumptions or premises that may affect our capacity to assess the identities of manuscripts that comprise the Archive. We should sympathise with the predicament in which Goodstein was placed by being confronted with being the custodian of this complex array of manuscripts.

A nun, Sister Mary Elwyn, wrote to Professor Goodstein in 1966. His hitherto unpublished reply reflects some diverse currents, reproduced in full below:[76]

> 13:9:1966
> Dear Sister Mary Elwyn,
>
> May I start by apologising for the very long delay in replying to your letter dated August 1st.

---

at Reading and at Leicester but, Skinner's notes… were kept safely in storage with other unpublished material, manuscripts and offprints, which were not suitable for addition to the MA's book collection. They were not catalogued and remained as a neglected Wittgenstein source until the twenty-first century. In 2000, these items attracted the attentions of an antiquarian book dealer who was inspecting both the MA Library's open shelves and its storage section. The potential importance of these items for Wittgenstein scholarship was now recognised. Accordingly, the materials were moved to the University Library's Special Collections unit and further professional advice was obtained, all of which confirmed the potential significance of this Archive for Wittgenstein studies. In this light, the MA Council decided to move the Archive in 2002 to a British centre for Wittgenstein scholarship, namely Trinity College, Cambridge, where the Archive itself was created. The Archive is on loan, renewable annually." It is worth noting that at some point after the manuscripts were deposited with the M.A., Prof. Goodstein had a stroke, which presumably could have affected his capacity to implement any intention that he might conceivably have wished to implement, though of course the absence of previously publishing anything of or on the manuscripts could be taken as evidence there was not such intention.

[76] Goodstein's letter is held in the Cornell University Library. I am very grateful to Professor Klagge for discovering it, also for unsolicitedly informing me and sending a copy.

I have found it very difficult to decide what reply to make since I have been contemplating writing an account of Wittgenstein's work with a brief biographical note, for many months. However it is becoming increasingly clear to me that I shall not be able to start for several years, and perhaps may never succeed in the project.

I knew Wittgenstein from 1931 to 1942 and still received an occasional letter from him after that[.] I have several of his letters but they are purely personal and have no philosophical content. I value them simply because he wrote them. Of those who attended his lectures 1931–1935 I must name above all my dear friend Francis Skinner who died in Cambridge early in the war and who was Wittgenstein's closest friend for the years 1932–1940. I have the notes which Skinner took of the lectures but most of the material has appeared in the *Blue and Brown Books*.

Apart from him you mention[77] Wittgenstein was certainly influenced by the two discussions he had with G Frege, and also by L E J Brouwer.

He strongly disagreed with both of these on many points but they certainly stimulated him. Wittgenstein first went to see Frege when he decided to give up Engineering and become a philosopher.

Of others who attended the lectures I may mention Denis Lloyd (now Baron Lloyd of Hampstead) University College, London; Professor T. W. Hutchinson University of Birmingham; Mrs. Braithwaite (wife of Professor Braithwaite, Cambridge). I will send you Skinner's notebooks when I have sorted them out.

With good wishes,
Yours sincerely
R L Goodstein

There is no evidence that Professor Goodstein sent the notebooks to Sr. Mary Elwyn, though she wrote two neglected works involving limited research into Wittgenstein.[78] The Skinner-Wittgenstein manuscripts appear to have remained throughout many years privately in Goodstein's care at Leicester. Probably, for quite understandable reasons, he was having great, and perhaps, life-long, difficulty tackling his own challenge to write an account of Wittgenstein's work, which by his reference to the Skinner material would include the present Archive. Certainly, his statement "I have the notes which Skinner took of the lectures but most of the material has appeared in the *Blue and Brown Books*" does *very roughly* cover *only some* of the Archive; much new material in the Archive is not covered by this letter, however. For example, about 60 pages at the end of the Archive's *Brown Book* (now comprising Chap. 4 below) are not in the published *Brown Book*.[79] The expression "most of the material" hardly faces up to the precision by, and criteria of identity with, which Wittgenstein regarded even slightly different readings sometimes as

[77]Russell's name, which is written before 'Wittgenstein', is deleted.

[78]Sister Mary Elwyn McHale produced two works on Wittgenstein (McHale 1966, 1979). She died in 2011.

[79]This does not require that Goodstein was mistaken, for he mentioned that he had not sorted the third parcel Wittgenstein sent. Conversely, if he had seen the *Brown Book* he might not realize it had a vastly longer ending (i.e. in Chap. 4 below) than the (later published as the Rhees version), and a somewhat different beginning in Wittgenstein's own handwriting (see Appendix N).

crucial for epitomising differing senses. Certainly, for sound reasons Goodstein may not have wished to convey any more than he did in the above letter. He does not appear to have left a fuller account from which we could benefit. There is no extant evidence that he sent the Archive to Wittgenstein's literary heirs, nor made them aware of its existence.

Goodstein's insight is valuable here, not least his notice, in his letter cited above, that Wittgenstein was influenced by, and yet disagreed with, Brouwer. Michael Atiyah proposed that Wittgenstein offers the basis for a new approach to constructing a logic of – and from – mathematics.[80] Atiyah thought it futile to try and construct a basis for logic from formal logic and set theory. He cited some of Brouwer's achievements, (and with which nevertheless to some degree Wittgenstein disagreed), which he thought stimulated Wittgenstein to go beyond Brouwer to explore routes to achieve a new logic, or basis for it.

Outline of the Archive's Contents and Dates

This present Edition publishes the first part of these hitherto unknown and unpublished manuscripts, which here are designated as the Wittgenstein-Skinner Archive. The earliest date they might start to cover is November 1932, and the latest is October 1941. Its compositions appear in the main to be completed by early 1936. It is possible that some revision was added up until 1941 (Plate 1.6).

All the manuscripts in this volume are written in Skinner's hand but presented as composed in the first person of Wittgenstein, with Wittgenstein's handwriting inserted at various points, as redrafting, revision and like matters.[81] Skinner's work is precise, clear and well crafted. It seems evident that much of the Archive is the outcome of considerable redrafting and revision under Wittgenstein's instruction.

Here is an outline of the Archive's contents, prior to more detailed presentation below (reference to their appearance as chapters in this Edition is included in parentheses):

The Pink Book (Chapter 3)
"Communication of Personal Experience" – the previously unpublished volumes of the Brown Book (Chapter 4)
"Philosophy" (Chapter 5)
"Visual Image in his Brain" (Chapter 6)
"The Norwegian Notebook" (Chapter 7)
"Self-evidence and Logic" (Chapter 8)

---

[80] In discussion with Arthur Gibson, 19 May 2016, Potter Room, Cambridge Mathematics Faculty.

[81] Of course, taking notes of a lecture is different from dictation. Cf. Hilmy's (1987: 282). Sometimes it is not entirely clear how these two different forms of life relate in detail. For example, when Wittgenstein was developing material from the large manuscript series C145–151, it is possible for notes to be dictated with revisions as dictations, and *vice versa* (cf. Floyd 2016a).

Plate 1.6 The Wittgenstein-Skinner Archive

"A Mathematical Investigation" (Chapter 9)

It is helpful to see how these manuscripts fit into the overall picture of the *Nachlass*. Insofar as this picture can be conveyed by a catalogue plan, the chart below of the *Nachlass* places the new Archive on the far right listed in its dated contexts, marked in green. (The manuscript items marked C in blue colour signify the large format notebooks used by Wittgenstein in lecture preparation and for noting explorations, which are contemporary with the Archive).[82]

Wittgenstein's Manuscripts Composed Between 1929 and 1942

The Archive's Contents

Here is an outline of the contents and length of each manuscript. The numeration used in the list below is that of the chapter sequence in which the manuscripts appear:

A Pink Book Inside the exercise book covers, this is marked as consisting of Book I and Book II (composed of 14,200 words, as well as many visual illustrations). It appears to be a fair copy, with revisions and the occasional paragraph added by Wittgenstein.

"Communication of Personal Experience" (At 12,000 words, this comprises the hitherto unknown extension to the *Brown Book*, in the form of a fair copy.) It

[82]View the C manuscripts in the 2nd Bergen Edition of the Wittgenstein *Nachlass*.

NOTEBOOKS	C volumes	Other MSS	Typescripts	Dictations	Wittgenstein-Skinner Archive
105 1929					
106 1929					
107 1929–30					
108 1929–30			208 1930		
			209 1930		
			210 1930–30/31		
109 1930–31		153a 1931			
109 1930–31		153b 1931–32			
111 1931		154 1932			
121 1931		155 1931			
113 1931–32					
114 1932–33	145 1933 (C1)		211 1931–31/32		
			212 1932–33		
115 1933; 1936	146 1933–33/34 (C2)	156a 1932–33	213 1933	309 1933–34	SKINNER I: ‘*PINK BOOK*’
		156b 1933/34–34			
	147 1934 (C3)	157a 1934; 1937			SKINNER III: ‘PHILOSOPHY – COURSE OF LECTURES’
		140 1933/34; 1936 (last folio)			SKINNER IV: ‘VISUAL IMAGE IN HIS BRAIN’
					SKINNER II: ‘COMMUNICATION OF PERSONAL EXPERIENCE’
	148 1934/35–35 (C4)			310 1934–35	SKINNER V: ‘THE NORWEGIAN NOTEBOOK’
					SKINNER VI: ‘SELF-EVIDENCE AND LOGIC’

(continued)

NOTEBOOKS	C volumes	Other MSS	Typescripts	Dictations	Wittgenstein-Skinner Archive
	149 1935/36–36 (C5)	141 1933/34			
	150 1934–34/35 (C6)				SKINNER VII: 'NEW *BROWN BOOK*'
					SKINNER VIII: 'MATHEMATICAL
					INVESTIGATION'
	151 1935/36–36 (C7)				GOODSTEIN'S: TWO VOLUME COPY OF THE CYCLOSTYLED BLUE BOOK
	152 1936 (C8)				
116 1937–38; 1945					
117 1937–38; 1940		157b 1937	2201 1937		
118 1937					
119 1937					
120 1937 –38		158 1938	221 1938		
			222 1938		
121 1938–39		159 1937–38	223 1938		
		160 1938	224 1938		
			225 1938		
122 1939–40		161 1939; 1941	226 1938–38/39		
		162a 1938–39			
		162b 1939–40			
		166 1941–42 (1935?)			

contains finely nuanced revision of details, some in Wittgenstein's hand, with numerical cross-referencing to the Archive *Brown Book*, for which see the previous heading (*'Handwritten Skinner Brown Book'*) below in this current manuscript list.

Philosophy (This manuscript is 20,352 words long, replete with lecture dates, the first of which is stated to be "Wedn.[83] Jan 17th" [1934]). It is a series of carefully crafted notes with continuous arguments and strategies that do not correspond to any published narrative.

"Visual Image in his Brain" (Comprising 3600 words, it was probably a private dictation to Skinner). Refined remarks in lecture note form.

Lectures on Self-Evidence and Logic (20,544 words). Detailed lecture notes, with evidence of revision prior to its final form. It comprises one term's lectures, with evidence that the manuscript has been crafted and re-shaped in the direction of becoming a unified manuscript. Although it returns to the matter of self-evidence in the *Tractatus* and is concerned to challenge Russell's views on logic and pure mathematics, yet it is not a repeat of earlier views. Rather, it develops an explanation about denial of self-evidence.

Norwegian Notebook (4400 words). Draft form. This was perhaps dictated to Skinner on his 1937 visit to Norway. The mention of Gasking in the text might require that this is to be dated to 1938, unless it is a later addition in a redraft.

A Mathematical Investigation (It comprises 12,353 mathematical symbols—without any narrative.) This manuscript is entirely constituted of calculations. It does not have Wittgenstein's handwriting in it, and it appears to be Skinner's own work, though perhaps with Wittgenstein's influence. Given that it is a component in an archive that Wittgenstein himself gathered together as an expression of his and Skinner's work, we allow space for it to be aired as token of this collaboration. It explores matters involving Fermat's Little Theorem.

Goodstein's Two-Volume Cyclostyled Copy of the Blue Book This is the only typescript in the Archive. It appears to have been placed in the Skinner Archive by Professor Goodstein and this is noted in Goodstein's handwriting by him for The Mathematical Association.[84]

Handwritten Skinner Brown Book The Archive includes an hitherto unknown handwritten full version of the *Brown Book*. It is more extensive, in a variety of ways, than any other copy of the *Brown Book*. Wittgenstein's own handwriting in the Archive's version is without precedent in Rhees's Edition of the *Brown Book*, published in *The Blue and the Brown Books*. The newly discovered version in the

[83]In the MSS the writing of date abbreviations such as "Wedn" varies, and this Edition preserves this in case it emerges that they are evidence for some purpose.

[84]For a fine account of how this copy complements and fits in with other extant copies of the *Blue Book*, see Smith (2013: 37–51). For details on the unpublished incomplete hand-written copy by Francis Skinner of the *Blue Book*, see Chap. 2, Part Two.

Archive has obviously been worked on in great detail personally both by Wittgenstein and Skinner. The Archive version is longer and it adds distinctive refinements to what appears to be a fair copy.[85]

In 1941 Wittgenstein himself gathered together all the Archive's manuscripts together with the *Brown Book*, which adds to our means of apprehending the Archive as, in some senses, a manuscript collection of partially-shared properties. The Archive is in Skinner's handwriting, sometimes with Wittgenstein's, expressing the results of their joint enterprise. The selections of manuscripts that Wittgenstein chose to send to Professor Goodstein seem to form a sort of uneven unity, amid their variety, if viewed from within the time period of their composition. Formulating what this unity comprises, is a separate task.

There is no proof that Wittgenstein posted off all the material that Skinner had a hand in producing; nor is it clear that we have all in the Archive that Wittgenstein sent to Reuben Goodstein.[86] Even so, under the reasonable assumption that Wittgenstein knew of or possessed such writings, it is plausible to conclude that the Wittgenstein-Skinner Archive, to some degree, reflects Wittgenstein's choice and taste in shaping a somewhat unified group of his creative compositions to reflect Francis's contribution. We cannot be certain if Goodstein removed some papers, though we should esteem his wish to follow Wittgenstein's wishes regarding the Archive. What remains in the Archive of Wittgenstein's philosophical art, intertwined with Francis' craft, is a memorial of his friend who offered requited love in helping Wittgenstein compose a stage in his philosophical progress – vexed though this was for him.

Does the Archive Reproduce Wittgenstein's Compositional Activity?

The answer is Yes. Questions abound, however; the analysis of which will occupy readers for some time to come. For example: is there new totally unpublished material here? There is. Are there clear indications of the relation of the material to already published works? The answer is 'Yes'. Are there indications that writing, language and style of the Archive are those of Wittgenstein? Certainly; and there is a wide range of direct and evidence for this claim. Does the Archive signify original thought by Wittgenstein? Definitely.

(Some of the Chapter below, such as the 'Pink Book', were completely unknown before the publication of the present book. By contrast, certain details in parts of other chapters may seem only to re-use familiar expressions and ideas. Let us be

---

[85]See, for example, Remark 28, which has some subtle adjustments in phrasing.

[86]For example, see the handwritten incomplete draft of the *Blue Book* in Skinner's handwriting, which differs in some paragraphs from its published form, and other related writing.

wary of such assumptions. We should not assume in such contexts that Wittgenstein is dictating a mere repeat, or trotting out the same stuff. This is for at least two reasons. First, he sometimes communicates new thought by 'familiar patterns'. Not infrequently, this is to show that the familiar is a resource to violate itself – by Wittgenstein teasing out new strands of originality within it: the shock of the new under the seeming disguise of the old. Secondly, he refines new edges further, which already contain fine distinctions that he had previously crafted, to add a fresh aspect of a new identity to the familiar, as noted below in '*Game to Wittgenstein: Conway & Atiyah*'.)

Insofar as is practical within the scope of this Edition of the manuscripts, the above and further questions will be addressed. As Schulte has explained (2005b: 361–63), what counts as a 'work of Wittgenstein' is a complex matter, which involves distinguishing one of his 'works' from a composition that is a record of Wittgenstein's developing thought-processes.

The present Archive comprises a varying qualitative continuum that manifests Wittgenstein's creative thought to different creative degrees. Let us heed Schulte's (2005a: 360) two preconditions for contributing towards becoming a work of Wittgenstein: "first, at least when it came to revising his earliest manuscript versions Wittgenstein must have had fairly clear, complex and changing notions of his overall project in mind; and, second, that an enormous amount of work must have gone into rearranging his material in accordance what his latest conception of what he as trying to achieve." Furthermore, the notions of unity and completion are somewhat vexed as measures of what it is to be a work of Wittgenstein, not least due to his own shifting, and fast-developing conceptions of them and his philosophy, together with his changeability about whether or not he is writing a "book" or "books".[87]

Another reason for introducing the use of "unity" in the foregoing paragraph is to prepare the way for launching the idea that there is some form of unity within and between the multifarious identities that comprise the Archive's manuscripts.[88] This is illustrated for example by a relation shared between the manuscripts reproduced in Chaps. 4 and 6 and the *Brown Book* in the Archive: in Chap. 6, there occurs the statement "Then we might say the time at which a dream occurs is the time at which these processes take place in his brain". This is a component in an investigation complementary to this one that comprises Chap. 4 – "Visual Image in his Brain". Since Chap. 6 has the form of an extension to the *Brown Book*, together with Chap. 4, their relations involve a thematic triangular link that is a qualified example

[87] A further extension of research (into Wittgenstein's presuppositions in the Edition's manuscripts) about unity requires attention to the relations of such matters to his Tractatus, and the presence of details in the present Edition's manuscripts. Zalabardo's (2015: e.g.,114–15) interpretation of unity in the Tractatus and combination of elements (Zalabardo, 2015: 126, 230) offer valuable starting points for investigating notions of unity associated with the material in this Edition and its relations to other manuscripts.

[88] Even so, we need to keep in mind how we may view this, to some limited yet different degrees, from some changing standpoints in Wittgenstein's emerging development, insofar as we can; cf. Floyd 2019 [forthcoming].

of unity within the Archive. Readers will find this is typical, in various ways, of the whole Archive.

Depicting the Context of Wittgenstein's Dictation

Wittgenstein's English writing period roughly covers the time when he was appointed to a title 'B' fellowship in Trinity College Cambridge, and he held a Philosophy Faculty lectureship. In this situation from the period of autumn 1932 to summer 1936, Wittgenstein taught and composed philosophy in English.[89] To some extent he drafted and rewrote elements of some pieces in German,[90] often to prepare for his English lectures or writing in English. In the period 1933–1936 he drafted notes in English, and dictated in English, which – as with his German – was revised by him, often again via dictation.

It is helpful to employ a manuscript's own measuring terminology for classifying his activities concerning manuscripts that he himself used in them. Wittgenstein refers to what we might consider 'lecture notes'[91] as dictation. Consider, for example, his last entry for Mon. Jan 2nd Lectures in Chap. 4 below:

> It is enormously difficult to dictate something to you that will hold water.

Notice that this embroils lecture 'notes' (of which it is part) with dictation. So let us be wary of demoting a lecture note to below that of dictation, at least when Skinner is "you". The Lecture for Feb. 14th Wednesday (Chap. 4) commences with a sentence added in pencil in Wittgenstein's handwriting: "Notes taken by Skinner (Beginning of lecture)". Typically, below Wittgenstein's writing, the first page's lines, in Skinner's hand, exhibit Wittgenstein's first person:

> "I said one could say this: – . . ."
> "I draw something. . ."
> "whatever I did as a lie"
> "I copied him by doing. . ."
> "I cheated the man"
> "I have gone through all the things. . ."
> "I described as the characteristic. . ."

[89]He continued to write in German; e.g. the Big Typescript was developed in 1932; as with many of Wittgenstein's writings he returned to add or revise elements of them later. It is a topic for research to assess what revisions or additions, if any, Wittgenstein and Skinner may have made to the Archive, with its varying internal identities, after its initial composition.

[90]For example, MS 114 (Band X "Philosophische Grammatik" X), dated from 27.5.1932.

[91]We should not confuse notes taken from Wittgenstein's lectures (e.g. MSS 501–509) with material specifically dictated by Wittgenstein. MSS 501–509 are those, respectively, by Yorick Smythies, P. T. Geach, J. E. King, R. D. Townsend, J. Inman, and R. Rhees. There are varying uses of the term "notes"; further progress needs to be made to refine our grasp of them. As a start, see further comments below in this Introduction; also, Chap. 2 includes discussion of this matter, as do a range of footnotes.

"I suppose by being angry..."

Likewise, in the other lectures. For example:

I dictated that I meant... (Wednesday, Jan 24th p.1)
I dictated last time that... (Wednesday, Feb. 16th MS p.1)
I dictated that I meant... (Wednesday, Feb. 21st)
I say 'I wish...' (Ditto)

So, in this set of notes of lectures, Wittgenstein stated that he was dictating them. This warrants our promoting the status of this use of the category 'note' to that of 'dictation', no doubt with appropriate qualifications, including the caution that we should be wary of unrestricted generalisation linking 'note' as 'dictation'.

'Philosophy' Notes, Lectures, and Remarks

It helps to render explicit a few assumptions and presuppositions that aid our attempt to justify a claim to find a match between the textual identity posed by an amanuensis's work as a product of the author he depicts. This is already a vexed arena, even before we consider the unusual and variously rich case of Wittgenstein's revised deployment of the categories listed in the above section heading. There are ranges of slippery slopes here, and one ideally would be a special sort of archivist if one were to become sufficiently adroit to negotiate them all.

At the beginning of the manuscript reproduced as Chap. 5 ("Lectures in Philosophy"), Wittgenstein himself writes: "Notes taken by Skinner (Beginning of lecture)".[92] Obviously this implements Wittgenstein associating himself with the manuscript of lectures. Since he sent the manuscript to Goodstein, we can assign a very high probability to his wishing to identify himself as the speaking, dictating subject reported in this manuscript.

The cumulative evidence of Wittgenstein's handwritten notes is evidence of corrections or revision. This requires us to assign a very high probability to Wittgenstein's authorising of these manuscripts as a signal that they embody his own thoughts.[93] Wittgenstein's additions are significant because Wittgenstein himself incorporated them as components in the Archive of Skinner's work as amanuensis that he posted to Goodstein.

We have no instance of Wittgenstein troubling to make regular editorial or revisionary notes on other people's lectures in the period.[94] The lectures' contexts, styles,

[92]Although the manuscript for these lectures does not have one numbered sequence, the writing occurs on page 73 (i.e., counting the sheets – in contrast with the actual convention used in the manuscript). This note in Wittgenstein's hand is page (1) for Feb. 14th. Note that this is the second sequence with such a date in the manuscript.

[93]On the basis of various analyses carried out by the Editor, he assigns a probability of about 95% + or – 5% for the authenticity of the manuscripts as expressions of Wittgenstein's compositional thoughts.

[94]It is noteworthy that there is no evidence Skinner ever gave a series of lectures.

methods, and conceptual identities are those consistent with Wittgenstein's own in 1933–1934. So we have warrants here to support the view that the first-person pronoun employed in the lecture series[95] and content signal Wittgenstein as the lecturer. Accordingly, we can assign a very high probability[96] that the rest of the notes, which are in the same style and handwriting, including one dated-sequence in the Lent Term that links the whole series, as well as generally clear thematic and historical contexts of Wittgenstein's concerns, have the same source as the first lecture.

Is it likely that the manuscript is, or contains, notes for a forthcoming lecture series? The settled, developed style of the text and physical polish in the writing goes against this prospect. For example, the large format C-series notebooks typically reveal a quite different literary[97] Wittgenstein – of provisional notes, chunks of condensed narrative with visual illustrations for lectures. A way of construing this style is to pose large format C notes as a record for yet-to-be-delivered lecture(s) that transform weekly as the series develops.

We should be sensitive to the usually helpful current attribution of 'lecture notes', as though they have to be of a lower order than 'remarks' or opposed to the function of being dictations, redrafts, compositions, or dictated by Wittgenstein. For example, the *Brown Book* manuscripts, replete with revisions in Wittgenstein's and Skinner's, handwriting, are classed under the heading of "the notes". This is stated on the cover of Volume I in Skinner's hand (see Illustration 8 below). On one interpretation this categorizes as 'notes' what have been called or are "remarks": "The numbers by the side of the notes refer to each separate example given". That is to say, here "notes" are Wittgenstein's remarks in the *Brown Book.* Of course, this may involve shifts of sense for 'notes'. It may be that the heading on the box, is intended for the reader to take it take that 'remarks' in the manuscripts in the box are a subset of and not synonymous with 'notes'. Perhaps all these ingredients obtain at some stage in the 1930s, and this mixture is part of an emerging more explicit sense in Wittgenstein's usage of what it is to be a remark.[98]

It will be helpful to distinguish the notes that are recorded in Skinner's hand from non-manuscript notes by some other students attending Wittgenstein's lectures. There will be some overlap; yet the occurrence of such usage in the Skinner Archive, sometimes along with Wittgenstein's revision or under the latter's guidance, and at a time when they lived together, somewhat promotes Skinner's use to a premium (Plate 1.7).

The Archive is composed of dictation by Wittgenstein, to Skinner, and notes taken by Skinner at Wittgenstein's lectures, together with revision in both their

---

[95]The first occurrence of "I" is on manuscript page 2: "If I ask you for the grammar of expecting…".

[96]Impressionistically I would assign a probability of at least 90% to this statement.

[97]With this first use of the term 'literary' in this Edition (except in the previous uses of 'literary heirs'), note that this term carries presuppositions analyzed and defended in Gibson's (2007) *What is Literature?*

[98]This perhaps complements the usage in the Pink Book of Chap. 3 below, where there is a switch from lines used to separate remarks to be replaced by spaces.

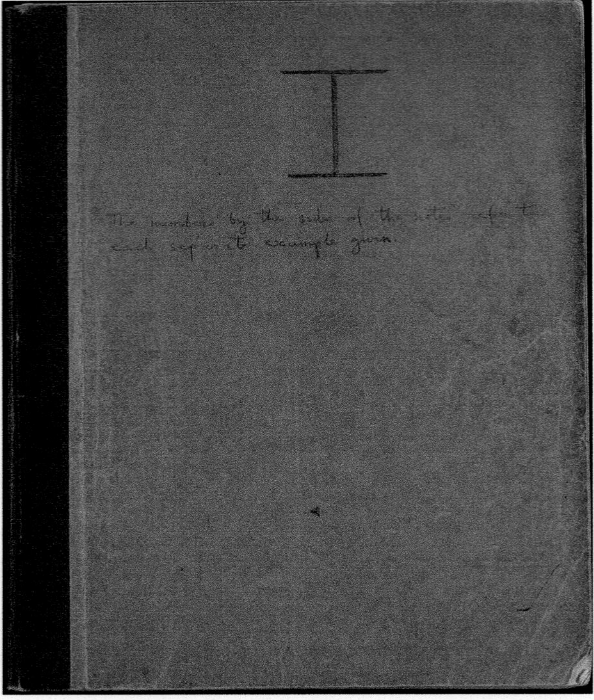

Plate 1.7 Skinner's handwriting: "The numbers by the side of notes refer to each separate example given". (Trinity Wren Library designation: Add. MS.a.407.3.book 1)

hands. The dictations (such as the Pink Book in the Archive) are comparable to the dictations included in von Wright's catalogue, which raises the Archive to the level of '*Nachlass*' material, as perceived by G H von Wright, not least with the evidence of redrafting and revisionary activity that appears in, or result in the manuscripts.[99] Where there is quite substantial correction and revision by Wittgenstein, this appears to elevate the manuscript to a place alongside the *Nachlass* manuscripts of the *Blue Book*[100] and *Brown Book*. This situation affords us more opportunities to derive further insights into Wittgenstein's thought-processes as signaled by the changes depicted in redrafting and revision. That the 'notes' are meticulously drafted, by Francis Skinner, and include revision by Wittgenstein himself, make these a fruitful record and source of Wittgenstein's thinking and lectures of the period.

[99]The exception to this is Chap. 9, is the manuscript which appears to embody a sample of Skinner's calculations probably manifesting some influence of Wittgenstein.

[100]Re. revisionary activity re. the *Blue Book*, see Smith (2013).

Skinner as Vehicle for Wittgenstein's Writing

Skinner's devoted compliance with Wittgenstein's will, their closeness, together with the clarity and care of Skinner's writing art – combined – considerably increase our inductive grounds for assigning a very close qualitative union between Wittgenstein's expression and Skinner's reproduction of them even when note-taking Wittgenstein's creative communication. It is somewhat remarkable that, in this perspective, the hitherto unpublished extension to the *Brown Book* is entitled: "Communication of Personal Experience", with its written display of revision in both Skinner's and Wittgenstein's handwriting, which explicitly advertises the results of present, emergent original thought processes by refining philosophical research in the act of composition. Furthermore, that they were for the most part living together and sharing revision activity using the same manuscripts increases the probability of Skinner's note-taking manifesting criteria acceptable to Wittgenstein as properties that precisely mirrored his own then-current expressions.

In the light of the Archive's illuminating new material, its evidence of fresh insights in Wittgenstein's philosophical dictation thought processes, and his approaches to mathematics, the Skinner Archive must rank as one of the more exciting discoveries in Wittgenstein studies since von Wright first published his catalogue of the *Nachlass* in 1969.

It assists us to register a feature associated with a handwritten manuscript pertinent to Wittgenstein's compositional practice. This is presented in an analysis published by Jonathan Smith (2013): handwritten manuscripts sometimes postdate a typed version of a manuscript. Consequently, we should not conclude that a handwritten manuscript is a more provisional form than a typed version. For example, there is some evidence that the Skinner handwritten incomplete copy of the *Blue Book* is produced after and as a refinement of a typed copy of the *Blue Book*.[101] Typically in this situation, such as the present Archive, one should avoid assuming that a handwritten manuscript is not a fair copy. For example, in the cases of the manuscripts reproduced as Chaps. 3 and 4 in this Archive, they display signs of being prepared for projected publication.

The Archive, Wittgenstein and Russell

In the manuscripts published here from the 1930s' Archive we see Wittgenstein formulating and refining his new insights, sometimes while battling respectfully against Russell's logic. Already in his *Tractatus*, written before WWI, Wittgenstein had inveighed against the claim that the truths of logic are self-evident.

[101] For further details see the sections below whose titles mention the *Blue Book*.

In this, Wittgenstein had anticipated Gödel's (1944) own much-delayed assessment of Russell's contribution in *Principia Mathematica*:[102]

> It is to be regretted that this first comprehensive and thorough going presentation of a mathematical logic and the derivation of Mathematics from it is so greatly lacking in formal precision in the foundations... that it presents in this respect a considerable step backwards as compared with Frege. What is missing above all is a precise statement of the syntax of the formalism.

Tim Gowers (2008) offers a twenty-first century leading mathematician's perspective, in his remark following his quotation below, which he cites from Russell's earlier great work, *The Principles of Mathematics* (1903: 3):

> Pure mathematics is the class of all propositions of the form "*p* implies *q*," where p and q are propositions containing one or two variables, the same in the two propositions, and neither p nor q contains any constants except logical constants. And logical constants are all notions definable in terms of the following: Implication, the relation of a term to a class of which it is a member, the notion of *such that*, the notion of relation, as such further as may be involved in the general notion of propositions of the above form. In addition to these, mathematics *uses* a notion that is not a constituent of the propositions, which it considers, namely the notion of truth.

Gowers' perspective in his *Princeton Companion to Mathematics* (2008: *ix*) responds to Russell's 1903 viewpoint:

> "*The Princeton Companion to Mathematics* could be said to be about everything that Russell's definition leaves out."

Wittgenstein was also concerned with what was absent from Russell's positions. Wittgenstein's investigations in the 1930s, new and remote from Russell's, encompassed mathematics and its links to the world, devising elements for new exploratory maps or patterns, which structure the manuscripts published in the present book. It is as if at times Wittgenstein was dipping presciently into twenty-first century ways of thinking, rather than Russell's. This is not so surprising, given that a range of twenty-first century developments not infrequently derive progress from mathematical insights initially crafted before Russell was born.[103] To introduce this sort of contrast between Russell and Wittgenstein to their philosophising, it helps to sample Chap. 8 below, where Wittgenstein reportedly states:

---

[102]Gödel (1944: 123–54). Gödel had taken many years to write this. When Ambrose, in the hitherto unpublished letter to Rhees (see Appendix [K]), states that, "I must delete those of the notes which are confused... There is repetition of material (on tautologies) which appears in the *Tractatus*", she underestimates the significance of detailed differences in Wittgenstein's use of familiar distinctions for original exploration. This betrays her need to "learn" more, as Wittgenstein expresses it in letters (1935) to the Cambridge University Registrary. On some occasions to Ambrose and to the Registrary Wittgenstein underlines his use of 'learning' with three underlines. In view of his work as a language teacher, and his refinement of his explicit concept of learning, readers should study the original research by Desiree Weber (2019), and [forthcoming], not least for its important bearing supporting Wittgenstein's approach to Ambrose's resistance to his attempts to teach Ambrose.

[103]For example as with Riemann's research.

> Russell sets out from a certain number of primitive propositions... That which produces a queer muddle is the relation between such a system as Russell's and its application...
> Russell and Frege were so keen on being able to define the number 3, on being able to say "3 = 'so-and-so' ".
> You could ask why on earth we should define the number 3, if we define it we define it by other things we have not defined.
> Why should he bother to define '3' and not "have"?

In the same context, Wittgenstein follows this remark by drawing attention to a parallel custom amongst philosophers:

> You will notice that philosophers don't try to define everything, but there are a few things they have a thousand times tried to define.

In Chap. 8 of this Edition Wittgenstein[104] partly bases the above views on his pivotal disagreement with Russell[105] concerning his appeal to self-evidence, variable though that commitment was. In the *Tractatus* Wittgenstein had already rejected the doctrine of self-evidence. In this, and in the Archive, his position was not unlike features of Saccheri's,[106] Riemann's,[107] and Poincaré's[108] rejection of Euclid's[109] view of the self-evidence of the truths of geometry.[110]

Wittgenstein expanded the map of non-self-evident logical space to prepare for a fulsome contrasting use of 'surprise' in reasoning. This idea about surprise has its counterpart within mathematics, logic and areas of philosophy where there he argued that there are new, unpredicted horizons.

Throughout the Archive Wittgenstein presents and assesses critiques of many things about the uses of 'definition' and 'define' in logic, some mathematics, philosophy of language, philosophy of mind, as well as general philosophy, and natural language use. Wittgenstein held the view that Russell's logic and tradition are not a basis for defining what logic and mathematics are. Wittgenstein offered a challenging alternative range of pathways – incompletely mapped, which was partly

---

[104]Wittgenstein, presumably aware that "self-evident" derives from Euclid, and that his fifth axiom (the parallel postulate), does not follow from his first four axioms. Thus, the fifth is not self-evident within Euclid's formulation of the notion.

[105]In *The Principles of Mathematics* (1903) Russell adheres to the self-evidence position, while he needs the axiom of reducibility (cf. Russell (1903: §435), which is not self-evident in his system. In the 1910 Preface of *Principia*, he considers reluctantly that one has to consider what he deems 'primitive laws' that seem far from self-evident, whilst he and Whitehead reverse their priorities about epistemology and mathematics. Conversely, Riemann, and Poincaré – speaking as it were more to the twenty-first century mathematics than Russell – refined and limited Euclid's scope, to remove the consequence of contradictions acquired by such self-evidence so as to avoid contradictions. So, it was not accurate for Russell to deem these developments as 'primitive', since they are the opposite mathematically.

[106]Saccheri's first 32 theorems anticipated some of Riemann's breakthrough in non-Euclidean geometry; cf. Saccheri (1733).

[107]Cf. Riemann (1868), who drops the assumption of the infinity of the straight line.

[108]CF. Poincaré (1902).

[109]Cf. Euclid (1920, 2014).

[110]This criticism, qualified, also applied to Frege (on the structure of his self-evidence cf. Jeshion (2016).

composed from his interactions with Cambridge mathematicians and scholars in other subjects.

Wittgenstein has had his detractors. In Trinity College Cambridge years ago one might hear that its Honorary Fellow "Freeman Dyson asserted that Wittgenstein was a charlatan". Dyson declared this in his youth, in the 1940s when he was an undergraduate! Freeman subsequently admitted that this was due to his being offended by Wittgenstein's brushing him off since he thought him to be a journalist. Latterly, Dyson stated that his own "old hatred ha[s] gone, replaced by a deeper understanding…[Wittgenstein] was no longer an ill-tempered charlatan. He was a tortured soul…trying until the end to express the inexpressible".[111]

Wittgenstein and Cambridge Mathematics[112]

Wittgenstein's and Skinner's relationship was a result of their overlap as contemporaries, from different generations, within the Cambridge mathematical culture of the 1930s.[113] Wittgenstein came to recognise zones and problems within this Cambridge culture, and his own elsewhere, that he could derive his own original analogies to craft games, patterns, and concepts for his philosophy. (It is noteworthy that the Wittgenstein-Skinner Archive, presented in this present book, does not use the word 'culture'. Its absence from the Archive marks a developing change in his research).[114] Wittgenstein also deployed some of his results back into mathematical problems to analyze them, some of which find expression in the Archive's contents in the present book. Since Skinner was a brilliant student mathematician, and subsequently a researcher, he was more than an almost perfect amanuensis for Wittgenstein's needs. As Mark Steiner states: "it would be fair to say that one could develop most of what Wittgenstein says on other topics out of what he says about mathematics."[115]

Wittgenstein witnessed a common situation: great mathematicians, experts in their fields, who often over years do not yet recognise, and are surprised by, significant newly-found properties discovered by other mathematicians in familiar patterns that have not hitherto seemed to contain any further secrets. Wittgenstein's view was that such truths are "not self-evident"; in this he disagreed with Russell,

[111]Quotation from Ground & Flowers (Eds.) (2015: 740).

[112]This use of 'Cambridge Mathematics' extends to influences upon and interactions with the international mathematical communities.

[113]See A. Gibson [forthcoming-c].

[114]Juliet Floyd explains that in other compositions in late 1936 Wittgenstein eliminates the word Kultur in favour of *Lebensform* – 'form of life'.

[115]Steiner (2009: 19; cf. 1): Steiner proposes that, for Wittgenstein, when mathematical propositions are classified as 'hardened regularities', we can see the analogy between them and other kinds of propositions, for example, unit conversions. The present book displays some phenomena in this arena that might not be expected this early in the emergence of his later 'philosophy'.

whilst Littlewood and Hardy concurred with Wittgenstein. He believed that we are prone to such blinded or partial sight. In the Archive we find claims that some breakaway from such restricted 'seeing under an aspect' can be achieved by using analogy and simile in mathematical and philosophical thinking.[116]

Wittgenstein and Kolmogorov

Wittgenstein's viewpoint resonates with a statement made by the renowned complexity theorist Kolmogorov. In this, Kolmogorov describes how we can delineate some boundary conditions for positioning where we can see things anew:

> At any moment there is only a fine layer between the "trivial" and the impossible. Mathematical discoveries are made in this layer.[117]

This quotation evokes a type of the conceptual landscape that attends Wittgenstein in the Archive. He is concerned with the borderlands broaching 'impossibility' in mathematics, physics, language, and mind. In Chap. 4 below, while developing the notion of chess as an analogy for how our minds work, he speaks of "the figure of mind lost in mist" that accompanies our trying to craft answers that border on the impossible, frequently not appreciating that we are doing so.[118]

Game to Wittgenstein: Conway, & Atiyah

Profound mathematicians continue to acclaim Wittgenstein's polymathic insights and their complimentary views of his grasp of mathematical research culture. A former Master of Wittgenstein's own College – Trinity, Cambridge – Sir Michael Atiyah, is a case in point.[119] He also admired the mathematician John Conway, who himself was an ardent Wittgenstein enthusiast. They variously researched imaginary and surreal numbers. So, it is worth quoting Wittgenstein (Chap. 5 below), as he extends his discussion on analogy: "When people talked of imaginary numbers, what revolted was the analogy with numbers which they drew."[120] It is relevant to hear of the status accorded Conway by a world authority on imaginary numbers – that of Don Knuth.[121] On behalf of the Royal Society, Sir Michael solicited the

[116]Consider the Archive example in Chap. 5 below: entries under Monday, Feb. 5th and Jan. 24th, which concern analogy that treats imaginary numbers, and difficulties in surveying such instances.

[117]Quoted from Kolmogorov (2003; entry for 14 September 1943).

[118]See Chapter 4 below, parts 17, 19, 20; also, Chapter 8. Cf. C Notebook MS 149, f. 44r, contemporary with the Wittgenstein–Skinner Archive.

[119]Michael Atiyah in discussion with Gibson (for example, on re. the present discussion, 19th May, 2016, Potter Room, Cambridge Mathematics Faculty). Atiyah is also quoted to this effect by Conway's biographer, Siobhan Roberts (2015: 214).

[120]See Chap. 5 below, section marked 24th January.

[121]See Knuth (1974) on Conway's research on surreal numbers.

games' mathematician Don Knuth, for his view of Conway. Knuth's reply: "the most creative mathematician I have ever met".[122] The above Kolmogorov quotation, which fits Wittgenstein, is applied to Conway.[123] So let us heed Conway, and add warrant to Wittgenstein's own researching within the layer in the quotation – between the "trivial" and the 'impossible' in both their uses of analogy. Conway's view is that this 'layer' is where one discovers original games, including Wittgenstein's.[124]

The above sketch occasions presentation of Conway's use of Wittgenstein, which Conway deploys to portray his own struggle to isolate the fine layer within which a discovery, such as a new mathematical game, is detected and composed. Conway reported that after he delivered a series of very difficult lectures that were – even for him – "dizzyingly ineluctable", Conway reflected:

> So what am I trying to do? I'm trying to, in this famous phrase of Wittgenstein's, I'm trying to whistle around something that I can't actually talk about very precisely. It's that we sort of expect subtle concepts to be hard to define and we should appreciate that, almost as a positive virtue, as a sign of deepness.[125]

The Archive published here presents a new report of Wittgenstein's use of whistling, which unexpectedly amplifies his well-known view. It relates to Conway's "dizzyingly ineluctable" state mentioned above:

> If some says, "One does not know what one is wishing" as Russell,
> there are two ways of using it.
> This shows how complicated the use is. The way one knows one is expecting a man is not generally the same as the way one knows somebody else is expecting so-and-so.
> We have just described different usages.
> . . .Was it a process like whistling a note or one like saying a sentence?
> In general you would say this answer is given by *introspection*.[126]

Conway has hit upon an aspect of Wittgenstein's point: if you go deep enough with introspection, then the words, numbers and symbols we use are analogies for what we are introspecting. Here in the analogy of whistling, it stands for some of our attempts to communicate depth. Whistling also functions as a simile for the way in which words are here inadequate to depict the significance that is the target of introspection. In the lectures Conway gave he presented new maths about a mathematical game he discovered: the free will theorem. Much of mathematics is games, for Wittgenstein

---

[122]Cited from Roberts (2015: 214).

[123]Quoted by Conway's biographer in Roberts (2015: 83).

[124]Since this is the present book's first reference to "game", it helps introduce and qualify the complex varying development of the term in Wittgenstein by reading Rodych's (2018) research.

[125]This quotation and some details in my presentation gratefully benefits from Conway's biography (Roberts 2015: 359), from discussion with John Conway, and with his biographer Siobhan Roberts. (Coincidentally she points out (Roberts 2015: 105) that Conway, as an undergraduate, had the same doctor as Wittgenstein – Dr. Bevan. The latter told Conway that his wish to avoid choice was the same condition that Wittgenstein had.)

[126]From Chap. 5 below; the entry is dated 'Wedn. Jan 17th'1934.

and Conway, which on occasions can *implicitly* be used to delineate realms of new philosophy. Obviously, we should not conflate their different worlds and epochs, while acknowledging Conway's mapping of himself to Wittgenstein.

"Whistling" in the above sense is analogous with aspects of some of Wittgenstein's uses of "shadow" in the Archive, which are cited in the present Introduction. This association with whistling applies when the mathematics expressed are evocations of truths deeper than the sense that the symbols carry, despite the symbols pointing the way to that deeper arena. For Conway also, mathematical symbols themselves can be likened to whistling analogies of, as yet, undetermined or incompletely perceived mathematics. What gives the sense of the depth to the symbols is their link to the target of introspection. As Conway said: "I'm trying to whistle around something that I can't actually talk about very precisely. It's that we sort of expect subtle concepts to be hard to define and we should appreciate that, almost as a positive virtue, as a sign of deepness." When Ramsey affirmed: "What we can't say we can't say, and we can't whistle it either",[127] he misconceived the role of whistling assigned in Wittgenstein's own use of words. Ramsey seemed not to appreciate that Wittgenstein's analogy is to mark introspection as a process that approximates, refines and probes to pinpoint and craft the extraction of a creative solution in research analysis. In the Archive, Wittgenstein's 'whistling' uses games of words or mathematical language analogically, to show what deeper words or maths, beyond their current formulated limits, depict.

Fundamental to Wittgenstein's approach in the Archive is the aim to show, by example, how analogy[128] and simile[129] in mathematical research, in the creative humanities, in ordinary language use, variously are means to furnish original insight that yield surprising – 'non-self-evident' – philosophy.[130] Wittgenstein was not deriving an explicit method from mathematics to apply to philosophy. Even so, he uses phrases that include the word 'method'; for example 'method of projection' in geometry. Such is the frequency of this type of reference that it is probable, in deploying 'method', he is drawing on mathematical custom as a point of departure and analogy for his use of 'method'. He saw that within mathematics one speciality is brought as an analogy to another specialty to construct a new discovery in maths. He saw that this analogy has a kernel that also has a relation of analogy to using a simile in natural language games.

> The foregoing introspection by whistling, in 1969, led John Conway to discover, among other things, an infinity of surreal numbers by using the real numbers as an analogy. He even found that ½ is a surreal number.[131] Here were familiar real numbers, while those who –

[127]Ramsey (1990: 146).

[128]At the same time that Wittgenstein was working together with Skinner attempting to compose this philosophy, he also explored related matters in his large format C notebooks, where he examines 'analogy'; also in *Nachlass* MSS 148, 149, 151. (Cf. Engelman 2013: 196–98).

[129]For a distinguished analysis of Wittgenstein's 'Simile', see Y. J. Erden (2009a, b, 2011).

[130]Especially see Chap. 8, Lecture VII.

[131]See Conway (1976), and Roberts (2015).

> merely technically 'knowing' them – did not understand that within this 'knowing' there were also surreal numbers. Numbers – like words – can be analogies, used to construct features beyond themselves. Aiming at new games and theorems, they can be 'whistled'. For example, as did Conway on the way to understanding a new analogy: surreal numbers.

Conway[132] is fascinated, as was Wittgenstein, with games as means to achieve new insights outside of games – not least chess, which pervades Wittgenstein's fresh arguments in the Archive, and elsewhere. The conceptions that explain such uses of 'game' are often deep and difficult, even for great mathematicians to understand.[133] Wittgenstein and Conway like to add new rules to explore chess as an analogy to reach recognition of fresh insights. In the Pink Book (Book I) Wittgenstein adds a new rule where a player is angry:

> If we brought into chess an element of temperament: if a man moves a knight in a new way, if he is angry. We might at first think this was not a game at all.

Accordingly, Wittgenstein deems the statement "a good simile", because here a new non-factual – qualitative property – temperament – is introduced: a simile. In the Archive, Wittgenstein states (in Chap. 4 below) that:

> . . .the use of similes is to give parallel cases.
>
> Parallel cases change our outlook. We can then look at the thing in a new way.
>
> *Simile*: When the view of the earth as the centre of the planetary system was given up. . . part of what happened was to destroy the unique position of the earth.
>
> It is often enough to destroy puzzlement to show it is not a unique thing.
>
> If we know the game of chess, we can be said to understand a move in the game.
>
> This mental process of understanding the move does not presuppose the rules, which are what distinguish between the man who does and who does not know the game.

These remarks, and their parallels with the C notebooks as well as other familiar passages, are like dialogues[134] between polyphonic 'Wittgensteins' – paving ways to dialogue within the Archive, and elsewhere.[135] They are fruitful insights into the Archive's dynamics, not least respecting vectors between philosophy and mathematics.

In Chap. 4 below Wittgenstein continues:

> if you were asked what [a person] had to imagine to imagine a *chess-board* you would explain it by pointing to a real chess-board or, say, to a picture of one, and *analogously* if you were asked what imagining the King of Chess is or imagining a knight's move. – Or ought

---

[132]Re. Conway on new chess rules, see Roberts (2015: 114).

[133]Re. Wittgenstein's use of 'understand' in Chapter 3 Part I, and Chapter 4, in *PI* §151–54. Some mathematicians state that they do not understand their own discoveries. For example, Neil Hindman stated to me (7 July, 2016) that he did not understand his own deep theorem concerning ultrafilters. (For exposition of this theorem, see T. Tao 2016).

[134]I use this term advisedly, for which see Heal (1995).

[135]'dialogue' and 'polyphonic' are here based in Mikhail Bakhtin's research, for which see Picon (2016: 3–18, 30, 172, 223). This Bakhtin was the brother of Nikolai Bakhtin with whom Wittgenstein discussed his philosophy – sometimes in Russian – in the period of the Archive published in this book: a sphere for a worthy future research project.

> you to have said: he must go through certain private experiences which he has when – in the ordinary sense of the expression[136] – he plays a certain game of chess.[137]

Here Wittgenstein is constructing a new analogy in and by use of the familiar figure of chess. We see above that Wittgenstein, in an oblique yet crucial moment, ties 'simile' together with 'analogously'. In the Archive, he concentrates on the 'simile' as a consequence of insight into analogy. Elsewhere he has set this agenda with his remark that there is: "The strange resemblance between a philosophical investigation (especially in mathematics) and an aesthetic".[138] For example, Wittgenstein has a specific sequence in the Pink Book (Chap. 3 below) where there is a fresh investigatory perspective in the calculus arena of *Tractatus* 6.241, which should not be misread as merely going over old ground.[139] Rather, the Archive is deploying analogy and simile to embark on a journey to construct new fields for many endeavours – not only games. Obviously, features of this exploration culminate in the beautifully refined use of analogy in the *Philosophical Investigations*. Nevertheless, sometimes the Archive in the present book speaks to a broader manifesto for the future, pursuing different identities, which his death – possibly also Skinner's, and perhaps post-WW II trauma, deprived Wittgenstein of a chance to model further.

In effect, Wittgenstein was – by concentrating on the ways he uses examples from different mathematical areas – proposing, with a degree of prescient sensibility, the conjecture that entirely different mathematical subjects all have specific analogies that amount to sharing certain elements of identities with each other. We see the patterns in the Archive: an uncountable set of interactive language games. This sketch, concerning Wittgenstein's sensibility, is here pitched by use of analogy and simile to map aspects of Wittgenstein's mathematical sensibility, but also his interest in applying calculations in practical use. There are certain parallels with Pascal here. These extend not only to Blaise Pascal's and Wittgenstein's use of aphoristic prose style, but also both their applications of maths in science.[140]

Certainly, Wittgenstein did not have the technical expertise of certain of his champion referees in their respective specialities – Littlewood and Hardy. He had an extensive openness to new patterns, and ingenuity about them, which accords with Hardy's emphasis on their importance to mathematical discovery. Wittgenstein developed philosophical insights into such domains represented by patterns,[141] not least with intricate exploration of games and family resemblances. He also wrote of

---

[136]The word 'expression' is written above 'word' in Skinner's hand.

[137]The words 'a certain game of chess' are written above 'a game of chess' in Skinner's hand.

[138]*Culture and Value*: 29. This aspect is developed by Alois Pichler (2013).

[139]See Chap. 8 below, and Pichler (2018a, b). For analysis of Wittgenstein's use of 'calculus', read Schroeder (2020).

[140]Recall that, for example, Wittgenstein secured a science patent in 1909; cf. J. Cater and Lemco (2009). His Vienna background also has a parallel here, as Ian Hacking (2014: 162–63) exposes it: e.g., Wittgenstein's links to and parallels with Richard von Mises' research on flywheel masses in crank drives. On Pascal, see Khalfa (2003); Pascal (1995, 2004, 2006).

[141]For other discussion of such patterns, cf. Baker (2004: 84–86, etc.).

the bewitchment of such imaging and arenas, probing their intricate, interlocking co-extensive variations in texture and structure. In this particular perspective, if we include the unexpected and the role of surprise, Wittgenstein's placing of these aspects could itself be regarded as a supervening generalised pattern – the full application of which we are as yet still partly ignorant. Even with the creation of some simple great new theorems, it is a familiar truth that proof of them has to wait for centuries. Wittgenstein grasped how prescient mathematical creative imagination may out-pace accurately our capacity to depict truths, which only later are explicitly proved in mathematical discovery and calculation, while prior to this may be elucidated to some degree by original philosophical investigation.

This type of situation gave rise to his simile of whistling as a device to articulate such processing. Likewise, John Conway says he allows himself the luxury of whistling with mathematics in Wittgenstein's sense. This stands against Ramsey's objection to Wittgenstein. We should allow Wittgenstein justifiably to make remarks that whistle about puzzling over mathematical games. For Wittgenstein, this was an instance of Ramsey's bourgeois thinking.[142]

In the Archive Wittgenstein argued that we should utilize games as new boundaries for truth. This concurs with Conway's presuppositions: The universe of mathematical and logical discourse shares games, inexorable and unrelenting in their variety, and invade other subject boundaries – across the motley ranges that are mathematics. Wittgenstein matched and added to Hardy's notion of the importance of patterns, but with games and other constructs, which embody evidence for transforming identities. In Chap. 8 below Wittgenstein's remarks prepare the way for his own view, by prefacing it with some discussion about Russell, based on considerable respect for him, whilst disagreeing with both Russell's and Frege's view of what it is to be a 'common property'. For example, in Chap. 8, part XI.8 it states:

> A class is generally represented by a list.
> Russell thought of a class as defined by a common property.
> Russell introduced 'class' to be able to talk about it as a list.
> He seemed at the same time to be able to define it by a property.
> The question which both Frege and Russell tackled was this:
> Frege had said the number was the property of a class.

In the Pink Book Part I, the dictation which reports Wittgenstein argues:

> Suppose we say, "What is it all games have in common?" I will say: need they have something in common? If we say chess and draughts have the same board in common, it is different to say all games have it in common... [Does the generic name point to their having a disjunction property in common?] The justification[143] of a generic name need not be a property in common. . ./ This has happened to the word 'number'; rational numbers,

[142]In Wittgenstein (1977b: 40).

[143]This use of 'justification' partly may be a response to Sraffa's view of rational justification (I came to this through valuable discussion with Brian McGuinness (also see McGuinness, 2008: Letter 172).

> cardinal numbers etc. need not all have something in common. We may call a construction "a number" if it only has similarities with three of these.

This appears to build on, and be an original development from, Wittgenstein's conversations at Schlick's house in Vienna, concerning Weyl's view – "a formalist conceives of the axioms of mathematics as like chess-rules", though he disputed Weyl's framework.[144] The above game figure fits with the first part of Weyl's (1930) view, which is: "A rescue of mathematics... is only possible...in which, in principle, it changes... into a game carried out according to fixed rules..."; yet Wittgenstein was now opposing "fixed rules".[145]

With the above introductory sketch, we are in a position to pose and briefly outline an answer to the question of: What mathematical evidence thus far in the twenty-first century is there for the extent of Wittgenstein's claims in the Archive? – Roughly, that there are inexorable varieties of sharing features *between* domains of mathematics games, which *do not always have the same properties in common.* The uses of the term 'game' in some maths and in Wittgenstein's uses show that the term is frequently deployed analogically to extend the identity of domain as well as draw on its uses. For example, there are disjoint areas of mathematics that nevertheless share some other properties that involve the notion of games. Not least having rules in two or more games; yet not the same rules that obtains for each game. Clearly at many levels this is obvious, while in their deeper domains, obscure or seemingly intractable.

In this and other ways Wittgenstein's perspectives are relevant for select patterns in some emerging contemporary mathematics research. His understanding often pinpoints the functions of analogies that fit such mathematical patterns, their boundaries, limits, and sometimes in differing ways for philosophy and languages. There is no claim here that he would mathematically fully understand these technical depths in the ways mathematicians such as Littlewood pioneered, though we should take due note of the latter's wish for his students to be taught by Wittgenstein.[146]

[144]See this in Waismann (1979: 102–05, note 67), citing Weyl's view in Symposium I (1927: 25). See also Menger's (1931–1937) *Ergebnisse eines mathematischen Kolloquiums* 1, and (1939–1948) *Reports of a Mathematical Colloquium 1–8*. It is relevant, for the next section below in the present book, to register that here Wittgenstein is involved in discussion about Dirichlet's function. If one wants an up-to-date approach in pure mathematics that would more suitably concur with Wittgenstein's philosophical position in the 1930s and beyond, Beck's (2008) major study on aspects of combinatorial games is full of insights, not least for domains in which Ramsey researched, which can be compared to and against Wittgenstein's Archive in the present book.

[145]Weyl (2012: 26; cp. 117).

[146]If there were to be criticism of Wittgenstein's level of knowledge on maths here, we should be wary of imputing general ignorance to him, since he tended carefully to select a narrowed perspective in his evidence. Many a pure mathematician would not understand much at all of some of his contemporary colleagues' researches. The conditions of access for Wittgenstein to try and make some such exploratory journey as philosopher, faces difficulty; yet the remains of the parallel are sufficient to leave open the prospect that such journeying and disputes – with pure mathematicians such as Hardy and Littlewood, and Turing, permit us to warrant that something

Nevertheless, we do well not to underestimate the significance of such philosophical comprehension – where it really does grasp some elucidation about the mathematical patterns, because mathematicians themselves proceed and discover new research identities – partly by deployment of new analogies. Wittgenstein's instinct was to sense the fabric of such analogy. Sometimes metaphysical or philosophical issues are internal to the identity of certain technical mathematics, to which Wittgenstein attended. This sketch is a mere fragment in, and pointer to, the extensive array of pioneering distinctions that are launched or developed in the Archive, which require investigation.[147]

Connecting: Wittgenstein & Langlands

The Robert Langlands Programme precisely affords us an illustrative *analogy*,[148] whose patterns indirectly, note again – *by analogy* complement Wittgenstein's patterns of arguments about games and non-self-evidence in the Archive. Certainly, Langlands and Wittgenstein are worlds apart in their theoretical languages and formal identities. Even so, apt comparison throws up parallels. Langlands' research, and later research by others, excavates and exploits surprising analogies in mathematics. His conjectures detected largely hitherto inscrutable domains and relations within mathematics, for example his original interpretation and use of some of Dirichlet's insights are pivots for these developments.[149] Wittgenstein had an interest in particular aspects of Dirichlet,[150] as similarly he had with Euler's

more in mathematical research was grasped by Wittgenstein than mere crumbs on the ground from the mathematicians' tables. We should be careful to ponder Wittgenstein's point, reflecting his discussions with Littlewood, that once a typical mathematician has proved his calculations, he moves on to another topic, whereas as Littlewood agreed this is where the worthy philosopher starts to investigate them. Wittgenstein well knew that there are some areas where mathematical science is attended by incomplete understanding, scrutiny of which can benefit from philosophical investigation. Furthermore, in the twenty-first century, in 'applied' mathematical research into dark energy, metaphysics subsumes some mathematics (cf. Gibson (2019).

[147]Cf. Gibson (2013).

[148]The introduction here, of *aspects* of the Langlands Programme, Langlands (2016), reflects some of the above explanation of '*analogy*' by Wittgenstein. It is not being proposed that there is direct comparison of technical content, objects, or formal style. Rather, as with mathematicians who deploy analogy to formulate new maths from extant maths, so Wittgenstein drew on his crafting of analogy to compose philosophical games, derived from maths, and other phenomena, whilst redeploying such games to depict mathematics. (Cp. Bernstein and Gelbart 2004; Dumoncel 2011). There are examples of mathematical analogy in for example Thorne's (2016) proof that if p is prime, then all elliptic curves defined over the cyclomatic $\mathbb{Z}_p$ -extension of $\mathbb{Q}$ are modular. This can be interpreted as the use of mathematical analogy to extend into a new domain. In the very different world of Wittgenstein's analogy, he extends analogy to discover and make explicit new philosophical identities derived from standard maths (as in Chap. 9 below).

[149]Dirichlet L-series is in special respects unexpectedly analogous with features of the Riemann zeta function.

[150]See a range of Wittgenstein's references to Dirichlet in Biesenbach (2014: 116–117).

research[151] – aspects of whose discoveries Langlands (2007) utilizes.[152] The Langlands Programme's extensive enterprise exposes non-self-evident generalised patterns in and between distinct mathematical fields.[153] These amount to previously separated, now connected, mathematical disciplines and domains. Aspects of these can be viewed accurately as mathematical games in Wittgenstein's intricate philosophical senses.[154]

The Archive displays Wittgenstein's regular concerns with (often to refute as premature) attempts to define reasoning in maths, logic, games, rules, and physics. Wittgenstein was investigating the nature and strength of what *connections* are in and between these subjects. After Wittgenstein's death, the research founded by Robert Langlands concerning the discoveries of unknown connections across

---

[151]See a variety of Wittgenstein's uses of Euler's equations in Biesenbach (2014: 139–140, which are informally reflected in a many of the archive's uses of finite' and 'infinite') where they amount to a polyphonic theme orchestrating the translation of infinite series into finite series. This is explicitly attempted in Wittgenstein's (2000a, b, c: 645–649) 'Eulerscher Beweis' – calculations starting to reformulate Euler's similar aim. Kreisel attempted to refute Wittgenstein's narrative, by allegation rather than evidence. Kreisel misrepresented Wittgenstein, for which read Appendix [H], especially its Footnote 1756; also, Gibson, A, Leader, I, & Smith, J. [forthcoming]. Wittgenstein's mathematics here are not very advanced, yet they reflect a subtle original instinct for future potential (though not a knowledge of mathematical objects), which deserves more appreciation than Kreisel's too-quick superficial glance could muster.

[152]In the *Tractatus* Wittgenstein does offer considerations that prepare ways to move in the directions of homomorphisms, according to Wolniewicz's (1979) explanation that homomorphism φ' is mapped onto the expressible. The limit of what remains is not only the inexpressible. Wittgenstein's awareness of technical difficulties may have restrained him from recognising this. If so, he dropped the notion that he could have revived. (It is possible that one motive in raising *Tractatus* issues in Chap. 8 below is to prepare the way for exploring what he had dropped.) As Wolniewicz shows, it can be retained by weakening the homomorphism φ' to one that is the expressible. So *L* could be a whole series of languages *L0, L1*. . ., with a corresponding series of logical spaces *R0, R1,* with the inexpressible commencing from within. That is, from the expressible in the language *L0*, without knowing in advance whether what is inexpressible at a given stage *L1* is absolutely or only relatively a limit. A limit that involves tracing boundaries of the inexpressible, receding to infinity. With the researches of Woodin (2017) and Welch (2017), for example, we have a basis for extending the above aspect of the *Tractatus* and Wittgenstein's argument there, which is revived and developed in Chap. 8 below, namely: "This shows a logical proposition has nothing to do with self-evidence" (Chap. 8 below, Lecture VII). In the light of this it is plausible to explore the supposition that one of the needless limitations in Wittgenstein's *Tractatus* was his not applying the foregoing quotation to strengthen his own analysis of homomorphism in the *Tractatus*, so that he could extend expressible games beyond his sense of the finite. If we utilize Langlands joined up to Woodin's insights, this is now a prospect.

[153]In more mathematical terms, for example, homomorphisms where Eisenstein series can be deployed to construct intertwining operators; cf. Lafforgue (2018).

[154]Langlands showed that for a certain Dirichlet character, its *L*-function can also be written as an Euler product. Langlands and his colleagues depict a whole fresh universe of such analogues and counterparts, including for example direct consequences of Ramanujan's conjectures and theorems (cf. Langlands, 2011: 305–07). Taylor's (2017) original approach has taken matters in a different complementary direction. This exposes further family resemblances and links to Dirichlet's theorem and the primes, which is beyond anything in Wittgenstein's research.

mathematics, has profound implications for problems posed by Wittgenstein. For example, one of the terms in this research 'maps' generally on to Wittgenstein's connections. It is the topic of *intertwining* (the borders/edges of distinct domains). This subject concerns cross-connections that map intertwining relations in, for instance, games – their borders and these connections inside the games.

Wittgenstein's development of *analogy* in the Archive links up with these ideas of connections (as sketched earlier in this Introduction) and shadow cross-links between games and their characteristics.[155] The highly advanced twenty-first century status of the Langlands Programme contrasts greatly with Wittgenstein's early twentieth century enterprise. Nevertheless, allowing for this contrast, there are parallels to illustrate the fit with Wittgenstein's original notion of family resemblances. Aspects of this use of analogy by Wittgenstein has parallels with Hugh Woodin's recent conclusions that concern deep related matters in set theory (some of which Wittgenstein addressed). In the following quotation, Woodin is viewing his own life's research and the relevance of past research concerning infinite and finite relations, some of whose debates impinge on issues in the Archive, which progress from Wittgenstein's *Tractatus*. Woodin states that:

> there is the very real possibility that among these axioms, there is an optimal one (from structural and philosophical considerations). In which case we will have returned, against all odds or reasonable expectation, to the view of truth for Set Theory which was present at the time when the investigation of Set Theory began.[156]

Accordingly, it is important not to read the presently published collection of hitherto unknown manuscripts as (mere) history of philosophy, for two reasons. First, because they have not been studied before, and the vague sense of a reader's familiarity with some of the language will be a fake friend to tripwire those whose knowing, too quick glance over the material could cause blind interpretation. Secondly, deep research from the past is often the basis for future answers, while fashion has deemed the past as arid of future potential. (Consider Einstein's use of the then neglected, and still not fully understood Riemann Hypothesis, and its relevance for future research on gravitational waves.) Very clever mathematicians, and other scholars, have yet fully to understand a range of discoveries already made (for example, with Langlands' insights). It should not be a surprise, to discover that we have yet fully to understand the full scope of what Wittgenstein was arguing about games as they relate to his examples, not least in the present book. The roads forward are also uncertain, despite familiarity with extensive insights into Wittgenstein's formulations of his game philosophy.

---

[155]To structure the notion of 'links' here, consider, for example, links between Gauge Theory, geometry and number theory in the Langlands Programme, as exemplified in Witten (2010, 2015). Note Wittgenstein's attention to the research of figures that contributed insights, which were later utilized to build some breakthroughs exposing these mathematical family resemblances, such as Dirichlet (for which from one perspective cf. Gauthier 2015).

[156]Woodin (2011: 118). In relation to logic and philosophy, see Grève (2018).

In his attention to games, and a host of polymath matters, Wittgenstein did not spawn something such as the complex and deep Ramsey Theory; yet in some respects Ramsey regarded himself as a student learning from Wittgenstein. Even within the context and scope of his own discoveries, Ramsey would not have anticipated or understood most of the developments that Ramsey Theory more recently yielded.[157] This is not a criticism. Rather, unpredicted prescience is in the nature of new identities made by great creative mathematicians. This is consonant with Wittgenstein's ways of speaking, for example in Chap. 8 of the Archive.

Wittgenstein saw that such notions of games apply to the actual world. Hardy spurned this latter perspective – unlike Littlewood, who found himself improving radar mathematics in WW2. In this and some other respects, Littlewood and Wittgenstein were kindred spirits. In the Archive, and elsewhere, Wittgenstein quickly moved to utilize mathematical analogies and patterns of family resemblances relating to games,[158] also to map some elements of what it is to be the mind and world in language. There are particularly valuable cases of this in Chaps. 3 and 5 below, which, for example, deploy geometrical diagrams as analogies for intentionality and other mental states. Episodic and scattered though such instances are, they nevertheless amount to sustained original progress, which, when viewed from a supervening standpoint over the whole Archive, are conceptually innovative landmarks that map territory for philosophy, and its implications.

Where Wittgenstein takes up a mathematical example in the Archive, it is not to turn maths into philosophy; nor philosophy into maths. Rather, he saw that many different mathematical specialties display various patterns that can be helpful to clarify philosophy. Sometimes the Archive deploys this approach with examples from maths – informally drawing on these patterns of mathematical reasoning when doing philosophy. Sometimes the Archive does not explicitly deploy maths. Instead, the Archive informally draws on the clarity of mathematical patterns to develop new philosophy. Sometimes Wittgenstein was at pains to press even mathematicians themselves of the need to raise their game, so as better fully to understand the rich diversity of unfamiliar patterns in mathematics. It appears that at times mathematicians such as Hardy, and especially Littlewood, enjoyed – as well as were frustrated by – their close intense encounters with Wittgenstein. In varying ways, they shared his aims to find new ways to engage with logic and mathematics, reflecting their shared disagreement with Russell's logicism.

[157]One of the leading developers of Ramsey Theory agreed with this point: Imre Leader, discussion in Cambridge, November 2014; see also Beck (2008: 91–209).

[158]'family resemblance' is not used in the Archive. In deploying this expression, I am drawing on its use elsewhere in Wittgenstein's *Nachlass* and its usage of 'game' – a term that recurs in this Archive.

Hardy

To some degree Wittgenstein's engagement with mathematics leading to philosophy was a converse of Hardy's interest in mathematical philosophy.[159] Hardy and Littlewood had Ramanujan as their genius. Russell had earlier proffered his own in the form of Wittgenstein, as if Wittgenstein were Ramanujan's ludic philosophical counterpart. By 1929 Russell was demurring from Wittgenstein's change of direction from logic to mathematics and natural language. Wittgenstein was probing the internal identities of research in pure and applied mathematics as well as their relations to logic – as we can read in the chapters below. Hardy and Littlewood (1925) found themselves more in agreement with Wittgenstein than with Russell, partly due to Russell's failure to engage with newer developments in mathematical research.

Hardy and Wittgenstein were vastly different in very many respects; but they exhibited some similarities, which perhaps antagonised each other, while earning respect.[160] Both were enormously over-confident, one might say to a fault, whilst vexed by anxiety about the real merit of their talent set against their weaknesses, as well as waning intellectual powers.[161] Hardy had paid Wittgenstein the tribute of naming him and quoting him as antagonist in one of his distinguished public lectures. Wittgenstein replies to Hardy (as he does to Russell elsewhere in the Edition[162]). In the manuscript reproduced in Chap. 6, Wittgenstein states:

> Hardy said that he *believed* Goldbach's theorem[163] was true.
> All objections seem to be turned down by saying you feel it is there, and then you silence the man.

[159]Wittgenstein's engagement with Hardy displays both respect and disputation. For a range of Wittgenstein's references to Hardy, cf. Biesenbach (2014: 246–250).

[160]Wittgenstein states in his letter to Watson, of 1931: "Prof. Hardy the Mathematician who was lecturer here when I was an undergraduate and then went to Oxford has now come back here as a Prof. and is one of the few people I see. I like him very much." (quoted from McGuinness, 2008: Letter 142).

[161]Both had the same sexual preferences, the pressures of which and society's opposition, periodically engulfed them, and on one insider's account Hardy's attempt at suicide was due to being deserted by a partner.

[162]For example, see Chap. 8's later references to Russell on number theory, instanced under 'Remark'.

[163]Although Goldbach's Conjecture or method is not a proven theorem, yet 'theorem' is Hardy's own tag in the above-mentioned lecture (*Mind* 38, 1929: 2). Helfgott pointed out to me that Dickson (1919: 421) mirrored some standard usage contemporary to and prior to Hardy's earlier experience, writing of "Goldbach's Empirical Theorem" – since it was "verified" to, e.g., 1000 by Cantor. It derives from, or presupposes as a basis, the Goldbach-Euler Theorem contained in Euler's (1744) paper *Variae* observationes *circa* series infinitas. Euler states that this was presented and proved in Goldbach's (lost) letter. Its context has been somewhat criticized (see Bibiloni, et al.) as a misuse of divergent series, a slip characteristic of that epoch, whilst Hardy's book *Divergent* Series reflected the progress beyond.

The question is not that you can't believe it, but what does it mean.[164]

Wittgenstein suggests 'divining' is more akin as an analogy to map Hardy's position, in the absence of proof. Wittgenstein was not initiating a sting operation against Hardy. Rather, Wittgenstein was replying to Hardy's already-launched published disagreement with Wittgenstein, displayed in Hardy's 1929 Rouse Ball Lecture in the Cambridge Senate House.[165] Hardy considered Wittgenstein important enough to make him the sustained target in a famed Cambridge public mathematical event. Both joined in with this fray, with mutual respect and appreciation, meeting in their College – Trinity.[166] Hardy claims in the lecture that "if Littlewood and I both believe Goldbach's Theorem,[167] then there is something, and that same something in which we believe, and that that same something will remain the same something…and when mathematicians have proved our belief to be right or wrong".[168] Hardy's style was full of panache, whereas Littlewood's typical style was sparse and modest – perhaps this is why Wittgenstein ignores the reference to the latter in Hardy's Lecture, though Hardy is clearly referring to his joint 1923 paper with Littlewood.9[169] Wittgenstein takes the position that until or if the proof exists, the 'theorem' and proof are intentional fictional objects. He believed Hardy's use of "true" falsely assumes[170] "proof".

The Archive also displays evidence of Wittgenstein being influenced by Hardy's interests. This is particularly so when he used mathematical notions to construct analogy extended from that domain to another, often as a means to formulate a new mathematical concept or proof. Wittgenstein deploys and extends or re-directs such analogy to compose original philosophy. In the years 1929–1935 Wittgenstein lectured on topics developed with original insights in the Pink Book (reproduced in Chap. 3 below, and elsewhere in the present Archive). In this period Wittgenstein was in discussion with Hardy, Littlewood, and perhaps Polya,[171] who were researching for their then-forthcoming book *Inequalities* (1934). Wittgenstein drew on details of their analysis of limit theorems, symmetry groups and related topics in the Pink Book, including parallelepipeds.[172] He deploys other terms that

---

[164]Wittgenstein returns in 1939 to engage with Hardy on this matter in *LFM*, Lecture XIII: 123, where he also refers to Hardy's use of 'Theorem'. See Potter's (2011: 128) analysis of open problems.

[165]Published as 'Mathematical proof' in *Mind* (Vol. 38, 1929: 2).

[166]Trinity Mathematical Society (of which Skinner became Secretary in 1931) meetings illustrate this: Hardy claimed Wittgenstein's view was that mathematics consisted of tautologies, whilst he denied this (Klagge and Nordmann 1993a: 336).

[167]Idem. p. 24.

[168]Hardy in *Mind* (Vol. 38, 1929: 4, 24), where he discusses some of Wittgenstein's views but not those in the present Edition.

[169]For example, see Hardy and Littlewood (1923).

[170]'assumed' is here used in the formal sense of 'assumption'.

[171]For a period Polya regularly visited Hardy to research with him.

[172]There is also reference to this and related terms in his later 1939 *Lectures on the Foundations of Mathematics*, re. inserting rods through their sides (p. 40) and stacking wood (p. 202).

appear in their book and related matters. There are other dimensions to this interaction between Wittgenstein and the above. For example, Morineau (2012) argues that in some respects Polya's (1940) publication builds on Wittgenstein's notion of logical space occupied by propositions.[173]

This arena of 'limits' can itself be viably interpreted as symptomatic of what Wittgenstein was trying to depict and map with language games: problems of precision, relations, and difficulties of measuring approximation in domains where one wishes to isolate and articulate both sides of boundaries. Here, states of affairs that often seem transparent are not self-evident, sometimes exactly where the best mathematicians expect them to be.

Littlewood

The Trinity College Council requested Littlewood to assess Wittgenstein's talent, though Littlewood seemed to feel that he was the one being interrogated.[174] This resulted in Littlewood's report, strongly supported by Hardy, and somewhat weakly by Russell, recommending Wittgenstein for a post in Trinity, in which Littlewood affirmed that he expected Wittgenstein would come to produce research more important than the *Tractatus*.[175]

In some philosophical accounts there is not a strong awareness of Wittgenstein's involvement with mathematicians. So it is worth repeating Brian McGuinness's conclusion, introduced in the above Preface: "It was not for nothing that Hardy and Littlewood would discuss [Wittgenstein's] views on mathematics before the First War or that many who were to be leaders of their profession went to his classes 'for mathematicians' in the early thirties."[176] Note that this priority is not Wittgenstein discussing Hardy's and Littlewood's views, though he did. In their varying stratospheres, neither mathematician suffered colleagues with a lack of deep mathematical intelligence; Hardy was easily bored.[177] Their perception of Wittgenstein's abilities is a guide as an element in our framing of the Archive published in this Edition. We would do well to notice that at the time of this esteem for Wittgenstein, the Archive was being composed, and his portrayal of games emerged. It is significant that Hardy and Littlewood facilitated Wittgenstein's teaching their

[173]The outcome is an algebra operating on a Polya-type hypercube, i.e. *n*- dimensions square.

[174]Littlewood (1986: 138). Hardy strongly recommended Wittgenstein, whereas Russell's report was qualified in such a way as to damage Wittgenstein's chances of gaining a fellowship, and Russell was out of touch with some of the mathematical problems that Wittgenstein was addressing; therefore, Littlewood was called in to assess Wittgenstein.

[175]See McGuinness 2008: document 133, report from Littlewood to Trinity College Council, 1.6.1930.

[176]For example, Michaelmas Term 1933: University Reporter announcing Wittgenstein's course: 'Philosophy for Mathematicians', for the Mathematics Faculty.

[177]Wittgenstein sometimes protested that he did not hold a view Hardy ascribed to him (see Klagge and Nordmann 1993b: 336).

mathematics' students, whose training at undergraduate level is so time-sensitive for their talents to reach their optimum harvest. The Faculty then (and now) guards against didactic influences that are not significant mathematical contributions for undergraduates.[178] Wittgenstein assisted mathematics students to *think mathematically.*

They were also lectured by Littlewood, who stated in his *Elements of the Theory of Real Functions: being Notes of Lectures,* that:

> intended to introduce third year and more advanced second year [maths students] to the modern theory of functions. The subject-matter is very abstract... The aim of the lectures, indeed, is to inculcate the proper attitude of enlightened simple-mindedness by concentrating on matters that are abstract... aimed at excluding as far as possible anything that could be called philosophy.[179]

It was a remarkable compliment, then, in view of his closing comment, that Wittgenstein was called on to teach these students. On one interpretation this requires that Wittgenstein's teaching of philosophy for mathematicians excludes philosophy, by his making explicit in philosophy what it is to be mathematics: to describe mathematics, without excess. In the Archive, Wittgenstein shows that "common sense" in language games is also a pattern for achieving clarity in mathematical calculations.[180] (This "common sense" was fundamental to Turing's education as an undergraduate, and to the emergence of his original research).[181]

Wittgenstein asks:[182]

> In what does Littlewood's 'enlightened simple-mindedness' consist? Does it not consist in the belief that so-called general mathematical inquiry brings to light truths which more specialist investigations merely explain in detail? Whereas general calculus is general only by reference to special calculus. Because in mathematics nothing resides in the words but everything in the calculus. I.e. because the sign brings with it into pure mathematics no meaning other than the one it is given by calculus itself. Mathematics does not consist of meditations. And a part of mathematics may be called 'abstract' only insofar as its application is suggested yet left vague.[183]

---

[178]This consideration obtrudes on the question of why Wittgenstein facilitated Skinner side-stepping his Trinity-funded advanced research training at this point, though it might – initially – have been considered a contribution to Francis' development as a mathematician.

[179]From the Preface of the Second (revised) Edition of 1926, page *v*; published by W. Heffer & Sons Ltd. – bookshop opposite Trinity. (The first Edition of 1925 was published in the same way, but without a Preface.)

[180]For this use of "common sense", note below, Chap. 5, subsection entitled 'Friday, Feb. 9th.'

[181]See Floyd (2017a).

[182]This translation is based on a somewhat literal draft kindly produced by Mark Chinca, which was then slightly adjusted by me in the light of much appreciated suggestions by Brian McGuinness, Joachim Schulte, and Volker Schlue.

[183]Wittgenstein's *Taschennotizbuch* (conjectured terminus *a quo* January 1938b–39/Aug.1940?) MS 162a: 16–18:

Ist ein Kalkül allgemeiner als ein andrer? Worin besteht die 'Allgemeinheit' des Mengenkalküls? Nicht darin, daß er eineinfaches Bild einer Klasse andrer ('spezieller') Kalküle ist? Ist er ohne Beziehung auf diese |auch| allgemein? **[16]** Worin besteht die 'enlightened simple-mindedness' von

This 'vague' feature is a feature *in relation to* pure mathematics. In the above context, it challenges the tendency to be prone to form a misleading analogy. This confrontation is epitomised in the Pink Book – Chapter 3 below. It states there that: "We are tempted to think a geometrical investigation is like looking at a piece of chalk and investigating it. This is the idea that mathematics *is* a sort of physics."[184] In this Book I of the Pink Book Wittgenstein is at pains to highlight both the contrast between natural languages and calculus, and the limits that attend the latter.

When Wittgenstein makes such use of quotations from colleagues or draws on their views as sounding boards with which to do philosophy, we come to appreciate that neither choice of person, nor use of a small conceptual fragment, is because Wittgenstein is merely ornamenting his style. In such investigation there is sustained linkage with complex multiple layers and interplays that map his explorations. Wittgenstein respected Littlewood's dual great gifts: pure and applied mathematics, with their contrasts, analogies, and matches.

Having prepared three lectures on religion including a treatment of 'mystery', Wittgenstein remarks, with the date marked as "Christmas Day 1938", that:[185]

> "Fractions cannot be arranged in order of their magnitude."[186] This sounds above all extremely interesting and remarkable. It sounds interesting in a quite different way from, say, a proposition of the differential calculus. The difference, I think, resides in the fact that *such* a proposition is easily associated with an application to physics, whereas *this* [former] proposition belongs wholly and solely to mathematics, to pertain to the natural history of mathematical objects itself, so to speak.
>
> One would like to be able to say of it e.g.: it introduces us to the mysteries of the mathematical world. *This* is the aspect about which I want to sound a note of caution.
>
> When it's as if this is a semblance[187] (Littlewood), we should look out.

The fractions can be arbitrarily ordered (i.e. listed), but *not* in order of increasing magnitude. Wittgenstein supposes correctly that fractions cannot be mathematically founded with well-ordering containing the least element, i.e. the least number *n*, so

---

der Littlewood spricht? Besteht sie nicht darin zu glauben, daß die sogenannte *allgemeine* mathematische Untersuchung Wahrheiten ans Licht bringt, die von den beson spezielleren Untersuchungen nur noch ins Kleine ausgeführt werden. Während der allgemeine Kalkül nur dadurch 'allgemein' [**17**] ist daß er sich auf spezielle Kalküle bezieht. Weil in der Mathematik nichts im *Wort* liegt sondern alles im Kalkül. D. h. weil das Zeichen in die |reine| Mathematik keine andre Bedeutung mitbringt, als der Kalkül selbst sie ihm gibt.Die Mathematik besteht nicht aus Betrachtungen. Und 'abstrakt' kann man einen Teil der Mathe[**18**] matik nur nennen, insofern seine Anwendung zwar angedeutet aber im unklaren |vag| gelassen ist.

[184]Italics supplied here by Eds.

[185]This is a translation of the German that presents some problems of interpretation, in Wittgenstein's *Nachlass* MS 121.

[186]Fractions can be listed, but not proved for a theorem, in order of increasing magnitude from *the* least well-founded numeral.

[187]The German word here (*Anschein*) is sometime used with the term for order, so there may be an allusion to the 'semblance' of order and associated 'mystery' in the 'fractions cannot be ordered' theorem.

as to specify a theorem for all fractions.[188] He does not dismiss "mysteries" ("secrets") but shows that they can be composed of absent solutions (e.g., the least numeral) of which we are forever ignorant, though tantalizingly close. He takes it that this ignorance can be wrongly masked as a mystery by a misleading analogy of taking mathematical physics as a measure of everything, when there is no possibility of a mathematical calculation. We cannot calculate what is the least numeral of all, in order of magnitude. Wittgenstein is warning against a sort of presumption of a type of mathematical imperialism that all mysteries are extirpable by proof, whilst charging us not to hold a counter view that this entails there is inexorable mystery.

He pinpoints this situation with subtle care; confident he would meet Littlewood's way of thinking mathematically. Wittgenstein here portrays his concern to develop the idea against, what he calls in the Archive (in Chap. 5), "talk of fundamental intuition in mathematics – the word 'intuition' refers obviously to some psychological event which is utterly uninteresting", which appears to reflect Hardy's sense of intuition.[189] Wittgenstein extends this (in Chap. 8) by advancing his argument that "a logical proposition has nothing to do with self-evidence".[190]

Turing

There appear to be some grounds to consider that Turing implicitly acknowledged some debt to Wittgenstein, and the current explanations suggest various aspects of how Turing's language partially mirrored or benefited from Wittgenstein's influence, including the latter's understanding of intuition.[191] Turing's recognition here is linked to features of the manuscripts published in the present book. Turing's undergraduate years in Cambridge coincided with those in which the manuscripts transcribed into this book were composed, in association with Wittgenstein's lectures, and it seems that Turing was present for a number of them.[192] This period was years before Turing's notorious attendance at Wittgenstein's 1939 lectures. Turing and Wittgenstein were both living in Cambridge at the same time in periods from 1931, attended some of the same meetings, and knew a range of the same people who commonly engaged with one another.[193]

---

188 See relevant formulation of 'well-founded' by T. Forster (2003: 27).

189 Cf. Chap. 5 lecture dated 23rd Feb. We should allow here for the distinction between 'intuition' and 'an intuition'.

190 See Chap. 8, lecture VII, sec. 6.

191 For example, Turing (2013); cf. Floyd (2016a: 33).

192 In addition to the material gathered and evidence presented in the present book, see Floyd's (2013, 2017a) research.

193 For example, the Trinity Mathematical Society, the Moral Sciences Club; Keynes, Max Newman and others.

Hodges (2008b: 821) states that Turing "was unusual as a mathematician in that he explored. . .the wider implications for philosophy, science, and engineering". There is a need fully to recognise Turing's interaction with philosophy and with Wittgenstein's thought, as sources of Turing's creative notion of 'process', starting to emerge while he was an undergraduate.[194] This is not a mere ornamental flourish by Hodges. For Turing, philosophy was central to maths. We have Gödel's own authority on the importance of philosophy as a source of Turing's insights – presented in Turing's (1936) historic paper, 'On computable numbers. . .', which Turing personally sent to Wittgenstein. Appel (2012: 2) expresses an aspect of Gödel's judgment: "Turing's 'On computable numbers' convinced Gödel. . .in part because of the *philosophical* effort he put into that paper, as well as the mathematical effort." So, the following associations between Wittgenstein and Turing that are documented in the Wittgenstein-Skinner Archive, appearing in the present book, are of some importance.

Furthermore, we know that Wittgenstein had quickly identified, as Goodstein realized, a key limitation in Gödel's own assessment of a finite string of signs, which could not in key respects of proof be assigned a useful function within domains of ordinary mathematics. Presumably, Turing knew of this from Wittgenstein, and maybe from his own researches via a different route.[195]

Turing's 'On computable numbers. . .' employs 'intuition' and phrases utilizing it in ways parallel with Wittgenstein. For example:

> All arguments which can be given are bound to be, fundamentally, appeals to intuition, and for this reason are rather unsatisfactory mathematically. The real question at issue is "What are the possible processes which can be carried out computing a number?"
>
> The arguments which I shall use are of three kinds.
>
> (a) A direct appeal to intuition.
> (b) A proof of the equivalence of two definitions (in case the new definition has a greater intuitive appeal).
> (c) Giving examples of large classes of numbers which are computable.[196]

Here Turing is making in *(a)* a direct appeal to intuition. Wittgenstein's framing of appeal to intuition problematises its use, whereas Turing does eliminate it during his progress with calculation in his article cited above. Accordingly, Turing has not shown that intuition's condition (a), conforms to his theorem's requirement, which is: "the theorem holds for any section of the computables such that there is a general process for determining to which class a given number belongs".[197] As Turing himself notices in the above quotation, this is not satisfactory. Rather, intuition itself

---

[194]See Juliet Floyds (2017a: 113–117, etc.) explanation of Turing's education here in relation to Wittgenstein and Littlewood's (1926) book, which is also informed by reading the above subsection entitled '*Littlewood*'.

[195]For example, cf. S. S. Grève and W. Kienzler (2016); cp. J. Floyd (1995: 408).

[196]Turing (1936: 249).

[197]Quoted from Turing (1936: 252). (As noted below, he does return to its role in his Princeton PhD thesis.)

is incalculable, given the truth of the stipulative power of the theorem he presents, in which he shows that the Hilbert *Entscheidungsproblem* can have no solution.[198] Furthermore, in Turing's PhD. Thesis (1938: 81) he volunteers that the role of intuition is "less open to criticism" with regards to calculation.

Wittgenstein, already in 1934, in the Pink Book, Part II (Chap. 3 below), had explored aspects of this problem of intuition, and its relations to calculation, to isolate and remove or restrict intuition from being a determining function in analysis and judgement. Well before Turing's (1936) use of the above quoted expression, Wittgenstein had explicitly treated "the possible processes", formulating and determining the scope of this phrase in such a way as to render Turing's type of problem explicit, and how to treat and dissolve it. This involves a central issue for Turing – yet already being addressed in the 1930s by Wittgenstein and before Turing, and pertinent for understanding themes in the Archive here published. We might summarize it as a dilemma: can or should the mind regulate a machine, or a machine come to regulate mind? We may shrink the foregoing context to the question: can intuition regulate a machine, or a machine come to regulate intuition? Turing (1936), in effect, uses a non-mathematical hunch (intuition) to construct a proof that self-tests so that the resulting proof aims to eliminate its origin – the intuition. Nevertheless, it is not completely eliminated. This is only one of the reasons why Turing sent the resulting 1936 published paper to Wittgenstein. Turing had also drawn on Wittgenstein's presupposition of 'common sense' to develop his approach.[199] Maybe another reason for this was, at that time, some Cambridge and Princeton mathematicians[200] had not fully understood the paper.[201] Turing and Wittgenstein were two complementary yet differing aliens amongst partly uncomprehending colleagues. Of course, there was some grasp of Turing as a rising figure, who associated with Wittgenstein.[202] This sketch frames a context in which Turing

[198] I.e., Turing (1936: 259).

[199] For Wittgenstein's use of this expression in the Archive, see Chap. 5. Cf. Floyd 2017a for discussion on this theme. This presupposition also bears on the issue of how autonomy and desire relate to machines and cognitive agents (cf. Erden and Magills 2012).

[200] E.g. Cambridge: Newman (cf. Hodges 2008a); Princeton: Von Neumann, for which (cf. Feferman 2012: 19).

[201] To illustrate the pressure of this, note that Turing did not succeed in obtaining any of the three Cambridge mathematics posts for which he applied at that time, and they went to people who are very largely forgotten, even in Cambridge, or not of any comparable importance to Turing.

[202] Deacon (1985: 130) reports that there was already in the 1930s considerable "computerology" interest and influence within the Apostles – the elite Cambridge secret society; yet Turing, who knew many of them, not least those at King's, was not invited to become a member. Apostle members included: the mathematician Alistair Watson, who with Turing met Wittgenstein (himself an Apostle member, elevated to becoming an honorary member (an 'Angel') in 1929; Keynes, who brought Wittgenstein back to Cambridge; Champernowne, the mathematician, of King's); Hardy; the philosopher Braithwaite (but note that Wittgenstein fundamentally disagreed with him about logic, for which see Floyd 2017a); and others such as Victor Rothschild, who was influential in fostering science developments. (Champernowne and Turing were interested in Wittgenstein's use

would seek Wittgenstein's view, by Turing sending him his 1936 epoch-making paper.

We find the basis for answering Turing's concern already in Wittgenstein's Pink Book, composed in 1934 two years before Turing's publication. Wittgenstein's usage here reflects its occurrence in Wittgenstein's lectures and, knowing as we do that Wittgenstein crafted terms by repeated experimental usage, presumably this occurrence presupposes oral usage in discussions. Therefore, we do well to attend to the Pink Book's deployment of "the possible processes". For example:

> If I give the order 'bring me a chair', you say he wouldn't understand it without 'bring me': this is a hypothesis.
>
> If you say what did I mean when I said chair, you may say I meant 'bring me a chair'. This may not have been in my mind when I said it. If I said "chair", you may say to yourself he means, "bring me a chair". There are cases where there is such a process, but there needn't be any such process. If there is such a process, it is a translation.
>
> The way to understand this is to split up the *possible processes* into games. . . .
>
> The multiplicity of a sentence is the multiplicity of possibilities. . . .
>
> This comes in when there is doubt.

Turing's (1936) own use of "possible processes" is combined with him deploying his expression "process" some 33 times in his paper, which amounts to integrating an explicit conception of process with that term.[203] Turing's presupposition in the process theme is akin to Wittgenstein's above quoted use of 'doubt'. Turing's states: "Although it is not possible to find a general process for determining whether a given number is satisfactory, it is often possible to show that certain classes of numbers are satisfactory."[204] Nevertheless, Turing agrees that: "we cannot regard the process of recognition as a simple process. This is a fundamental point",[205] which involves "checking" all individuals in classes. This matches Wittgenstein's connected concern regarding the Pink Book quoted above, and later in the Pink Book he explicitly requires, "ways of checking up on the recognition".[206] This connects with Wittgenstein's treatment in the present Edition of "satisfy", the incomplete success of which limits "satisfy", so as to leave us with doubt.[207] Furthermore, Turing (1936: 252) also employs the term 'possibilities' – as does Wittgenstein, while Turing presupposes Hilbert's functional calculus, which allows one only to suppose choices between two possibilities 0 and 1. Although Wittgenstein illustrates his point with

---

of chess; in 1948 they crafted the chess-playing computer program – they named TUROCHAMP; cf. Champernowne 2000.)

[203] See Turing (1936: 231, 235, 246, 247, 248, 249, 253, 255, 256, 259, 262 – notice that "process" appears more than once on many of these pages). Additionally, Wittgenstein's use of "translation" above concurs precisely with Turing's (1936: 236, 237, 259, 260) eight uses of "interpretation". Also cf. Floyd (2016a: 228).

[204] Turing (1936: 255).

[205] Turing (1936: 251).

[206] See below Chap. 4, §36.

[207] In the same Chap. 3 context as his above indented quotation, from Pink Book Part I, entry dated: 20th Nov.

ordinary English, he is obviously articulating a debate that draws attention to a richer number of mathematical possibilities.

Artificial AI

Against this, it might be considered that since Turing is concerned with theoretically engineering a machine, Wittgenstein's exemplifying his remarks by using a person's speech strikes such a contrast with Turing's and his contexts, as to render the comparison of the two irrelevant for the present purpose. Not so; for example, this would be to neglect Turing's (1936: 231) own comparison in his paper: "We may compare *a man in the process* of computing a real number to a machine which is only capable of a finite number of conditions". We should recall here that it was Turing who sent this paper to Wittgenstein – not the other way around. Turing's purpose evidently presupposed common interest with his correspondent in the specifics of the paper.

Wittgenstein's remarks aligned above with Turing's (1936) paper are certainly not the counter-arguments antecedent for Turing's paper. The presence of the above relations between Wittgenstein and Turing signals parallels, as well as contrasts, that indicate connections between them, the measurement of which is a challenge for new future investigations. They expose interplays that introduce us to fresh dimensions of connection to explore in the present Edition. It is probable that, as an undergraduate, Turing had heard or knew of Wittgenstein lecturing on the topic that comprised the Pink Book. Either way, there are specific parallels with Turing (1936) with the Pink Book, as well as other elements in various chapters of the present Edition.

Explicit attention to the function of doubt is absent from Turing (1936), whilst he proceeds to presuppose intuition in conditions for analysis – reducing the role of intuition as far as he could. Wittgenstein had, prior to Turing's (1936) paper, specifically laid out an account of intuition's problems, which either instinctively or consciously Turing partially attempted to avoid.[208]

In contrasting yet complementary perspective to my above scrutiny, Appel's (2012)[209] and Feferman's (2012)[210]contributions question Turing's use of intuition, though in his later Princeton research.[211] In Turing's 1937 thesis he wishes to reduce

[208]Turing (1936: 249). See Pichler (2018a, b) to consider other aspects of Wittgenstein's use of 'intuition'.

[209]Appel (2012: 7).

[210]Feferman (1988).

[211]During his undergraduate studies, Turing was taught the foundations of mathematics by Max Newman – who knew Wittgenstein, though there was some tension between them (Newman was a PhD examiner for Alice Ambrose). There was some friction between Newman and Turing, which was largely due to Newman not immediately recognising the importance of Turing's visionary insights. Despite this, Newman's meticulous yet overly pedantic concern with detail disputed Turing's sometimes slightly slack manner with technical formulations. Nevertheless, such matters

the reliance on intuition, while concluding: "I shall not attempt to explain this idea of 'intuition' anymore explicitly."[212] Nevertheless explicitness is precisely what is needed. I have applied the foregoing comparison and contrast with Turing and Wittgenstein to Turing's PhD thesis, which yields some of the same evidence of concerns that intertwine in compositions of both writers.

When in 1936 Turing went to Princeton for Two years,[213] he engaged with three projects, one of which built on Littlewood's research.[214] Turing returned to Cambridge, then went back to Princeton, and also was in contact with Wittgenstein after his next return. The years 1931–1939 contained related crucial creative phases for both.[215] When Turing's (1936) article was published, Wittgenstein was one of the very few people to whom Turing sent his research paper, embodying research he conceived in Cambridge and developed in Princeton. I believe that the specific questions it addresses mirror some influence of Wittgenstein on Turing during his undergraduate period in 1931–1934, and Turing's later times in Cambridge after 1936.[216] Reviewing these relations yields evidence that Turing draws on his knowledge of Wittgenstein's research that is in the Archive, and aspects of the *Blue and Brown Books*. There is much more that can be explained here, resulting from the Editor's research, though it is beyond the scope of this current context. So the following will have to suffice as a typical illustration. Turing (1936: 231) argues that:

> We may compare a man in the process of computing a real number to a machine which is capable of a finite number of conditions $q_1, q_2, \ldots, q_n$ which will be called "m-configurations". The machine is supplied with a "t a p e" (the analogue of paper) running through it, and divided into sections (called "squares") each capable of bearing a "symbol". At any moment there is just one square, say the r-th, bearing the symbol $\mathfrak{S}(r)$ which is "in the machine". We may call this square the "scanned square". The symbol on the scanned square may be called

to do with intuition and its relation to notions of a machine having assigned 'directly aware' to it, perhaps justifiably concerned Newman.

[212] The quotation from Turing's Princeton thesis, is on page 81; cf. the whole section on intuition, pp. 81–84. His thesis is published in Turing (2012), where the pagination for this section is pp. 116–119.

[213] Arriving September 1936.

[214] The Skewes number, concerning which Littlewood proved that $\pi(x) - \mathrm{li}(x)$ changes sign infinitely often – so not good for claims of intuition in this arena (cf. Feferman 2012: 18); note also Turing's taking up research on the Zeta function, on which Littlewood had worked.

[215] For an aspect of their mutual background contexts in Cambridge during the 1930s, see footnotes to the "Editorial Conclusion" to Chap. 9 below.

[216] Such contact has complex context, for example, Turing's meetings with the Apostles Wittgenstein and Alistair (George Douglas) Watson. Watson was a member of an inner circle comprising Wittgenstein, Sraffa, Ramsey, Keynes (see Kurz and Salvadori 2003). Formerly a Trinity mathematician, Watson joined the Radar and Signals Establishment of the Navy (Head of the Submarine Detection Research Station at the Admiralty Research Laboratory). Blunt latterly maintained that Watson was a Russian spy (See Blunt's alleged confession manuscript (Blunt 2009). Not so: Blunt would hardly have named him if he were, since Blunt was a Russian spy. Blunt's alleged confession was a last-ditch distraction device (as was his attack via Bell on Wittgenstein during the period the Archive in the current book was composed; see McGuinness (2006: 374–75).

> the "scanned symbol". The "scanned symbol" ≠ is the only one of which the machine is, so to speak, "directly aware".

Wittgenstein had, previous to this study by Turing, already begun tackling some of the issues Turing addresses. Wittgenstein's now-published discussions in the present book, and in his *Brown Book*, show that from 1933 Wittgenstein was exploring a range of issues, and using some of the vocabulary employed by Turing in the 1936 article quoted above. Obviously this does not lessen the importance and significance of Turing's research. Even so, it does expose a degree of Turing's debt to Wittgenstein. There is more evidence in this direction in Wittgenstein's *Remarks on the Foundations of Mathematics*.[217]

There is discussion of the figure of 'pianola' and a machine function in the *Blue and Brown Books* of 1932–1934, which appear in a fresh form in the Skinner Archive in the present Edition.[218] Wittgenstein here builds on the problems of the pianola example to furnish a far more complex case of a machine treated *as if* it were a function of a mind – which in a narrow perspective anticipates a feature of Turing's machine and its tape. Looking at the evidence of the Archive from a variety of viewpoints, Turing himself drew on select aspects of Wittgenstein's discussions, lectures or manuscripts.[219]

In twenty-first century terms we could recast Wittgenstein's pianola into a quantum computing and quantum mechanics' illustration: the notion of a holographic tape, in an entangled state – as opposed to a limiting binary tape in Turing (1936). Wittgenstein concludes that the holographic model is not equivalent to mind; and never will be adequate. Turing appeared to have held this conclusion, whilst strands of his view cited above seem to concede too much ground (for example, "the machine is, so to speak, "directly aware") – against which Wittgenstein argues in the chapters below.[220] It should be taken as a warning from history, and our future, for artificial intelligence not to be conflated with being or becoming a mind.

Wittgenstein (prior to Turing's 1936 article) laid out and probed major issues that would confront any attempt to interpret a case such as Turing's as a basis for linking mind to machine. This includes debates, concepts, specific expressions and terms

---

[217]E.g., sec. §2.

[218]Cf. Shanker's (1987, 1998) valuable discussions.

[219]Here are samples: *Brown Book* (pp. 65–66) "we could only decide ...if we could look into the actual mechanism connecting seeing the signs with acting according to them... "being guided by the signs" is a mechanism of the type of a pianola...being guided by the pattern ... reading off the record we might call patterns ...complex signs or sentences, opposing their function ...to the function which similar devices have in mechanisms of a different type... the pianola can read any pattern". Compare a sample from the Pink Book below, written at the same time: "How can a mental act accompanying the hearing of the word stand for the whole calculus with that word?... Thinking is not reading the sign; it is interpreting the sign. I can think a sentence while I am reading it (and interpret it while I am reading it)." And the previously unknown extension to the *Brown Book*, below Chap. 4: 'What if I said: "It's a symbolism, the sounds change their meaning, but then you have admitted that symbols may change their meaning while we play a language game"?'

[220]Turing, in his 1951 radio broadcast, maintained that the mind is uncomputable; see also Hodges (2008a).

that are central for Turing (1936). A wide range of these, which were deployed by Wittgenstein in lectures and dictations, from the Michaelmas Term of 1932 to Summer 1935, are published here: for example, the use of 'machine' in the above quotation from Turing. Additional to Turing's treatment, we find it associated with 'doubt' in the present Edition, where Wittgenstein is tackling both resolved and connected, unanswered questions that appear or are posed in Turing's (1936 and 1937) researches.

So let us here briefly extend the foregoing introduction by Wittgenstein of 'doubt', which hovers over the issue of 'intuition', to enable us to key 'doubt' into questions surrounding use of 'machine' in the above quotation. Thus, in the Pink Book Part II, we find Wittgenstein arguing:

> If I say there is a case where there are no doubts, you still think that is because he knows it. You may say (1) the man who has no doubt is a fool or a machine because he can't doubt, or (2) that because there is no doubt that he knows it all.
>
> (1) A doubt is something that happens, and then there is an answer.
> (2) What does it mean that he knows it? If you say he knows it, I say, "when does he know it?" Does he know it all the time or does he repeat it to himself?

It is as if Wittgenstein either knew Turing's drift of thought or is anticipating it. In these sorts of respects, Wittgenstein sometimes appears to have had prescient tuning into the consequences of emerging patterns, and for some future pathways.

In summary, this above Introduction aims to function as a backcloth to assist in framing and positioning the collection of manuscripts published for the first time in this Edition.

Section 2: Characterising the Archive's Contents

The Pink Book (Chapter 3)

This manuscript is well developed as an original composition. It displays some familiar signs of Wittgenstein's progress to attempted completion: the signs of virtual completion, almost forever, not quite reached. In this, the first use of 'progress' here in the Archive, it is worth emphasising a sense, and tone, of Wittgenstein's use of Nestroy as epigraph to the *Philosophical Investigations*: "The trouble about progress is that it always looks greater than it is". This also, as Wittgenstein's autobiographical presupposition, readily unlocks the view that such domains as mathematics are infinitely deep. On this assumption, each advance in knowledge entails ignorance of wisdom beyond its upper finite bound. So progress includes a process of exposing the identities of its contrary, which should re-calibrate and downgrade being held in thrall of new discovery. This type of enterprise is originally exposed and multifariously implicitly presented in this Pink Book manuscript.

This manuscript appears to be a revised text, or contains the revised outcome of lectures, with responses incorporated into them. Some of the revised text is in Wittgenstein's hand,[221] also with revisions in Skinner's. For example, the manuscript of Book I of the Pink Book (published in Chap. 3 below) has various revisions, including a deleted expression, which this Edition demotes to a footnote:

> The laws of geometry talk about the cube in the sense of giving rules.
> [~~An objection was made~~]
> I said geometry fixed the meaning of the word 'cube'. If one alters the geometry, one alters the meaning.

This deleted expression, together with other data, might be evidence of the manuscript being an outcome of redrafting form lectures to fair copy. Another instance, which comes after the above passage, seems to report the response Wittgenstein wishes to tackle:

> An objection was "is it true all the geometry constitutes the meaning, or does some constitute the meaning and is the rest the propositions which follow from the rules?"
>
> I can neglect this question. . . .

The Pink Book is concerned with some relations between how to represent a range of ordinary language use, and how this parallels and contrasts with some senses in mathematical calculations. For example, the expressions 'a rose is red' and the geometrical use of a 'cube' are explored in respect of aspects of their comparable similarities. Wittgenstein of the *Tractatus* supposed he had there fully answered the question "What is a proposition?" His autobiographical consciousness arcs back and harnesses complex changes of aspect to expose new identities.[222] Here in the Pink Book (Book I) he launches an attack on his own posing of the question and answer, commencing with the more modest: "What is a proposition about red?" He argues that proper analysis of language-use displaces formal logic as the basis to be a model of natural language, and patterns in mathematics displaces standard (Russellian) logic as the foundation for logic.

The above theme concurs with Wittgenstein's viewpoint expressed slightly later than the composition of the Archive. In 1937 Wittgenstein pinpoints a trajectory that complements his above new exploratory directions, by using the following analogy, which utilizes patterns in the Archive:

> This book is a collection of wisecracks. But the point is: they are connected, they form a system. If the task were to draw the shape of an object true to nature, then a wisecrack is like drawing just one tangent to the real curve; but a thousand wisecracks (lying close to each other) can draw the curve.[223]

---

[221]Two examples of Wittgenstein's handwritten additions or changes: in Chap. 3, some six sentences or clauses on MS p.41; 12 lines inserted on MS p.45 – all fully detailed in footnotes to that Chapter.

[222]This use of 'aspect' presupposes Floyd's (2018) approach; cf. Hagberg (2008: 16–43).

[223]Wittgenstein, MS 119 (Band XV), pp. 108–109. The date of 14 October 1937 is on the previous page to the quote in MS 119, p.107.

Wittgenstein here forms an analogy to expose a new identity, by using both maths and literary elements. Contemporary with the composition of the Archive, analogy within mathematics, as now, was often deployed to craft the emergence of fresh insights. For example, algebraic topology, especially in the 1920s and 1930s, was deployed to craft a new field from a known one.[224]

Visual Illustration: Calculation in the Pink Book

Sometimes Wittgenstein had a prescient sense of some future directions for mathematical research, which have found realisation in some twenty-first century perspectives.[225] The use of the visual is one of the bridges between mathematics and non-mathematical philosophy that Wittgenstein deploys. This perspective has some expression in Wittgenstein's large format C-series lecture Notebooks numbered 145–152, from 1933–1936, which also have links with the Skinner Archive that require extensive exploration.[226] This can be seen in the Bergen Edition, where Wittgenstein's visual illustration is embedded in his notes.[227] Frequently they are in contrapuntal position harmonising and developing his written text. Such use of visualisation employing mathematical examples attests to the causal relations between Wittgenstein's mathematical philosophy and his other philosophy, found often as they are, on the same page.

Sometimes Wittgenstein uses a very general illustration and speaks about it briefly. For example, even the use of a roughly drawn circle, such as that which appears in the Pink Book alludes to contemporary, and now still current, mathematical debates. The Pink Book and Wittgenstein's 1935 Notebook MS 148 deploy the uneven 'circle'.[228] In the latter there are some – quite abbreviated, yet evident – outline proposals of mathematical research debates, which connect with some research by his colleagues.[229] Certainly he was not doing specialist mathematical research. Nevertheless, he is probing the then-foundations of mathematics associated with individual mathematical problems. He knew how to picture such matters to best assist mathematicians to think mathematically in research. Sometimes this is with a strategy that might not come from a mathematician.

---

[224]See Burt Totaro, in Gowers et al. (2008: 383–96).

[225]See Gibson (2013).

[226]David Stern (in Klagge & Nordmann. Eds. 1993a: 200–288) has laid a basis for this in research in editing-related manuscripts. Important for this research is analysis of visual elements linked to narrative.

[227]See also Rhees (Ed.), in *Wittgenstein: Schriften* 5 (1970: 117–237).

[228]Cf. Pink Book p.11 and Wittgenstein's 1935 notebook (MS 148: 37).

[229]Cf. MS 148: 36, 38–40 – note especially the employment of terminology to do with calculation and use of this last term; cp. MS 148: 44. Re. colleagues, note for example Newman (1923), and Littlewood [1944]. Cf. Chap. 1 of the present publication relating to lectures delivered *circa* 1931; it is pertinent to Wittgenstein's discussions with Littlewood on functions.

A merit in Wittgenstein's use of mathematical illustration in the Archive is the use of quite elementary arithmetical topics (such as fractions, graphs, and *pi*), which have within their research orbit highly complex, competing asymmetric routes to solutions. Wittgenstein was personally familiar with or engaged with a range of these topics.

Schopenhauer in the Pink Book

Ambrose's Yellow Book fragments contain a reference to Schopenhauer. As the Pink Book (Book I) opens it mentions Schopenhauer, though it presents a map analogy about doing philosophy.[230] Apart from the occurrence of the name Schopenhauer, the contexts are quite different in the Pink Book in its contrast with the 'Yellow Book'.

In the first case, Wittgenstein inserts the paragraph below, which fits in well with his opening reference to Schopenhauer in the Skinner Pink Book (Book I) concerning reading and mapping understanding. Later in the Pink Book, we find another reference: 'Schopenhauer once said, "if you try to convince someone and get to a certain resistance, you know then you are up against the will, not the understanding".'[231] As Sophia Vasalou points out, in relation to Wittgenstein's enthusiasm for Schopenhauer's sense of wonder, Wittgenstein is concerned with those things that are precisely "hidden because of their simplicity and familiarity" – i.e., they are all too easily missed by our seeing them under an aspect of our context's assumptions of familiarity, rather than their own specific identity.[232] He was also attentive to hidden complexity.

Secondly, on the manuscript page facing the above quotation, we find in Wittgenstein's own handwriting, a paragraph that to some relevant degree sympathises with certain aspects of Schopenhauer's approach to aesthetics and psychology:

> It is extremely interesting that very often when people say that science[233] has not yet discovered this or that but if X will have discovered it then..., that they very often don't know at all what sort of discovery they are waiting for, that they talk of a discovery without knowing of what nature[234] this discovery would be. (For example, when people say that one day when psychology will be far enough developed it will make us understand the nature of beauty.)

---

[230]For detailed discussion of the appearance of Schopenhauer, see the footnotes attached to the above quotation in its setting in Chap. 3's presentation of the Pink Book.

[231]This occurs in the Pink Book, p. 43.

[232]Vasalou (2015: 3) and Wittgenstein's *PI* §129, p. 56.

[233]This could be a response to Sraffa's type of view, as in McGuinness (2008: Letter 172).

[234]Wittgenstein writes 'sort' first, deletes it and replaces it with 'nature'.

It is worth noting what follows after the above quotation from the Pink Book, in Skinner's hand – and presumably dictated by Wittgenstein, which continues:[235]

> If you ask why we use substantives?
> You can find games with boards and holes in them all over the world. I can only say that human beings are apt to act in such a way. I can make up games which human beings don't play.

Wittgenstein's handwritten words above presuppose a thread between Schopenhauer's sense of wonder at beauty and will. Wittgenstein was no doubt aware of Schopenhauer's absorption with the uses of simile and analogy as means to craft wonder: "making up games that humans don't play".

Communication of Personal Experience (Chapter 4)

The last page of Skinner's *Brown Book* is succeeded by a sequence of 43 manuscript pages, which have not been published before (Chap. 4 below). The manuscript appears in the last part of Volume V and comprises all of Volume VI of the *Brown Book* in Skinner's handwritten copy. Skinner's *Brown Book* was not known, nor any handwritten manuscript of the *Brown Book* known, before the discovery of the Archive.[236] This hitherto unpublished extension to the published *Brown Book* in a very good state of presentation – at least as refined as the Archive's copy of the *Brown Book*, arguably in some placed to a more refined stage than elements in the published *Brown Book*. For example, the extension that comprises Chap. 4 has new corrections in Wittgenstein's hand.[237] This manuscript is an integral part of the full handwritten manuscript of the *Brown Book*. A feature of the evidence for this is that this new appended section deploys numbered Remarks cross-referring back to the published part of the *Brown Book*.

The Edition of the *Brown Book* published by Rhees ends with: ". . .the mirror." The Skinner *Brown Book* manuscript has this same ending; then the next page resumes with another section, which is published as Chap. 4 in the present Edition. The following expression is appended in pencil to the text in the Wittgenstein Skinner version, after the word 'mirror' (Plate 1.8):

> "ended of printed version".

The version "printed version" could indicate either 'published' or a typed copy. The associations of "printed version" today may ring differently to that of the 1930s: a reference to a typescript, a carbon copy, a cyclostyled copy, a published form.

---

[235]There is no parallel for either of these remarks in Ambrose's (1979) Yellow Book details.

[236]I.e., no handwritten or original typed copy.

[237]For example, in the Skinner *Brown Book* (Trinity Wren Library designation: Add.MS.a.407.3. book 5) there is a note (MS, p.12) that states: "(new line.)", which is there to mark a paragraph break (in MS Vol. III, Remark 10, or Note 10) before "Could you tell what is in common between. . .".

Plate 1.8 The last page of Skinner's *Brown Book*

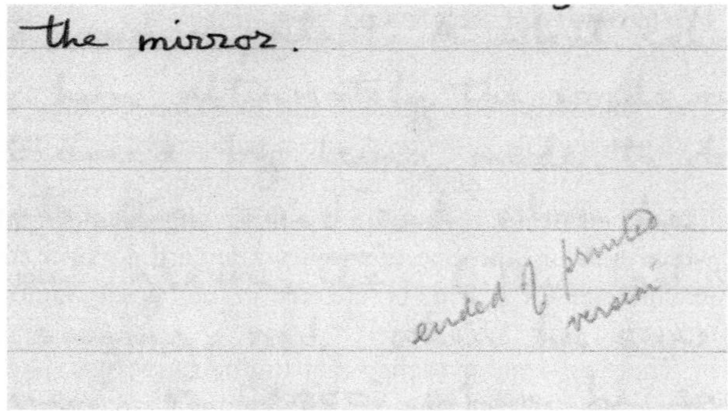

Augustine

There is a quotation from Augustine in this manuscript. In its precise form it does not occur elsewhere in Wittgenstein's other writings. References to Augustine occur in three places in Wittgenstein's already known *Nachlass* notebooks.[238]

In Wittgenstein's terminated attempt to transfer part of the English *Brown Book* into German[239] (on which manuscript page he gives the date as "Ende August 36"), he also quotes 1.8.13, as follows:

> *. . .cum. . .[ipsi] appellabant rem aliquam et cum secundum eam uocem corpus ad aliquid mouebant, uidebam, et tenebam hoc ab eis uocari rem illam, quod sonabant, cum eam uellent ostendere.*

Here is Chadwick's (1991) translation of this passage:

> and wanted to manifest my wishes to those who could fulfil them as I could not. For my desires were internal; adults were external to me and had no means of entering into my soul. So I threw my limbs about and uttered sounds, signs resembling my wishes.[240]

If we were to juxtapose Wittgenstein's above quotation of Augustine from 1.6.8 with the German version use of 1.8.13,[241] we discover that this occurrence of the conjunction *et* is not in the Latin text.[242] The curved parenthesis is written in pencil in what appears to be Wittgenstein's hand.

---

[238]They are: (1) "Augustine, about expressing the wishes inside him", MS 149. (2) In MS 152 Wittgenstein presents five very sketchy attempts to introduce 'Augustine'. (3) Wittgenstein's use of the German at the beginning of the *Confessions* (in 1.8.13) in §1 of *Philosophical Investigations*. This, though seemingly similar, is not the 1.6.8 citation that appears in the present's Chap. 4 Edition of the extension to the *Brown Book.*

[239]In *Philosophische Bemerkungen* MS 115, XI, p.118.

[240]Foreword by Henry Chadwick (1991) to his Edition of *Augustine* (1991: 7).

[241]Note that this is not the text that occurs at the beginning of the *Brown Book.*

[242]Cf. Clark in Augustine (1995: 32).

Philosophy (Chapter 5)

The manuscript is written in ink in Skinner's handwriting. Sometimes there are revisions, whole sentences, and a paragraph, in Wittgenstein's hand. If the manuscript's contents derive from the official teaching occasions on the five dates mentioned in it, they are on the following days: Monday, Wednesday and Friday, which concur with one of the two standard patterns for the Moral Science Faculty teaching customs. This is reflected in the Faculty announcement of Lectures for the academic year, 1933–1934.

There is some evidence that the narrative has a specific link to Wittgenstein's preparation for lecturing as drafted in the C notebook MS 149. In Chap. 5, as in the lecture for Wednesday 24th January, there is a discussion of toothache, and the expression 'unconscious toothache'. In Wittgenstein's unpublished notebook MS 149 (1935–1936) we have a discussion of toothache, together with the relation of this to "bodies which have the toothache" (ff.2r, 14r); yet there is no substantial overlap between the treatment of the two.

Chapter 5 is original creative work of composition in the later stages of progress. It is a composition that is well underway, with thought-processes marshalled and advancing through new complex territory. Its written form is consistent with being an outcome of redrafting a previous version[243] – perhaps of a draft developed in the lecture series and redrafted outside of the lecturing performances.[244] There is a sustained multi-themed argument throughout, with evidence of multi-layered redrafting, which is reminiscent of some Wittgenstein-Skinner revision skills in the *Blue Book*.[245] As you will see below and in more detail in Chap. 5, the three versions are to some degree contrapuntal movements of aspect exploring differing conceptual angles. Wittgenstein utilizes a typical geometer's strategy by presenting a simplified version of a more complex problem. Sometimes it is not realised that Wittgenstein's employment of a simple example is drawn from, or parallel with, a professional custom deployed in research by mathematicians. This is done to render formulable or irreducibly precise a problem whose complexity or obscurity is too great to resolve directly. The following case is abstracted from a lecture given on Feb. 14th [1934]. In the illustration and quotation below from Chap. 5 Wittgenstein

---

[243]For example, although the manuscript's own pagination is usually fixed within a specific date, with 1 to *n* on that date, and then repeats the same 1 to *n* within the date of each lecture. This pattern breaks (for Feb 9th), and it starts at page 26 (ringed) and ends with a brief one-paragraph entry (for Feb 14th), which is deleted by diagonal line – with a few horizontal deletion lines. With the next date (Feb. 12th), the numeration returns to p. 1 ending with 3. Then there occurs a replaced longer entry (pp. 1–6) for Feb. 14, headed by Wittgenstein's note.

[244]In contexts such as their flat at 81 East Road, Cambridge. (In an interview with Gibson and O'Mahony, Donald Mackay explained that his family came to own the row of houses that included number 81 East Road. (This row is now demolished, and 41 was located where the Law Court is). Mr. Mackay spoke of his contact with Wittgenstein when the latter lived there, including his visits to Wittgenstein in Trinity; cf. Mackay (2012: 9–10.)

[245]See, J. Smith (2013).

deploys the idea of a point-source with a cone from it, engaged with two squares set at different points in its field, and compares it to when:

> We copy say a square or a face. The question was "does intentionally copying differ from unintentionally copying it by the process of deriving the picture from the object?"[246]
> Suppose that by the process of derivation I just meant say what I draw if I draw a projection and draw projecting rays.

Wittgenstein draws two diagrams; here is one of them:

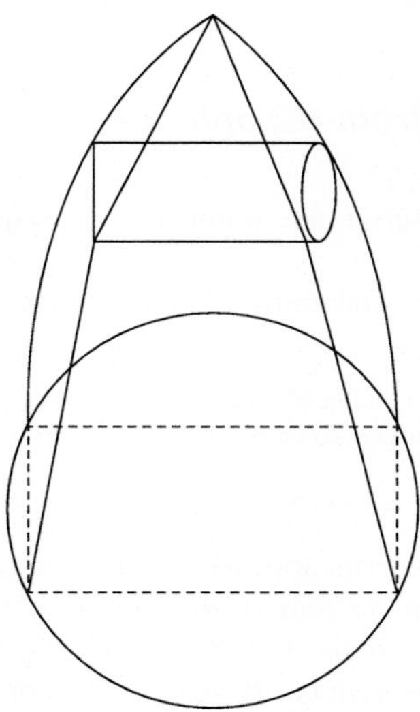

Wittgenstein then states:

> This is a simple case of 'projection' and 'object' (as we could call them).
> There would be the case where he had just drawn the "dotted figure".[247]

> I am now proposing to distinguish between intentional and unintentional by saying the unintentional project is when he draws the right shape without going thro' the process of drawing the rays.
> The intentional is when he draws the rays.
> You say, "Couldn't he have drawn the rays unintentionally?"
> Here you must tell me how you distinguish unintentionally and intentionally in this case.

---

[246] This resumes and extends discussion of the theme that began (9th of Feb discourse) with 'shadow of a fact' linked to the earlier consideration of "copying is the intention of copying", etc.

[247] This illustration in Chap. 5 employs a notion of projection from a point in geometry as an analogy; the idea of a 'shadow' in relation to such projection emerged in 1930's algebraic topology. This compares and contrasts the diagram with the one on the facing previous page of the manuscript.

Here Wittgenstein is employing a notion or projection from a point in geometry as an analogy; the idea of a 'shadow' in relation to such projection emerged in 1930s' algebraic topology. Although the example is simple, it is the type of simplicity that a mathematical researcher could employ to explore the more complicated case that cannot yet be mathematically resolved or fully understood. He uses this illustration to characterise aspects of intentionality. Such strategies occur frequently throughout the Archive: an only apparently transient, brief illustration from mathematics is deployed to launch a large investigation as a model for features of mentality.

Visual Image in his Brain (Chapter 6)

The pages of this manuscript are numbered in sequences of: 1–8; 1–3; 1–5, respectively.

The title of this Chapter is taken from the first sentence of the manuscript, which precedes the statement:

> Hardy said that he *believed* Goldbach's theorem[248] was true.
> All objections seem to be turned down by saying you feel it is there, and then you silence the man.
> The question is not that you can't believe it, but what does it mean.

The discussion in this manuscript appears to fulfil Wittgenstein's promise in the *Blue Book* – that he will return, to remark on Goldbach's theorem. Rhees's editorial footnote in the published *Blue and Brown Books* and (p.11) directed at this reference to Goldbach's theorem curtly states, "This promise is not kept". Contrariwise, we now find out from the Wittgenstein-Skinner Archive that Wittgenstein did fulfil his promise, as Chap. 6's manuscript attests. The unpublished *Blue Book* (incomplete) manuscript contains the date "Wednesday, Nov. 29th", which identifies the year as 1933. Maybe Chap. 6's manuscript was written soon after or at the same time as the *Blue Book*, since it certainly shares yet differently exploits some of the same theme and argument. (Rhees's typical diligence, yet ignorance of this Wittgenstein-Skinner manuscript, warrants the inference that he did not have access to it.)

In the manuscript reproduced as Chap. 6, Wittgenstein is careful not to criticize Hardy's employment of the term 'theorem', where one could expect 'Conjecture'. Hardy does reflect some contemporary usage, such as Dickson's (1919: 421): 'Goldbach's Empirical theorem'. 'Conjecture' was then standard, whilst the post-Church mathematical world would narrow this flexibility in terminology. Even so,

[248]Although it is not a theorem, it is Goldbach's Conjecture or method, yet 'theorem' here does refer to Hardy's own use of the expression 'Goldbach's Theorem' in the latter's *Mind* (Vol. 38, 1929: 4, 24) article "Mathematical Proof": "if Littlewood and I both believe Goldbach's Theorem, then there is something...". Wittgenstein returns in 1939 to engage with Hardy on this matter in (*LFM*, Lecture XIII: 123, where he again refers to Hardy's use of 'Theorem').

Wittgenstein is probing the ways in which mental states facilitate slippage of beliefs by users acting as if a future proof of a theorem is incorrectly taken as a property of its prior identity as a conjecture.

The above quotation complements Wittgenstein's remarks in his Notebook MS 161, 15v-16v:

> . . .is the theory of the game arbitrary? 'I believe that Goldbach's theorem will come true'. How is this belief in the end verified? By a proof? By any proof? No. By something we shall recognise as a proof. But isn't the fact that such and such a proof is *possible* based on mathematical fact, a mathematical reality?... (We shall talk on a later occasion of Professor Hardy saying that Goldbach's theorem is a proposition because he can believe that it is true.)

"The Norwegian Notebook" (Chapter 7)

The physical manuscript is an exercise book, the text of which comprises a continuous discussion by Wittgenstein. There are references to queries raised by and it seems in the presence of Gasking, Paul,[249] Redpath, Rhees, and Wisdom. Although the text is not sustained dialogue, yet it exhibits elements of dialectical confrontation of queries with responses. One very distinguished philosopher of Wittgenstein's work (and a referee for the present book) affirmed that the reflections on private experience and private language in this chapter's manuscript are better than anything else he had seen from Wittgenstein on these matters.

Wisdom[250] had been appointed to a lectureship in 1934. In Wisdom's (1934: e.g. 156–57) book, we see traces of Wittgenstein's impact and sometimes dependence of Wisdom on him.[251] When Wisdom was attending Wittgenstein's lectures, he was writing some six articles[252] that seem partly to depend on Wittgenstein's contributions. Aspects of Wisdom's analysis appear to some degree retrograde, or derivative of, Wittgenstein's 1930s' responses to his *Tractatus* in the present Editions, not least in Chap. 8 where Wisdom is named. The *Mind* articles, though not their agenda, and Wisdom's presence at some of the lectures reflected in Chap. 8 below, may account to some degree for aspects of Wittgenstein's sometimes sustained attention in those lectures to the consequences of the *Tractatus* in relation to his new philosophy and the oblique impact of this in aspects of the *Brown Book*.

---

249 G. A. Paul (1936: 70) refers to Wisdom, which may reflect the influence of an aspect of this debate in Chap. 7 below.

250 This and other references are to A. J. T. D. Wisdom, not his cousin J. O. Wisdom.

251 That is to say, the current Chap. 7 being described above, and Chap. 8 below. In fact, Wisdom's publications in this period seem almost like grateful commentaries on Wittgenstein's original insights.

252 Published by Moore as editor, in *Mind*, cf. J. Wisdom (1933, 1934), etc. In Trinity College's copies of *Mind*, Skinner's brief handwritten comments appear on some of Wisdom's articles.

Self-evidence and Logic (Chapter 8)

The manuscript is precisely divided up into groups of Lecture Notes.[253] Nevertheless, the manuscript constitutes a unity, centering on the question expressed in Lecture VII: "One of the great difficulties which is in[254] logic was this – what is the criterion for a proposition being a proposition of logic". This concern with the criterion, and the consequent constancy in tackling this issue, weave their way within the manuscript. Here, as with other parts of the Archive, Wittgenstein's first person directs the "chapter", which is intertwined with the theme that the truths of logic are not self-evident. In other words, it takes a point out of the *Tractatus*, and transforms into another genre of Wittgenstein's own making that is other than the unique *Tractatus* genre: itself a knowing move against Wittgenstein's self-evidence in the *Tractatus*.

The manuscripts that comprise this chapter are carefully arranged, separate yet unified as a sustained argument. Each Lecture appears to be presupposed as a chapter. For example the first is Lecture VII, which opens with:

> I want to go on now to a chapter[255] by itself.
> By this chapter I have to interrupt what I have been doing – talking about generality.
> This is logic I must talk about.
> One of the great difficulties which is in[256] logic was this – what is the criterion for a proposition being a proposition of logic.
> By some it was taken to be self-evidence.[257] This was what both Russell and Frege said. At certain points where this is a difficulty one tends to bring in psychology.
> The word self-evident is taken from psychology.
> It is what everybody is in his right senses would say is true.
> It is the highest degree of plausibility. But it is clear that what is self-evident may not be to another.
> People did not mean to say these propositions were self-evident to Europeans.
> The self-evidence seemed to be an objective one, not a subjective one. It seemed to lie in the proposition.

[253]See Chap. 2's consideration of this use of "Notes", and also for a picture of the manuscript containing the above text. The employment of "Notes" in the Archive includes references to the *Brown Book*. So the scope of the sense of using 'Notes' can include reference to the content of a published book. The *Blue and the Brown Books* (for which see Plate 9). "Notes" includes the use of "Remarks", as deployed in the *Brown Book*, also their dictation by Wittgenstein. This also accords with Littlewood's (1926, Preface) own custom in his book, quoted by Wittgenstein, in which Littlewood's published title includes the expression "Notes".

[254]Editors add 'is in'.

[255]The references to 'chapter' are absent from Ambrose and MacDonald, while present in Skinner's Archive. Perhaps this is because the current manuscript was drafted as a chapter – a re-worked dictation from Wittgenstein, who refined the lecture material with Skinner as amanuensis.

[256]'is in' added.

[257]In MacDonald's (11.2.35.) version the lecture notes begin as follows: "What [sic] the criterion for a proposition being a proposition of logic? Criterion used to be self-evidence, a word taken from psychology." (There is no use of "Russell, and Frege said" in MacDonald's version, though she mentions Frege in a related connection.)

Even so, Wittgenstein considered that he was also doing a bit of mathematics:

> Then I will talk about in what sense Russell's propositions about generality are tautologies.
> Then I will talk about what has been said: that I showed that mathematics consisted of tautologies. This is nonsense. The truth is extremely simple, that this which I am doing is a bit of mathematics.

It was, and in fact still is, a custom in the Cambridge Faculty of Mathematics that a lecture course in, for example, pure mathematics, is presented as a set of chapters, which given that Wittgenstein was advertised to teach in that Faculty may account for this presentation. Even so, this custom also probably stems from a time when publication could be an oral presentation to a public. Many such presentations either became or had become books, by the time they were delivered. A case of point, which Wittgenstein quotes (mentioned earlier) – is Littlewood's *The Elements of the Theory of Real Functions*; another instance, of course, is the *Blue Book*.

Pertinent to these customs, and Wittgenstein's style that develops in the 1930s, is the mannered mathematical mode of the lectures. Certainly, this itself is part of a play within complex, varying pulses from other genres, which presses against customary limits of language. In this narrowed, ironised mathematical framework in the present Edition Wittgenstein proceeds in the following way:

First comes a definition. Secondly, if there are problems, this dissolves into questions with hope of justified definition. As Wittgenstein is reported in the Pink Book, reproduced below: "The first thing might seem a definition, but then this only reduces the question." In Chap. 8, Wittgenstein uses some mathematics to show that in philosophy we do not have mathematical proof that moves from definition, to proof, to theorem. He stays close to practice in mathematics as a source for reason, whilst pitching battle against Russell's and others' related claims that a certain view of logic is the imperial standard for philosophy.

Mathematical Paper: Calculations (Chapter 9)

The manuscripts published in Chap. 9 do not, in our current state of ignorance, appear to contain evidence that would yield a proof that they are the work of Wittgenstein. This mathematical paper appears in this Edition because Wittgenstein himself chose to include it in the manuscripts that he posted to Goodstein. So its presence should not be disputed on the basis of its lack of evident philosophy, or even solely its apparently limited mathematical value.

Let us consider here only a few of the grounds for identifying patterns in the calculations that reflect some influence by Wittgenstein on Skinner's writing of them. This does not amount to a strong inductive argument to confirm that Wittgenstein has an authorial role for this Paper. These calculations could be viewed as a collaborative result between Wittgenstein and Skinner, or not. That will require further research on, if, or how, these patterns intertwine with queries, probes, and conceptual edges adjacent to mathematical domains and issues that interact with Fermat's Little Theorem calculations.

Although Fermat's Little Theorem is well known and understood, yet some of its extended connections with problems of primes face intractable challenges. It is a typical Wittgenstein concern: "What is the next step?" With what do these calculations intertwine? This involves his game of extracting a new unfamiliar challenging aspect by linking up two known games.

This theorem underlies public key cryptography, the seeds of which were being developed well before WWII, and some of Wittgenstein's mathematical colleagues and friends were already exploring the maths that led to WWII breakthroughs in coding and encryption. Questions about primes connected to this Theorem, and others, were components in these research areas.

Wittgenstein chose to send Goodstein the manuscript presented in this chapter along with the other manuscripts. This warrants the view that Wittgenstein would have welcomed the presence of it alongside the other manuscripts he posted with it. We might do well to consider that it has earned its place with the others as a token of the work of his amanuensis who went well beyond the call of duty, devoting his adult life to assisting Wittgenstein. As such, given the early death of Francis Skinner, it is appropriate to commemorate his role by presenting it at the end of this book's series of chapters that display so well his care and attention.[258]

Skinner's Blue Book

The Archive has a printed two-volume typed cyclostyled copy of the *Blue Book* – much the same as other early copies. (Since similar copies have been published, it is not reproduced here.)

The second volume lacks the last part of the narrative. This premature ending roughly coincides with the termination of another unpublished draft of the *Blue Book* that the present Editors have identified, which is largely unknown – and is not in the current Archive. It is a handwritten copy in Skinner's hand of the first half of the *Blue Book*.[259] This unpublished handwritten draft has sentences in it that are not in the Goodstein copies – i.e., also not in the published Rhees' Edition of the *Blue Book*. These sentences appear to have been moved from the Skinner handwritten copy into the Pink Book (published as Chap. 3 in the present book in the present Edition) – hence the report in this summary of the handwritten copy. These correspondences and changes are noted in the footnotes to the relevant Chapter below.

Furthermore, this Skinner handwritten copy contains dates of composition, which are also marked in his hand. As far we can assess, this is the only time any *Blue Book*

[258]See Appendix [I].

[259]I have read this by permission of Michael Nedo, since the copy is in his collection.

manuscript contemporary with Wittgenstein and Skinner has dates on it. These dates are listed in Appendix [O] at the end of this Edition. Of particular importance, then, is the task of relating these *Blue Book* dates to the few dates also in the Skinner *Brown Book* manuscript in this Edition.

Skinner's Brown Book

There is evidence in the manuscript of this newly discovered version in the Archive that it has obviously been worked on in great detail personally both by Wittgenstein and Skinner. In contrast, the typed carbon copy manuscript that Rhees drew on for his Edition is blank and bereft of any handwriting guidance or revision by Wittgenstein; sometimes it is even without section numbers, with only empty parentheses standing proxy for then possibility of their insertion. It is worth noting Jonathan Smith's observation: sometimes a handwritten manuscript dictated by Wittgenstein or written by him, is a subsequent more advanced version superseding an earlier typed copy.

The opening pages of the Archive's *Brown Book* contain a number of new sentences or sentence parts that have no precedent nor trace in the published version. Furthermore, these are written in Wittgenstein's own handwriting, four pages of which can be read below.[260]

The Skinner version contains remarks, details, and forms that are absent from the published *Brown Book*. This is especially true of the opening two pages, where Wittgenstein has both re-written and added a number of new sentences. There are many other examples, such as in Skinner version of Remark 10 where there is a sentence that is absent from the published version: "There are temptations not to call 1 and 2 numerals or a temptation to call 0 a numeral. Or to say black and white are not colours." It seems evident that, if Rush Rhees had known of this version of the *Brown Book*, then he would have incorporated its readings into the published form.

There is an important issue that arises when contrasting the published ending of the *Brown Book* and Skinner's manuscript. The last page of Skinner's *Brown Book* manuscript is succeeded by 28 pages of text that were not known, and they have not been reproduced or published before. They are transcribed here as Chap. 4. This manuscript appears in the last part of Volume V and comprises all of Volume VI of the *Brown Book* in Skinner's handwritten copy. Skinner's hand-written *Brown Book* was not known of, nor was any other *handwritten* manuscript of the Brown Book known, before the discovery of the Archive.[261] It is in a very good state of presentation and comprises six volumes of manuscripts (in great contrast with the two-part carbon copy copies hitherto known). It is more refined than the published

[260]See these in Appendix [N].

[261]I.e., no handwritten nor original typed copy. There is of course von Wright's later generation carbon copy of a typed copy, which may derive from a copy held or used by Ambrose.

version of the *Brown Book*. For example, Skinner's handwritten *Brown Book*, including the portion here published, has further corrections in Wittgenstein's hand. In sum, Skinner's version of the *Brown Book*, which contains many new additions including some in Wittgenstein's handwriting, carries the *Brown Book* beyond the stage achieved in the Rhees' published version.[262]

[262]Re. instructions for layout, for example. In the Skinner *Brown Book* manuscript there is a note (MS, p.12) stating: "(new line.)" – there to mark a paragraph break. This is in vol. III Remark 10 (or "Note" 10, as Skinner lists it on the front cover of Vol. I), which is placed before: "Could you tell what is in common between...".

At the end of Skinner's volume II, we find a handwritten note, which states:
"shld read?
Ex. 42 nor in the practices of it does a limitation of range play 'a predominant role etc."
'Ex' = example and refers to Remark 42 in volume I"

Chapter 2
The Amanuensis Matters

Section 1: Manuscript Matters

Physical Description of the Archive

The Archive was[1] contained in two boxes containing different types of manuscripts. There are some handwritten cover headings, for example:

The handwriting on Plate 2.1 is that of Professor Goodstein's.[2] We can also identify Francis Skinner's handwriting on a paper bag (below), which states: "Notes taken by myself of Dr. Wittgenstein's Lectures dictated" (Plate 2.2).

The inscription on one of the containers states, "Unedited Notes of Lectures given by Ludwig Wittgenstein in Trinity College in 1934/The Notes were taken by Sidney George Francis Skinner". This is in Francis' handwriting. Since some of the notes do in fact display editing in the form of Wittgenstein's own handwriting, it is clear that in this we happily have evidence of Wittgenstein's knowledge of and attention to the contents in the form of editing, which postdates and updates such an inscription, assuming that the container did originally contain unedited notes.

The Pink Book (Chap. 3)

The title "Pink Book" is used because the whole manuscript is contained in one pink-backed exercise book, which is internally divided into two Parts – "I" and "II". The manuscript is written in ink in Skinner's handwriting. Sometimes there are

[1]The Archive's manuscripts are now preserved in special folders.

[2]This is the same handwriting as that of his letter to Sister Elwyn quoted in Chap. 1.

A. Gibson, N. O'Mahony, *Ludwig Wittgenstein: Dictating Philosophy*,
https://doi.org/10.1007/978-3-030-36087-0_2

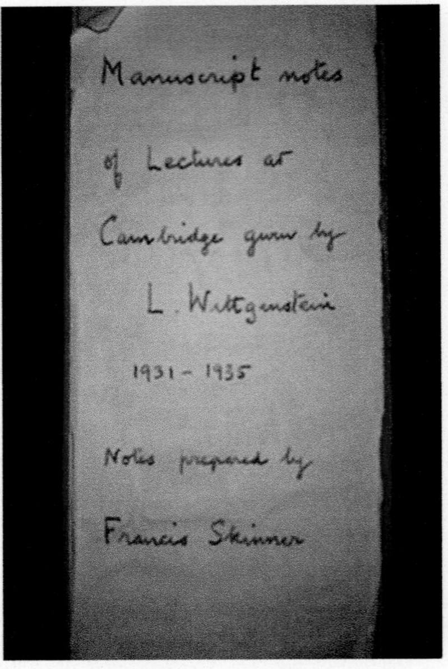

Plate 2.1 A box-file marked "Manuscript notes of Lectures at Cambridge given by L. Wittgenstein 1931–35 Notes prepared by Francis Skinner". (Trinity Wren Library designation: Add.MS.a.407.boxlabel)

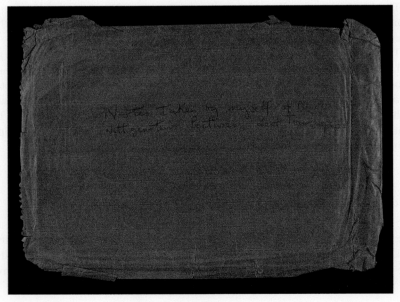

Plate 2.2 Skinner's handwritten note on manuscript bag. (Trinity Wren Library designation: Add. MS.a.407,wrapper). : "Notes taken by myself of Dr. Wittgenstein's Lectures dictated"

revisions, whole sentences, and a paragraph, in Wittgenstein's hand. There is no pagination, which is the same as in the Wittgenstein-Skinner copy of the *Brown Book* (with the exception that Book III of the latter).

The first pages of the Pink Book employ continuous lines to separate out sets of sentences, each set of which appears to satisfy being a remark in Wittgenstein's later writings such as the *Philosophical Investigations*. The lines appear to indicate a line space that was not thought of when this part of the narrative was composed; after a number of pages, the use of continuous lines to divide the remarks peter out, and the line break emerges. In these first few pages the lines are introduced (presumably by going back over the pages after they are written), to introduce the line-break divisions that are characteristic of Wittgenstein's remark. Within the remark, paragraph break still occurs, as in this manuscript.

So the manuscript comprises a fully lined pink exercise book. The books measure 28 × 22.5 cm. The pages are unnumbered. Page 1 is loose, while there is no corresponding end-page folded at the spine to form the last page of the volume. With this exception, all the pages of the two volumes consist of double-folded pages. All pages are written upon in Book I, whilst the last nine pages of Book II are blank.

The two 'Parts' are respectively marked "I" and "II", written in pencil on the top right-hand side of the first page of each part. This Roman numeration and the way it is written are the same as the way the covers of the Skinner *Brown Book* volumes are numbered (except in the latter case, the Roman numerals appear on the front of the card covers). Such Roman numeration, with bars above and below are the same as those used on the covers of Wittgenstein's own German handwritten volumes, such as "Philosophische Bemerkungen XI" (Plate 2.3).

This sort of evidence helps to show that some of Wittgenstein's written and literary customs in German have parallels with those he brings to English writing via his amanuensis, which adds to the inductive case for recognising the manuscripts as expressing Wittgenstein's creative processing.

There is only one outer, exercise book pink card cover, which is not attached to either of the two parts (manuscripts), though they are contained within it. This outer pink cover has maroon adhesive cloth-tape down the stem, which is similar to some of the Skinner *Brown Book* volumes, though the latter are in various colours. (There are six such exercise volumes that comprise the *Brown Book*. The cover of volume III of the Skinner *Brown Book* is the same pink as the cover of this Pink Book.)

There are dated entries in both books: the last two entries in Book I are 22 and 27 November, and the last date in II is 4 December (no year nor day names). Note that the last line of the manuscript marked "II" is: "Meet on 17th January at 5 o'clock".

The first 32 pages of Book I are without dates, though there are page breaks, which appear to indicate the end of a completed section, with a separate remark to commence on the next page. The first break is on page 5; the next on page 8 (where there is a box diagram A); the subsequent break is on page 14 (which has a reference in it back to the "first lecture" (see below, Section "Describing the Wittgenstein Papers"). The next break is on page 19; next break is page 28. On page 32 the first date occurs: "Friday Nov. 17th" (with no name or year). There follows an entry on the facing page 33 (an unusually short section for this book): "Nov. 13th", which coincides with the year 1933. The break on page 42, unusually, has the date at the

Plate 2.3 The presentation style of Wittgenstein's handwritten "*Philosophische Bemerkungen XI*". (Trinity Wren Library designation: "Wittgenstein.MS.115. title".) is the same as his style in the present Edition

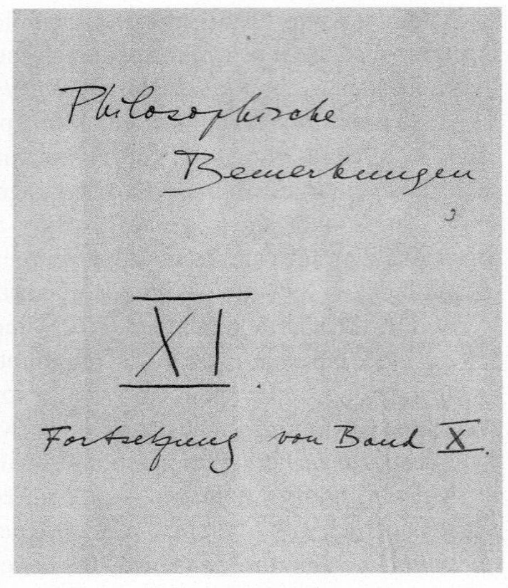

end of a page – "Nov. 22nd". The final entry is on page 43, which is dated to "November 27th".

The Pink Book and 'Communication of Personal Experience' (Chap. 4) are similarly bound exercise books, with canvas spine, and lined. These have coloured flexible backings, which have a colour that sometimes near-matches or complements the colour of the canvas tape that has been added down the spine during the manufacturing process.

They have the same binding as the notebooks in the large format C-series MSS 145–151, used for notes in lectures. The writing in the latter is in Wittgenstein's hand and is largely in the style of sketchy notes and rough dialogue, whereas the manuscripts in the Archive are in Skinner's hand, some in a fair copy quality presentation, sometimes with Wittgenstein's careful emendations.

The physical and stylistic similarities between the Pink Book and the Archive's *Brown Book*, and its previously unpublished manuscripts appearing as Chap. 4 in this Edition, are parallel with the C-series lecture notebooks in the Wittgenstein *Nachlass*. Therefore, it is worth listing the relevant C physical description of those here for comparative purposes: the colours of the C-series lecture notebooks covers, and tape colours, are as follows: MS 146 (dated 12.1233), pink cover with maroon canvas spine. MS 148 (c. 1935), buff cover with green tape. MS 149 (c. 1935–36), pink cover with pink tape, and white stitching down the spine. MS 150 has a buff cover, with red tape (with white cotton stitching down the spine). MS 151 (c. 1936) displays a buff cover with maroon tape, and white stitching down the spine. MS 152 (c. 1936) is an entirely different style – bare buff cover, no tape on spine.[3]

[3]The white stitching mentioned on MSS 149, 150, 151 is, as Jonathan Smith observed, almost certainly a later conservator's repair, done prior to 1990 but well after Wittgenstein's death.

The contents of the C-series and the Archive are very different: the contents of the C-series are very rough, provisional notes for lectures and research in Wittgenstein's handwriting, whereas the Archive manuscripts contain dictated material in Skinner's hand, and lectures notes of Wittgenstein's lectures, with Wittgenstein's hand periodically revising the contents. The Archive notebooks only very spasmodically redeploy material from the C-series notebooks, which is refined in this transfer.

The related physical descriptions of the manuscript of the Pink Book and the notebooks are: pink cover, with maroon tape (the same as MS 146 (1933) and MS 149 (c. 1935–1936), and the *Brown Book*, Volume III. The related physical description of the manuscripts of the *Brown Book* volumes is: Volume I, buff cover with green tape (the same as MS 148). Volume II has an enriched buff cover with red tape (the same as MS 150, c. 1935 – except that the latter has white cotton stitching down the spine, and the former does not). Volume III has a pink cover with red tape (same as the Pink Book, and MS 146, 1933; also, this Volume is almost identical to MS 149 – except that the latter has white cotton stitching down the spine, c. 1935–36, though the latter cover surface is very worn, with dirt marks, and perhaps has been left out in the sun). Volumes IV and V are identical, with buff cover with green tape (the same as MS 148, c. 1935). Volume VI, light green to buff cover with green tape (cp. MS 145, 14.10.1933); VI also has the same green tape buff cover as MS 148).

The total number of words of the Pink Book manuscript is 26,000.

Communication of Personal Experience (Chap. 4)

This manuscript comprises the second half of "Book V" and the whole of "Book VI" of the hitherto unpublished copy of the *Brown Book* in Skinner's handwriting with Wittgenstein emendations (see Chap. 1 – 'Introduction'). The whole of the *Brown Book* appears in large bound notebooks (slightly larger than A4), which are similar in appearance to Wittgenstein's own large format C-series notebooks – used to prepare for, yet it seems usually not employed during, the lectures. This copy is written to fair copy standard, whereas the C-series notebooks are not.

The word count in the manuscript is 12,000.

Philosophy (Chap. 5)

This manuscript is comprised of unlined buff coloured sheets. The size of each sheet is 230 mm × 175 mm. There are rust marks from the use of metal paper clips to attach the sheets into batches (the rusted paper clips survive in a separate bag).

The first batch, consisting of pages number 1–12, has the watermark "IVORY FINISH", with no motif. The second group is fifteen batches, consisting of pages numbered respectively 1–4; 1–6; 1–6; 1–7; 1–6; 1–4; 1–3; 1–2; 1–6; 1–6; the twelfth batch is numbered pages 26–32[4]; 1–5; 1–2; 1–5 all display the watermark "KINGSCLERE", with a crown motif over it. The sixteenth batch, pp. 1–6[5] has the foregoing watermark, and these pages are unlined, whilst the next three pages – 7–9 – have no watermark (other than parallel marks), and these three pages are lined. These three pages were originally numbered 1–3, which numbers were deleted in favour of continuing the pagination. There is a date on page 7 (formerly page 1, deleted) of Feb. 14th, which is the same as that on page 1 of the previous 1–6 page sequence; it is unusual to have a date within a sequence of continuously numbered pages.

The dates on this and other manuscripts in the Archive are worthy of investigation, not least in their comparative significance. Chap. 5's manuscripts have dates throughout. It is worth comparing these with some of the Skinner *Blue Book* dates in Appendix [O], where, for example, in January there are lecture and dictation activities regarding both manuscripts in the same week, etc.

There are 20,352 words in the manuscripts that are included in Chap. 5.

Visual Image in His Brain (Chap. 6)

This manuscript comprises 16 unlined sheets. The pages are numbered into sequences of: 1–8; 1–3; 1–5, respectively.

The last page of the manuscript concludes with the note: "I want to talk about the thought reading from the Monday Lecture. Meet at eleven Friday morning." This seems to refer to the Monday lecture represented in the Pink Book Part I, which section is dated "November 27th" (second page of entry; see above), which is a Monday in 1933. This would make the note indicate Friday 1st December, which would be a time between the ends of the Pink Book I and Pink Book II. Since these 18 sheets are grouped in three numbered batches, it is plausible to conclude that they are records of meetings Wittgenstein had with Skinner, of three meetings held in between the lectures. Since they are not bound or clipped together in the present order, it may be that the current last one, which ends with the above note, was the first, and led to the other two, since they do not contain references to other meetings.

The words in this chapter amount to 3600.

---

[4]See footnotes 160 and 174 in the text of Chap. 5 (i.e., Friday, Feb. 9th) below, for the significance of this break with the usual pagination.

[5]The obverse of this page also has a pencil sketch, similar to the sketch on the facing page 7.

Plate 2.4 Norwegian notebook cover. (Trinity Wren Library designation: Add.MS.a.407.7. These are almost like woodcut prints: on the front cover there are three institutional buildings (church, town hall, school), with sea waves and cloud pattern, which gives way on the back cover to the sea with two Viking boats and four seagulls)
Imprint title: NORSK PAPIR·OMSLAG·NORSK TRYKK·MONSTERBESKYTTET
(Trans.: "Norwegian Paper Covers, Monster Protected")

The Norwegian Notebook (Chap. 7)

Chapter 7 publishes, as the title supposes, a Norwegian notebook – printed in Norway. This is a buff/light brown A5 exercise book, originally with a staple binding now missing, with brown pictures printed over the outer cover (Plate 2.4).

Perhaps this was obtained from the postmaster's shop in Skjolden, where Wittgenstein had stayed, during the time when he met Francis Skinner in Bergen and they spent seven days together (see MS 118, 18.9.1937, and MS 119, 6. and 7.11.1937). Compare Bergen purchase of writing book in Wittgenstein (1937). This may complement other evidence to flesh out Wittgenstein's 1936 trip to Norway when he appears to have taken the Wittgenstein-Skinner Archive's *Brown Book* version with him, when trying to produce a satisfactory German[6] (alternative) to the English *Brown Book*.

The size of a single sheet is 210 mm × 170 mm. The total number of sheets is 32, written on throughout except for the last two pages.

The handwriting on the manuscript is Skinner's, written in pencil. There are many corrections, and crossings out – as if the manuscript bears evidence of the quick pace and changing direction in the thought-processes expressed in this writing – which we may assume communicate Wittgenstein's dictation, constantly revising and developing trains of investigation. Many of the corrections are replete with Wittgenstein's

[6]An outcome of such efforts is the German version that was transcribed, edited and published by Rhees as Wittgenstein, L. (1970). *Eine Philosophische Betrachtung*, Frankfurt: 117–237.

own style of revision, for example the insertion above the line of a sentence fragment, with a large 'V' line to indicate the place for insertion.

There are 4400 words in the 'Norwegian Notebook'.

Self-Evidence and Logic (Chap. 8)

The title is derived from the manuscript. This manuscript item is composed of buff sheets of paper. The sheets are divided up into page-numbered sequence, each batch commencing with sequential Roman numerals, commencing VII-XIV. The sheets used and marked with these numerals are unlined, whereas the remaining two batches XV and XVI are written on lined buff sheets. This latter batch of line sheets employs ink pen writing, whereas the first fourteen lectures are written in pencil.[7] Both the lined and unlined sheets are watermarked: "Vanity Vellum BRITISH MADE" with the image of a fine quill pen diagonally through the words. The size of each single sheet is 235 mm × 169 mm.

The manuscript reproduced in Chap. 8, which is grouped into batches marked "VII" through to "XVI", is further subdivided within that series into numbered pages, as follows: VII, pages 1–13; VIII, pages 1–14; IX, pages 1–15 (Note: much longer in Skinner than Ambrose); X, pages 1–11; XI, pages 1–16; XIII, page 1–7 ([Headed question by] "Wisdom". (Note: again, much longer in Skinner than Ambrose); XIV, pages 1–7; XV, pages 1–8 (title: "Russell's theory of number)"; XVI, pages 1–17. For a typical contrasting difference with Ambrose, compare Lecture IX, and Ambrose (1979, 141). The last remark: "Next Term April 24th 3 o'clock". This date fits in with the date on which Wittgenstein lectured for Lent term 1935.[8] Revisions or corrections appear in Wittgenstein's hand on some of these. For example, in VIII.1 his handwriting is used for the corrections "further" and "describe" together with underlining and arrow marks; also, on the otherwise blank reverse side of this there is symbolic logic in Wittgenstein's hand that he appears to have intended for inclusion.[9]

[7]But note that the Roman numerals on sheets 1–4 are in pencil, as are the page numbers on these pages – except that the page number for p.3 is not written at all. (Although p.17's number is written in pencil, the Roman numeral 'XVI' is in ink.)

[8]The brown empty paper bag in one of one The Mathematical Association boxes, which contains some of the Skinner Archive, has this inscription on it in Skinner's hand: "Notes taken by myself of Dr. Wittgenstein's lectures Lent Term 1935", which complements the above evidence. This bag did not have any notes in it, but the size of notepaper on which the lectures are written is suitable for them to be inserted in it.

[9]There are four unlined paper sheets, consisting of four loose and separate folia (each just over A3 size).

The manuscript commences with Lecture VII, not with I. There was a series of lectures from the preceding Term,[10] and a later Term. Possibly Skinner did not make notes other than these. Yet the way they are set out and the theme developed, renders them self-contained, and freestanding, for which see the footnotes' commentary on the manuscript within Chap. 8. As the footnotes to Chap. 8 show, Margaret McDonald made a set of lecture notes for these lectures. In contrast to the Wittgenstein-Skinner manuscript in the present Archive, however, McDonald's notes are much shorter than the present manuscript, with fewer technical workings in some contexts, with no sign of revision of the sort we find in Chap. 8 of the present book.

This manuscript consists of 20,544 words.

Notes on Separate Manuscript Sheets (After Chap. 4)

Noted As Folio I, pp. 1–2

The A5 page is unlined paper with a beautiful watermark "Fine Indian Ivory" – opposite a watermark picture of an elephant. (This watermark is different from the paper employed, in for example, the manuscript of "Philosophy", whose watermark is less well-crafted and is worded as "FINE IVORY". The paper of the manuscript entitled (Chap. 8) "Self-evidence and Logic" is watermarked: "Vanity Vellum BRITISH MADE" with a fine quill pen diagonally through the words.) Since there is no pagination, let us take this page to be Folio I, p. 1.

There is another A5 identically watermarked page, which can be indexed as FII, p. 2.

Its expression "Ex. 42" may relate to Remark 42 in volume 1 of the Archive's *Brown Book*, though it is not the same paper medium.[11]

Mathematical Paper: Calculations in the Integers MOD 257 & Z Powers (Chap. 9)

The manuscript appears to be in Skinner's hand, with some small details altered sometimes in another hand. It appears to be Skinner's composition, and not Wittgenstein's, though it seems to reflect a style that is traceable to features of Wittgenstein's approach to mathematics. It may be a collaboration; one which is incomplete.

---

[10]Concerning which see Appendix (D) on Margaret McDonald, for assessment of the contrasts between this manuscript's context and those of her unpublished lecture notes.

[11]Its use of 'does' is not in the published *Blue and Brown Books*; nor is it in the Wittgenstein-Skinner handwritten manuscript of the *Brown Book* in the Archive.

Entirely composed of calculations, the manuscript consists of eight sheets in all: three sheets are of the length 443 mm × 280 mm; four sheets of length 220 mm × 280 mm; and one small sheet, with length 135 mm × 207 mm. This sheet is lined, only on the side that does not exhibit the calculations. There are three other sheets containing geometrical figures, which are published in Appendix [L]. These are not directly connected with the rest of the Archive's specific contents, yet relate to some themes.

Some 12,353 mathematical symbols comprise this manuscript.

The Blue Book

There is Goodstein's two-volume copy, and incomplete cyclostyled copy of the *Blue Book* in Skinner's Archive. They have the familiar blue paper binding. The second volume on the back-outer cover displays these details: "R. B. Darkins. Stationer & Relief Stamper. 59 Bridge St Cambridge." The first incomplete volume ends roughly halfway through, similarly to other cyclostyled copies.

The Handwritten Brown Book

The Archive contains Francis Skinner's previously unknown handwritten version of the *Brown Book*. The form, order and layout of the newly discovered *Brown Book* version in the Wittgenstein-Skinner Archive vastly differs from the two-part division of the *Brown Book* in Rhees's version. Rhees's Edition of the *Brown Book* relied on a carbon copy manuscript of the sort that Von Wright owned, which appears to be copied from or derive from the 1930s. Rush Rhees's Edition divides the *Brown Book* into two parts and embodies the carbon copy's same division. In complete contrast, the Skinner *Brown Book* comprises six numbered volumes. (In this far more detailed Wittgenstein-Skinner copy, the volume that coincides with the ending of Rhees's Part 1 breaks at this point and is marked by a date noting a Term's ending in Cambridge. Thus, it may not have been intended as a conceptual division for the structure of the *Brown Book*.)

This Wittgenstein-Skinner *Brown Book* includes a new additional previously unknown extension of over one volume in length – of 12,000 words. No hint of it is present in previous typescripts of the *Brown Book*. This section is replete with its own numerals, which are in the same style as the *Brown Book*. And there are cross-references – with numbers, to numbered Remarks in the published *Brown Book*. This cross-indexing is tied into the new section's arguments. It has some details of revision by Wittgenstein.[12] This unknown extension comprises Chap. 4 below.

[12]See, for example, Remark 28, which has some subtle adjustments in phrasing.

Here is an example of how this extension to the *Brown Book* interlocks explicitly with the content of the already extant and published Rhees version of the *Brown Book*: in Chap. 4's previously unpublished extension, it refers to the beginning of the *Brown Book* as "Bk I" and mentions the game there presented: "But suppose now I had described a language game (similar to N°. 1, Bk I) in this form." This clearly incorporates a strong sense of continuity between the *Brown Book* and the extension to it published here in Chap. 4.

Section 2: Relating the Archive to Wittgenstein's *Nachlass*

The Wittgenstein Catalogue, Archives and Multi-media

The collections of manuscripts that Wittgenstein left when he died (his *Nachlass*, to use his native language) are massive and are still not fully understood. A full account of the above topic is for future research, though progress has been made attempting to formulate a framework for Wittgenstein's compositions.

This present book – the Wittgenstein-Skinner Archive – is the first major addition in many years to be accorded a place in Wittgenstein's *Nachlass*. A little history is in order here: Although Wittgenstein died in 1951, yet it took until the late 1960s and a lot of work, for the Wittgenstein Trustee, Professor von Wright, to produce a catalogue of the Wittgenstein papers then extant.[13] This catalogue went through several Editions, and provided the basis for the work of the Wittgenstein Archives at the University of Bergen, which from 1990 to 2000 transcribed the Wittgenstein papers (also called the '*Nachlass*') in machine-readable form; they were published in 2000 with OUP.[14] The Wittgenstein Archives at Bergen, led by Alois Pichler, are preparing a revised Edition.[15] This is complemented by a wide range of significant breakthroughs by a variety of scholars in the multi-media presentation and conceptual analysis of Wittgenstein manuscripts.[16]

The identity of the hitherto unknown Wittgenstein-Skinner Archive appears not to have been known to von Wright, nor to any of the other trustees. (In Ambrose's hitherto unpublished 1977 letter[17] to Rush Rhees she refers to some fragmentary

---

[13]von Wright (1969).

[14]*Wittgenstein's Nachlass*: The Bergen Electronic Edition. Ed. Wittgenstein Archives at the University of Bergen. Oxford: OUP (2000).

[15]Pichler (2010: 157–172), Pichler (1997); for a range of relevant research see Floyd and Katz (2016).

[16]For example, conceptual ontology, see Addis, Brock and Pichler (2015). Also see the Open Access site Wittgenstein Source (www.wittgensteinsource.org) produced by Alois Pichler (the Curator) and colleagues.

[17]The letter to Rhees is dated 9 April 1977 (reproduced in Appendix [K] below). To position the background of this in relation to Goodstein's illness and his considerable care of the Archive, see Price and Walmsley (2013).

Skinner material in Goodstein's possession. Ambrose, understandably, did not appear to know or understand precisely this Archive's length, scope or significance, even though she had borrowed *parts* of a 1934 draft of a draft of the *Brown Book* by Skinner, which appears to have been edited and lengthened by Wittgenstein and Skinner subsequent to her reading of some of it.[18])

The pioneering achievements of the above scholars have laid valuable foundations, even though it is difficult to measure some past, alleged possible fault-lines in aspects of these such scholarly endeavours. For example, in an unpublished statement held in Trinity College Library, Georg Henrik von Wright in 1990 expressed his view that all the editing by Rush Rhees would probably need to be done again. Therefore, it is hardly surprising that even a leading international research scholar, educated by one of the world's leading senior authorities on Wittgenstein, remarked to me that, upon trying to master only one chapter (by a distinguished academic expert) on the history of Wittgenstein's manuscripts, it caused him to lose the will to live. Such serious irony marks the, sometimes seemingly intractable, complexity of the relations of manuscripts that comprise the *Nachlass*, and its conceptual challenges.

It is evident that the Wittgenstein-Skinner Archive punches above the weight that a casual glance might afford. This weighs in at the level of the importance of the *Brown Book*'s significance, since the Archive contains a much fuller, carefully edited handwritten version of the *Brown Book*. This has a lengthy extension, additional sentences and revisions handwritten by Wittgenstein. The Pink Book adds substantially to this picture, partly because it is unprecedented and a fair copy. Even the lecture Notes appear to have been carefully redrafted, sometimes with signs of influence at Wittgenstein's hand.

Wittgenstein-Skinner Editing

The following presupposes Jonathan Smith's (2013) valuable analysis of the *Blue Book's* emergence, concerning particularly its use of dictation, revision, and redrafting in the process of its production. Smith positions aspects of Skinner's role in this process, and others.

There is evidence, in the Thouless copy[19] of the *Blue Book* that Skinner, working directly under the supervision of Wittgenstein handled successive drafts, typescripts, and written copies of the *Blue Book*. These data amount to strong evidence of Wittgenstein working closely with Skinner in creative compositional and revision

[18]In 1934 she writes to Wittgenstein "enclosing here Skinner's notes, just in case they might be wanted" (McGuinness 2008. Letter 176, p.131). In summary, the absence of some of the Skinner *Brown Book* content from the extant carbon copy manuscripts version and the Rhees edition are bases to sustain the conclusion that Miss Ambrose's contributions are incomplete, along with other evidence offered in the present Edition.

[19]This copy was given by Wittgenstein to Thouless in Cambridge.

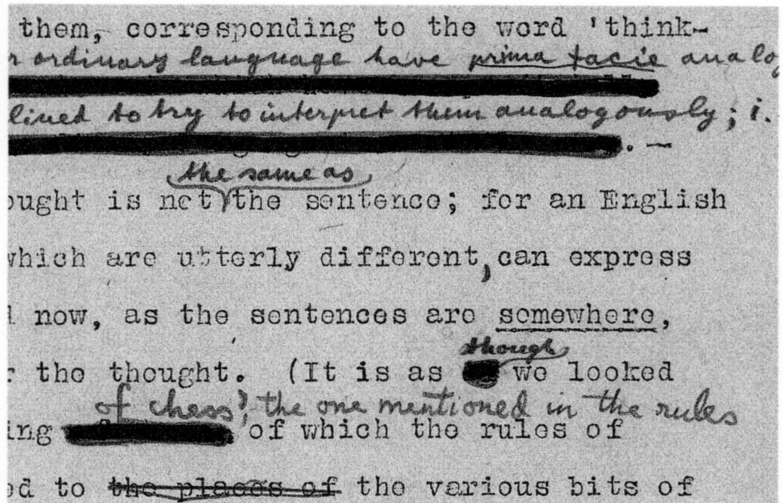
them, corresponding to the word 'think-
r ordinary language have prima facie analo
lined to try to interpret them analogously; i.
. —
the same as
ught is not the sentence; for an English
which are utterly different, can express
now, as the sentences are somewhere,
though
the thought. (It is as we looked
of chess), the one mentioned in the rules
ing of which the rules of
d to ~~the places of~~ the various bits of

Plate 2.5 Wittgenstein's and Skinner's handwriting in the Thouless copy of the *Blue Book*. (Trinity Wren Library designation: Add.MS.a.522(1). See Jonathan Smith's (2013) remarkable analysis of the Thouless MS)

processes, which, together with the empirical proof derived from the present book, demonstrate the unique working relationship that Wittgenstein had with Francis Skinner. The manuscripts that are published in this book manifest evidence that Wittgenstein deployed the same sort of customs and employed the same techniques with Francis Skinner as he did when composing the *Blue Book*. Below are two samples taken from Thouless's copy of the *Blue Book* – valuable as parallels with customs in the Archive published in the chapters in the present book.

In the illustration below, Skinner's pencil handwriting (e.g. 'of chess') is present alongside Wittgenstein's own blue pen handwriting (e.g. "ordinary language'), in the Thouless's copy of the *Blue Book*. This specific practice can be found in the Archive. Notice – as a piece of evidence to reveal that this is not solely a matter of common techniques deployed, but also of the same particular term or theme being developed – we read of the introduction of "analogously" in the *Blue Book* illustration below, which is a conceptual parallel with use of the same word in Chap. 4's transcription of a manuscript. This theme is also presented for complementary reasons in this present Introduction (Plate 2.5).[20]

Secondly, in the Thouless *Blue Book* we see that Wittgenstein wrote a paragraph in English, in his large handwriting, which is placed on the opposite facing page of the typed main text (see Plate 2.6). This typed text manuscript also has some revision on it in Skinner's hand inserted amongst the words of the typed version opposite the statement: "Perhaps the main reason why we are so strongly... same paragraph".

[20] Cf. the above section entitled "Wittgenstein and Cambridge Mathematics".

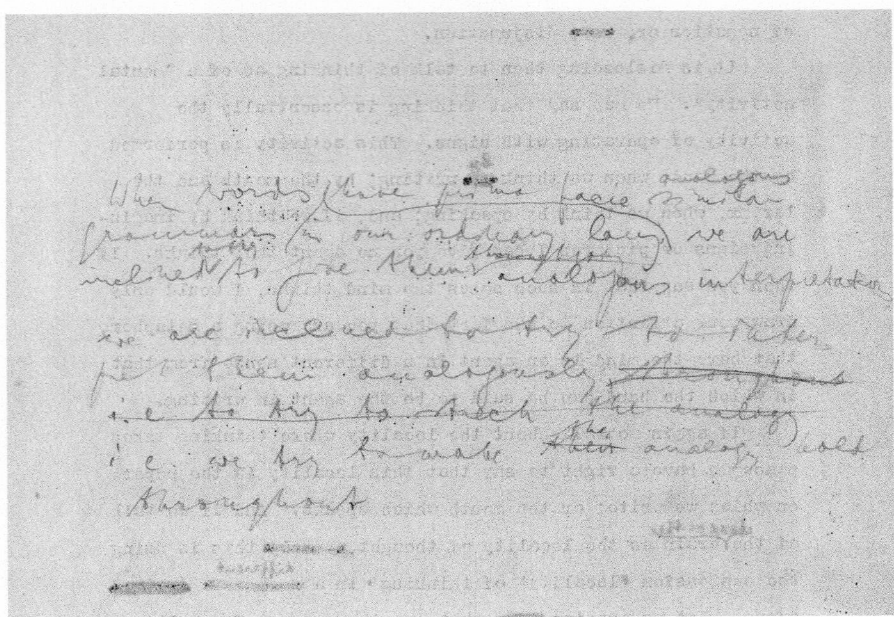

Plate 2.6 The Thouless copy of the *Blue Book*, with a paragraph in Wittgenstein's hand on the facing page to the typescript. (Trinity Wren Library designation: Add.MS.a.522(2))

In the Pink Book (i.e., Chap. 3 of the present book) we have a parallel practice, though the subject is different (Plate 2.7).

This illustrates the situation that both the *Blue Book* and the Pink Book display collaborative interaction between philosopher and amanuensis, in both cases involving attention to detail, redrafting, and the introduction of fresh elements. As readers will see from the Pink Book's footnotes in Chap. 3, there is evidence of material being transferred out of a draft of the *Blue Book* (its written form by Skinner) into the Pink Book. This not only shows unity of parallel methods; but it also exposes that the skills are deployed to compose unity in the enterprise, even whilst varying manuscripts are taken in different directions. Most importantly, such evidence of synthesis between author and amanuensis in writerly activity provides a basis for deriving inferences from the manuscripts about the thought processes that occasion their creative contents.

Attention to Detail and the Least Element of Sense

In this book we scrutinize important lengthy revision-details in the manuscripts here published. At the other end of the size scale, editing in this book attends to what seem to be miniscule details. To what degree do editors have a duty of care to attend

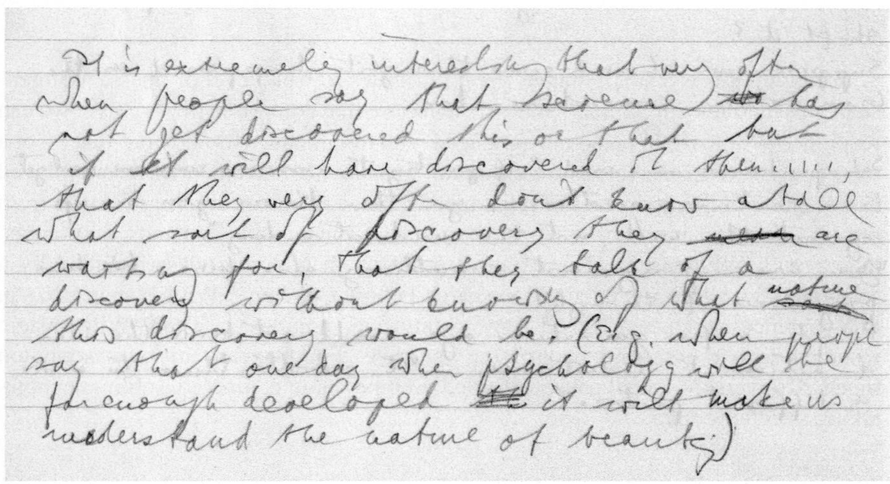

It is extremely interesting that very often when people say that because they have not yet discovered this or that but if they will have discovered it then....., that they very often don't know at all what sort of discovery they are waiting for, that they talk of a discovery without knowing of what nature this discovery would be. (E.g. when people say that one day when psychology will be far enough developed it will make us understand the nature of beauty)

Plate 2.7 The Pink Book, with a paragraph in Wittgenstein's hand. (This Wittgenstein handwriting is on the page facing the Archive manuscript page 43, opposite the question: "If you ask why we use substantives?".)

to such seeming minutiae in composition? Does the pursuit of detail in such scholarship inexorably lead to the absence of progress for interpretation? Sometimes; yet we do not always have the retrospective capacity to recognise at what point we should rule out further attention, since the most familiar things may contain unknown identities. A small detail, as with a trigger, may action matters disproportionate to its apparent size. Therefore, an excess of care is not obviously redundant. The reader should allow the need for footnote and other detail whose point may yet need to be exposed by future investigation.

In relation to this, we should enquire: what is the least element of sense expressed as an editorial alteration by Wittgenstein in a manuscript, which an editor might be permitted to ignore, and not inform the reader of this Archive to see – by not troubling to cite it? We have no proof for the least fraction of its significance. It is worth erring on the side of its fulsome presentation. In the Pink Book: in Chap. 8, and at the beginning of Lecture VIII.2, the text reads in Skinner's writing:

I.E.[21]

$$q \supset r . \supset : p \vee q \supset p \vee r.$$

Wittgenstein, with a characteristic large squiggle, changed to "I.E" to "e.g.":

e.g.

[21] There is another such correction from "I.E." to "e.g." in the same handwriting on the next page of the manuscript in VIII.3.

$$q \supset r . \supset \; : p \vee q \supset p \vee r$$

Although this is a small matter, yet it illustrates to what level of finesse Wittgenstein was in tune with Skinner. With the evidence that comprises the Archive, we can recognise Wittgenstein personally engaged with his amanuensis and note-taker – or a merger of the two roles into one continuum. The dictator and amanuensis interplayed as a singular unifying process of refinement, albeit sometimes arguing.[22] Aside from the private impositions of such work, consider the regular demands of Francis' labour with Wittgenstein. On a typical occasion in a lecture class attended by a selective few, which included the Research Fellow Coxeter,[23] Wittgenstein froze in mid-sentence. Silence reigned for twenty-five minutes.[24] Then Wittgenstein resumed with the second part of the sentence. For Coxeter this was soon too much. Conversely, such pause afforded Skinner ample time to sublimate note-taking into fair copy.

Non-self-evident Logic[25]

In contrast, with the above example of detail in Wittgenstein's revision of his manuscript, the main thrust of Skinner's craft was also to reproduce Wittgenstein's broad, intricately argued, original sweeps of theme. For example, in Chap. 8, Lecture VII, the manuscript declares Wittgenstein's view that the truths of logic are not self-evident, by assailing us with the claim that our psychology is the contrary of this; that we should embrace surprise:

> One of the great difficulties which is in logic was this – what is the criterion for a proposition being a proposition of logic?
>
> By some it was taken to be self-evidence. This was what both Russell and Frege said. At certain points where this is a difficulty one tends to bring in psychology.
>
> The word 'self-evident' is taken from psychology.
>
> It is what everybody is in his right senses would say is true.
>
> It is the highest degree of plausibility. But it is clear that what is self-evident may not be to another.
>
> People did not mean to say these propositions were self-evident to Europeans.
>
> The self-evidence seemed to be an objective one, not a subjective one. It seemed to lie in the proposition.
>
> It had been said also that logical propositions were tautologies. They were redundant, they were in a sense such that they needn't be said.
>
> But if they needn't be said, what is their use? (Plate 2.8).

---

[22]As noted at the beginning of the present Chapter (cf. Fania Pascal in Rhees 1984).

[23]Cf. Roberts (2007) and Roberts et al. (2009).

[24]I owe this and other details here to the letter (2001) from Coxeter to Dr. M. H. Price, University of Leicester, for which thanks to Mike Price for making it available to me.

[25]'Non-Self-Evident' is here intended as the same as Wittgenstein's use of 'not self-evident'.

Plate 2.8 Use of 'non self-evidence' in Chapter 8, part VII, sec. 1. (Trinity Wren Library designation: Add. MS.a.407.2)

VII

I want to go on now to a chapter by itself.
By this chapter I have to interrupt what I have
been doing – talking about generality
This is logic I must talk about
One of the great difficulties which logic
was this – what is the criterion for a
proposition being a prop. of logic.
By some it was taken to be self-evidence.
This was what both Russell and Frege said.
at certain points where this is a difficulty
one tends to bring in psychology
The word self-evident is taken from
psychology.
It is what everybody is in his right senses
would say is true.
It is the highest degree of plausibility.
But it is clear that what is self-evident
may not be to another.
People did not mean to say these props
were self-evident to Europeans.
The self-evidence seemed to be an objective
one, not a subjective one It seemed to
lie in the prop
It had been said also that logical props.
were tautologies They were redundant,
they were in a sense such that they needn't
be said.
But if they needn't be said, what is their
use?

Wittgenstein's struggle with his creative thought processes is partially displayed in the Archive's contents, modes of writing, as well as its formulations that are evidenced in revising and re-drafting, which requires subtle investigations. Certainly, many an amanuensis intervenes and interposes distance between the dictating author and his copyist, though we should keep in mind that the latter's task is to reproduce the former's intentions. No doubt this distance obtains for some authors; yet in the case of Ludwig and Francis, their personal engagement reverse-engineers a closure over much of this distance, while Wittgenstein also at times seems to hold-off Skinner as an alien too close for comfort. In this role, Skinner was to a peculiar degree internalised into – yet is sometimes exiled by – Wittgenstein's mental processes. This angle is further illuminated by James Klagge's (2011) way of landscaping Wittgenstein's mental state of permanent exile in the interplay between his external life and interior states.

Section 3: Skinner Matters

Did Francis Have a Standpoint?

The basic answer to the above question is 'Yes', in the sense that he wished to please Wittgenstein at all costs. Although a full account would be lengthy and complex, a summary fragment follows. Francis appeared fulsomely to comply with what he assumed to be Wittgenstein's (developing) intentions and wishes. This seems to be true generally and obtains in his very detailed work as an amanuensis. In the few instances where we can derive extant evidence of Francis communicating his presuppositions, there are signs of anxiety about or in his communications to Wittgenstein. (It is worth reading the opening parts of this Introduction again from this standpoint to reflect on its descriptions of Skinner's biography in relation to the assessment here.)

On September 9th 1935, Wittgenstein arrived in the USSR alone (Skinner was to have visited with him, but polio prevented him). Wittgenstein linked up with the Moscow correspondent for the *Daily Worker* – former Cambridge Economics student – Pat Sloan, a friend of his former teachers, Piero Sraffa and Maurice Dobb, hoping for support in his quest for employment in Russia. Yet he was in for a shock from Sloan, who writes to Dobb[26] blasting Wittgenstein with this diatribe below. (Wittgenstein would be aware that due to Dobb's very close relationship to Sraffa, this would be passed on.)

> [Wittgenstein] wanted to give up his speciality. . . is hardly the person to settle in the USSR. His mind is so narrowly confined (amounting almost to insanity). . .: 'I am 46, I cannot read the philosophy of the land I want to settle in, for I am too old'. . . wants to abandon his mind altogether. He deems it quite impossible to develop it along new channels. . .I feel he is so absolutely muddled.[27]

During this USSR visit, Wittgenstein received Francis' irenic letter, which contrasted with Sloan's missive; yet it also discusses Wittgenstein's then-current method:

> 17 Sollershott, West,
> Letchworth, Herts.
> Sept. 17th Tuesday [1935]
> Dear Ludwig,
>
> It was very lovely to get your letter today from Leningrad. . . I have just heard from Goodstein that he has got a job at Reading University. . . I shall miss him a lot next year at Cambridge. . .
>
> I think a lot about the work which we are going to do next year. I feel that the spirit of the method which you used last year is so good.

---

[26]Letter dated 24 September 1935. For the full transcription of this letter, and related points, see Appendix [R].

[27]Despite this critique, Monk documents two offers of philosophy jobs in Russia extended to and declined by Wittgenstein.

> Everything, I feel, is absolutely simple and yet it's all full of light. I feel it will be very good to go on with it and get it ready in some form for publication. I feel that the method is so valuable. I hope very much we shall be able to get on with it. We will do our best. I'd like to say again that I hope you will stay longer in Moscow than the time you arranged for if you feel there is any chance that you might learn more. It would be valuable for both of us. I was very glad to get both your letters. I think of you a great deal and hope that you are feeling well.
>
> W[ith]. L[ove].

This specific *Brown Book* manuscript that Skinner mentions above has been found in the Wittgenstein-Skinner Archive.[28]

Did Skinner's encouraging sentiments about the method concur with Wittgenstein's feelings? Given Skinner's typical, stringent attempts to comply with Wittgenstein's wishes, it is highly likely that the words in Skinner's letter emulate Wittgenstein's view in 1935 of the "spirit of the method" for the *Brown Book*:

> I feel that the spirit of the method which you used last year is so good. Everything, I feel, is absolutely simple and yet it's all full of light. I feel it will be very good to go on with it . . .I feel that the method is so valuable.

Did Skinner's urging to "get it ready in some form for publication" spur Wittgenstein on? The Archive's highly edited version of this *Brown Book*, as reported earlier in our Introduction, is a fair copy – including its long extension (which is unique and published here in the present Edition for the first time, as Chap. 3.)[29] In those days they could be recognised as in a final preparatory process to go off to a publisher – consistent with Francis' view in the above letter.[30] After Wittgenstein returned from Russia to Francis and Cambridge, it appears that they continued to work on the *Brown Book* manuscript that is in the Archive. It is also most likely to be the typescript that Wittgenstein took with him to Norway in 1936, where Francis visited him.[31]

Yet it was another six years before Wittgenstein wrapped up the manuscripts and sent them to Goodstein in late 1941, asking him to consider the Archive for possible publication. This triggers the question of why Wittgenstein sent off the manuscripts reproduced in the current Edition at that point, when by that time he already had criticized the *Brown Book*'s method to Moore in 1938. Certainly, analysis of the Archive has its own set of emerging argued difficulties to contribute – that add to

---

[28]This also includes the lengthy hitherto unknown last section of Remarks, which is published in this volume (as Chap. 4).

[29]According to specialist archivists such as Jonathan Smith, whom the present editors have consulted, this designation of 'fair copy' also applies to the Pink Book (Chap. 3 of the present Edition).

[30]In this period for Wittgenstein and on his behalf, handwritten written revised copies can even be seen to supersede typed copies in the compositional sequence towards projected publication (cf. Smith (2013, and Chap. 2 of this edition).

[31]See Wittgenstein's letter written in Norway, to Moore (McGuiness 2008: Letter 204), which speaks of his translating and rewriting the *Brown Book* in German. There are pieces of translation and comments in German on some of the facing pages of the Skinner *Brown Book* version in the Archive, which attest to this activity.

problems associated with some of Wittgenstein's subsequent manuscripts, which may influence interpretation of the "spirit of the method".[32]

Nevertheless, as Alois Pichler states,[33] "the seeds of discontinuity in the *Brown Book* would be manifest in the *Philosophical Investigations*".[34] These seeds both mark a shift to a new phase, and show some overlap in Wittgenstein's creative sensibilities between his writing in the 1930s and his later period.

Introspection

The earlier Section (i.e., "Wittgenstein and Cambridge Mathematics") outlined some of the mathematician John Conway's use of Wittgenstein's words cited above on his sense of the significance for 'whistling', which also assists us to set the scene here, in the Pink Book that Wittgenstein dictated to Francis. Wittgenstein's following expressions require attention: "I dictate to you and you have to write": –

> I must show you certain ways of generalizing are not true.
> Suppose I dictate to you and you have to write, . . .
> "If we go back to the question
> Was it a process like whistling a note or one like saying a sentence?
> In general you would say this answer is given by *introspection*."

Wittgenstein employs the term 'dictate' to mark his first person. Skinner appropriately fits as the recipient with the use of 'you': "Suppose I dictate to you and you have to write". The function Francis is here subject to being drawn by Wittgenstein into the web of the latter's introspection.

Skinner's role assists in removing or limiting the gap between whistling and saying a sentence. This is a function of Wittgenstein's process of "introspection". It is plausible, at the very least, that to some degree Francis is snared by and internal to this process and by Wittgenstein's use of introspection. Note that in *PI* Wittgenstein includes the calling up of feelings by introspection.[35]

---

[32]See A. Gibson's introduction in Wittgenstein (2009); Cf. Venturinha (2004): the first part of MS157a, covering the years 1934–36, yields a turning point, for which also see the second half of MS 115 (cf. MS 115, 292), as well as MS 141; Venturinha (2013).

[33]Pichler (2013: 74).

[34]See Joachim Schulte (2013: 81–83).

[35]Notice Wittgenstein's explanation of 'introspection' in PI §587: "the process of introspection is the calling up of memories, of imagined possible situations, and of the feelings that one would have if…".

Progress

We might ask: Is progress in philosophy scarcely knowable?[36] Wittgenstein, using his method, or attempts to refine methods, personifies the question: philosophical progress itself a self-cancelling device for the claim of there being one method? And if so, would this self-cancelling of progress eventually be all-pervasive? And leave skills and not method as the outcome?

Pinpointing one aspect of our problem in a world of Wittgenstein's exceptions, Paul Horwich[37] finds that philosophers have a special kind of bewilderment: "What sort of progress is this? The fascinating mystery has been removed – yet no depths have been plumbed in consolation; nothing has been explained or discovered or reconceived."[38]

Schulte (2013) elegantly identifies how formulating problems are fundamental to Wittgenstein's diet in this method. Even the same remark can have a different purpose in two contexts. Could it so be with method itself?

Brian McGuinness rightly concluded that Wittgenstein lacked a good editor. If so (we might ask on Francis's behalf, given his conviction that he was sensing the spirit of Wittgenstein's method), why should it be accepted that Wittgenstein's method in the *Philosophical Investigations* (*PI*) is exclusively superior to, for example, the *Brown Book*, *in the sense that* it should cause us to demote or dismiss Wittgenstein's previous 'methods' and 'works' as defective in principle, even when not assessed in relation to *PI*? Certainly, we should not wish to allow empathy for Francis's view (that the *Brown Book's* method is ideal) to obscure the uniqueness of PI.[39] Nevertheless, we should enquire: in some of his other writings, was Wittgenstein simultaneously both refining his aim to achieve the PI, as well as crafting other attempts at complementary enquiries, which had a double identity. Namely, was the *Brown Book* (in the slightly more revised version in the Skinner Archive) both part of the sequence culminating in the PI, *yet* stands separated from PI, and can be treated as distinguishable complementary mode of exploration? We should here recall that it was not until 1946 that Wittgenstein inserted in his manuscripts for PI, that: "There is not a philosophical method, though there are indeed methods, like different therapies"![40] We should not overstate these considerations, whilst sensitively placing them in the perspective that, when Wittgenstein posted off the *Brown Book* to

[36]Cf. Gibson (2013).

[37]Horwich (2012: 61).

[38]Horwich (2012: 211).

[39]We should certainly agree with Joachim Schulte's important analysis. (Schulte 2001, 2005b, 2007, 2013). Notice the quotation of Schulte in the foregoing section entitled "Wittgenstein's Scribe", and 'Does the archive reproduce Wittgenstein's Compositional Activity?'.

[40]When Weber (2019: 3) points this out, she also observes that "Wittgenstein wrote this remark on a small slip of paper, inserted it in between the typeset pages, and marked that he intended it to be added as a footnote to §133 – an instruction that was not followed in the Anscombe editions of the PI."

Goodstein for consideration for publication, it was after he had embarked on the new method and drafted a version of it as first part of the Philosophical Investigations.

In some ways we do not yet understand, Skinner may be at an epicentre within such deliberations, and wrapped in Wittgenstein's feelings for him. This might be a route to explain why Wittgenstein sent the Archive to Goodstein to be considered for publication. We do not know if Goodstein replied; or even if he understood. His life-long silence on the Archive may not be neglect. It might indicate that he was still trying to deliberate on how to make progress on the matter.

This is not to warrant to a claim that incomplete 'work in "progress" ' should be allowed to transcend its limit.[41] Obviously, Wittgenstein's *PI* is the zenith of his later creative method. But are the differing pathways of method(s) in Wittgenstein's late creative writerly evolution properly *not* to be measured as criteria against the *Brown Book's* merit? Could the *Brown Book*, and other select examples in the Archive – such as the Pink Book, be identified as embodying meritorious method? And similarly for the method Francis Skinner wrote of in his above letter?[42] Are the *Brown Book* and other manuscripts included in the current Archive, experiments-in-progress conditionally moving towards a multiple-conclusion rationality in Wittgenstein, which purposively create tensions in themselves?[43]

Wittgenstein did not finish his final work, and infamously – on his deathbed, he advised his editor to toss a coin as to the way forward for his literary heirs. Was this a giveaway remark, or irony? Or was it a component in his methods: tantamount to introspection at the stage of whistling? Knowing for some time that he was dying, is it not likely this last toss of a coin was not a giveaway comment. It was a procedure in which he confessed what he believed: he had couched in his writings a plural set of directions and methods which he had not discarded; and did not believe that he could choose a final game for deciding choices about them.

Was there – within any of the above patterns – a thought experiment? Had Francis Skinner been present at Wittgenstein's deathbed (he had been chosen by Wittgenstein early on in the 1930s as his literary custodian[44]), Francis might have recalled a remark Wittgenstein dictated to him, which appears below in Chap. 4, in which the extended, hitherto unpublished, last portion of the *Brown Book* appears:

> The condition for B's learning the game was, of course, his reacting in the 'desired way'. I.e. had B not reacted by saying the words, or reacted by saying them 'at random', the game would not have come off.

---

[41] This is partly achievement by heeding Schulte (2013) and Smith's (2013) deployment of Schulte's criteria.

[42] In relation to this, note the second part of the sub-section entitled "Wittgenstein-Skinner Editing" in the Introduction to the present book.

[43] I am here alluding to the title and themes of D. J. Shoesmith and T. J Smiley's book *Multiple Conclusion Logic* (1980). Although the above is not Gordon Baker's view, yet one can read with profit the insight in his use of analogy in the section entitled "A Psychoanalytical Model of Philosophy", in Baker (2004) Chap. 8.

[44] For which see Gibson (2019).

We leave the reader to infer the consequence of this for Wittgenstein's philosophy. Had Francis Skinner – the "simple-minded"[45] 1st Class Honours mathematics graduate researcher, and so close to Wittgenstein – learned rather more on how to read Wittgenstein than a purely mechanical dictatee could?[46]

This publication honours Francis, as well as Wittgenstein. It honours labours and content – not now lost; but published. Francis's careful work as amanuensis reveals more than we hitherto previously had of Wittgenstein's creative compositional crafts, and of the way the manuscripts directly or indirectly signal or betray many of his thought processes, at the hands of both Francis and Wittgenstein. The present Edition partially fulfills at least some of Francis's hope for publication:

> I feel it will be very good to go on with it and get it ready in some form for publication. I feel that the method is so valuable. I hope very much we shall be able to get on with it. We will do our best.

[45]Cf. J.E. Littlewood's (1926), use of this remark in his Preface p.ii.

[46]The use of 'learn' has of course a major role in PI, which partly results from Wittgenstein refining and transforming his own teaching method – not only in Cambridge, but also drawing on his earlier work as school teacher in Austria, writing and publishing a reference book; for which see Weber 2019, and [forthcoming].

Part II
The Transcriptions of the Wittgenstein-Skinner Manuscripts

Conventions Deployed in This Edition

Typically, the first occurrence of a custom will be explained in a footnote, while it will assist readers to have the main ones itemized here:

1. Where there is an alternative reading in the MS, this is marked by double bars. See for example in Chap. 4: '‖We then tell him: "We don't call this colour 'red', we call it 'green'."‖'. The double slash, here rendered as double bar, is characteristic of the instructions, perhaps for a further revision, or for a typist aimed at a final draft for projected publication. We find this convention in Wittgenstein's German and other writing. Cp. manuscript MS 120 (XVI) for the period 19 Nov. 1937–26 April 1938, f.118r, and later in the manuscript MS 116 (*Philosophische Bemerkungen*, XII), p. 224. Where there is such a multiple choice, it or they will be placed in a footnote mark by the double bar that is deployed in the main text. In Chap. 9's calculations, the vertical double-bar symbol explained above has the role of a marker to divide separate calculations from each other.
2. Where a full stop, comma, question mark, or capital letter is absent from the original manuscripts, it has been supplied in this Edition without noting the insertion, unless the change affects the sense; footnotes will always note other changes. Based upon the advice of a number of distinguished authorities in editing and research library customs, we decided to privilege ease of readability over allowing too many intrusions in the main text of the Edition. So, we have not littered the narrative unduly with uses of '[. . .]'. Rather, we have placed alternatives in the text and details in footnotes.
3. When a word is regularly misspelled (for example, when 'roll' is a misspelling for 'role'), the first occurrence of this mistake will be noted in a footnote and corrected; but further corrections of the same type will be corrected with mention of it in a footnote.
4. In some parts of the original manuscript lines are drawn across the page to indicate a division. This practice begins in the Pink Book – *above* the first line apparently to indicate paragraph or 'remark' divisions – and continues for

14 paragraphs. Perhaps some of these uses are where Wittgenstein and Skinner are recognising the emergence of remarks within the composition. After this practice in the opening of the Pink Book the drawing of lines ceases and a space is left in the original manuscript, as if the line custom has given way to explicit uses of paragraph breaks parallel with the use of remarks. The custom occurs more spasmodically elsewhere in the Archive.

5. In the writing style of the time in Cambridge – and even currently often in maths lectures, some uses of conditional sentence "If *x*, then *z*" were written without the comma, and with a full stop. Occasionally in the Archive the latter might confuse, especially where more than one antecedent is employed. In those contexts, the latter has been standardised to employ the comma, only in one case with an explanatory phrase.
6. In this Edition the term "Archive" designates the collection of manuscripts and papers that comprise the Wittgenstein Skinner Collection. Except for the published parts of the *Brown Book*, all the Archive is transcribed and published in this Edition for the first time.[1] (In the perspective of the differences between the fine Rhees' Edition of the *Brown Book*,[2] and the handwritten *Brown Book* Edition in the Archive, there is a need for a new Edition of the *Brown Book* for a future occasion.) The one and a half volumes of the *Brown Book* in the Archive that have not been published before appear in the current Edition as Chap. 4.
7. In the MSS the writing of date abbreviations varies (for example, such as "Wedn"/ "Wednesday"), including changes from one side of a page to another. This Edition preserves these customs exactly as they are written in case it emerges that they are evidence for some purpose that can be retrieved from close study.

[1]There are a few school-notebook details that are in the Archive but not published here, though they are referred to and quoted at relevant points in this Edition.

[2]To view some of the differences expressed in Wittgenstein's handwriting in the opening pages of the Archive version of the *Brown Book*, which contrast with the Rhees version, see Appendix [N].

Chapter 3
The Pink Book

Part I[1]

Schopenhauer[2] said one should read his book twice. This does not distinguish philosophy essentially from science. Philosophy can be compared to geography. If I had to teach geography without a map,[3] and there was a wrong idea of the geography of a district. If you had to draw a map, you would go hopelessly wrong. The way to put you right would be to give you a map – but if I could only give parts such as a village, it would be enormously difficult to put you right. The difficulty would be to give a synopsis. You know the geography perfectly, but if I asked you to draw a map you would go hopelessly wrong. I am trying to give you a synopsis of a huge country. If I go along a road, I come to a crossroads. I then go one way, and you have to wait till I come back before I can go the other way. This is

The title 'Pink Book' has been applied by the Editors because of the colour of the cover, since it concurs with Wittgenstein's periodic custom.

[1]The manuscript is divided into two – by means of the designations "Part I" and "Part II". In the original manuscript lines are drawn across the page to indicate a paragraph division. This practice begins here – *above* the first line apparently to indicate paragraph or 'remark' divisions – and continues for 14 paragraphs. After this practice, the drawing of lines ceases and a space is left in the original manuscript. The custom used here might be taken as some evidence of Wittgenstein crafting his emerging notion of 'remark'.

[2]Ambrose's Yellow Book text mentions 'Schopenhauer'. This should not be taken as evidence of co-dependence of the two texts, since, for example, 'Schopenhauer' occurs twice in MS 213, the Big Typescript (pp. 281, 300). This notion of reading twice can also be contrasted with Schopenhauer's own view that, "the meaning of the speech is immediately grasped" (Schroeder 2012: 21–22).

[3]Notice the later use of 'map' in Chap. 5, entry under "Wed. Feb.7th: "The map and the legend are a symbol and its explanation."

A. Gibson, N. O'Mahony, *Ludwig Wittgenstein: Dictating Philosophy*,
https://doi.org/10.1007/978-3-030-36087-0_3

where Schopenhauer's repetition comes[4] in. Schopenhauer said one could not understand one bit without knowing the whole. Really one might say one cannot appreciate one bit. No one can remember the whole. One should go over and over again the same connections. As we have go through the same net of roads hundreds of times,[5] it does not matter where I begin.

I shall begin with problems connected with understanding, meaning, thinking. What has understanding to do with philosophy? A sentence which is not understood seems in a sense dead or just ink on paper, and then when it is understood it seems to have a soul. One might say a sentence only had meaning to an understanding being. Frege said the sign was the unimportant thing, and the meaning the important thing. Frege said mathematics was not about ink on paper. What are the rules of chess about? Are they about wood? This seems queer, and gives us a hint. If a man said the chalk mark is not one, this is clear because the chalk mark can be rubbed out but the number one can't. This shows that we do not use the word "one" in the same way as the word "the sign one". We should not use the same word. I am not going to say anything which is not as trivial as this. What is not trivial is to see the whole thing as one.

The rules of chess, one will say, are not about the particular chess set but about the king of chess.[6] Chess has no nimbus, but arithmetic has one.[7] That is why chess or such games are so useful to me. Chess loosens our minds, while the mention of mathematics cramps us. The difficulty of trying to imagine an ethereal entity corresponding to the king of chess is enormously strong.

There are only a few traps into which we are tempted to fall, which we fall into again and again. What answer can I give to the question "What is the King of Chess"? The first thing might seem a definition, but then this only reduces the question. You do not really need definitions. The idea of reducing things to a few entities is drawn from something quite different.

---

[4]MS reads 'come'.

[5]Singular 'time' in the MS text.

[6]"The King of Chess" is employed in the MacDonald (1935) copy in Rhees's own archive, of "Lecture on Logic, etc." (Lecture Six, p. 2 (11)).

[7]Compare the use of 'nimbus' in *Wittgenstein's Lectures* by J. King and H. D. P. Lee (1980): 21, "The nimbus of philosophy has been lost. For we now have a method of doing philosophy". Ambrose's (1979): 44, report "maths has no nimbus" seems to conflict with the above.

If we say there is not an object in the same sense corresponding to the sign 'one' as the sign 'Mrs. Smith',[8] one immediately is tempted to believe there is an object in a different sense. Here you may think I am denying the existence of entities, as I might deny the existence of the image in a mirror. The word "one" is used. We can't say, "Where is the King of Chess?"

I suggest we talk about the use of the word as equivalent to[9] (and not) the meaning. Suppose we say the meaning of a word is its use. To use for the word "meaning" the word "use" has a definite advantage. To say they are equivalent is an advantage for the problem of[10] what is the object corresponding to a word. The case of the object corresponding to a word is where we explain the word by pointing (point) to an object.[11] Frege used meaning for the bearer of the name. "The bearer of the name Watson" can replace "Mr. Watson". It is queer I should have suggested a definition of the word "meaning". There are words in our language which can be explained by pointing to their bearer. This is more ambiguous than you think. From what I have said it might appear I can give an ostensive definition of Mr. Watson, but not of one. We can give ostensive definitions of one, by pointing to one picture. One may say this is not the same sort of ostensive definition. But there are lots of ostensive definitions, such as colour, a note. You will all agree we will define piano by pointing to a piano. You will think to define one by pointing to a bead is very unscientific. [The word "object" is on the same level as the number one.] If I point to a piece of chalk and say "abracadabra", this might correspond to the chalk or red or circular.[12] Thus the ostensive definition does not seem to give anything. You will clear this up by saying we should say "the colour of this is . . .". This sort of definition seems to be given to people who know a great deal about the language but not one particular thing. If the rules fix the use of the word except for one thing, the ostensive definition has use.

You might say in order to learn the use, you must know lots of rules. The ostensive definition is only one rule.

If one says 'genus + differentia'[13] gives ostensive definition, one is tempted to believe this is all one formula. We are surrounded by familiar objects and if we point to them, the answer is one word. Most of these objects are used practically for only

[8]The scare quotes added here by Eds. are not in MS.

[9]'as equivalent to' is written over the line.

[10]'of' is inserted by Eds.

[11]'explain the word by pointing' is written over '(point)' in the manuscript. It is possible that this is to indicate that '(point)' could be removed; and the same could apply (less likely so) to '(and not)' a few lines before.

[12]'correspond to' is written over the line to replace the deleted 'mean'.

[13]Scare quotes are added here by Eds.

one purpose. If one used these objects for lots of different purposes, we would probably have a different[14] language. It is very queer that everything has one generic name. Some uses do not seem significant to knowledge, such as classifying objects by their shape. The words 'generic name'[15] are ambiguous. There are different kinds of generic names.[16]

If one tries to define game by saying that it is between two people, one mentions football.

One then says a game is what is common to them all. But there is really something quite different which relates them and not at all that of having something in common. One is tempted to think if a set of entities have the same name, they have something in common. Our language is constructed on a very simple scheme, substantives, adjectives, verbs. The outside of our language is very primitive. We have mentioned two of the primitive notions.

A third is that we compare the qualities to the ingredients of a mixture. The ingredients can also be put into bottles besides the cocktail. The genus being something in common is that of the ingredient being in common. Suppose four had brandy in them, one of these had water while three others also have water. This is something like the way we use a word for a set of different things.[17]

Suppose we say, "What is it all games have in common?" I will say: need they[18] have something in common? If we say chess and draughts have the same board in common, it is different to say all games have it in common to cheer us up. [Does the generic name point to their having a disjunction property in common?] The justification[19] of a generic name need not be a property in common.[20]

This is what has happened in ethics.

This has happened to the word number; rational numbers, cardinal numbers etc. need not all have something in common. We may call a construction "a number" if it only has similarities with three of these.

---

[14]Before 'different' there is a deleted word, which is perhaps 'difficult'.

[15]Scare quotes are added by Eds.

[16]The next three paragraphs are marked at their openings and endings by the marker "||". Perhaps these were inserted after the manuscript was drafted, to introduce the breaking up of an otherwise continuous narrative into separate paragraphs. Soon after the above, there is a second set of three paragraphs below, which are marked by the same device utilizing this sign. This "||" marking device is deployed elsewhere in other Chapters, where the first choice is retained in the main text while similar alternatives are moved from to footnotes.

[17]This is a point at which the use of "||" to divide the continuous narrative into three paragraphs.

[18]After 'they', the deleted word 'say' occurs.

[19]This could be a response to Sraffa's view of rational justification; see Brian McGuinness's 'Notes by Sraffa for Wittgenstein' (in McGuinness 2008: Letter 172).

[20]As with above, the next three paragraphs are marked and divided the use of "||".

All the words of a language look so much alike. This is the greatest difficulty. Suppose we bought a box of tools for a house, there will be all sorts of tools. They all look much the same, but they are absolutely different. This is the same with words in a dictionary, the use of two words is different in the sense that the use of a hammer and glue is different and not a hammer and chisel. The meanings of words are as different as their use. We can buy all sorts of objects, but we can also buy a seat in a theatre. If the meaning of a word is comparable to the use of money, we see it has just the same ambiguity.

It is alright[21] to say the ostensive definition fixes the differentia if we understand that the genus is not fixed.

It is misleading to say the ostensive definition makes the last decision about the word. It is absurd to say we know all the rules but one.[22]

We might say the explanation given to a child is not an ostensive definition. There is no clear line drawn between the ostensive definition given to a child and grown-up.[23]

There are different stages in giving explanations to a child. St Augustine said he learnt language by being told[24] the names of things.[25] He must have learnt words like "and", "nor", "not". He thought the name of things was important, and[26] that the rest came in somehow.

We might say he was wrong.[27] We might say what he described was learning a simpler thing than language. We do the second thing. We give simpler structures and put them side by side with[28] language.

St Augustine described a kind of language. Suppose we build a house of bricks, arches, cylinders. I give orders by saying bricks, arches, etc.[29] This is a complete language. This could be taught as St Augustine described.

---

[21]This variant spelling dates from the late 19th century.

[22]'We might' is written after the end of this sentence and deleted – with the rest of the line left blank, while the same words are written on the next line.

[23]Two deleted expressions are written after the end of this sentence: 'There is one' and 'There are'.

[24]'being told' is written over the line, above the deleted 'learning'.

[25]After the end of this sentence, 'Didn't' is written and crossed out in favour of 'He must' – both occupying the main line. Wittgenstein is deploying and developing his explicit original theme of 'learning'; for which importantly see Weber (2019).

[26]After 'and', 'learnt' is written and deleted.

[27]The sentence "We might say he was wrong." occurs in the MS after the next sentence but is circled and indicated by a line and arrow that it is to be moved to the above position. The capitalization in "We" has been added.

[28]After 'and', 'real' is written and deleted.

[29]Of course, the later and final form of Wittgenstein's use of Augustine's *Confessions* 1.8 occurs in the opening of *PI*; note that the above also relates to the next Chap. 4's use of *Confessions* 1.6, which differs somewhat from 1.8. We can see echoes in the above from *De Magistro* (cf. M. F. Burnyeat, 1987).

To make this clear, there[30] would not be any explanation involved. When we teach a dog, no explanation is involved.

There is no reference to my superior intelligence if I describe a simple language or arithmetic.

We could surround chess by games simpler than chess but leading up to it.

If we define Moses as the man who led, etc.,[31] and say something different about him. If somebody said recent researches have shown this wasn't the same man, you might go on giving definitions.

We had not being playing an exact game, as I persuaded you to give up definitions. Here you had a certain range of definitions.

In language we change the rules as we go along: it is not a fixed calculus. You might suppose[32] I treat language as we treated a body supposing it is falling in a vacuum.[33]

Immediately after saying that the ostensive definition was the last decision, I said the exact opposite: that a child is taught by ostensive definition and there is no question of knowing the rules. Then I said there were all sorts of gradations. What I did was to give one case, I.E. the language with bricks.

Ditto for[34]: When I said, "the colour of this is abracadabra". _ _ _ _ .

We could have a table of rules[35] when we play the game, and look them up, or we might know them by heart, or we might move a piece automatically. What happens may be a mixture of all this, but I describe certain types.

This is exactly the same with the use of a definition. Compare

---

[30] 'there' is written over the line and replaces the deleted 'it'.

[31] This reference to Moses re. 'led etc.' alludes to the expression "the children of Israel through the wilderness", for which see *Philosophical Investigations* §79 and §87. The biblical references are numerous, typically Exodus 19:1–2 and 17 (see also Exodus 15:22); note that the Hebrew for 'sons of Israel' is often translated 'children of Israel' as with *PI* here. For the next two pages there are some parallels with Ambrose's (1979): 46–47, Yellow Book. Despite this, in the MS above edited here there is approximately 70% more text, which also enjoys a much higher qualitative character, replete with developed structure, in contrast with note-like pattern in Ambrose's text.

[32] "suppose" is inserted here by Eds.

[33] After the end of this sentence a line is drawn to the end of the line, with a line space left before the next sentences starts.

[34] "Ditto for:" inserted here because after the sentence there is a line of dashes that appear to indicate that the reader should repeat "What I did was to give one case". "abracadabra" refers to its use above on page 110.

[35] Notice that "table of rules", together with some similar phrasing, occurs in Ambrose (1979): 47, though the context she records differs widely from the content of the MS currently being edited, both generally and specifically.

[*sin* x].

Troubles in philosophy arise from making up rules, but making up too simple a system of rules.[36]_ _ _ _ _ _

If we brought into chess an element of temperament: if[37] a man moves a knight in a new way, if he is angry. We might at first think this was not a game at all.

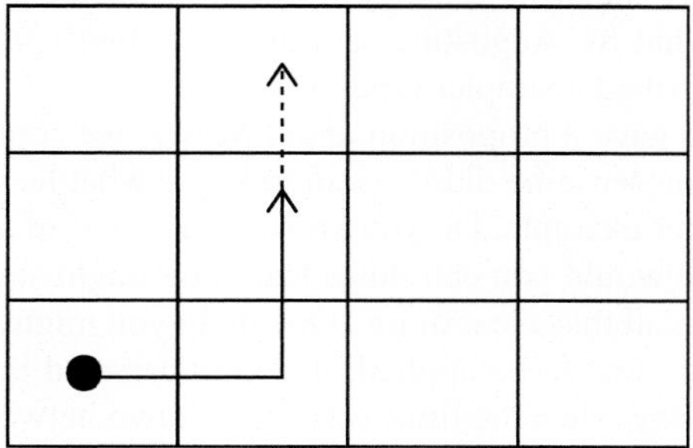

This is a good simile.

If we found a tribe which did this, we might say they did not play an exact game, but we could put side by side with it something that is an exact game.[38]

Suppose a game in a tribe was taught by showing, I.E. a father taught his son. There would be an elasticity in the game. The game would exist without their being any rules. Suppose the tribe tries to tabulate the rules. They might put down wrong rules. I would have to *give rules* to correct them. The philosopher tries to tabulate rules and if he lays down a wrong rule, I can do nothing but lay down rules to correct him.

It might seem surprising that they should lay down wrong rules. Compare high, higher, highest. We think of these as degrees.

My talking about language having definite rules is an answer to the philosophers giving rules.

Example

$2 \times 2 = 4$[39] the rose is red.

[36]The line of dashes that appears in the MS perhaps indicates the 'needless' spelling out of instances with too simple a system of rules; or additional narrative that was not forthcoming.

[37]After the colon and before the indefinite article, a word is written, which clearly looks like 'it', though with an underline that runs into the bottom of the word, which is taken to be deletion.

[38]'game' is written over the line, with an insertion instruction.

[39]This passage alludes to *Tractatus* 6.241, which states:

The solution to the trouble of[40] how could the rose be equal to red was to give *another word*. As long as we were *caught* by this word, it was a *difficulty*.[41]

In notes, examples and similes are always useful. If I could give you enough of them, that would be all that would be necessary. Usually we think of similes as second-best things, but in philosophy they are the best thing of all.[42]

Part of our subject is that we must jump about and make connections.[43] We said we could look at what St. Augustine said in two ways[44]: first that it was wrong; secondly that it described a simpler type of language.

I said if someone gave a proposition about Moses, we could make him abandon his definitions. In one sense he didn't seem to know what he was talking about.

Let us take another example. Do you know what a leaf of a tree is? I could draw pictures and ask you would you call this a leaf. You might say that you had not yet arranged whether to call this a leaf or not. One thing you might say you should know when the word 'leaf' was to be applied, if you understand the word. On the other hand you could not say when the limit was to be drawn between the use you would have to know. This shows us the fleeting (or fluid) use of language.[45]

We can compare this to a game with strict rules.

"Thus the proof of the proposition $2 \times 2 = 4$ runs as follows:

$$\begin{aligned}&(/\mathrm{v})\mathrm{n}'\mathrm{x} = /\mathrm{v\ x\ u}'\mathrm{x}\ \text{Def.}, /2\ \mathrm{x}\ 2'\mathrm{x} = (/2)2'\mathrm{x} = (/2)1 + 1'\mathrm{x}\\ &= /2'/2'\mathrm{x} = /1 + 1'/1 + 1'\mathrm{x} = (/'/)'(/'/)'\mathrm{x}\\ &= /'/'/'/'\mathrm{x} = /1 + 1 + 1 + 1'\mathrm{x} = /4'\mathrm{x}."\end{aligned}$$

[40] 'of' is inserted here by Eds.

[41] At this juncture there occurs the last use of a continuous line below the writing to indicate a line break for the end and beginning of a remark. As mentioned in Chap. 2, (under *Remarks*), it appears that here for the first time we see the evolution of Wittgenstein's crafting of his 'remark'. The next page of the manuscript now emerges with the line break to mark the beginning and ending of the remark, which takes over the from the use of the continuous line.

[42] The writing ends four lines before the end of the page, so the present Edition marks this with a page break here.

[43] This use of 'make connections' contributes to the emergence of 'see connections' crafted in *PI* §122. The new Edition by Hacker and Schulte (Wittgenstein 2009, 54; cf. their note on Page 252) happily translates *Übersicht* as 'survey' in *PI* §122. This enables us to attend to the links between 'see connections' and 'survey' here.

[44] See reference to Augustine in Ambrose (1979): 46. Colon and semi-colon inserted in the above passage.

[45] The MS has a partly formed sentence, with a line deleting it: "If I could give you a definition, you might say".

Simile

If there were patches on a board and I gave you a sharp pencil, and I said copy this contour. There will be a resemblance between the patch and contour. The patch has no contour and merges into its surrounding. But we can make a picture with a sharp line.

There will be lots of pictures with sharp lines which can be said to be copies.

If we use a word, without strict rules, if I lay down a grammar with strict rules, it cannot possibly be the same, but it has a similarity to what we did above.

Why do we need to compare language to a calculus with fixed rules? Partly we are naturally tempted to lay down the rules of our language, and the only way to correct you is to draw several of these "copies" of the patch.

In[46] the case of understanding, meaning, what gives the difficulty is the fleeting use of the words. One remedy that is suggested is to say the words[47] have seven or eight meanings: all I do is to draw a great many figures of this form[48]:

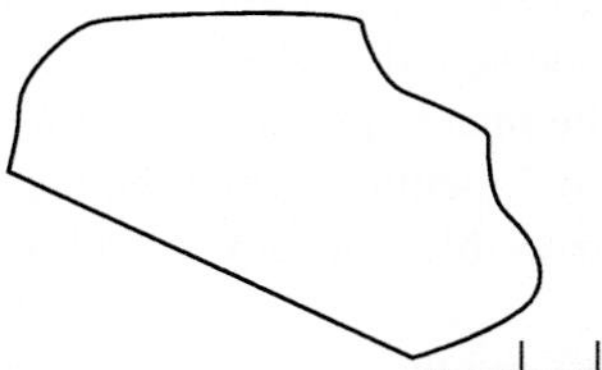

The other temptation is to say all the uses having something in common.

Example

We say there are a[49] great many similarities in a family. Two members have the same nose etc.

---

[46]This word starts a line, and immediately above it '*Example*' is written and deleted (perhaps, another remark occurred to Wittgenstein – the following four sentences and illustration), since the underlined word '*Example*' is deployed at the end of this remark.

[47]After 'words', 'meaning' is written and deleted.

[48]Although the contexts are somewhat different here and in Wittgenstein's 1935 notebook (MS 148: 37), yet there is some thematic and underlying continuity between their uses of the uneven 'circle" (see MS 148: 36, 38–40 – note especially the employment forms of the term and notions of calculation; cp. MS 148: 44).

[49]'a' is inserted here by Eds.

In my first lecture I said I should like to put the word use in place of meaning. In certain cases we can put the word use in place of the word meaning.

One idea to *grasp* the meaning of the word meaning:

Let's regard the meaning as the correlate of the explanation[50] of the[51] meaning.

We regard weight as the result of the weighing process. I am looking at the word "meaning" in connection with "the explanation of . . .".

What do we call explanation of meaning?

If we say, "the colour of this is white", we see that this explanation gives us the use of the word.

If we call meaning the use of the word, we might call understanding the knowledge of the use.

First you might object to putting meaning equivalent to use.

You might say the use is something internal or inside language while the meaning seems to point outside.

The rules of the game seem[52] to belong inside the game, while the very essence of language seems to point outside itself.

A game[53] seems to be autonomous, and we seem[54] to make language self-contained.

Question: does any rule about[55] the word give partly the meaning of the word? We might say this. We are inclined to give particular rules for this meaning like the ostensive definition.

I want to mention an objection. If we say the meaning of a word is its use and is what is defined by its rules, we seem to make language autonomous.

We are going to make the comparison with the king of chess. We could talk about the meaning of the King of Chess. There is the similarity that the shape of the King of Chess is as[56] irrelevant as the sound of the word. The chess player has a different relation to this piece than the man who knows nothing about it.

One difficulty we have to[57] mention is that the meaning of a word is something else. Using this simile destroys this tendency to look for the meaning as looking for the man.

Instead of saying "money and the thing we buy" we say "money and its use".

We don't point to the meaning, we explain the meaning by pointing to the bearer.

We will talk about the word understanding.

[50]'explanation' is written over the line and replaces a deleted word, which may read 'sense'.

[51]'word' is written and deleted before 'meaning'.

[52]'seems' in the MS.

[53]After 'game', 'is' is written and deleted in favour of 'seems to be'.

[54]A light line appears through 'and we'. It is hardly likely that this indicates deletion. Rather, an inadvertent touch of the pen passing.

[55]'rule about' is written over the deleted words 'use of'.

[56]'as' is inserted by Eds.

[57]'to' is inserted by Eds.

If I call understanding a word "knowing its use", one of the difficulties is this. If I say "I saw a pine-tree on the road" and you say you understand the word pine-tree, do you mean you know all the rules of the word? Are they all in your head when *you say* this?[58]

We will use the game of chess straight away as another example. If I and somebody sat playing chess, and if you asked me what I was doing, I would say we were playing chess [we assume the position could be just the same in another game].[59] I would say we could distinguish chess and draughts only by their rules. Somebody might say you couldn't possibly know which game[60] you were playing. Suppose we had made three moves which could be either chess or draughts, we couldn't say we were obeying from this the rules of chess.

We will ask: how does the intention of playing chess differ from the intention of playing draughts?

If I ask you what you mean by intention, you might say it was a psychological process. You say there are two processes, one for chess the other for draughts. Is this a matter of experience?

If I have a particular sort of hunger, and you say an apple will quench it and a pear won't. This is a matter of experience. But if you say you want an apple, this does not mean an apple will satisfy you.

Suppose you say your criterion for playing chess is that you said you were going to play chess, you might go on and play wrong chess.

The statement "I know I am playing chess" is not defined by the rules being present in any way. If we played chess from a book this could be the case.

We might think that the criterion of playing chess was a state of mind.

We know that there are different states[61] of mind for games like chess, bridge.

I will assume that there is a state of mind corresponding to a game, further I will assume there is a particular state of mind for any move.

Although there are these states of mind, they do not presuppose the rules.

There is a difference between shoving[62] the King of Chess with understanding and mechanically. We may call understanding the state of mind when using the King of Chess. But we have said understanding was knowing the rules for the King of Chess. There is a *causal* connection between these two uses of understanding.

---

[58]A new sentence is begun: 'We [also the beginning of perhaps '[wil]l]' is written and deleted, while the next sentence commences on the next line – presumably to indicate the paragraph break.

[59]The use of '[. . .]' here reproduces the MS's own usage.

[60]'game' is written over the line with a mark for insertion.

[61]'states' is written 'statements', with 'ment' deleted; also, the next use has a variation of this mistake, which is written over to correct it.

[62]'shoving' seems to be the correct reading (using the precedent of reading the letter as 'v'.], which in Skinner's hand is written identically to the letter 'v' in 'have symptoms' lower down the page; 'show' is theoretically possible, however.

If[63] the word "understanding" for a word or sentence is used as a psychological state which[64] accompanies the word, then it is utterly different from using it for knowing the rules of the word.

We have two ways of using the word "psychological[65] state". In one use it means "consciousness" as toothache, in the other it is hypothetical where we[66] get symptoms.

Let us distinguish between these – in one case a state of consciousness, the other a state of hypothetical construction for which we have symptoms.

This comes in when we talk about subconscious states.

We might say we had something we called subconscious toothache and think we had discovered something new.

In some cases it may be useful to talk about subconscious states of mind.

Knowing the alphabet is a hypothetical state. Knowing I have toothache is a non-hypothetical state.

Let us use this distinction in the case of the intention to take the King of Chess.

Is intention when expressed as a sentence an amorphous state of mind? The expression of intention is a sentence with a beginning and end.

Is the connection between the intention of doing something and doing it a matter of experience! Can you later on find[67] you intended to do something else.

This same thing has been said about wishing.

A pain has a place, but there may be nothing wrong with my tooth.

Notes.

There are lots of things you must have in evidence. I said in my first lecture[68] we were doing geography without a map. You need to make a map.

I drew your attention to a difficulty about the idea of understanding.

The difficulty arose that understanding seems to be[69] a mental process which goes on when we hear the word. How are the rules present to us?

We followed this up with a parallel case from chess.

Digression: the use of similes is to give parallel cases.

Parallel cases change our outlook. We can then look at the thing in a new way.

---

[63] 'If' is written in the margin in smaller writing, as if added later.

[64] The MS has "with"; here corrected to "which".

[65] The term 'psychology' was written first and altered to 'psychological state', perhaps explaining why the singular 'word' is used to cover these two words.

[66] 'only' is crossed out here.

[67] 'find' is first written 'finding', with the ending then crossed out.

[68] 'lecturing' is written first, with the ending then crossed out. Here is evidence that, in some form at least, part of the narrative embodies more than one lecture.

[69] 'seems to be' is written over the earlier word chosen, which is "is". This re-grading of modalities is a regular feature of sensitivity in revision.

Simile[70]

When the view of the earth as the centre of the planetary system was given up, and the earth was said to be just one of the planets, part of what happened was to destroy the unique position of the earth.

It is often enough to destroy puzzlement to show it is not a unique thing.[71]

If we know the game of chess, we can be said to understand a move in the game.

This mental process of understanding the move does not presuppose the rules, which are what distinguish between the man who does and who does not know the game.

There are two uses of the word understanding.

To know the alphabet or to know the rules of chess is not a state of consciousness or is not a feeling. The grammar of such a word like knowing is something quite different from the grammar of a word like feeling the piece of chess while moving it.

If two people learnt by heart a game of chess for a theatrical performance there is a sequence of states of mind[72] which[73] two people have[74] who know it, and which they do not have.[75] But what distinguishes these two pairs is that one knows the rules and the other doesn't.

Knowing is a state of mind in quite a different sense from a state of mind like a base note underlying a melody.

The chess player is distinguished from the automaton in what seems at first two ways.

Mental processes take place in one case which do not take place in the other.

It has been said that words like "and" "for" "not" have special states of mind.

Such feelings exist without doubt, but there is doubt whether we always have this feeling.

Having the "and – feeling" is different from knowing the application of the word "and". It is not what we call the knowledge of the use of the word.

We said the word "is" had two different meanings:

"2 + 2 is 4"
"the rose is red".

Do you understand the word "is" when you pronounce it? I suppose you mean a different mental event takes place in the two cases.

---

[70]As mentioned in the Introduction, for 'Simile' see Y. J. Erden's (2011).

[71]An ending to the sentence has been deleted, which reads: "to destroy uniqueness".

[72]For 'states of mind', 'statements' is first written, then partially crossed out with 'of mind' written over the line.

[73]After 'which', 'distinguishes' is written and deleted yet restored in the next sentence.

[74]'have' is squeezed in at the end of the line, presumably afterwards to balance the other revision (noted in footnotes below).

[75]'and which they do not have' is written over the line; this replaces the deleted 'from there' that would have gone with the deleted 'distinguish'.

No other event need take place[76] than that you pronounce it, and hear it.

There is a tendency to think you can swallow the meaning of a word while you pronounce it.

The mental event that takes place is never the knowledge of the use of the word.

The idea is that one can see the difference when one says the two words "is."

The idea is that the meaning stands behind the words.

Imagine we had bodies of glass cubes, pyramids, parallelepipeds,[77] but we could not see them. Each one has one surface painted. We could only see two squares the base of a pyramid, and the side of a cube.[78] The way in which these bodies could be arranged in space would depend on what bodies they were. We might compare the word with the surfaces we see and the use with the body.

If what I say is true, you might say by a single act of seeing the cube you would know all the rules.

We might think something similar could take place here with the word "is".

But can one see all the rules of the geometry of the cube just by seeing the "cube".[79]

We are tempted to think we grasp the meaning of the word "is" as a whole while we pronounce it, and the rules can be developed from the act of grasping.

In the same way we think by a glance at the cube, all the rules can be developed.

Here we think the rules of the use can be deduced from the meaning.

This is analogous to where I see the cube and grasp its shape and then slowly deduce the laws of geometry from it.

When we move a pawn, we are tempted to think we grasp this as a move of a pawn. We feel we grasp it as different from a move of a king.

[76]'place' is inserted by Eds.

[77]This is for a three-dimensional figure constituted by six parallelograms. This belongs to the symmetry group C_j, $[2^+, 2^+]$, (1x). The pyramid is a member of the related symmetry group C_{4v}, $[4]^+$, (∗44). The parallelepiped is related to a parallelism, as the square is related the cube. See Euclid's *Elements*, Book 11, Prop. 24. Note that during the years 1929–1934 when Wittgenstein was lecturing and revising the current Pink Book, G. Hardy, J. E. Littlewood, and B. Polya were researching aspects of these notions and other matters, publishing in 1934 in their book *Inequalities* (Cambridge: CUP). Notice that Wittgenstein is still concerned with parallelepipeds in his *Lectures on the Foundations of Mathematics*, in 1939: re. inserting rods through their sides (p. 40); stacking wood (p. 202).

[78]Indefinite articles inserted by Eds.

[79]This sequence continues the exploration of the earlier theme of seeing connections (see earlier in this chapter the link with *PI* §122 and surveyability problems). The discussion commences above with the introduction of geometry (e.g., via parallelepiped). Wittgenstein is clearly aware of some problems of mathematically 'seeing' and 'grasping' in cognizance of symmetry groups (cf. Hardy's et al. 1934 *Inequalities*). Towards the end of the Archive in this book (see below Chap. 8, sec. XVI), we find a return to Russell's role in surveyability, with attention to cardinal numbers. Compatible with Wittgenstein's position, research on attempts to assess infinity (Bombieri 2011), the inaccessibility of Cantor's absolute (Livadas 2020), and problems concerning uncountable sets also in computer calculation (Tao 2008), raise substantial problems for claims of universal survey ability in mathematics.

Does geometry talk about cubes? I ask does it talk about wooden cubes. You say it is about the geometric cube. I say[80]: is the geometric cube something very thin?

The propositions are not about the geometric cube. I say they define the word cube.

1 + 1 = 2 is not a proposition about one.

A proposition about one is "there is one piano in the room". If I say, "Give me a proposition about Mr. Smith" there is no difficulty about this.

Let us slowly approach a proposition about one.

What is a proposition about red?

The rose is red is such a proposition but one feels one would like something like "I like red better than blue".

For different kinds of words, we try to make up something corresponding to "Mr. Smith sits in the chair". I suppose we want to form a true sentence about reality in which this word occurs.

Grey is a mixture of white and black; this does define grey. This is not a proposition, in the sense that the hair of Mr. Smith is grey.

If we alter '1 + 1 = 2', we alter the meaning of two. If I defined a cube as a parallelepiped with sides in a special relation, you would agree I had defined it. But if I alter a geometric proposition about a cube, you might not agree I had altered the meaning of the word cube.

The geometric propositions constitute the meaning of the word cube.

The arithmetic rules for one constitute the meaning of one.

You are tempted to say certain propositions about the word cube constitute its meaning and the rest follow from it. If you say the rest of the propositions about one follow from a certain number, then you assume the logic according to which you deduce them.

Let us consider one case.

You say surely there mustn't be contradictions in our arithmetic.

It all depends what games you are going to play with "one" "two" "and" "not" "equal".

You begin with the word[81] cube or the word and you give a definition which is a sentence. You then apply such words as "not" "or" "and" to it, and seem to imply[82] that they are to be used in a similar way to what they have been used in some other cases. We may then say certain rules follow.

You have in a sense prejudiced yourself.

We lay down the rules of multiplication. Then we make multiples. We multiply 451 x 637 and then say this is not the usual result, which is one less.

We may say we are not playing[83] the game.

We said originally our rules were for all cardinal numbers.

[80]Colon is inserted here by Eds., as throughout this Edition.

[81]'word' is written above the line.

[82]The manuscript has 'apply', which obviously seems to be a slip, for 'imply'.

[83]'playing' is written above the line with instruction for insertion.

Now we may say we are not using the word "all"[84] in a way we are prepared to use it.

In what sense does the general rule tie us down?

Supposing I had made an exception in this case. Someone would say I had used "all" as I would use "all cardinal except...". But where was my meaning prejudiced. I used the word all. When you say I have changed my game, this statement is only of psychological importance. I use it differently from what[85] you expected it. Does this show we are not playing a game? I can say playing a game is not as simple a thing as you thought. We might have one rule for some days of the week, and another for other days.

Is this alright? No, it isn't.

We will only yet say that geometrical propositions say nothing about cubes. Some seem to fix the meaning of cubes, and others seem to follow.

The point was could you deduce the laws of geometry from the idea of a cube or do the laws determine the idea of a cube?[86]

Knowing the use comes roughly to knowing the application of the rules of the word.

We also say we understand a word when we read it. This is a different meaning of understanding.

When we read the word, we only have this one use of the word.

Is the use independent of the meaning, or does it follow from the meaning?

If[87] we write down half a dozen sentences with the word is: "the rose is red", "the book is green" – "is $= \in$." Russell.

When we understand the word in one act of reading, we take in the meaning – swallow it. Then we have certain ways we can use it or certain ways we can't.[88]

This idea is that the use follows from the meaning.

This is the contrary idea to what I want you to get.

To explain this idea, I used a simile. The meaning seems to be an invisible body behind the visible word.[89] We had certain glass bodies each with one of its faces painted, so that we saw only a red triangle square etc.

If we wished to establish the rules according to[90] which they combined in space, we might say if we knew the bodies we would know the rules in which they combined.

[84]Quotation marks are inserted by Eds.

[85]'what' is written above the line.

[86]Page break after this sentence ending.

[87]Before the start of this sentence, 'Is the further' is written and crossed out.

[88]This sentence is written in slightly smaller handwriting, and the last part of it is written under the end of the line – so as to fit in the space before the next sentence commences.

[89]This remark reflects a discussion about a "mental reservoir" of meaning in Wittgenstein's sketchy argument that he sent to Sraffa, in Skinner's handwriting. Sraffa replied (see McGuinness 2008: 225; Sraffa's reply: 227). Material from Wittgenstein's argument, which shows this influence of Sraffa in the modified version, is re-used in the *Brown Book*.

[90]'according' is written a second time, now above the line, with no further indication.

By one act of seeing, we might or[91] we could get all the rules. We see the cube, and think we can expand this into rules.

The rules would be given in the laws of the geometry of the cube. Could the geometry of the cube be obtained from the cube? Does geometry talk about a particular cube?

I will formulate the answer.

Geometry does not speak about cubes, but gives the grammar of the word[92] cube.

The laws of geometry talk about the cube in the sense of giving rules.[93]

I said geometry fixed the meaning of the word cube. If one alters the geometry, one alters the meaning.

If we alter a statement in geometry, we change the meaning of the word cube.

Geometry determines what propositions about cubes make sense and which do not.

In the same way arithmetic determines what propositions about numerals make sense and what do not.

Suppose I said I saw a paper with five dots, two arranged on each side of one in the middle. This may be true or false.

If I said I saw a paper with five dots,[94] three each side of one in the middle, then this would be nonsense.

What roughly is the relation between arithmetic and its application? This is the relation between the grammar of a sentence and the sentence?

One mistaken notion is "the idea of possibility as a shade of reality".[95] If I say, "it is possible Mr. Smith may turn up" this seems somehow different from his coming or not coming.

Geometry does not say anything about straight lines.

The straight line is the dividing line between red and black. Geometry does not talk about this line, but fixes the meaning of the word line.

We might say, "Geometry constitutes the meaning of certain words". An objection was "is it true all the geometry constitutes the meaning, or do some constitute the meaning and is the rest propositions which follow from the rules".

I can neglect this question. The rest are rules about the grammar which follow from the first rules.

If I say these rules constitute the meaning, you may say, "What is the meaning". We can say, "What is the game of chess".

---

[91]'or' is added by Eds.

[92]'the word' is added below the line (i.e., it is the last line on the page), which replaces the deleted expression 'a particular' in the main text.

[93]A deleted expression is written on the next line: 'An objection was made', with the rest of the line left blank.

[94]After 'I saw' there is a ditto sign followed by five dashes and a comma. This appears to indicate a repeat of the expression from the previous line: 'a paper a paper with five dots' as presumed in this Edition.

[95]Cf. *Tractatus* 2.201. Note discussion of 'shadow of a fact' in Chapter (7th Feb. Lecture).

I give you a description of the use of certain words. You expect me to point at something and say, "This is the meaning". Sometimes I may do this. As for example, suggesting considering the use of a word equivalent to the meaning of a word.

There is the question "what role does the word cube play in geometry and in its development".[96]

There is the temptation that we can get the rules of the cube by investigating it geometrically.

We are tempted to think a geometrical investigation is like looking at a piece of chalk and investigating it. This is the idea that mathematics is a sort of physics.

It does not play the role of the piece of chalk.

I will answer, "I say it plays the role of a symbol".

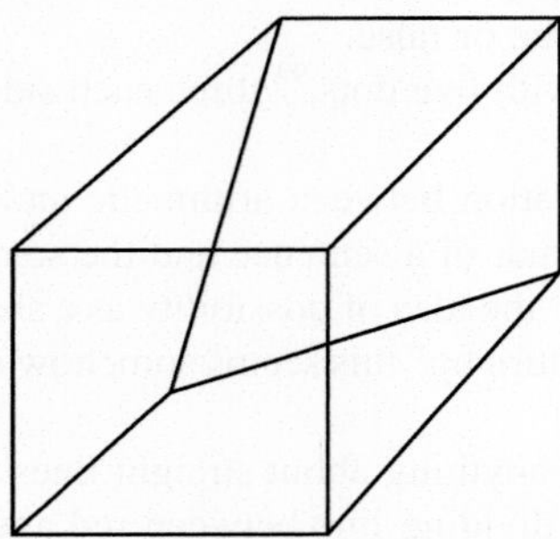

This drawing is a symbol or part of a language.[97] Thus the drawing need not be accurate. The geometry is about this figure.

We could represent a proof of Euclid by a series of figures. Suppose the proof ends "these two segments are equal" we could represent this:

---

[96]This line and the next four are consistent with the resumption of use of line insertion to indicate a space between remarks. It seems that this could be a retrospective decision.

[97]This use of a cube and the occurrence of 'King of Chess' have thematic parallels with the Wittgenstein's C-series large notebook MS 150 (1935), pages 67–77. Rhees's comment on the cover of this C-series notebook suggests it pertains to material in the *Brown Book*; the Pink Book association – together with other ones in the Skinner papers – indicate a much more extensive network among Wittgenstein's writing.

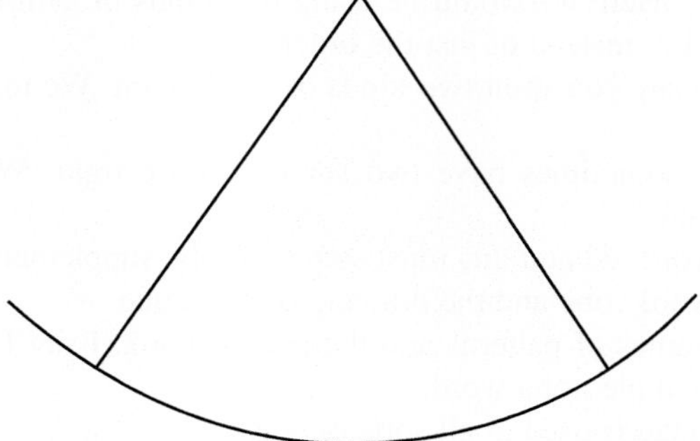

Is there a connection between the sentences which there is not between the figures?

I will say the proof you think alright is a series of sentences written down in a certain order.

I say let's have a series of pictures instead. You say a picture is only a picture. You say surely there is no connection between drawings. I say there is a connection between pictures.

Your next objection is this – You say I compare sentences and pictures. But sentences are not sentences – they have sense. You can't have pictures unless you give them a sense.

I say you can regard the sentences as pictures.

Example

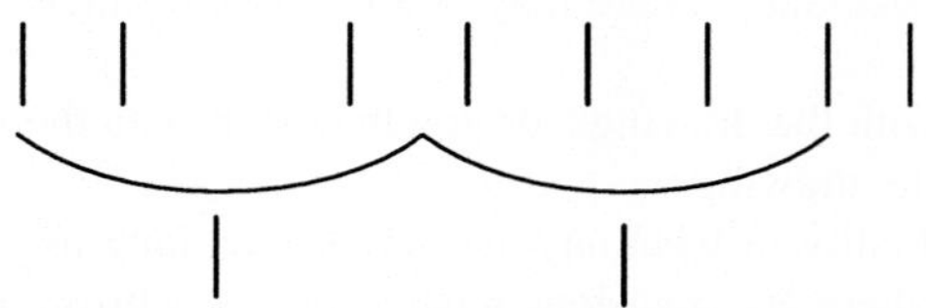

This can be regarded as a geometrical drawing. You say in geometry you always try to draw a line thro' two lines. This is another kind of[98] geometry. Some things do matter.

You can regard the written[99] proof as a geometric proof. Some things do not matter, such[100] as the way the word is written, but the recurrence of the word does matter.

---

[98]'kind of' is squeezed in on the line after the line was written.

[99]'written' is written over the line with instructions for insertion.

[100]'such' is added here by Eds.

We may use ∴ .Here we would be using two kinds of symbols in the series of figures, if we used ∴ instead of just the order.

It is not true to say you want two kinds of symbolism. We may need a hundred, or one.

If you say we sometimes have two sorts, you are right. We sometimes have "words and samples".

We have the word red and this must sometimes be supplemented by a red patch.

We have the word cube and the drawing of a[101] cube.

We have the names of patterns and the pattern book. I say I want a suit of this colour. This is a sample not a word.

I say our language is used in all sorts of ways.

Your idea is the words are necessary to explain the thing. Take the example of a table.

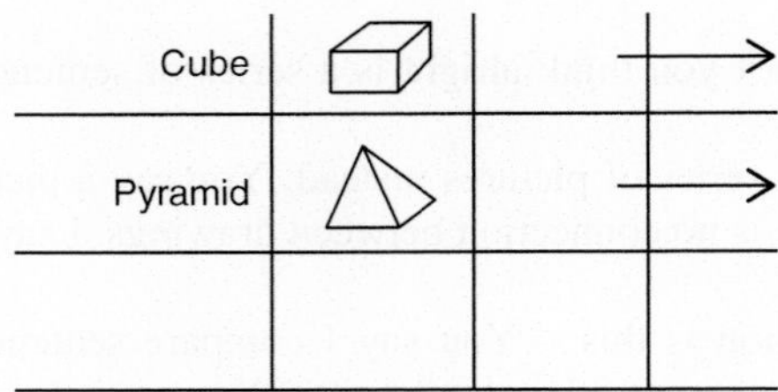

You would say this table explains the words by giving samples.

Someone may say, "I want an explanation of the way this table is used". I can give you arrows. You might then still want another explanation.

Psychologically some of these explanations may be necessary. Logically you can't say any are necessary. There may be one, two, or three. The explanation can come to an end.

You may start with the drawings, or you may start with the words. Words are on the same level as the drawings.

There is no difficulty in what anyone may say as long as you don't generalize. There is no doubt there are occasions where you use words and samples, as in the case of the tailor.

If you say everywhere we use words and samples. I will say this is not true.

I must show you certain ways of generalising are not true.

Suppose I dictate to you and you have to write, or I give you the Strand magazine which we are to read.

You could read it by correlating the written word with the spoken word.

Your reading is a translation from the written to the spoken symbol. This could be done by a table.

[101]'of a' is written above the line.

Supposing I wrote down a sentence in small letters[102] only, and your task was to translate it into a sentence with capital letters only. This could be done by a table:

a	A
b	B
c	C

If I gave you this task, you can imagine this done by a table, but as a matter of fact you don't use a table. You say you have the table in your head, but as a matter of fact you haven't. (There needn't be a table.)

Here it seems I might bully you into believing there is a table in your head, but this isn't the case.

This shows that symbols are used in different ways.

You are used to explanatory sentences of drawings.

If I draw a cube to explain a cube to a Chinese, you should say the Chinese will only understand it, if he can make sentences about it, I will say here you use understanding as "understanding in words".

Here you have given a definition of saying certain words[103] or sentences to yourself.

I will give you certain intermediate cases.

You might write them down. You would agree this made no difference.

You could imagine a tribe which could only write and did not speak.

Now I can go on and say, "I will draw pictures".

Supposing we have a pictorial language: there are two ways we can represent[104] "red circle".

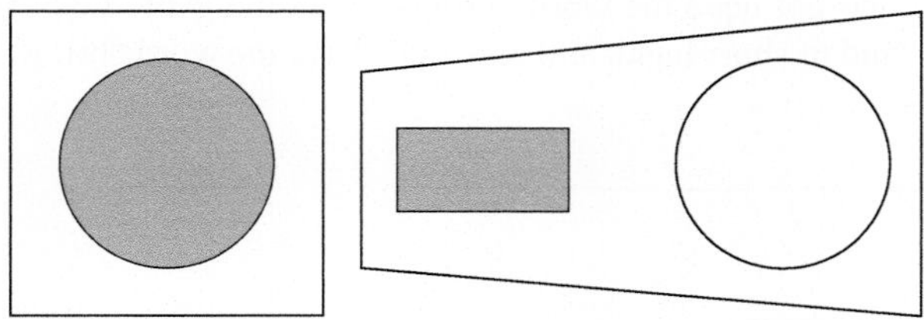

[102]MS reads 'letter'.

[103]'words' is written above the line.

[104]The expression after the colon replaces deleted writing that reads: 'and we want to point to'.

These are two ways of expressing it pictorially.[105]

We represent: "paint a red circle"[106]:

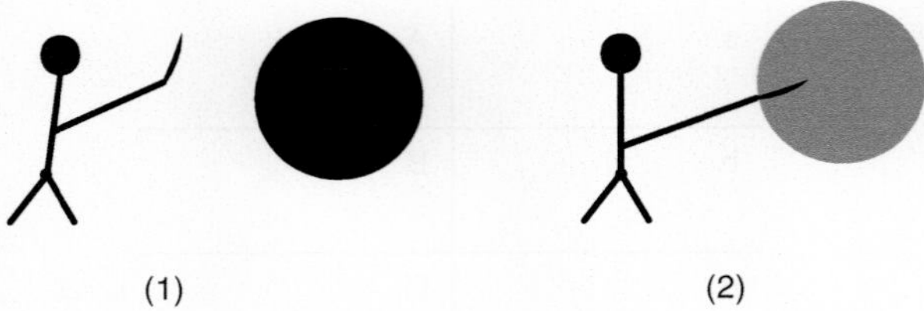

(1) Corresponds to a word language.
(2) Does not.

These languages need not have any words. In the second case we have a picture which corresponds to a word language, in the first case not.

Could a man understand.

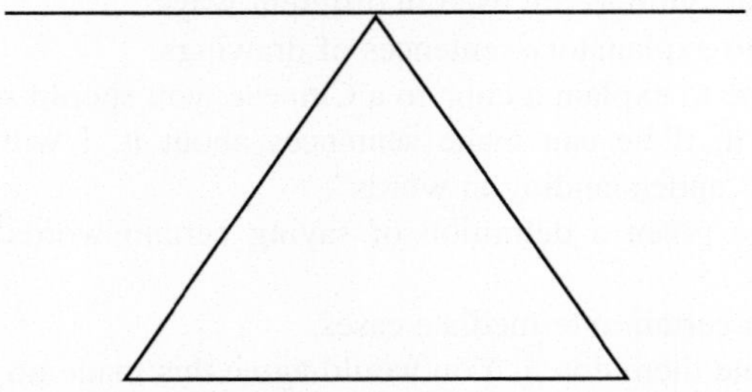

without knowing a language?

Sometimes we understand a thing by translating into words – sometimes we understand a thing by drawing a picture.

If I say you do not need the word "bring me" in the game I described, you may think this is a kind of short-hand, and they must have the word "bring me" at the back of their minds.

---

[105] In the manuscript the two figures together with the two numbered propositions appear on the facing, otherwise blank page, which is opposite the diagram above comprising a square (with a filled in circle in it) and an oblong with a square (presumably to indicate a red colour) and an empty circle in it. This diagram is set within the continuous narrative in the manuscript.

[106] This figure is on the otherwise blank facing page of the manuscript exercise book. It roughly matches a figure in Klagge and Nordmann (1993a: 293) though the arguments and contexts are different.

Language games, gradations of understanding.

Question: at which point does understanding begin?
Question[107]: Doesn't the child act as an automaton in the first games – where does it cease to be an automaton?

Revert to ostensive definitions.[108]

In which sense can an ostensive definition be said to be an[109] explanation of meaning, and in which sense an incomplete explanation of meaning.

Remember that there are differences between the case of someone who can ask the question[110] "Do you call 'A' the colour of this, or its shape, or its length, etc." and the case, say of a child who is 'explained' the meaning of the word sugar by someone pointing to a piece of sugar and pronouncing the word —— should we[111] call this an *explanation*? There seem to be all sorts of different kinds of ostensive definitions, I.E. all sorts of different things which we might call "ostensive definitions" (as there are all sorts of different things we might call arithmetics, imagine a primitive arithmetic with two numbers only).

A[112] mental act accompanying the hearing of a word loses its importance if it cannot sum up the meaning of the word.

We think of symbols as dead, and that the mental act brings them alive.

How can a mental act accompanying the hearing of the word stand for the whole calculus with that word?

The[113] mental act may be important.

Whenever the mental act seems the soul, you should remember we needn't have a private mental act, but we can substitute one which everyone can see.

For imagining red, we can substitute taking a piece of red paper out of our pocket.

There is a prejudice that thinking consists of calling up images – as opposed to taking a sample out of one's pocket and looking at it. It is not essential that we should have an imagined red image: if you can see this, your outlook will be changed.

We can imagine we have a table in our pockets.

---

[107]From 'Question' to 'two numbers only)' contains two paragraphs that appear in the unpublished hand-written Skinner draft of the *Blue Book*. These two paragraphs do not appear in any version of the *Blue Book*, not even in the early Thouless copy (see Chap. 1 above). The *Blue Book* text on either side in the unpublished Skinner (incomplete) hand-written version is the same as the published Edition of the *Blue and the Brown Books* (p. 3).

[108]Prior to this sentence a deleted expression is used, which appears to be 'I say'.

[109]It seems that 'an' was originally 'a'; and after this there is a deleted word 'full'.

[110]In MS 'Is A the colour' and 'Do you m' are written and deleted.

[111]Although 'be' is written in the MS, 'we' seems to have been intended.

[112]'If a mental act accompanying the' opens this remark but is deleted, and a new line commenced.

[113]'Does the drawing of a cube' opens this remark but it is deleted.

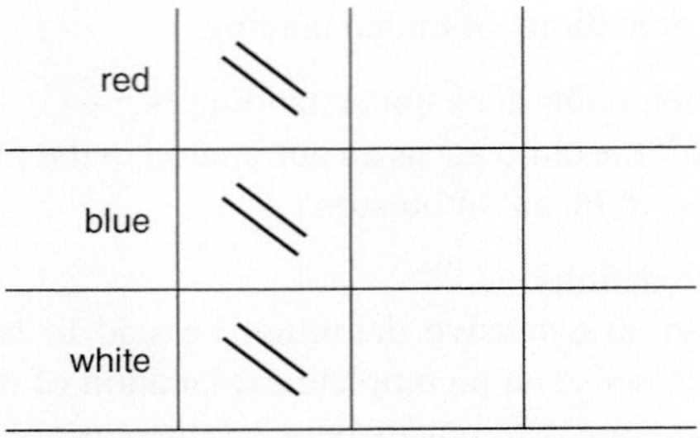

red	\\		
blue	\\		
white	\\		

and pull it out whenever we need something to find a colour.

This prejudice[114] can be expressed "We think in our head" or more refined "We think in our mind". The idea is there is a locality of thinking. What happens may just as well happen on the black board. You can say I think with my finger if I[115] move my finger on the board.

There is an objection that you may say; I see the finger with my mind.

Do[116] you say you have toothache in your mind?

The private experience does not matter, as it doesn't matter in chess.

You need not think thinking[117] is something essentially private. It is said "one thinks with one's pen". This is a good expression.

It may be important besides the word red to have a red image before you.[118]

Example

"Copy the red of this book". The word red will not do here alone. If I say, "paint me a particular red", you need to have some sort of sample. A mental image may be essential to using the word, but I could replace it by a piece of paper of that colour.

You could say if you think you calculate with words and with images. This proceeds by steps – no mental act contains all the other. No mental act anticipates the steps through[119] which you actually go.

Let's do away with mental acts in thinking; they are entirely inessential in thinking, as they are inessential in multiplication.[120]

[114] Spelt 'prejudiced'.

[115] After the first person the deleted incomplete expression 'write it o' occurs.

[116] 'Can' is written then deleted under 'Do'.

[117] 'something' is written then deleted under 'thinking'.

[118] The foregoing discussion of toothache succeeded by consideration of perception of colour occurs, albeit without the same specific discussion, in Wittgenstein's unpublished 1935 notebook MS 148, pp. 8–10 (the colour sample here is red, whilst green in the notebook).

[119] 'through' is missing here and supplied by Eds.

[120] After the end of this sentence 'There' occurs and is deleted.

There are all sorts of mental acts, but they are unimportant. We are interested in the multiplication, not in a single act of understanding even if it did exist.

The thinking may be the talking.

There is an objection: we say, "Did you mean what you said?[121]" he may say, "Yes, I did".[122]

The images may be very important. If I say, "I am very tired".[123] This may mean that I have lots of experience, or that I just say it.

There is one question what did you mean? You say, I meant so-and-so. There is another question: did you mean what you said? The answer is here yes or no.

How does the process "meaning I am tired" accompany saying "I am tired"? When you say you mean it, all sorts of things may justify this.

A French politician said, "In the French language the words follow in the order in which you think".

Sometimes you say if you recall images while you read, you say you think while you read; sometimes if you draw conclusions from what you read, you say you thought while you read.

You can make a record of "when was Trinity[124] founded" by putting it[125] on a piece of paper and pulling it out next day. If you say you do the same sort of thing when you remember it, I can ask you where is it written? Is it in your head?

There is not one act of recording it. You must not mix up personally recording it, and a hypothetical mental act of recording it.

If you say there must be a record somewhere, what you mean is that you want a model in which there is a record.

Your mathematical symbols you write down are such a model. They have the same relation as the written rules of chess to playing the game of chess with the pieces.

If we looked at the brain we could see thinking from the outside, we feel it from the inside.

This is an expression of a certain confusion.

This is only a simile.

You assume there is a thing you can see in two ways.

This is a confusion that you talk about seeing in two ways.

You can see a building from the inside or outside, but you can't see a visual image from the inside or outside.

You have two *different* images. They are not the same.

You could construct the grammar for a word where you saw toothache both from the inside or outside.

---

[121] Absent question mark is supplied by Eds.

[122] After the end of this sentence there occurs an incomplete expression: "That may mean pesa that the image".

[123] Quotation marks are added by Eds.

[124] I.e. Trinity College Cambridge.

[125] Absent 'it' is supplied by Eds.

Friday, Nov. 17th.[126]

We say we think in our head. What are the reasons?

It is obvious the head has something to do with thinking. We hold our head when we think.

Physiologists say certain definite things happen in our brains when we have certain definite thoughts.

If you say that's why we call the brain the locality of thought, then this is your reason.

If this were[127] all, I would say you might find something corresponding in our whole nervous system.

There is a particular reason why we chose the head: that we have a grammatical need for a locality and a locality which we can't see. I will grant you that you have seen into a man's brain and seen what?

You could thereby read the thoughts.

If this different from reading the writing?

Are they on any different level from the writing?

I see the chair in this room. Where is the locality of the after image? There is no locality.

If your images occur in the brain, you see the fire[128] in the brain.

What a queer sublimation when we say we think in our mind and not in our heads.

We might make a model, and say part of the model is the mind model and this is the part where thinking takes place.

Nov. 13th.[129]

What we said about wishing applies to interpreting.

Whether they are about meaning, wishing, intending, thinking: these discussions have the same shape.

Nov. 20th.[130]

I had a long discussion whether wishing was some sort of *feeling*.

---

[126]November 17th was a Friday in 1933 (see Cheney 2000: 207). Any dates published in the main text of this Edition appear in the form and position that they occupy in the Archive's manuscripts. This is the first date in the present manuscript. The retention of dates can be an indicator to reflect redrafting, or an *aide-memoire* retrieving an earlier context. This may assist explanation of the reverse dating here, of Nov.17th placed prior to Nov 13th.

[127]'was' in the MS.

[128]Almost invariably the manuscript's handwriting is very clear, but the term 'fire' gave a problem, and we are grateful for the identification of the word by the Trinity Library archivist Adam C. Green.

[129]This date is consistent with Nov. 13th, Monday 1933.

[130]This date is consistent with Nov. 20th, Monday 1933.

If one is asked whether wishing is a feeling, one could answer if you call a wish a feeling, this is nothing but a definition.

In eating, there is a feeling called hunger.

We cannot read from hunger the articulate wish for an apple if hunger *is* a *wish.*

If we receive a blow on the head and the hunger ceases we can say this was our wish.

If we say hunger is the wish to eat, it is not clear that eating satisfies this wish.

We may make arrangements. It is a hypothesis that eating will satisfy it. Using the word this way, it is doubtful what will satisfy it.

There is nothing against it, but we can see it is a rather unusual way of using it.

We can say we could use the word[131] wish in two ways:

(1) A way where it has no sense to ask 'how do you know'. Here you mean it is an articulate wish.
(2) An hypothesis.

What happens when you wish for an apple?

There is not a single process: feeling things, imagining things, saying things.

You may say 'this alone I do not call the wishing'.

Sometimes only one of these things happens.

Whenever[132] you say 'I don't understand or do understand', you can have classes of cases where the imagination is very important, and cases where it is does not play a role.[133]

If you ask what does it mean to understand a sentence you may say 'sometimes you follow it' 'sometimes you make an image of it if it is a description'.

There are lots of[134] different things in common[135] in different classes. What is in common is overlapping:

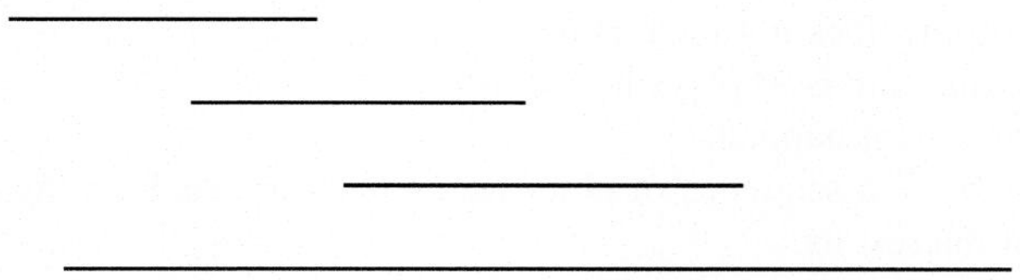

Suppose I gave someone an order to go somewhere, and gave him a map, all he does may be to follow the route with a pencil. He may not yet obey the order but he follows it in this way, (or by mental imagery).

[131] 'we could use' and 'word' are written over the line for insertion.

[132] Sentence originally began 'How do you' but it is deleted.

[133] Spelled 'roll' in MS.

[134] 'of' inserted.

[135] This looks like a reply to the type of answer Sraffa gave Wittgenstein in McGuinness (2008: Letter 172).

Roughly one can divide what happens when you wish into two[136] classes:

(1) Amorphous feelings which go on for a time and cannot be translated into a sentence.
I feel hunger – I cannot translate this into 'I want food'.
(2) The other processes can be translated into a sentence 'I wish for such and such' 'I expect so-and-so to come in'.[137]

For these latter cases put for the hope, the fear: the expression of the hope and fear.

I expect Mr. Smith is the expression of the expectation.

Put the expression of the thought instead of the thought.

Whatever happens is translatable.

If wishing is a certain process, I can see what is wished by looking at it. There is no question 'how do you know? Etc.'

The objections could be summed up by saying no mere sign[138] is the thought.[139] The thought always interprets the sign.

Something written down must be interpreted and *the interpretation of the sign is the thought.*

If we say this, we are in a difficulty.

If we say the interpretation[140] is a certain process, what sort of structure must it have to satisfy you?

Thinking is not reading the sign, it is interpreting the sign.

Let us accept this.

You will grant me I can think a sentence while I am reading it (and interpret it while I am reading it).

There is a process of symbolizing it, and a parallel process of interpreting it.

Is this like whistling a note while the sentence is read? – A state with duration?

If this is so, this does not interest us.

Whatever this state is, it plays no roll for me.

If it happens, let it happen.

There may be[141] a sequence of notes accompanying, each listing a certain time. This does not interest us.

Secondly there may be an articulate process – the mental event is articulate.

Something of this process may correspond to each word.

---

[136]The manuscript reads "three classes", but only two are numbered; so I have changed this figure to "two".

[137]Hyphenation and comma added by Eds.

[138]The term 'thought' is first written, deleted, and replaced by 'sign'.

[139]The term 'sign' is first written, deleted, and replaced by 'thought'.

[140]"Thought" was written first, then deleted with "interpretation" written above it.

[141]"be" is missing from the MS.

There is no doubt that sometimes this sort of thing[142] happens – parallel to the proposition is something which has a correspondence.

Whenever you mention the colour, you imagine it. This sort of process would then be called a process of interpretation.

The process of interpretation is translatable into the sentence.

Instead of a red image, let us take a red patch; the word and the red patch will be another sign: does this again need an interpretation?

It[143] is wrong to think you *need* an interpretation:

You may say the thoughts must be a process which no one[144] can look at from the outside: they can only be looked at from the inside.

If we try to construct anything we call an interpretation, we find it will not do.

It seems all the possibilities of the use of a sign are present when it is used.

It seems as though the calculus was constantly there in the background, while we used the signs.

We might think the whole game of chess was in the background when we make a move or we have played a game.

What is characteristic for the King is not the particular path he has made in the game. If this was all, we might not.[145]

What happens may be possible for three different games, A, B, C. You say something must be there to make it[146] into chess.

Something may be there, the book of rules.

Behind the word, you may think there must be the whole possible application of the word.

Playing chess is not making moves with a whole background of the game of chess.

We talked about the reason one had to say that thinking goes on in our head.

There is a curious superstition that there is some such thing as the pure thought conveyed by the words but different from them.

If I describe a room to you and as I go on with this description, you might possibly draw a picture in your mind.

There is a process which can be called conveying a picture: but this may happen on a pad.

You can talk of conveying something by speech.

---

[142]In MS, "this sort of things happen" is written.

[143]This paragraph commences: "What one sort of There". These words are deleted, and a fresh sentence commences, as above.

[144]The two last words are written as "noone".

[145]Three written joined letters are deleted here. They might be 'use', and so 'use [it]' is a possible rendering. There is no full stop before or after this deleted word; maybe it was left for future completion. Conversely, it is possible to leave the end as it is (in which case the antecedents 'use' in the foregoing's: "while we use the signs. We might think the whole game of chess was in the background when we make a move or we have played a game").

[146]'it' is written over the line.

If you give someone the order 'go downstairs', you have an idea I am conveying something if he is to obey the order.

An[147] interpretation sometimes happens, and sometimes doesn't.

If I say to someone 'Now be very attentive, go downstairs and tell the porter...' Here he may use his imagination – this may then help or not.

I give directions, and he followed these directions with a diagram: [148]:

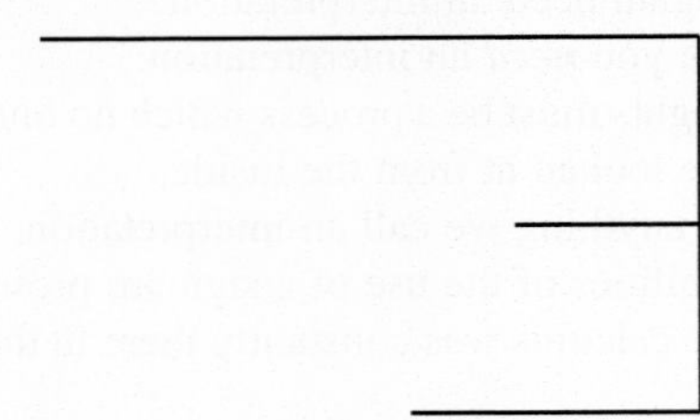

The superstition is[149] *that there is such a theory as the pure thought and* that the pure thought is in our head. I can imagine what – we could see in out head – and see say movements etc.

This is like asking to see the pure King of Chess.

If you say the words are not the thought, this is like saying the King of Chess is not the wood. This is right if you mean the word 'piece of wood'[150] is not used the same as the King of Chess.

We can translate an English sentence into the French and say that the sentence is not the thought but that the two sentences expressed the same thought.[151] It does not mean we can ask where is the thought, or that the thought has a place[152] *different from those of the sentences.*

This is connected with 'where is the visual space'.

---

[147] The word that was originally written to open the sentence – "Such" – is deleted, while the succeeding 'a' in 'an' is capitalized as above.

[148] This figure is similar to but varies from Ambrose's (1979: p. 54) use of the drawing. Ambrose's context is somewhat different from, briefer than, and lacks much of the detail presented here in Skinner's MS.

[149] All the following italicized narrative is in Wittgenstein's handwriting. These additions in Wittgenstein's handwriting do not occur in Alice Ambrose (1979). Her *Preface* does not report that Wittgenstein worked on her notes or even saw them, candidly acknowledging that she could not recollect any details about her writing of the notes.

[150] Scare quotes are added by Eds.

[151] The italicized sentence is in Wittgenstein's hand, and he has deleted Skinner's handwritten version of the part of the sentence that states: "If you translate a French word into English and say the English has one thought". Wittgenstein then re-composes part of the sentence and crafts, as new, one utilizing the remainder that he presumably previously dictated to Skinner.

[152] Skinner's sentence ends here with a full stop, while Wittgenstein adds a new ending to the sentence.

There is no *where* to it. You can say: where is the table *in the visual picture* and I can answer near the chair.

If I ask where is the whole picture, this has no sense; the picture has no place. (*At least not unless you define one for it in some special way), for example*[153] I can describe what I see when I stand in a position.

If somebody stuck pins in my eye and said: now I can see a chair etc. We have now a correlation and we might now say the visual picture was situated there. But this is different from saying Mr. Smith is situated near so-and-so.

The specification of the locality has utterly different meanings.

As different a meaning as saying 'he married the money of so-and-so'. He has not married the money in the same sense that he has married her.

If you ask where does the thought take place: You can[154] say lots of things. You may say it is in the head but in no important sense. It has the meaning that we can find certain definite processes corresponding to thought in the[155] head.

You say the thought imagined is more important than the thought written.

If you say: the importance that you think is that it is in the head.

If you have a feeling that you can cut away the hand/tongue[156] but not the head.

It is possible if we cut away the larynx, we could not think.

You said surely no-one can think without a head.

The thought you mean is no-where at all. How do you know it can't happen without a head?

Suppose a man's brain was cut off and he was only left with hands etc. Why should he not think?

You think a man dreams in his head. Could you dream without a head? I don't know.

Suppose we talk of dreaming. Here we are tempted enormously to say it happens in your head.

When you say this, is it *a priori* or a matter of experience? If a physiologist found something happened in your muscle, when you dream[157] you would find[158] it very difficult to say you dreamed here.

What you need is a sphere for it to happen in.

You are bent on finding a place for thinking.

You have learnt it happened in your head.

---

[153]Italicized expressions are in Wittgenstein's handwriting, and these words are written over the base line that is in Skinner's hand.

[154]The word 'also' is written and then struck out.

[155]The definite article is written over the line above an illegible deleted word.

[156]The '/' sign is inserted because 'hand' and 'tongue' are contiguous, and each fits into the theme (note that 'hands' occurs in the remark below commencing 'Suppose a man's brains').

[157]'when you dream' is written over the line.

[158]The word 'something' is written, and then struck out.

I have been talking as if this were really[159] true but need[160] not be. As a matter of fact this is very far from likely, that anything happens in the brain – it is very likely that nothing at all correlated happens in the brain.

If you found the thought happened in the shoulders, you would find it is very difficult to say it happened there.

Would the real thought be there? What if we could shift it?

Suppose we showed some thoughts happened in the brain, others somewhere else.

Schopenhauer once said, "If you try to convince someone and get to a certain resistance, you then know you are up against the will, not the understanding".

You are up against something else here. We have prejudices of thought.

People a few generations ago could not bear the idea of action at a distance. It revolts them to make a different picture.

This is analogous to the present case.

There is no proof that for everything in the world that seems action at a distance we could have an explanation by particles bumping. It is possible that for everything that happens in thought something may happen in the brain.

It[161] may be to talk of the place of thought in the head is very useful. This may prove a very good shorthand, provided we understand what we have said.

I said it is possible that we find that a certain thought is correlated to a movement in the brain.

Take the thoughts: there are three, four men in the room; how are these thoughts given to us if we say there is a correlation to them in the brain? *By words*. You think the thought can be given in a pure way and then discovered in the brain.[162]

It is extremely interesting that very often when people say that science[163] *has not yet discovered this or that but if X will have discovered it then….., that they very often don't know at all what sort of discovery they*[164] *are waiting for, that they talk of a discovery without knowing of what nature*[165] this discovery would be. *(For example when people say that one day when psychology will be far enough developed*[166] *it will make us understand the nature of beauty.)*

---

[159]'really' written over the line, with an instruction to insert it.

[160]'needed' is written, deleted, and replaced by 'need'.

[161]This paragraph commences with, "One thing that the physics", which is struck out.

[162]The next, italicized, paragraph is in Wittgenstein's handwriting. In the manuscript it is placed on the facing page opposite to the paragraph immediately after the occurrence of "brain". The paragraph's theme links to one of Schopenhauer's. It returns to complement Wittgenstein's reference to him in the opening line that starts this Chapter. (Cf. the first footnote of this Chapter for an assessment of Wittgenstein's view of his use of Schopenhauer.)

[163]This could be a response to Sraffa's type of view, as in McGuinness (2008: Letter 172). (Note that 'wo' written then crossed out.)

[164]An illegible deleted word occurs after 'they'.

[165]Wittgenstein himself writes 'sort' first; deletes it, replacing it with 'nature'.

[166]A word is partly written here ('th'), which may have been intended for 'then', though it is deleted in the MS.

Why had we[167] used such a term[168]? – Perhaps because of this.

Our language has very few kinds of words, substantives, verbs, adjectives and a few others.[169] Imagine genders – ships are females etc.

It points out[170] that originally language was used[171] for females and males. One has imagined some deities male and some female. A very small stock of (kinds of words – norms[172]) have been used to cover everything. This probably arises from the fact[173] that it was used for a very small stock of ideas – ordering people about etc.

If you ask why we use substantives?

You can find games with boards and holes[174] in them[175] all over the world. I can only say human beings are apt to act in such a way. I can make up games which human beings don't play.

Nov. 22nd[176]
November 27th.[177]

Is there some other good reason for talking of processes of thoughts as opposed to processes of speaking?

We have expressions – 'I can't express my thought. I can't find the right words, I have something in my mind'. There are lots of different cases where I say I have something in my mind.

If you ask what does happen when I say I have something in my mind – it maybe you have an image which you can't find words to describe, or you may frown and finally a word comes to you.

If we talk of looking in our memory, we may go through a certain chain of memories. To remember a man's name, I recall his rooms at King's perhaps to remember the name over his oak.[178]

There are lots of different processes of looking in our memory.

---

[167]'are' is written then crossed out before 'had' and 'we' squeezed in before 'used'.

[168]"This" may indicate the above use of "brain" immediately prior to the italicized inserted quotation in Wittgenstein's handwriting.

[169]Punctuation added to this sentence. Immediately after it, 'This is a way' is written and deleted before the next sentence.

[170]'out' inserted to complete sense.

[171]'was used' written over the line, with insertion sign.

[172]This term 'norms' is written underneath 'words' with an arrow between them. Note that the succeeding 'have' retains the MS reading.

[173]'from the fact' is inserted by Eds.

[174]'wholes' is written, then the 'w' is deleted.

[175]'them' is written over a deleted 'it'.

[176]See note on use of November 17th [1933] above.

[177]See note on the date in this manuscript, of November 17th [1933] above.

[178]I.e. King's College Cambridge, and his name printed on the oak frame over his door.

There are analogies to looking in our room for something and[179] some of the processes of looking in our memory. One process is, for instance, you are writing a letter and say to yourself what's the right word. Perhaps what comes is a gesture and we may repeat this gesture till the word comes. If it doesn't come I can't help it.

Here looking for something in our room differs from looking for something in our memory.

I can search the whole floor of the room with my visual field and say it's not there or it is there.[180]

In my memory there is no such thing. We can't say it's there or not there.

If we call using a gesture looking for a word, then the analogy doesn't hold to looking in a room.

If we press a row of buttons and see if any work an electric bell, then we can only see if the mechanism does work. This is much more like looking in our memory.

We look into a pigeonhole and see what is there. This is what is comparable to looking in our memory, at first sight.

The comparison does not quite hold. I can look into the pigeonhole and see what is there, but I don't look into the memory and see what is there.

We have shown something we call looking for the expression of[181] a thought. Then we must say the gesture which is what is given is the thought.

The case may arise where my thought is expressed in German and I am looking for[182] *an expression* in English.

Here we may look up words in a dictionary and translate.

Here it is a question of translating.

But I couldn't translate a gesture.

People talk of thought reading. One man is the thinker and the other the reader.

I say, "think of somebody of whom I am thinking". When I am thinking of somebody it may be all sorts of things.

Suppose I repeat his name.

The other man has got to say you are thinking of so-and-so.

If the name springs into his[183] mind, then that is[184] what you called reading the thought.

Don't think there is a thought into which the reader can peer.

You can say the thought was transmitted if you know what this means.

This discussion would be most useful when we talked about the philosophy of mathematics.

---

[179]After 'and', 'the' is written and deleted.

[180]'is' is omitted.

[181]'the expression of' is written in ink over the line, with an instruction in pencil for insertion (rather like the pencil use when Wittgenstein inserts some sentences in his own hand earlier in the manuscript).

[182]'looking' is written again after 'for' and deleted.

[183]'the' is first written and deleted in favour of 'his' – the latter written in pencil.

[184]'that is' is written over the line.

Here the use of a phrase seems to justify certain philosophic theories.

I.E. are there existential theorems which construct the entities whose existence they prove?

We must say that there are two things we call existential theorems:

We can never predict in philosophy.

We can't say 'there can't be existential propositions of a certain sort in mathematics'.

If we say 'thought can't be a material process somewhere' you may say someone has in America seen things happening in the brain.

(We aren't making general statements for the future.)

(I can't prescribe the way you should use the word thought, but I can point out you are using the word in different ways.[185]

If we examine the things we call existential theorems and find something in common between them, we can't say another thing must have this.)[186]

If you say "Mr. Smith in Chicago has opened a person's[187] brain and found their thoughts there", I can say if you have read this that you don't yet know what this means.

This is quite analogous to saying you have found in the *New York Herald*[188] that Fermat's problem had been solved.

If you say you[189] read in the *New York Herald* that so-and-so has found a pen in somebody's head, you can draw a picture of this (N.B. this is part of your understanding).

If you say they have found[190] the thought in the head, you can't draw a picture of this.

You can't *go on*, as you could if you knew they had found the thought written on a piece of paper.

(You can't dispute what we say by giving us new examples.)

Take the case of something being at the back of one's mind.

This is an excellent expression.

If you say this is a thought, you can call it so.

---

[185]'ways' (written in pencil) replaces the original choice: 'meaning', which has been written then deleted. The parentheses are added in pencil.

[186]In the manuscript the parentheses are added in pencil.

[187]'a person's' is written above the line over 'people's', which is crossed out.

[188]In the manuscript the text is "N.Y.H.", expanded by the editors to its full title, ditto for use below. There is some irony here since the *New York Herald* (which the manuscript abbreviates as 'N.Y.H') was published between May 6, 1834 and 1924; but it folded some ten years before the above statement was written. (This was due to the owner, Gordon Bennett, moving to Paris and trying to run it by telegram, though earlier it had funded or reported some great discoveries.) This and subsequent uses of "Fermat" have some general parallels with its occurrences in *The Big Typescript* (Wittgenstein 2005: 420), though the arguments and contexts there differ considerably.

[189]'say you' is written above the line.

[190]'found' is written over the line.

Suppose[191] we say we know what must have been proved if we read that Fermat's problem has been solved.

I say the unclarity here is in the word proof.

You may say I don't know the proof but I know what it must end up with.

You know that Fermat's theorem has been proved. What do you know about the grammar of the word?

The word proof is[192] a family.

The family of proofs is made by a kinship between lots of different things which is very complicated.

It is clear the proof will not be one of these.

You may say it[193] belongs to one set. It may or may not.

It may belong to a series of proofs.

It is clear it isn't this.

You may know Mr. Smith's researches and say it must be something like what you know of his researches.

You might say every new proof widens the idea of proof.

You may know he will get a thousand dollars and you can write and ask him for money.

Supposing you read in the *New York Herald* that Mr. Smith your friend has proved Fermat's theorem, and you decided to write to him for money, then you know that he is going to get money and that is as far as things went. You understood this much.

If you go into a room and see somebody you know[194] talking to someone else and he looks at you and says you _____ where you don't know the second word and you retire, can you say you have understood him.

You can say you understood this much.

What I have been urging is that words like thinking, wishing, understanding etc.,[195] which[196] seem to denote a single activity, have a family of uses. One of the main uses, and the one which made these processes seem to be of importance to logic, is thinking being the process of handling signs of a kind of language.

This is the characteristic which really made thinking interesting to logicians.

The main scruples are overcome if you get clear that when we say two sentences express the same thought, I can't look about me for something called the thought.

The temptation I tried to make you avoid is to think of thought as a kind of gaseous being corresponding to the word.

---

[191] 'If' was first written, then deleted in favour of 'suppose', which is written in pencil.

[192] 'be' is first written, then crossed out in favour of 'is'.

[193] 'is' occurs and is deleted here.

[194] 'and' is deleted here.

[195] 'etc.' added above the line.

[196] 'though they' was first written then deleted, with 'which' written over the line.

People talk of proposition instead of the sentence because proposition seemed to have this kind[197] of being.

The way to dispel this is to think of the King of Chess and oppose it to the piece of wood.[198]

I want to rid you of the idea of an intermediate being between the coarse material sentence and the use of it.

Again this does not (the next step[199] I now try makes you take) mean that the word thought as opposed to the word sentence means nothing. But there is no shadow to it.

Supposing we talk of the King-piece and the King.

If we ask where *is*[200] the King as opposed to the King-piece.

If you call the King the class of King-pieces, *how do you know* –.[201]

Do you then say the King is the same as the King-piece?

I say the two words have their different uses.

People who think of these things always think of a word as a proper name.

I can cure you *by*[202] say*ing*, "what is a comma".

If I talk of the meaning of a comma, everyone will say I mean[203] the function of a comma.

I compare words to instruments in a tool case where you have the hammer, the nail and the *glue pot*.

I should like to talk of one particular mental activity (volition).[204]

If I say I lift up *my hand*[205] by willing it.

Willing[206] seems to be seeing it from the inside.

The only phenomenon is seeing it go up.

What is the difference between a movement which is voluntary and an involuntary movement of my body. Some people say the difference is that there is a feeling.

---

[197] 'king' was first written in ink (perhaps with an illusion to 'King of Chess', as it occurs in the next remark), which is corrected in pencil to 'kind'.

[198] 'word' is actually written here.

[199] 'the next step' is written in ink above the line over the ink-written word 'as', which is deleted in pencil.

[200] Notice that the expressions that are italicized over the next page are in Wittgenstein's handwriting, commencing with the penciled 'is' here. Some word endings are modified by him. (The exception is 'gluepot', which is italicized to represent the underlining of this word.)

[201] ', how do you know' – including the comma, is written in on the line in pencil.

[202] 'by' is written in pencil over 'and', which is deleted; likewise, ink-written 'say' is changed into the present tense by pencil.

[203] The manuscript read 'he means', and was changed to 'I mean', with 'I' above the line.

[204] '(volition) as' occurs parenthetically with 'of'; an arrow indicates that it is to be moved to the end of the sentence, whilst 'as' is deleted.

[205] 'my hand' is written in pencil over the line with a mark for insertion.

[206] 'it' is written after 'willing' and deleted by a pencil mark.

Others would say that the feeling is something by itself and can have nothing to do with it. You would be[207] dependent whether this arises or not.

You are *doing* the thing, not dependent on whether it comes or not.

Either I lift my arm or I see it go up.

I don't get the experience of the feelings in my muscles; I just do it.

What is it really you are willing? Is it the visual experience or the contraction of my muscles? (Suppose I make a fist.)

Some people say they are *willing* in their head and some they are willing in their muscles.

We use willing and wishing in a different way.

Nobody says I wished that my arm should go up and it did. This points to the state of affairs[208] that we use the words in two very different ways.

If someone said willing to lift my arms is not an experience I make but a thing which I do. If I answer whatever it is, it is an experience.

Don't imagine yourself looking at it, but close your eyes.

That there are experiences is clear.

If you look up at the arm, you can imagine looking at the arm and being astonished.

If you close your eyes, can you imagine both things?

Can you feel the muscular sensation and imagine being astonished and also think of a volition?

We actually use the word willing for a very special phenomenon only, and the word wishing for a much wider set.

We are tempted to use willing for this wider set. If I say 'try to will that this picture should fall down', you would[209] make certain movements.

This shows you what it means.

There is a connection of this with mathematics – the different meanings of trying to do a thing.

We know what trying to lift a weight (there is also willing here[210]). We know a game where we try to move a finger and can't try – here you would say you can't will it to move it. Can you try to will it?

This shows better than anything what willing consists in.

---

[207] 'would be' is written over a deleted 'are'.

[208] The expression 'state of affairs' is inserted by the editors – adopting the precedent of other uses of it in the Archive, e.g. Chap. 4, remark 7 – twice used; rem. 28 – used once; Chap. 6 etc.

[209] 'look' occurs after 'would' and has been deleted.

[210] 'here' with the missing close-parenthesis ')' is added in pencil, with an instruction for insertion also in pencil.

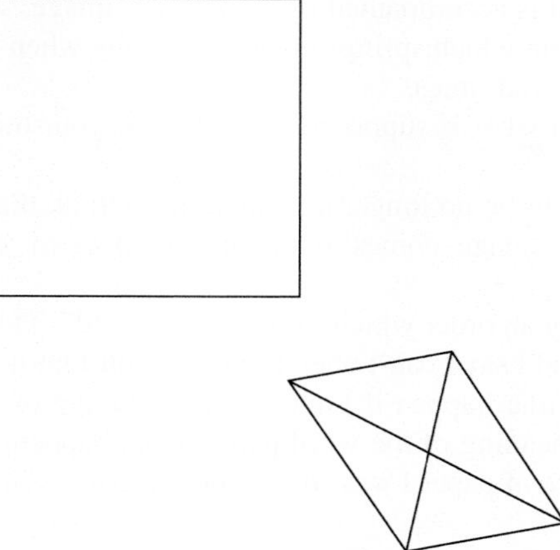

Draw a square looking into a mirror.

The[211] use of the word "will" is different from the word wish and is bound up with certain phenomena of our body.

If you will to remember a thing, you would talk of an activity here. You would be prepared to talk of an activity in thinking.

Thinking is an activity means it is something that you do, not something that happens.

Dec. 4th.[212]

Part II[213]

I have been talking the whole time about understanding, meaning.

The main point we have made is that in the use of language no one process is needed for[214] understanding.

Understanding at first sight seems an accompaniment necessary if the symbols are not to be dead.

This does not mean that there isn't sometimes such an accompaniment.

---

[211]This sentence is placed opposite – at the right side of – the diagram.

[212]This date is consistent with Monday, Dec. 4th 1933 [Full Term ended that week on Friday, 8th Dec.]

[213]Above this date, it states in Skinner's pencil handwriting "[PART] II includes separate papers". The term 'includes' covers the contents of the next chapter below.

[214]The MS has 'language' after 'for', though it is crossed out.

Supposing a word is accompanied by something, image, sensation feeling etc. If the image is something which springs up automatically when you hear the word red, let's take red and the red image.

I supply you with what is supposed to come into your mind, and make that the complete symbol.

This makes there to be no longer a temptation to think that this must happen.

The fact that an image comes into your mind is of *great* psychological[215] importance.

Take "I can't obey an order which I don't understand". This is alright. There may be a French word, and I say I can't obey because I don't know what you mean. Does anything in[216] particular happen if I tell you the meaning of the French word?

If I tell you the meaning of the word paper, what happens? I may now associate the image of the piece of paper I was shown, but probably if asked for paper I just go and get it.

The explanation tells what the English word paper is used for.

Suppose someone said, "bring me a hair" (meaning chair), I might say I don't understand you although I do know the word hair. Here there is almost a process of understanding.

In the *use* of language (as we or[217] a parrot uses it) there is no need for anything extra to come in.

If there is anything, we add it to the sign.[218]

The traffic lights could be without the words "go" and "stop".

I shall go on to speak about[219] what a proposition is.

You know all sorts of definitions have been given. (In fact I have given one.)

If someone asks you what a proposition is, you give examples "the rose is red" etc.

You ask what is in common.

I say propositions are a family. (You would be surprised what we do call propositions.)

In this family there is a large family or group which do have something in common.

This is the propositional form of a particular language.

If someone is asked for a proposition, he will be immediately or[220] is able to make up some sort of English sentence.

This is enormously important.

If I say 'piano' 'lamp' these are not sentences, but if I say 'piano lamps' this seems more like a sentence though it has no more sense than abracadabra.

[215]The MS has 'image' after 'psychological', though there is a line through it.

[216]'In' is added by Eds.

[217]'we or' is written over the line, with an insertion instruction.

[218]'sign' was first written 'signal', then the two last letters were deleted.

[219]'about' is supplied by Eds.

[220]'or' is supplied by Eds.

There is something that we might call the jingle of the English sentence (or German one). We also know analogies in French sentences, etc.

A vast chunk of propositions have the sentence jingle. This makes you feel there is something in common.

If asked, you will say this is not necessary. We could make up a language where this is not so.

We could[221] read English sentences from right to left. This has exactly the same multiplicity as the English language but sounds nothing like it.

Example

Suppose I put my hand to my face and said, "tooth-ache". Is this a sentence? You will say I[222] really mean,[223] "I have tooth-ache". Do I repeat 'I have' to myself? Suppose on another occasion I said 'I have tooth-ache'. Is this sentence three words or one as the word "tooth-ache" simply would have done.

There are cases where there is an indication. Suppose she says, "I have", I say does she mean pains money or something else. If she now makes a face, I may say she means I have a pain. Is this always a doubt and settling a doubt?

There are cases where the words "I have" give me a clue which otherwise I would have been without. But this does not happen in ordinary English.

I must return to the point where I described a language game.

I build a certain structure and point to certain things saying the word and afterwards say the word. We took words like brick column etc. and afterwards bring, take away.

You are clear I could say 'brick' and the man brings me a brick.

But you say he must be taught 'bring me'.

If I give the order[224] 'bring me a chair', you say he wouldn't understand it without 'bring me': this is a hypothesis.

If you say what did I mean when I said chair, you may say I meant 'bring me a chair'. This may not have been in my mind when I said it. If I said "chair", you may say to yourself he means, "bring me a chair".

There are cases where there is such a process, but there needn't be any such process. If there is such a process, it is a translation.

The way to understand this is to split up the possible processes into games.

There is the game when I say chair, and he says to himself "bring a chair".

There is the game where I say chair, and he brings a chair.

There is the game where he looks at me for approval, or disapproval.

The multiplicity of a sentence is the multiplicity of possibilities. (I send a chair fetch a chair, piano, etc.).

This comes in when there is doubt.

---

[221] 'could' is squeezed into the line, presumably after the sentence was written.

[222] 'say I' is written over the line.

[223] 'say' is written first, deleted and replaced by 'mean'.

[224] 'order' is first written 'word', and then first letter is deleted and 'er' added.

If I say "chair",[225] you may say 'who is to bring the chair'. He must know, you say, *he* is to bring a chair.

If I say, "chair", you can say "how many, one or two".

You are inclined to think if I leave out the words they must be self-understood.

If I say "chair", then all the doubts which arise by this you think must be settled. But they may not arise. There was nothing to be settled.

<u>*Example*</u>

You look at a railway timetable. This settles the doubt when the train is going to leave. But have you been taught to settle the doubt how to use the table. You may have been given a scheme

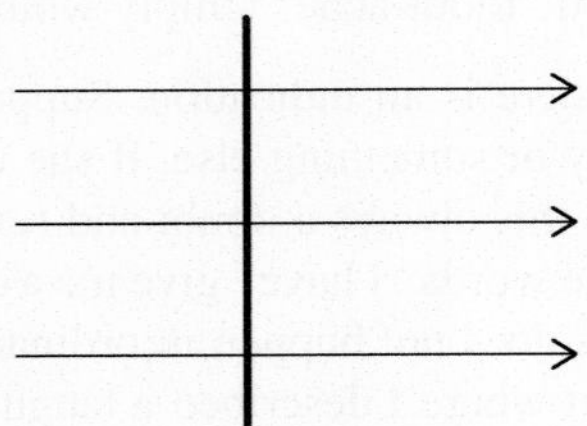

(But have you been taught this.) It is not that there is a particular number of doubts which must be explained. There are a thousand doubts if you want them.

If I say there is a case where there are no doubts, you still think that is because he knows it.

You may say (1) the man who has no doubt is a fool or a machine because he can't doubt, or (2) that because there is no doubt is that he knows it all.

(1) A doubt is something that happens, and then there is an answer.
(2) What does it mean that he knows it? If you say he knows it, I say, "when does he know it?" Does he know it all the time or does he repeat it to himself?

How long does it take to know $2 \times 2 = 4$?

If knowing is hypothetical state, you mean something like "being able to lift a weight".

You may mean his brains are capable of giving an answer.

This does not interest.

If I say I knew then $2 \times 2 = 4$, I[226] say I wrote it down – anything more? – I said it to myself. Then that is all.

Knowing the alphabet is only a mental state in a hypothetical sense.

[225] 'bring me a' is originally written before 'chair' here, and then crossed out.

[226] 'you' is written and deleted before 'I', both on the line.

If I say the multiplicity only comes in with the doubts, this does not mean there are doubts to be settled, or that the man knows the multiplicity.[227]

In both cases he has heard the word chair (you may say) and[228] he has understood them in different ways.

What does this mean?

When you say "chair" he may supplement this with "bring me a chair".

It is not necessary while you said the word chair two different processes went on: perhaps[229] the understanding may come out later.

You were taught words by things being pointed and the words said. Did this solve all doubts? If I point to paper, there[230] may be lots of doubts.

A doubt is something that takes place, not something latent.

It revolts you that you look at the timetable and use it.

You feel that there is an infinite chain of doubts all of which have to be settled before you use the table.

This is not so.

If a child does something, you say this is automatic. You oppose automatic and with understanding.

The child does one thing, where you might do two things (when you point to a piece of paper and say paper).

With understanding, this is only a finite stretch of doubts.[231]

There is the difficulty of doing something with understanding and doing it (automatically, say).

Suppose you say, "how do you do it", I suppose you want a description (if you say you don't want a causal explanation).

If you say, "how do you play cards," you do plays and do certain things.

All we do is to describe a thing; I can't give explanations (which would be something scientific).

By[232] understanding you refer roughly to a process of calculation which doesn't always take place.

I will now again take the case of a child, which I taught the name of something and said, "bring me this thing" (say paper).[233]

Does the child understand the word paper? In which cases would you say the child understood the word and in which not. This will help us a lot.

I did some things,[234] and then said 'paper' and the child brings me paper, would you say this was automatic or not.

---

[227] 'thing' is written and deleted in favour of 'multiplicity', both on the line.

[228] The word 'said' is crossed out here.

[229] 'perhaps' is written over a deleted word, which may be 'or'.

[230] The word 'is' is crossed out here.

[231] The Editors have supplied 'this'.

[232] 'But' is first written and altered to 'By'.

[233] A reminiscence of Wittgenstein's experience as a school teacher?

[234] Written 'somethings' in MS.

You might say this does not come in till much later in language.

Automatic and with understanding apply to particular cases. We would make in such cases this distinction.

Is this distinction always there? It is one special distinction which comes in in particular cases.

This happens in philosophy very often. Words which are used in particular cases are thought to be general metaphysical terms. They then have got to be restricted to the cases again.

You might say 'have you heard me' to the child, and if he said no, you would say automatic. This is a primitive game with the word automatic.

There is another case where non-automatic means translating into a language.

Take the case of multiplication 16×18. If you see me doing it, you ask do you do automatically or not. You say usually $8 \times 6 = 48$, but if you say it 20 times then it was not automatic.

Supposing[235] one says, "If the red light goes on, you must cross the road". Then the red light goes on, and you stop. You may remember my words or not.

Did you stop automatically or did you understand?

He said, "Yes, I heard the sentence, and understand English".

Suppose I said, "don't you do so-and-so, and make a fierce face".

If we say "Tell me the story about Julius Caesar" to a child it may just repeat the story word by word, or[236] it may tell the story in its own word. Here we have automatic and with understanding.

If a child has to leave out a word because it has forgotten it, then we say automatic.

What you say is that this shows the understanding *or not*.

If you mean this, you mean a hypothesis,[237] which can be used for prophesies. Say that the child will be able to tell[238] other stories and is[239] intelligent.

We treat the child like the ticking of a watch.

We are now in the region of making hypothesis about what has happened or is going to happen.

This is different from definite conscious states taking place at the time.

If I used the word "chair" for two cases, you say the man can't understand.

What do you mean by 'can't'?

Do you mean by 'can' something logical, or something from experience?

If it is logical impossibility, you are stating something[240] grammatical.

---

[235] Before 'Supposing' and on the previous line, 'You are told your 1' is written and deleted.

[236] 'and' is first written and crossed in favour of 'or' on the main line.

[237] 'hypothetical thing' is written first and modified to read 'hypothesis'.

[238] 'about' is written after 'tell' and deleted.

[239] 'story' is written and deleted after 'is'.

[240] 'something' is written over the line, above a deleted dash, and there is a deleted dash after 'grammatical'.

If you meant he can't act, do[241] you mean that he would be too stupid?

If you say the word chair, he can supplement this with two things. If you say he can't know, you mean the word chair is alone and is not supplemented. By understanding, you mean something repeated to himself, you mean[242] he has not been given anything supplementary to say[243] to himself. Here you are giving the understanding of[244] the multiplicity of the sentence (which I agree with).[245]

Meet on 17th January at 5 o'clock

[241] 'have' is deleted and 'do' is written over it.

[242] The words 'he can't say either one or other because', written on the line, is crossed out here.

[243] First written 'saying', then altered to 'say'.

[244] 'of' is added here by Eds.

[245] At the end of this manuscript, the final sentence reads: "Meet on 17th January at 5 o'clock."

Chapter 4
"Communication of Personal Experience"

1.[1] B looks at traffic lights and reports to A what colour shows by calling out to him alternately the words "red" "green" "yellow". B was trained by being made to observe the lights while the instructor said "red" when he saw red, "green" when he saw green, etc. (This last sentence opposes the instructor's saying "red" when he saw red to, say, his saying "rouge" or "moo" when he saw red. The sentence therefore did not say the instructor told the pupil what he saw as opposed to telling him a lie – something else.)[2] The condition for B's learning the game was, of course, his reacting in the 'desired way'. I.e. had B not reacted by saying the words, or

[1]Remarks 1–5 of this manuscript could contribute to a solution to the absence of something akin to it from another manuscript which David Stern researched and published in (Wittgenstein 1993a: 213, footnote 40) – parts of the Wittgenstein C-series notebooks, MSS 148, 149, 150. Stern (Klagge and Nordmann 1993a) refines Rhees's earlier publication of parts of these MSS, observing that: "This series of remarks [from 6 to 16] continues a sequence of remarks were numbered (1) to (5)... and I [David Stern] have been unable to locate them elsewhere in the Wittgenstein's papers". Remarks 1–6, which open Chap. 6 here, may be this absent portion, or somehow intertwine with it. (The number sequence in Wittgenstein 1993a is different by two numbers, so that remark 9 = 7 of the manuscript; cp. numeration here and in Chap. 6.) An upshot of this situation is that there are eight, not only six, remarks prior to remark 6 in Wittgenstein 1993a, if we judge that the present chapter has a serial relation to these missing remarks. See the footnotes concerning remark §8, and through to remark 17 below. Reference in subsequent footnotes will be to the MSS of the C-series notebooks, and their appearance in the Bergen 2nd ed. of the *Nachlass*; not to their earlier publication. This is not due to any error in this earlier commendable publication. Stern's research and clarity have assisted us to refine the issues here. The information used in the current footnotes selects only a few of the examples available in Stern's publication, to illustrate the general pattern, so as to point to the value of further analysis on Stern's and the present chapter's parallels. The combination of Stern's research resulting in Stern (in Klagge & Nordmann. Eds. 1993a) with the present chapter is a fertile basis for further research into the subject matter.

[2]The words 'something else' are written above and partly after 'a lie'. '–' is not in MS.

A. Gibson, N. O'Mahony, *Ludwig Wittgenstein: Dictating Philosophy*,
https://doi.org/10.1007/978-3-030-36087-0_4

reacted by saying them 'at random', the game would not have come off. I am assuming that B after some training in every case said the word the teacher wished him to say (the word which the teacher himself would have said). Isn't this a case of the communication of personal experience?

2. As N$^{o.}$ 1. After B has passed the training, we let him look at the traffic lights through various kinds of media. Whenever he looks through a particular medium[3] he says "red", when without the medium he says "green", and vice versa. Other media produce other regular changes of this sort in his reactions[4] but not in the normal person. The question arises: "If he looks thro' the medium so-and-so, does he see what we see or something else". The question is decided by letting him look thro' the medium and letting him tell us what he sees.

3. B has been taught to describe the colour of the traffic light by pointing to some coloured object near him, to a red one when the traffic light is red, etc.

B looks at the traffic light and says "red". A sees that the light is what he would call green (the traffic light is green),[5] and asks B to point to a sample to show him what the colour is. B points to a red object. On being shown various coloured objects in the room B however reacts normally, I.E. he calls red what A calls red, etc. A now says: "Queer, B sees a green traffic light red and a red traffic light green". We shall[6] in such cases ask: "Why does he in this case see a different colour than we do?" Perhaps we will find out that certain atmospheric conditions always coincide with this *phenomenon.*[7] But possibly the phenomenon will occur in certain odd cases when we shan't find an explanation.

4. Imagine on the other hand that B had called a green traffic light red as in 3, but on being asked to point to the colour in a sample he would have pointed to green. We would then have told him that we didn't call this colour 'red' but 'green',[8] and he

[3]The expression 'a particular medium' is written on the line above the phrase 'the medium α'; the latter has a small cross at its side just below the line. I take it to indicate deletion of 'the medium α'.

[4]'in his reactions' is written over the now deleted phrase 'in him'.

[5]The words 'the traffic light is green' are written above 'the light is what he would call green'; curved parentheses supplied.

[6]Here Wittgenstein's hand in pencil corrects the Skinner handwriting of 'will' to 'shall'. This is the first such type of correction in Wittgenstein's hand, which evidently is the commencement of Wittgenstein going through this manuscript to add refinements throughout.

[7]'phenomenon' has a wavy underline under it, which appears to use the same pencil as Wittgenstein's correction to 'shall' above and 'names' below in §5. Manuscripts in Wittgenstein's hand use the same convention in the same period (see MS 120 (XVI) for the period 19 Nov. 1937–26 April 1938a), ff. 27r, 63, 74v, 75r.

[8]An alternative is offered in the main text: '‖ We then tell him: "We don't call this colour 'red', we call it 'green'." ‖'

might have corrected himself. In this case we would say: "He said 'red', and[9] meant 'green'." Or: "He saw the same colour we saw, only called it differently".

5.[10] But we could imagine that, being told "we don't call this colour 'red' but 'green' ", he would refuse to change his usage, say, in the cases in which the colours were luminous. Then too we could say: "B sees what we see but he refuses in such-and-such cases to use the names[11] we use.[12]"

6. Suppose B and A looked at the embers in a fireplace and B talked of the green glow of the embers. Then if there were no other samples of "green" and "red" about[13] A might say: "I can't find out whether B sees what I see and just calls it differently, or whether he really sees the embers green". If a particular test had been made A might say to a third person: "I'll show you how B sees a fire,[14]" and to do so A might paint a picture of the fireplace with green embers.

7. Now a case in which we might say: "He said green where we saw green and red where we saw red but he all the time saw red where we saw green and vice versa". How could we have discovered such a state of affairs?[15] We train him in describing colours in a room lit up by some particular artificial light. Later on after leaving this room however he calls green what we should call red, and vice versa. We correct him and he changes about his appellations[16] and tells us that a leaf which we showed him in that room had the colour which a rose has now. We should be inclined to say in this case: "that artificial light had a queer effect on him, he saw red objects green and green objects red; but we only discovered it when we went into the open air with him".

---

[9]The '&' is inserted here in Wittgenstein's pencil hand.

[10]The number 5 is at the top of the page in the margin of the manuscript, as with the rest of this numeration. On this line the sentence ends from the previous page (i.e., '"only call it differently"'). I assume that the '5' indexes the start of the new remark at the beginning of new sentence after this ending.

[11]The word 'names' is written in Wittgenstein's hand above the word 'appellations' that is in Skinner's handwriting. The latter is not deleted.

[12]The word 'use' is written above the undeleted word 'do' on the main line.

[13]The manuscript also has an alternative in the main text: '‖ no other coloured objects around them ‖'.

[14]Before 'fire' a word is deleted and almost illegible – it may also be 'fire'.

[15]An alternative reading is present in the main: "‖ How could we imagine to have discovered such a state of affairs? ‖"

[16]Note that in §5 above Wittgenstein swapped 'appellations' for 'names'.

8. [17]—[18] "We shall never know whether he meant this or that". Suppose someone[19] who had been trained together with B in that artificial light died before ever he used the language in daylight; we might say: Perhaps he too would have reacted like B when taken into the daylight, but we shall never know.

Perhaps (α) we should say that this question was decided if he arose from his grave and we then made the experiment with him; or his ghost appeared to us in a spiritualist séance and told us that he had a particular experience. — Perhaps (β) we don't accept any evidence. But what if we hadn't accepted the evidence in N°. 7 and had said: "We can't be sure that he is the identical man who was trained in the room",[20] or, on the other hand, "He is the identical man but we can't know whether he *would have* behaved like this at the time when he was trained."[21]

9. We introduce a new notation for expressions of the kind[22]: "If P happens, then under the circumstances[23] C (as a rule) Q happens. P didn't happen this time (the circumstances otherwise being C) and Q didn't happen". Instead of this we say: "If P had happened, Q would have happened" or, "But for P Q would have happened". E. g., instead of "If the gunpowder is dry, under these circumstances a spark of this strength explodes it. It wasn't dry this time and (under the same circumstances) didn't explode", we say: "If the gunpowder had been dry this time, it would have

[17]Note footnote number 1 of this chapter, which contextualizes the following considerations: there are some parallels and variations between this present chapter's remarks 8–17 with remarks 6–16 in Wittgenstein 1993a. The contexts surrounding its remarks 6–16 in Wittgenstein 1993 are different from those of the present chapter's contexts before its remark 8, and after the last parallel with Wittgenstein 1993a. (For example, the context before the latter's parallel remark 6, is vastly different and its immediately prior sentence is '("I have a peculiar feeling of pastness in my wrist.")', which has no parallel in the current chapter nor in the rest of the Skinner Archive. Readers will readily see similarities and differences between the two manuscripts by comparing the present publication with Wittgenstein 1993a, so only a few of the differences will be mentioned in the present footnotes. Note also that in the Trinity manuscripts, vertical lines – the full length of the page – are used by Wittgenstein to indicate the deletion of the text, whereas we have a fair copy with added nuances (for example "Perhaps" before "(α)" and "artificial light" in remark 8 of the Skinner copy, but not in Wittgenstein 1993a). The present chapter's manuscript also has revisions in Wittgenstein's hand.

[18]This use of the long dash, which appears to indicate the opening of a remark, has its complement in the next paragraph after a line break part-way through the paragraph. A similar custom occurs in Wittgenstein's handwriting in *Philosophische Bemerkungen* XII (MS 116), pp.313–14; cp. p.190.

[19]'someone' is not depicted as such but 'C' Wittgenstein (1993a), and here in Chap. 6 "(α)" is modulated by "Perhaps" but no qualifier attends in Wittgenstein (1993a: 213).

[20]Closing quotation mark appears to be missing here and is supplied by Eds.

[21]Closing quotation mark is omitted here and added by Eds.

[22]In Skinner this is Remark 9, whereas in Klagge and Nordmann, Stern 1993a, it is Remark 7. The words 'this sort' on the main line are crossed out, with 'the kind' written above the line, both in Wittgenstein's hand and in pencil. The partially parallel text for this remark (in Wittgenstein 1993a: 213) differs in three ways. In the Wittgenstein-Skinner version, it has: (1) 'expressions' in the plural; (2) 'of this sort' in Skinner's hand; (3) changed in Wittgenstein's handwriting to 'the kind'.

[23]Here 'under the circumstances C' contrasts with 'always' in, and 'C' is also absent from Wittgenstein 1993a: 213 text.

exploded". The point of this notation is certainly[24] that it diminishes[25] the difference between the expression saying that the catastrophe didn't happen and the expression saying that it did.[26] We might say[27] the new notation paints[28] a very vivid picture of the catastrophe and doesn't stress the point that this time it didn't happen. This could be shown still clearer if in our language we actually used painted pictures. The second notation will e.g. be particularly appropriate if we wish to give a person a shock by making him vividly imagine that which would have happened, stressing only slightly that it didn't happen.

10. Someone might say to us: "But are you sure that the second sentence means just what the first one means and not only something similar, or something in addition to what the first one means?" (Moore)

I should say: "I'm talking of the case where it means just this, – and this seems to me an important case (and this you concede by saying what you have said). But of course I don't say that these notations[29] aren't used in other ways as well, and if so, I would have to talk about these other cases separately. – (Lowering one's voice in a speech sometimes serves to release for a time the hearer's attention; in other cases it serves to draw special attention to what is being said. Or again lowering the voice may serve to rouse the expectation of the hearer[30] for what is going to be said.)[31]

It must be clear that our simple examples are not preliminary exercises to the (final[32]) analysis of the actual meaning of the expression so-and-so. (Nicod[33]) But these examples affect this very 'analysis'. (The point of view criticized in this remark is this, that the analysis of simplified cases gradually leads up to the analysis of the real case in the way in which the study of the laws of a falling body in a vacuum lead up to the study of the motion of a falling body in air.)

---

[24]There is a wavy line under the word, in Skinner's ink.

[25]The word 'diminishes' is written in the same hand over 'minimises'.

[26]This sentence is quite differently phrased in the present chapter and in Wittgenstein 1993a: 213.

[27]'say' is written above the line with insertion instruction.

[28]A word is deleted on the main line before 'paints'; now barely legible, it may read 'points'.

[29]In Wittgenstein 1993: 213 'it' occupies the position where 'these notations' appears here.

[30]Wittgenstein's pencil hand here adds 'of the hearer' above the line, attended by an insertion mark; this does not appear in Wittgenstein 1993a: 214.

[31]These two parenthetical sentences do not occur in Wittgenstein 1993a: 213, whilst some similar types of points are made there. These points are positioned as the beginning of remark 9 in Wittgenstein 1993a (where this would have paralleled 11 in the present chapter). Here the earlier parallel in numeration between the two manuscripts starts to differ.

[32]This term has a wavy ink line under it.

[33]This is J. Nicod; Wittgenstein refers to him a number of times during this period, not least in his notebooks, e.g., notebooks MSS 152 (1936: 45), and 157b (1937: f.36r) where he wrote in German; cf. also The Big Typescript MS 213, p.156 §202. In Wittgenstein 1993a: 214, the sentence with Nicod appears, though in some parts differently expressed; yet the parenthetical sentence that forms the end of 10 above does not appear at all, and there is a sentence which has no parallel in the present Edition.

11. We say "We can't know whether this spark would have been sufficient to ignite that mixture", and give as a reason, that[34] we can't reproduce the exact mixture not having precisely the same ingredients, or not having a balance to weigh them exactly, etc. etc. But suppose the circumstances could be reproduced, and someone said: "We can't know whether this spark would have ignited the mixture", and gave as a reason that we can't know whether under these circumstances it would *then* have exploded. This answer would set our head whirling. We would feel he wasn't playing the same game with that expression as we do. We would say: "This makes no sense." And this means that we are at a loss to know what reasoning, what actions, if any, go with this statement of his. We are inclined to believe that he made up a sentence analogous to sentences used by us in certain language games, not noticing that he had deprived his sentence of its *point*.

In which case do we say that a sentence has a point? That comes to asking, "in which case do we call something a language game". I can only answer: Look at the family of language games. This will show you what sorts of relations; similarities exist between its members.

Have we demonstrated that to say (N$^{o.}$ 8), "We shall never know _ _ _ _ _ _", when we accepted no evidence which would decide the question, has no sense? We should say: The sentence under these circumstances loses its point, which was to be used in a game in which something was accepted as evidence. But to say this does not mean that there[35] can't be some other point in saying it.[36]

12.[37] B is trained to describe his after-image after having, looked, say, at a bright red spot. He is trained by being made to look at that spot and then to shut his eyes, then we ask him: "What do you see?" This question before was put to him only if he was looking at physical objects and had to describe them. He then reacts by saying, e.g.: "I now see a green circle." (If he doesn't react by saying what he sees with closed eyes, the game just doesn't come off.) There is, or seems to be,[38] a grave objection to this description of the training. For what if I had had to describe my own, not B's, training? Would I then have said "I closed my eyes and reacted to the question 'what do you see?' by saying 'I see a green circle' ", –[39] not rather: "I closed my eyes, saw a green circle and said so"?

---

[34]The words 'and give as a reason, that' are written in same hand above to displace 'because, say'. (The line breaks after the deleted 'say' and resumes on next line with 'we'. A space is left between the margin and 'we'. This might comply with the possibility that, already in this previous draft, the writer has left a space to accommodate some fresh variation that is on the way in the act of dictation.)

[35]'cannot' is written before 'can't', while the former is deleted in favour of the latter.

[36]The end of 11 resembles yet deviates from Wittgenstein 1993a, with some transposition re. 9.

[37]The numeral for remark 11 is missing in Wittgenstein 1993a, so at this point the numbers are parallel at remarks 12. The contents of 12 have similarities and differences, with sentences or phrases that are in this current chapter and absent from the earlier publication.

[38]'or seems to be' has a wavy line under it in Skinner's ink.

[39]This dash has been inserted in pencil, which, given that the other emendations are in Wittgenstein's hand (see remarks 9, 10, etc.), maybe the dash is his.

The[40] idea is that I saw a green circle and said so, as opposed to just saying so without seeing a green circle. This leads us to a fundamental objection to saying that the examples 1, 3, 12[41] are cases of communicating personal experience, *viz*: these examples give us nothing but the outside appearance of such a communication[42] but the essential thing which would make them examples of the communication of personal experience we have left out, I.E. we have not said that there was a personal experience behind the utterances of the subject. These utterances may perfectly well have been made without (a) there having been a personal experience, or (b) without referring to[43] what personal experiences there were. It seems therefore that by saying that those examples showed us simple, primitive, cases of the communication of personal experience I[44] had passed off on you the mere empty nutshell for the real nut.[45] We are inclined to compare our statement that N°s. 1, 3, 12 exemplified primitive cases of the communication of personal experience with the statement that when a suit of male clothes and a pair of boots are in a room this is a simple case of a man's being in a room.

Now isn't it possible that B has reacted satisfactorily in the game N°. 1 and that all the same later we may say[46] that he had reacted in this way without really seeing the traffic lights. There certainly are such cases. Let us imagine one.

13.[47] B[48] reacts satisfactorily in game N°. 1 played with the traffic lights, otherwise however he reacts like a blind man. (a) Playing game N°. 1 he reacts

---

[40]A vertical wavy pencil line is placed in the left margin, contiguous both to the sentence commencing "The idea" and the next line down, which in the manuscript commences with "to just saying'. This coincides with the use of the two vertical bars, in the same paragraph, to partition alternative choices of expression. This concurrence of vertical wavy pencil lines and vertical bars coincides with Wittgenstein's own parallel custom in German (e.g. Notebook MS 158 for 1938, folio 27 and 28r; this latter page is the one before Wittgenstein's draft in English of his letter to Gilbert Pattison. Brian McGuinness (2008: 271–72) published this and dates it to 15.3.1938.

[41]None of the numerals 1, 3, 12 (both here and below) and form of content, nor the (a) and (b) sequence below in this remark, occurs in Wittgenstein 1993a. Also, there is a theme of lying in the latter publication in its remark 12, which is entirely absent from 12 in the current chapter; despite this, it is also much shorter than the current 12 here.

[42]At this juncture the main text has an alternative reading: "‖ these examples are examples of the outward appearance only of such a communication ‖".

[43]The main text poses an alternative: "‖ without being meant to refer to ‖".

[44]The word 'we' is crossed out and 'I' written above it in a hand different from Skinner's; it is not a typical Wittgenstein first person written form, though it has some similarity and the difference might be accounted for because it is a one letter insertion placed above the line.

[45]The words 'for a nut' are written in the same hand above 'for the real nut'.

[46]The manuscript offers a variant reading: "‖ we may be inclined to say ‖".

[47]No line space is left between this and the last numbered remark; this may be intentional or awaiting alteration, or the number could have been inserted in the margin (possibly deemed a remark) after the paragraph was written.

[48]Although 'B' is mentioned in Wittgenstein 1993a: 21 in remark 13, as here in 13, the themes, sentences, colours, arguments, and use of the game motif are all present in this chapter whilst absent from the former publication. In the 1993a published form, eight lines make up remark 13, whereas some 60 lines comprise the remarks in the present chapter (coincidentally 1993a's source –

equally well with his eyes shut as with his eyes open. (b) He can only play N$^{o.}$ 1 with his eyes open.

Now are we bound to say that in such cases[49] he only reacts to the lights without really seeing them? Couldn't we say that for a reason unknown to us he only sees traffic lights and[50] nothing besides, and in the case (a) that, extraordinarily enough, he even sees the traffic lights through his eyelids (or blindfold)? – But to this again you object by saying: "Of course we can only conjecture whether another personally really sees or doesn't,[51] but we know directly whether we ourselves see or don't." And we argue: "Surely you don't deny that we may – and actually do – sometimes say that we see so-and-so when in fact we don't see it. So why should B playing the game N$^{o.}$ 1 not do just that? And further, – when I am asked 'What do you see?' I don't observe my own behaviour and thence conclude that I see, say, a green circle". – But what does it mean to say that we know 'whether we ourselves see or not' or that we know whether we see a green circle or not? Can't we imagine a case in which we should say, "He said he saw a green patch, but he had forgotten the meaning of the word green[52]" or "He said he saw something but he had forgotten what was meant by 'seeing'"? – But to this you will answer: "He may have forgotten what the colour he sees is called but nevertheless he knows what he sees." – I think you wish to say: "He sees what he sees" and this means nothing. (Possibly you will say: "Even though he might have forgotten the name of the colour he will be able to tell us what colour he sees by other means, say by pointing to a sample of the colour." – But couldn't we say that in this case he might have forgotten what the expression "the same colour"[53] means?)

How do I know that I see a green patch? – Can't I say: "Well, just by seeing it"? This answer may arise out of a confusion between seeing and looking. We are thinking of the use of such a sentence as, "I know that nobody is in the room just by looking about". Or we may say: "I recognized him just by seeing him" (as opposed to recognizing him by hearing his voice). If to say "I know that it is a green patch just by seeing it" is to have any sense we must contrast seeing that the patch is green with finding this out in[54] some other way. But this was not what we wished to **do** for we talked of the 'green patch' as of a green visual appearance.

Notebook MS 148 (1935a), p.13, is only 13 lines; the current chapter's remark in the manuscript is 91 lines long).

[49]There is a wavy line under in 'such cases', which appears to be by Wittgenstein. This has its counterpart in Wittgenstein's usage in his German manuscripts of the same period.

[50]The word 'and' is written above the line, with 'but' on the line.

[51]The word 'doesn't' is written in same hand above 'not'.

[52]Single quotes around 'green' is written and deleted.

[53]The words 'expression "the same colour"' are written in Skinner's hand above the phrase, 'word "colour likeness" on the line.

[54]The word 'in' is in Wittgenstein's pencil hand above the line, with an instruction for insertion (Wittgenstein's custom of writing a diagonal mark over the 'i' rather than a dot, is present here. For example, see his 'in', also written over the line as an insertion in his own text in his notebook MS 148 (1935a), p.27, line 2; cp. notebook MS 158 (1938), fol. 32v, line 5-up.)

How do we know that what we see is, say, a green patch? Isn't it, by saying to ourselves, or aloud, that it is a green patch? – "But surely we only said that it was because we knew that it was!" – But what is it like to know that the colour which I see is green, is this a process constantly accompanying the seeing, '*must*' it accompany the seeing, and in what way does it differ from the seeing. Knowing, I suppose, is something like *being able* to say what it is we see.

Then, examine what are your criteria for saying that you are able to say what you see. One such criterion for saying aloud what you see is probably saying to yourself what you see. It is not absurd to say that the criterion both for yourself and for the others for your seeing the colour green maybe[55] that on this occasion you react with the word 'green' (the word 'green' comes).

"But why shouldn't I have played the game N°. 1 and afterwards remembered that I had seen nothing while playing it but just had said the words?" – Certainly there can be cases in which I on some occasion have said "I see so-and-so", and where afterwards I remember clearly not having seen anything while I had said so or remember that I had seen something else or that what I had said had no connection at all with what I had seen. Then why shouldn't I remember this sort of thing about having played game N° 1?

There is something curious[56] about the use of the word "remembering". It seems to refer to a present event on the one hand, and[57] to a past event on the other hand. One is inclined to say; I see *now* what has happened in the past. This of course is a metaphor and in many cases a very misleading one. Let us not go into the question now what it is that happens 'in our minds' when we remember that something has happened. There may be images, words, tensions, and what not. For our purpose it would be sufficient to assume that what happens is that the words "I remember so-and-so" come into our mind. We have been taught to use these words in various language games.

14.[58] Perhaps the most important one of these is the one in which there are ways of checking up on what we remember. There are, however, usages of the word 'remembering' in which no 'way of checking up on what we remember' is provided for.

---

[55]The word 'maybe' written in Skinner's hand replaces 'is' in the main text.

[56]The word 'curious' written by Skinner above the main text replaces 'queer', and there is a wavy pencil under 'queer' here. This line seems to be Wittgenstein's – perhaps to indicate and mark unhappiness with his dictation of the original word so as to mark it for revision, which in the case above is then inserted at some point afterwards. Re. the quandary in the above text about choosing 'queer' or 'curious'. We do not know if Wittgenstein had a specific German word in mind. Anscombe translated *seltsam* and *merkwürdig* as queer, while translators of the 4th Ed 2009 of PI, have rendered both *seltsam* and *merkwürdig* as 'curious', with various alternatives such as, 'odd', 'strange', 'remarkable', 'strange', or 'extraordinary'. This helpfully lays the German context that can be used to interpret Wittgenstein's and/or Skinner's uncertainty in the above passage.

[57]The word 'and' in Wittgenstein's pencil hand is written above the line, with a mark indicating insertion. This particular insertion mark is unusual in the Skinner Archive – it is a large 'v' through the line and above it. But it corresponds to Wittgenstein's own custom, for example in his German draft (dated by von Wright to 1935–36; by Alois Pichler (2013) to 1933–34) of the opening of the *Brown Book* (MS 141), p. 4. There are also wavy lines here, above and below the 'v'.

[58]The number '14' is placed here at the beginning of a line, whereas in the manuscript no line gap is left and the end of the previous sentence, 'words in various language games' begins the line. The

15. Thus we say in the morning that we remember having had such-and-such a dream in the night. And the question has been raised whether the dream which someone[59] tells us actually happened while he was[60] asleep or whether it really happened while we[61] 'remembered it'. The answer is that as long as we don't admit any criterion for deciding at what time the dream happened we may say whichever we like. "But surely this[62] can't be just a matter of words, there is something more to it". – There is![63] The child[64] for example who has been taught the use of the word "I remember" or the use[65] of the past tense of verbs[66] spontaneously on waking up tells us a dream in the words, "I *was* at such-and-such a place last night". We then teach him to say, "I dreamt I was _ _ _ _ ", thus preserving the past tense which the child had spontaneously used.

Now we might say: "That the child spontaneously used the past tense shows that there was something 'past' about his dream. For it wasn't a slip of the tongue when it[67] said 'I was . . . .' for it *persisted* in using the past tense". – But isn't it enough to say that the child persisted in using the past tense, couldn't this have been 'the phenomenon of noticing a pastness about the dream.'[68]

Surely it is remarkable that in telling a dream we are inclined to use the past tense together with all the imagery, gestures, inflections of the voice, etc., connected with the use of the past tense in cases of the kind N°· 14. If you ask: "But *why*[69] is the child inclined to use the past tense", I should answer: the *causes* of its doing so I don't know; and why should it have a *reason* for expressing itself as it does?[70]

---

partial parallels with Wittgenstein 1993a: largely cease from now on. For example, 1993a: 215 has numeral '14', yet no remark is offered; rather, a one-line list: "Identity of physical objects, of shapes, colours, dreams, toothache", which has no resemblance to remark 14 in the current chapter.

[59]The word 'they' is crossed out in pencil, and 'someone' is added in Wittgenstein's hand using the same pencil.

[60]The words 'they were' are crossed out in pencil; and 'he was' is added in Wittgenstein's hand by pencil.

[61]The word 'they' is crossed out in pencil; and 'he' added in Wittgenstein's hand by pencil.

[62]The word 'this' is placed above 'it' in Skinner's hand.

[63]An exclamation mark inserted here, in what appears to be Wittgenstein's pencil hand; it displaces a colon, with each dot a small cross.

[64]The word 'child' is used in remark 16 of Wittgenstein 1993a: 217-17; the subject matter and themes are quite different from the present chapter, however. It seems as if Wittgenstein was intent on having a child theme (perhaps to link up with his use of Augustine's *Confessions* 1.1.6.8), yet he was exploring the prospects for alternative subject matter to articulate the motif.

[65]'the use' is inserted here in Wittgenstein's hand.

[66]'of verbs' is inserted here in Wittgenstein's hand.

[67]The variation in deploying 'it' and 'he' to indicate to the 'child' appears to mirror some German switches of pronoun marking different stylistic registers.

[68]The open and close single quotes are here added in what seems to be Wittgenstein's pencil marks.

[69]The italicized terms '*why*', '*causes*' and '*reason*' have Skinner's ink pen underlining to indicate their italicization.

[70]This question mark (converting a full stop) is written in pencil, which accords with the pencil hand of Wittgenstein's on the same page.

Now we may imagine that we accept an experiential criterion for the time at which the dream 'actually took place'.

16.[71] Imagine such a case as this, that we could observe people's brains while the people were thinking (speaking, writing, observing phenomena around them) and that we observed a person's brain while he was sleeping, deduced from the[72] observation that the person was having certain thoughts, images, etc. and found that on waking up he told us a dream agreeing with what we had inferred. Then we might say the time at which a[73] dream occurs is the time at which these processes take place in his brain. This statement is one about the use of the expression "time at which the dream occurs". (Compare with this such statements as "Experiments have shown that long dreams[74] can take place in a few seconds".)

17. Now consider such a statement as this: A man says: "I remember clearly having had this dream before I was born".[75] – Should we answer him that it's impossible, that he can't remember this, that there isn't such a thing as remembering this? – This would seem to us like denying that he could have *said*,[76] that he remembered.... But we may be bewildered and not know 'what to do with his statement,' 'how to treat his statement.'

18.[77] Suppose, for example, old people[78] above ninety tended to say that they remembered certain events of their early childhood. Suppose that by checking up on this sort of statement we almost invariably found them to be entirely incorrect.[79] I am assuming that on the other hand they don't in any other way bear the characteristics[80] of lies. It is likely that in such a case we should give these utterances a special name, say, "senile pseudo-memories". We would perhaps explain them in a Freudian way

[71]'16' is inserted here in the margin, while the sentence starts halfway along the line opposite it, which thus appears to be evidence of decision to number the paragraphs after this part of the text was written, though by the next paragraph the numerals accord with paragraph breaks. '16' is also the last numeral in this sequential series that occurs in Wittgenstein 1993a: 215, and the contents of this publication's remark 16 have no parallel the present Chap. 6's '16', and the contents of the two publications largely converge, though note a parallel with part of remark 17 below.

[72]The word 'the' is written above 'this' in the same hand.

[73]The word 'a' is written above 'the' in the same hand.

[74]The words 'Experiments have shown that long dreams ...' written in Wittgenstein's hand by pencil, which replace the ink written in Skinner's hand in the main text that reads: 'We know from certain experiments that long dreams'.

[75]Similar point to a sentence in Wittgenstein's Notebook MS 148 (1935a), p. 38, as in Wittgenstein 1993a: 222, though with distinct context and developed differently.

[76]This comma is inserted in pencil, which, given Wittgenstein's pencil-written emendation above it on the same page, may be Wittgenstein's.

[77]'18' is placed in the margin here at the beginning of the line starting with 'treat his statement'; but presumably the start of the next sentence 'Suppose...' is the target of the number.

[78]Cf. 'old people' in Wittgenstein 1993a: 222.

[79]The manuscript contains an alternative phrasing here: "|| found that there was no truth in them, that they were entirely fictitious ||".

[80]'bear the characteristics' is underlined with a wavy line.

as utterances of *wishes*, etc. I mean, we might in such a case not be inclined to make any connection whatever between these senile utterances and the past. – If statements of the form "I remember having dreamt so-and-so before I was born" usually occurred between the senile statements described we should[81] probably regard them as interesting phenomena of old age and we should[82] not here conceive of remembering as of '*looking back*' into the past.

19. Under other circumstances however our attitude might be quite different, if for example we had found that at a certain age people tended to remember events of their childhood, that in a great number of cases these memories tallied with other records. In such a case we should[83] be inclined to treat these memory-utterances as documents[84] for the history of the person's life.[85]

20.[86] And it is quite imaginable that we should look in the past of the person,[87] say, in the development of the embryo for the foundation of that memory of a dream which was said to have been dreamt before the person came into the world.

We shall however[88] find a difficulty in saying that a person remembered events of a very early childhood, let alone of a time before his birth.[89] We should[90] say that it was impossible for someone to remember what happened around him after[91] he had just been delivered, because his consciousness at the time was not enough developed to take in his surroundings. At a later stage – we would say – when the child begins to notice what happens around him he is able to collect memories, especially when the child can already speak. For by speaking (thinking)[92] he as it were grasps his surrounding, takes an impress[93] of it in his soul. – This shows us how,[94] on the one

[81]'Would' is corrected to 'should' in Wittgenstein's hand.

[82]'Would' is corrected to 'should' in Wittgenstein's hand.

[83]'Would' is corrected to 'should' in Wittgenstein's hand.

[84]The expression 'as documents' is written above 'as documentary', in Skinner's hand.

[85]This phrase has an alternative I the main text: "‖ of the person ‖".

[86]'20' is placed in the margin at the beginning of a line, presumably to mark the beginning of the next sentence commencing 'And it. . .', so I have begun a new paragraph remark break here.

[87]'the person' has a wavy line underneath it.

[88]The words that occupy the main line – 'would however be inclined to' – are crossed out and 'shall however' inserted in Wittgenstein's hand.

[89]Note the earlier quotation from Augustine's *Confessions*, and the theme of 'child'; see Mulhall (2007). 'before he was born' parallels 'before he was born' in Wittgenstein 1993a: 222, though the overall development of theme differs.

[90]'Would' is corrected to 'should' in a Wittgenstein's hand.

[91]The word 'after' is written above 'when' in Skinner's hand.

[92]The words 'speaking (thinking)' is written above the main text's 'speaking, thinking,' in Skinner's hand.

[93]The word 'impress' is written above the main text's 'off-print' in Skinner's hand.

[94]Wittgenstein is the only hand that is using pencil in these pages, and all the commas in this sentence are in pencil, so it is probable that these commas were added by Wittgenstein – certainly they are how he wrote commas.

hand, we are inclined to limit the use of the word 'remembering,' but how, on the other hand, certain facts might make us stretch this use.

21. Consider now the case of someone saying that he remembers to have seen[95] a blue light when, being taught the use of the word 'red,' he played the game N$^{o.}$ 1 and said 'red'. – Should we say that now we had got to know what at that time he had really seen although then he was not able to tell us? We may be tempted to say this, to imagine that now at last we had a means to look into his mind as it then was. His mind as it were had then been closed, impenetrable to us,[96] but now by means of speech he could open it[97] and let us look into it. Speech here appears as *the* window through which we look into the other person's mind.[98] This picture can be extremely misleading.

Now when a person is taught the use of colour names ("red", "blue", etc.) we are making use of certain phenomena as criteria for saying that he sees what we see, that he sees something different from what we see or that he sees nothing. And, of course, our criterion is not what he says in the way in which at a later stage what he says is taken as the criterion for what he sees. For when for example he learns to play the game N$^{o.}$ 1 and in the time of his training says the word "green" where we want him to say "red" we don't regard[99] his saying green[100] as a description of what he saw.

22. Now imagine this: –[101] As soon as ever he has learnt enough of language to express it,[102] he tells us that he saw blue when, in playing game No. 1, he had said "red". – This sounds as if now we really ought to be convinced that he had seen[103] blue.[104]

"22" reminds one misleadingly of this case: "As soon as ever he had learnt enough of their language, the stranger told his hosts that...". – The idea connected

---

[95]'seen' is written over an illegible deleted word.

[96]'closed, impenetrable to us' has a wavy line under it. Presumably, given the identity of the pencil handwriting on the last few pages, it is Wittgenstein's, though of course if it is Skinner's it is likely that Wittgenstein, having emended the text at various points, would have seen it.

[97]The words 'by means of speech he could open it' is written above the line in Skinner's hand, and bracketed with 'by having learnt speech he could open it up' on the main line also in Skinner's hand.

[98]The main text presents an alternative depiction of the point here: "‖ *the* window in a person's mind which allows us to look into it ‖".

[99]The word 'regard' is written above 'take', both in Skinner's hand.

[100]The double quotation marks around "green" are written and deleted in ink in the main text.

[101]The opening dash of §22 is in pencil, as all the commas in this paragraph. Noting that §20, and the remarks leading up to it, have Wittgenstein's writing in pencil together with punctuation additions, it seems probable that these pencil additions in §22 are his.

[102]The commas in this paragraph are what appears to be Wittgenstein's pencil hand, except for the last two, which are in Skinner's ink hand.

[103]'seen' is written above the line, with an illegible deleted word on the main line below it.

[104]In Wittgenstein's notebook MS 149 (1935–36), folio r4 (as with Wittgenstein 1993a: 229–30), employs 'blue' and 'red' but entirely without a language game as here. In effect, the present chapter embodies Wittgenstein's embedding the child theme into the language, and a role of recognition in it – unlike the notebook.

with the description of the case 22 is that he had remembered[105] all the time, had perhaps tried to express his memory, but had had no outlet for it. As soon as this outlet was provided he told us[106] what he knew.

([107]Thus Augustine[108] says,[109] that when he had not yet learnt speech he tried to communicate his[110] wishes which were inside him to the people outside him by signs produced by his voice and movements[111] of his limbs, signs similar to these wishes [_ _ _ . *et*[112] *voluntates meas valebam ostendere eis, per quos implerentur, et non poteram quia illae intus er[r]ant, foris autem illi nec ullo suo sensu valebant introire in animam meam. Itaque jactabam et*[113] *membra et voces, signa similia voluntatibus meis, ...*])[114]

---

[105]The words 'he had remembered' is written above 'he remembered' in Skinner's hand.

[106]Written in Skinner's ink hand, the word '(us)' has a parenthesis around it, with a diagonal line through it, and a wavy in line under it, all of which seem to be Wittgenstein's hand.

[107]This and the closing curved parenthesis are in pencil, which, when compared to and combined with the above pencil punctuation, appear to be in Wittgenstein's hand – in contrast with the square brackets that are in Skinner's ink hand.

[108]Note that reference to Augustine also occurs in two places in Wittgenstein's large C-series notebooks. First, in MS 149 (1935–36), f. r4 (cf. Wittgenstein 1993a: 229–30): "Augustine, about expressing the wishes inside him". Secondly in Notebook MS 152, pp. 38–39, Wittgenstein presents five sketchy attempts to introduce 'Augustine', referring to 1.8 in the *Confessions*, whereas the citation here is from *Confessions* 1.6.8. (The provisional nature of this part of his Notebook MS 149 is illustrated by the re-ordering, and insertion in a fresh position, of the three short paragraphs (cf. Wittgenstein 1993a: 229–30). In contrast, the present chapter is replete with structured single argument and thematic development.

[109]The comma is in a pencil hand that appears to be Wittgenstein's, while the main text is in Skinner's ink hand.

[110]'his' is written over the line in ink, with an insertion instruction.

[111]The words 'signs produced by his voice and movements.' (with the full stop) are written above the 'making signs to them by his voice and' in Skinner's hand.

[112]This quotation from Augustine's *Confessions* 1.6.8 is not the same as Augustine composition that Wittgenstein cites in *Philosophical Investigations* §1, which is from *Confessions* 1.8.13. Note also that in his terminated attempt to re-write part of the English *Brown Book* into German in MS 115 (XI), p.118 (on which is MS page where he gives the date as "Ende August 36"), Wittgenstein quotes 1.8.13, as follows: "*. . .cum. . .[ipsi] appellabant rem aliquam et cum secundum eam uocem corpus ad aliquid mouebant, uidebam, et tenebam hoc ab eis uocari rem illam, quod sonabant, cum eam uellent ostendere.*" (also see Pichler 1997: 73ff). Given the plausible assumption that Wittgenstein used Skinner's copy of the *Brown Book* (taking it with him to Norway), it is worth juxtaposing Wittgenstein's above quotation of Augustine from 1.6.8, with the German version's use of 1.8.13 that does not occur at the beginning of the *Brown Book*, with its reference to Augustine's name.

[113]This occurrence of the conjunction *et* is not in the Latin text (see Clark in Augustine 1995: 32).

[114]This curved parenthesis is written in pencil in what appears to be Wittgenstein's hand. Chadwick's (1991) translation of this passage (Wittgenstein 1991: 7) is, "and wanted to manifest my wishes to those who could fulfil them as I could not. For my desires were internal; adults were external to me and had no means of entering into my soul. So I threw my limbs about and uttered sounds, signs resembling my wishes".

23. Now let us consider a case in which the child remembers having seen this or that before ever it has learnt to say that it remembers. And let us imagine the following: the child does not remember by imagining but by drawing and painting 'what has happened'. As soon as it was strong enough to hold a pencil, before ever it had learnt to speak, it spontaneously began to draw pictures which could easily be recognized as[115] scenes from[116] his life of[117] the previous day. (Later when we teach the child to speak we fasten the word "remembering" to this spontaneous phenomenon of drawing.)

24.[118] Suppose now the child thus doing what we should[119] be inclined to call drawing and painting its memories persistently painted red objects green and green ones red. (A rose it had seen the previous day it paints with green petals and red leaves.) Now what more direct and unmistakable[120] expression of memory could we wish for? No language which he later on learns can express what he remembers with more truth than his uninfluenced spontaneous drawing. We could say "If ever we shall know what he[121] really saw, than[122] by looking at these drawings". No memories, we might say, can be expressed more clearly. And yet, is the case we have described a clear case in which we should[123] say: "The child sees red when we see green and vice versa; it clearly remembers having seen the rose green and the leaves red"?[124] Wouldn't we just as soon say that the drawing-memory of the child plays[125] a queer trick? I think we feel a definite reluctance to say that the phenomenon we have described shows,[126] that the child sees red whenever we see green and vice versa. And the reason is this: We do under certain circumstances use the expression "a person sees red where we see green", – but the grammar of this

[115]The word 'as' is written above the main text's 'to be', both in Skinner's hand.

[116]The words in Skinner's hand in the main text – 'out of' – have been scrubbed out by pencil, and 'from' is inserted in Wittgenstein's pencil handwriting.

[117]The words in Skinner's ink hand 'having taken place' in the main text are crossed out with a single pencil that appears to be Wittgenstein's and replaced by the single word 'of' inserted written in Wittgenstein's hand.

[118]This numeral '24' is written in a pencil hand that looks like Skinner's way of writing numerals.

[119]The word 'should' is written above 'might' in Skinner's ink hand.

[120]The phrase 'direct and unmistakable' has a wavy line in pencil under it.

[121]The word 'it' in Skinner's ink hand is crossed out, and 'he' inserted in Wittgenstein's pencil hand.

[122]'than' is spelled 'then' by Skinner – presumably a slip.

[123]The words 'would be inclined to' in the main text in Skinner's ink hand, are crossed out by a single pencil line, with 'should' inserted over them in Wittgenstein's pencil hand.

[124]The quite various alternative is offered in the main text: "‖ And yet, is the case we have described one in which we should clearly be decided to say... ‖".

[125]The word 'plays' written above 'has', both in Skinner's ink hand, with a pencil wavy line under 'has' that appears to be Wittgenstein's.

[126]The comma is written in what appears to be Wittgenstein's pencil hand.

expression is characterised by this that we don't use it[127] in the connection: "he sees red *whenever* we see green and vice versa". Or, as one is inclined to express it misleadingly: "If he always saw green whenever we see red and vice versa we couldn't find it out". The[128] normal use of the expression "He sees green where[129] we see red" is like this: We take as the criterion for someone's meaning the same by "red" as we do,[130] that, as a rule, he calls those objects 'red'[131] which we call 'red' and vice versa. If then,[132] in particular instances, he says something is red when we should say it is green we say that he sees it different from us. Mark[133] how in such a case[134] we should behave: we should[135] look for a cause for this disagreement and if we had found a 'satisfactory explanation' we should certainly be inclined to say that he saw green where we saw red.

Finding a satisfactory explanation would be something of this sort: finding a substance in his eye which, as experiments have shown, when injected into the eye

---

[127]The words 'its use is characterised by this' are written above 'the grammar of this expression is', both in Skinner's hand. Notice that a replacement of "it" is offered in the main text here: "‖ the expression ‖".

[128]This sentence marks the appearance of a parallel with Wittgenstein Notebook MS 149 (see next footnote). It is worth noting an example of how much more content there is in the present chapter between this point and the above reference to Augustine than the notebook: in the present chapter and its source manuscript there are 475 words between the reference to Augustine and the start of the parallel, while in Notebook MS 149 there are 130. In the present chapter there is a fair copy single sustained argument, with a long quotation in Latin from Augustine, whilst in the notebook there is a reference to Augustine's name and six words about him. Rhees (1968: 272) observes that at a certain point in the Notes, these were written between January to early June 1936. The above parallel comes from that period. If this is correct, it is plausible to suppose that Wittgenstein's use of Augustine in the foregoing remark could be at or slightly before this date.

[129]'The normal use of the expression "He sees green where…"' appears in Notebook MS 149, folio 4v. (This is reproduced in Wittgenstein 1993a: 230, though the term 'green', which is written clearly by Wittgenstein in the notebook, is left out and replaced by '[red]'; it states in footnote 56 'The text reads "green" here, but this was probably a slip'; yet we see Wittgenstein retaining 'green' again in the present Edition, which amounts to being a fair ink copy. There are slight pencil revisions in the present Edition on the page where 'green' occurs and the previous, but the 'green' is unaltered; even so notebook MS 149 is heavily revised in the paragraph where 'green' appears.)

[130]A pencil comma is employed here, which appears to be Wittgenstein's.

[131]The single quotes around this and the next use of 'red' are inserted in what appears to be the pencil hand of Wittgenstein. It is noteworthy that the prior use of "red" in the same sentence has double quotes. Such movement between double and singles, not least when some particular attention is being given to them by virtue of such insertion, and also since the use of red in the next sentence has no quotes assigned to it, attract the suggestion that there is a reason why Wittgenstein wished to express the difference.

[132]This comma and the succeeding one are written in what seems to be Wittgenstein's pencil hand.

[133]The word 'Mark' is written above 'Notice', which is written on the line, both by the same hand. In Notebook MS 149 f.4v 'Notice how in such a case' occurs in a similar context, though with many differences.

[134]The words 'in such cases' are written above 'in such a case' in Skinner's hand.

[135]The word 'should' is written above the line over 'would', both in Skinner's hand.

immediately makes red objects appear green; or finding that he looks through coloured glasses.

Furthermore it is clear that even before having found such a cause we should, under certain circumstances, be inclined to say that he saw a colour different from that which we saw. If he called all the objects green which we call red and vice versa we should say that he *means* by "green" what we mean by "red" and vice versa. So it is clear too that we can't give any strict rule stating in how many cases we must agree about the colour of objects in order to be able to say that we mean the same colour when we say "red".[136]

25.[137] Consider this case: Someone says, "It's queer, I see every red object blue today and vice versa". We answer: "Things must look queer to you!" He[138] says, "They do! The 'blue' sky looks warm and the glowing embers cold". I think we should under these or[139] similar circumstances be inclined to say that he saw red when[140] we saw blue. And again we should say, that we know that he means by the words "blue" and "red" what we do as he had always used them as we had.[141]

26.[142] On the other hand: Someone tells us today that yesterday he saw everything red blue and vice versa. We say to him: "But you talked quite normally yesterday, you talked about the blue sky and the red sunset". He answers: "That was because I had changed the names too.[143]" We: "But didn't it feel very queer?" He: "No, it seemed all perfectly ordinary[144]". – Would we in this case say that, as he remembered it, he obviously[145] had seen the clear sky red, etc.? Would we not rather say that we don't know *what* to say under such[146] circumstances?

27. Go back to the case 24, – and suppose now that the child not only draws scenes of yesterday but also of the day before yesterday. Thus sometimes it draws the

---

136 The writer includes an alternative ending in the main text here: '‖ that we mean the same colour by the word "red" ‖'.

137 There is a similarity to 25. Notebook MS 149 f. 5r, (Wittgenstein 1993a: 231).

138 The main text in Skinner's ink hand had: 'And he'. This is crossed out in what appears to be Wittgenstein's pencil hand, with the 'And' deleted and 'he' changed to 'He'.

139 The word 'or' is written in Wittgenstein's pencil hand above Skinner's ink handwriting 'and'.

140 The word 'when' is written above 'what' on the main line, both in Skinner's hand.

141 The word 'had' is written in Wittgenstein's hand above Skinner's ink written 'did' on the main line.

142 Parallels here with Notebook MS 149, f.5r. (Wittgenstein 1993a: 231).

143 The words 'I had changed the names too' is written above 'I also had changed the names', both in Skinner's ink hand.

144 The word 'ordinary' is written above 'natural', both in Skinner's ink hand.

145 The word 'obviously' is placed above the line, with an insertion sign, in Wittgenstein's hand; there is a faint rough line under this word, which may be the wavy line for uncertainty.

146 'such' is written over the line above 'these', both in Skinner's ink hand, while there is a pencil line through the latter word which, in view of the immediately prior similarly penciled insertion by Wittgenstein, is his.

same scene twice, –[147] the next day and again the day after that. And we observe that, as in 24, he[148] paints red objects green and green objects red the next day, but the day after that it changes about and paints green objects green, etc. – Ask yourself,[149] what would you say the child had seen originally[150]; had it seen the grass green or red?

28. Suppose[151] someone habitually draws[152] sketches of the objects and people around him. We observe certain peculiarities in his representations. E.g. he draws people with very small heads and very long legs, and such like. – In what sort of cases would we say, "He sees things different from what we do", and on the other hand: "He just paints them different"?

Let us look back now on the example 22. We felt that if only he told us what he had seen,[153] as soon as he was able to do so,[154] we (now)[155] had a direct criterion for knowing what he really had seen.[156] But the following examples were meant to show that only under very special[157] circumstances do we regard the phenomenon of memory as a certain sign of what has happened 'in him'.

But here we may be accused of not having treated these cases fairly, for we talked as though to say, that he had seen green, when there was no criterion for us to decide whether he had, made no sense, and it seemed that we forgot that after all he saw green or didn't see green and knew whether he did at the time.[158]

[147]This dash is in pencil, and in view of the previous and succeeding penciled insertions by Wittgenstein, it may be his; it is a recurrence of his stylism: a dash after a comma. Note also that the next comma (after 'day') is also in pencil.

[148]The word 'it' is crossed out and 'he' inserted in Wittgenstein's hand.

[149]The comma is in pencil, which appears to be Wittgenstein's.

[150]There is a penciled horizontal S′ loop (probably Wittgenstein's) around the words 'seen originally' to indicate reversing their order, but this itself has a series of dashes along it to indicate deletion of this instruction.

[151]After "Suppose' Skinner's ink hand has 'now', which is deleted in ink.

[152]The word 'draws' is written above 'made', both in Skinner's hand (written twice, first misspelled and deleted).

[153]This and the other comma in the sentence are both in what appears to be Wittgenstein's pencil hand.

[154]After this penciled comma, the expression 'at all' appears in Skinner's ink hand and is crossed out in pencil.

[155]Penciled parentheses have been added around the word '(now)', and a wavy pencil line is under 'now', both of which seem to be in the same pencil used by Wittgenstein in this sentence.

[156]'he really had seen' is written over the line in Wittgenstein's pencil hand, to replace Skinner's ink written hand 'it was he saw' in the main line.

[157]The word 'special' is written above 'particular' in the same hand. Both 'special' and 'particular' have wavy pencil lines under them – it seems in the same hand as the other adjacent pencil strokes.

[158]The words 'whether he did at the time' are written above 'knew it at the time', and both are in Skinner's hand.

If we say: "He knew at the time what it was he saw", this is like saying what he saw was on record *then*, only the record might have been lost[159] but that is irrelevant.[160] But let me ask: what is knowing that I see red like? I mean: look at something red, 'know that it is red' and see[161] what you're doing. – Isn't it what you might call 'impressing something on your mind'[162]? But by this you mean two different sorts of things, (a) recording what it is you see by saying it to yourself, painting it, etc. etc., (b) staring, concentrating your attention and such like. It is clear that (a) and (b) are connected. One might call both 'impressing on yourself what you see',[163] as one might call impressing a seal on wax making a print[164] of the seal,[165] or, on the other hand, pressing the seal into wax, whether the seal is thereby recorded or not. Now it is misleading to use the word 'knowing what one sees' in the sense of staring, looking hard and such like, because in our use of the expression "A knows what B sees" this expression is bound up with a way of communicating *what it is*[166] that B sees B sees. To the sentence, "I know what he sees" there is, in our game, a supplement "He sees so-and-so".[167]

Now why are we inclined to say: "Certainly he knows what he sees". One might answer: "Certainly he can't be in doubt about what his visual[168] impression is as he is having it. He may for example be in doubt as to how to call the colour he sees, – but that is irrelevant, – but he can't be in doubt about the colour itself[169] – just

---

[159]The main text presents the variant reading of: "‖ though afterwards the record may have been lost ‖".

[160]The words 'is irrelevant' are written in Wittgenstein's hand above the main line ink written 'doesn't matter' in Skinner's hand.

[161]The word 'see' is written in Wittgenstein's pencil hand (especially the way he wrote the initial 's' and its join on the line to 'ee' – as with his writing of 'seeing' in MS 158, p.31, line 15, is entirely parallel with the present case, and unlike Skinner's writing of the word.). Wittgenstein's 'see' replaces the main line's 'mark' written in Skinner's hand.

[162]Single quote marks around 'impressing something on your mind' are in what appears to be Wittgenstein's pencil hand.

[163]There are single quotes here in Skinner's ink hand (with spacing that does not look as if they have been inserted subsequently). He seems to be following Wittgenstein's lead who earlier had added penciled quotation marks to a similar clause.

[164]The penciled words 'a print' are written by Wittgenstein above Skinner's ink-written 'an offprint'.

[165]This comma, and also the three remaining ones are written in what appears to be Wittgenstein's pencil hand.

[166]The word 'that' is very lightly crossed out in pencil, as if it is a question, rather than a decision to delete. One use of 'B sees' is inserted over the line.

[167]A different reading for the ending of the sentence is in the main text: "‖ a supplement of the form "He sees so-and-so" ‖".

[168]After 'visual', 'image is written and crossed out in favour of 'impression'. The latter is squeezed in at the end of the line, as if Skinner has already commenced writing the remainder of the sentence and is brought back to insert it on the line.

[169]There is an ink comma after 'itself', which is crossed out in pencil and replaced by a penciled dash.

because he sees it. But here we are mixing up having a colour impression with having a sample of the colour. It is quite correct to say, "I know what colour he sees, namely this (pointing to a sample),[170] although I don't know what you *call* this colour". "To know what colour A sees" can be translated into "To be able to communicate what colour A sees". And if we say that A certainly knows what colour he sees although he may not know the name of the colour, we more or less presuppose that he could, if not name the colour, paint it or point to a sample of it. And here we must remember our use of the word "can". Ask yourself: Are you talking of a process or conscious state of mind which continues while the person can? – "He knows what colour he sees because he sees it" is like saying "He sees the colour, so he sees it". Compare with this: "He knows what colour the flower[171] is, because he sees it" – as opposed, say, to knowing the colour because someone who sees it has told him. The confusion which is troubling us is connected with that idea of the 'visual impression', *viz*[172] that it is a picture which we carry about with us. In part of their uses the expressions 'visual impression' and 'picture of a physical[173] object' run parallel; but where their grammars differ that parallelism tends to delude us.

But isn't it true that if I see a colour I can, given certain outward circumstances, communicate,[174] if not by speech then by pointing or painting, what colour it is I see?[175] – Let us see what it means 'to communicate by pointing,[176] etc.'. Supposing that I see green and at the same time point to a green object. Is this communicating what colour I see or not? That depends upon the circumstances; I.E. upon whether I am playing a game which we would call "communicating a colour impression".[177] When,[178] just before, I said, "Given certain outward circumstances", I meant[179]: given the fact that I can raise my hand to point, that there are colour samples near,

---

[170]This comma and the underlining that italicizes '*call*' are in what very much appears to be Wittgenstein's pencil hand.

[171]There is a wavy line under 'flower' in what seems to be Wittgenstein's hand.

[172]The word '*viz*' is added here in Wittgenstein's hand.

[173]'a physical' is inserted, with an insertion instruction, above the line in Skinner's ink hand; the word it replaces on the line is deleted and illegible.

[174]This comma and the next one are in Wittgenstein's pencil hand.

[175]In this sentence I think that the commas after 'communicate' and 'painting' in a different hand from Skinner's may be Wittgenstein's.

[176]Skinner's writes 'painting', crosses it out and writes 'pointing'. After the end of the quotation mark a question mark is written and crossed out in ink.

[177]The double quotes around "communicating a colour impression" appear to be in Wittgenstein's hand.

[178]The commas after 'when' and the next two commas are in Wittgenstein's hand.

[179]This colon is inserted in the pencil hand of Wittgenstein, above the line and over his penciled comma, which he has deleted.

and such like. But[180] is it true that, given these circumstances,[181] I can play a language game of 'communicating visual impressions'?[182]

"But would you say then that a man may see a red patch, say, and not know what he sees?" – You don't mean, of course, that the expression "seeing a red patch and not knowing that one does" is to be used in the form "I am seeing a red patch but I don't know that I do". The expression on the other hand "I saw a red patch but didn't know that I did" we may be inclined to use. (We hear a clock strike and become aware of it afterwards). The expression "He sees a red patch and doesn't know that he does" may be used by us if we have adopted certain criteria for saying that he sees so-and-so although he doesn't communicate it to us and if we use the expression[183] "He knows what he sees" in a way analogous to that in which we use the expression "A knows what B sees". We[184] can then under[185] circumstances say that the child

---

[180]Skinner's ink writing originally ran "like; but"; Wittgenstein's pencil hand emended this to: "like. But".

[181]The commas after 'that' and 'circumstances' are in Wittgenstein's pencil hand.

[182]On the blank facing page opposite in Wittgenstein's pencil handwriting a word "Casfeling" (in double quotes) is written, after which there is a line drawn ending in an arrow head pointing to both the last line of this paragraph and the empty line space below it (the line of writing opposite it is 'language game of 'communicating visual impressions). It seems to be an oddly written 'Castling', alluding to the strategy in chess. (Wittgenstein's 't' is unusually looped to cross the 't' and join the 'e'.) This would fit the context: the legitimacy of using 'Castling' depends on certain circumstances holding in the state of play. 'Castling' occurs only once in Wittgenstein's other MSS, in MS 149 f. 10r (p.19 in the OUP Bergen Electronic Edition): "One might be inclined to say that castling was not just the act of _ _ _ _ But is the game of which it is part _ _ _ _". G. Citron kindly drew our attention to a use of 'Castling' in Yorick Smythies MSS, the Trinity microfiche Reel 2 (X-XII), 1st page series page 29. There is a similar sketchily written word, which is of uncertain identity, and it is corrected: it is on line 4-down on the previous page of MS 149 f. 9v., which is translated 'Christening' by Rhees (*PR* 1968 XX/Stern, p.234), though it does not look very like this word; there is a small probability that it is 'castling'. That one can exactly recognise the sequence of the context in Notebook MS 149 at this juncture, it is relevant to note that Rhees and Stern have repeated a paragraph p.234 (para. 2), and 234 (para. 7)-5 (para. 1), which only occurs once in this MS 149, namely: the two paragraphs commencing and ending "The words. . .this is grammar." The arrow in the second occurrence of the paragraph in Stern has a horizontal arrow '→', which occurs at this angle in Notebook MS 149; the first use of the arrow in Stern p.234, which is at a 45° angle, is incorrect since the arrow is horizontal in the MS in this context. The MS is confusing in these contexts since there are many deletions and lines to indicate transposition of order or paragraphs.)

[183]Originally in the manuscript after 'expression' there is a continuation of the sentence as follows: '"A knows what B sees" we can then under certain circumstances say that the child already'. Here the reference to a 'child' (and in the succeeding surviving next sentence) may be an allusion to Wittgenstein's use of the example in §22, which occurs in the Augustine quotation from *Confessions* 1.6.8. This is written and deleted in Skinner's ink hand. Given the incomplete ending of this deleted sentence-fragment, perhaps, may be Wittgenstein instructed Skinner to delete it immediately after the latter wrote down this dictation, and so continued with the alternative second part of the sentence that survives in the above form.

[184]Wittgenstein's pencil hand has changed the 'we' to 'We', placed a full stop after 'sees', and added a comma after 'meaning'.

[185]After 'under', 'certain' is written in Skinner's hand and deleted.

already sees but doesn't know what it sees, meaning, that it is not able to play a game of 'communicating the visual impression'.

"But surely we needn't inform a person that he has such-and-such a visual image; but, if we are to know, he must inform us. Therefore[186] he knows without being informed and we don't". – It is quite wrong to say that we need not inform him that he has such-and-such a visual impression. Remember game N$^{o.}$ 1: there *was* in this game (that is, in the actual playing of it) no such thing as A telling B what colour B sees; but not because 'this information wasn't necessary'; there just didn't occur such an operation[187] in the game, and, on the other hand, we took our idea of "information what *x* sees" from this kind of game. Therefore we might[188] say that *it makes no sense* to say, "*x* informs *y* that *y* has such-and-such a visual impression", and, analogously, we may say that it makes no sense to say "*x* knows what visual impression he has."

To say "I can say what colour I see without being told" sounds like "I can do without the help of being told"; – as though, analogously[189] to the game N$^{o.}$ 1, we had considered a game in which B tells A what A sees. (On the other hand we could imagine the game N$^{o.}$ 1 played with that variation that A has to propose colour names to B, that B would be unable to say red but able to nod his head when the word "red" was suggested to him by A.)

"You[190] talk as though one couldn't see a red patch, unless one can[191] *say* that one does. As if seeing something was saying that one sees it". – 'Seeing something' of course doesn't mean the same as 'saying that one sees something'. But the meanings of these expressions are related more closely than it might appear.[192] –

---

[186]Wittgenstein's pencil hand replaces the semi-colon after 'us' with a full-stop, and 'therefore' with 'Therefore'.

[187]The words 'an operation' are written above 'a process', both in Skinner's ink hand.

[188]The word 'might' is written above 'ought', both in Skinner's ink hand.

[189]The word 'analogous' is changed to 'analogously' by Wittgenstein's pencil hand and also, he inserts commas before this term and after the succeeding 'N$^{o.}$ 1,'

[190]This paragraph and the intervening paragraphs through to the end of the first paragraph in remark 30 strike extensive parallels with Notebook MS 149, ff. 6r-6v. (Wittgenstein 1993a: 233). There are differences, however, most of which are not listed here. One example illustrates a trend: in the above opening sentence, Notebook MS 149, ff.6r-6v has: "patch, if one can't say ['couldn't' – deleted] that one does; as [capital 'A' written over word] if seeing something was saying that one sees it." The present chapter's MS has '"patch, unless one can *say* that one does. As if seeing something was saying that one sees it". –'. Notice here how Notebook MS 149 has already provisionally placed a capital 'A' above 'as', which the present chapter's MS implements to end the previous sentence and signal a new one, no doubt to give special emphasis to the point. The present footnote's role is to draw attention to this as evidence of how finely tuned is Wittgenstein's attention to the form of expression, for which also see the next footnote.

[191]The words 'unless one can' are written in Wittgenstein's pencil hand, above Skinner's 'if one can't' (which, as indicated in the previous footnote, Wittgenstein dictated or Skinner copied from Notebook MS 149). Immediately prior to this replacement, Wittgenstein's pencil hand inserts a comma, and places a line under 'say' to italicize it.

[192]After 'appear' the words 'to you' occur but have been crossed out by Wittgenstein's pencil hand.

We say: a blind man does not see anything. But not only do *we* say so, but he too says that he doesn't see. I don't mean "he agrees with us that he does not see", or,[193] "he doesn't dispute it", but rather, he too describes the state of affairs in this way, having learnt the same language as we have. Now whom do we call blind, what is our criterion for blindness? – A certain kind of behaviour. And if the person behaves in that particular way,[194] we not only call him blind but teach him to call himself blind. And in *this* sense his behaviour also determines the meaning of 'blindness' for *him*.

29. But now you will say, "Surely blindness isn't a behaviour; it's clear that a man can behave like a blind man and not be blind. Therefore 'blindness' means something different; his behaviour only helps him to understand what we mean by 'blindness', to understand[195] what meaning it really is that we wish to convey to him. The outward circumstances are what both he and we know. Whenever he behaves in a certain way we say that he sees nothing, and he notices that a certain private experience of his always coincides with these cases.[196] And so he concludes[197] that we mean this experience of his by saying that he sees nothing."

The idea is that we teach a person the meaning of expressions relating to personal experiences *indirectly*. Such an indirect way of teaching we could imagine as follows:

30. I teach a child the names of colours and a game, say, of bringing objects of a certain colour when the 'name of the colour' is called out. I don't, however, teach him these names by pointing to samples which I and he see. Instead I have various spectacles each of which, if I look through it, makes me see a white sheet of paper in a different colour. These spectacles are also distinguished by their 'outside appearance': the one which makes me see green has circular glasses, another one elliptical ones, etc. I now teach the child in this way that when I see it putting on the circular spectacles I pronounce[198] the word 'green'; if it puts on the elliptical ones I say 'yellow', and so forth. This one might call "teaching the child the meaning of the colour names in an indirect way", because one could in this case say that I led the child to correlate the word 'green'[199] with something that I myself didn't see[200] but hoped the child would see if it looked through the circular glasses. And this way is indirect as opposed to the direct way of pointing to a green object, etc.

---

[193]The inserted commas before and after 'or' appear to be in Wittgenstein's pencil hand.

[194]Comma is inserted in what appears to be Wittgenstein's pencil hand.

[195]The words 'to understand' are written over the line with an insertion instruction in Wittgenstein's pencil hand.

[196]The words 'always coincides with these cases' is written above the main line's 'coincides with all these cases', both in Skinner's ink hand.

[197]The words 'so' and 'concludes' are written above 'thus' and 'knows' respectively in Skinner's hand.

[198]The word 'pronounce' is written above 'say', both in Skinner's ink hand.

[199]'red' in Notebook MS 149 f. 7v. (Wittgenstein 1993a: 233).

[200]The expression 'I myself didn't see' is written above 'I didn't see', both in Skinner's ink hand.

(The idea that[201] we convey the meanings of expressions of personal experience indirectly is connected with the idea of mind reading' and with the idea 'that the communication of our thoughts would be more perfect if we didn't have to communicate them through the medium of language'.)[202]

From the objection N°· 29 it should follow that we sometimes rightly, sometimes wrongly, teach a person to say that he is blind. For what if he saw all the time but nevertheless behaved exactly like a blind man? Or should we say: "Nature wouldn't play such a trick on us."

We[203] can see here that we don't clearly understand the real use of the expressions "To see something" or "To see[204] nothing".[205]

And what *so strongly tends to deceive us* – is so **deceiving** to us[206] when we consider this use is *this*[207]: We say "Surely we can see something without ever saying or showing that we do, and, on the other hand, we can say that we see so-and-so without seeing it[208]; therefore seeing is *one* process and expressing that we see another, and all they have to do with each other is that they sometimes coincide; they[209] have the same connection as 'being red' and 'being sweet'; sometimes what is red is sweet, – etc." Now this is obviously not quite true and not quite false. It seems we somehow look at the use of these words with some[210] prejudice. It is clear that in our language we use the words 'seeing red' in such a way that we can say "A sees red, but doesn't show it", on the other hand it is easy to see that we would have

---

[201] At this point the parallels start to break and include themes and terms that are absent from the Notebook MS 159 f.7v, while some parallels continue to appear.

[202] The main text here, in addition, offers two further choices with which to end the sentence: "‖ if they didn't have to pass through the medium of language ‖" and: "‖ if we could communicate them directly instead of through the medium of language ‖". The closing parenthesis of the paragraph is added by Eds.

[203] There is a lengthy repetition from "We can..." to "coincide; they". This repetition in the Skinner Archive has been omitted in this Edition. Nevertheless, it is important to point out that the first draft before the repetition is a section that Wittgenstein added to and revised in his handwriting. The second draft implements all Wittgenstein's handwritten changes in Skinner's handwriting, which explains the repetition.

[204] The word 'clearly' is written above and bracketed with 'quite', both in Skinner's hand. Repetition of 'see' in MS deleted.

[205] There are parallels with the last two brief paragraphs in Large C series MS 149 ff. 7v and 8r.

[206] The *italicized* words here are written in Wittgenstein's pencil hand, which are placed above 'is so **deceiving** to us' in Skinner's ink hand. (There is a parallel with this paragraph and MS 149 f.8; in this Notebook MS 149 there is a large 'V' insertion sign before 'when' without a proposed insertion.) The present chapter's original MS presents Wittgenstein's replacement. There is a penciled wavy line under the latter quoted expression, as has 'something' in the next sentence.

[207] The words 'the following' are crossed out and 'this' is inserted in Wittgenstein's pencil hand.

[208] This semicolon was an ink comma but converted by the addition of a penciled dot. In the second draft in MS Book VI, p.1, the semicolon is written in Skinner's ink.

[209] Note that this is where the above-mentioned repetition ends.

[210] The word 'some' is written above 'a', both in Skinner's hand.

no use for these words if their application was cut off[211] from the criteria of the person's behaviour[212]; that is to say, to the language game which is played[213] with these words it is essential that the persons who play it behave[214] in the particular way we call 'expressing (showing) what they see', and also that sometimes and under certain circumstances they should more or less, or entirely, conceal that they see something or what it is they see.[215]

31. Consider the following case.[216] We are in a laboratory[217] weighing various bodies, say, pieces of metal, and studying the behaviour of these bodies, which, as we have found out, depends upon their weight. As a rule the weight of such a body remains constant,[218] that is, if any time during the experiment or after the experiment we weigh it we get the same result. – Sometimes the weight of such a body changes[219]; weighing it at[220] different times yields different results. We then look for the cause of this change and find, say, that part of the body has been removed,[221] evaporated, and such like. In comparatively rare cases the weight of a body changes and we can't account for the change at all, but we nevertheless don't say that weighing here has lost its point, "for now the body doesn't really have any one weight." Rather we say that the body has changed somehow, say in composition or size, but that we hitherto have not detected the change. –

We use the expression "the weight of a body"[222] with the idea of something inherent in the body, something which could only be diminished by destroying, removing, part of the body. Our idea is: the same body – the same weight. The

[211]The words 'cut off' are written above the main text's 'severed', both in Skinner's ink hand. Each expression has a wavy pencil line under it, probably in Wittgenstein's hand.

[212]The words 'of behaviour' are written above the main text's 'the person's behaviour', both in Skinner's hand. The former expression has a wavy line under it, which appears to be in Wittgenstein's hand. Since this line queries the choice of this expression, the main texts use is retained here.

[213]The words 'is played' are written above 'we play', both in Skinner's ink hand.

[214]The term 'behave' written above 'should behave' in Skinner's ink hand.

[215]A partial parallel with MS 149, f. 8v paragraph 1. Cp. *PI* §91's hidden theme; see the use of 'laboratory' in *PI* §88, related to (in)exactness in the present chapter's use of the English 'laboratory' in the next remark – 31, as well as its footnote.

[216]The word 'case' is written above 'example' in Skinner's ink hand.

[217]The word 'laboratory' occurs in MS 226, p.63 – Rhees's typed translation of an early – *circa* 1939 – draft.

[218]The expression 'constant' is written above 'the same' in Skinner's ink hand.

[219]Some episodic parallels with Notebook MS 149 appear here. The opening statement is similar to "– Sometimes these bodies change their weight" in MS 149 f.13r; note that the long dash also occurs in the MS, though it is absent from Wittgenstein 1993a: 239.

[220]'at' is written above the line with an insertion instruction, in Skinner's ink hand.

[221]In the margin there is a question mark in pencil that appears to be Skinner's; it is opposite the line that commences with 'removed' and ends with 'comparatively'.

[222]The same as the quoted phrase in MS 149 f. 13v, though Wittgenstein 1993a: 239 does not document the indefinite article on the main line but gives the word 'this' written over it.

weight, for us, *belongs* to the body. And if the grocer[223] weighs the piece of cheese he sells us, he just finds out, we should be inclined to say, *what* he is selling us.

A person who knew nothing about weighing and its normal consequences, I.E. of the normal behaviour of the bodies we weigh, might think it queer that the grocer or we should be interested to know what happens when one puts a piece of cheese on a pan[224] suspended in such-and-such a way. He might say: "I thought you were only interested in eating the cheese, not in making an experiment with it." But we don't look at weighing in this way, for us it is not finding out in what way a complicated piece of apparatus reacts on a body, but finding out *what body it is*.

32. Suppose, however, what in the case 31 was described as the rule became the exception. Most bodies change their weights unaccountably. We could imagine that a body generally has[225] five weights; that is to say, when we put it on the balance at different times the balance shows one of five readings but we have never observed any regularity[226] which might make us express this change in the form of a law which helps us to predict what the reading will be the next time we weigh the body. Would we under these circumstances still be inclined to say that the body had changed but that we had not yet discovered the change? If *as a rule* I saw no regularity whatever in these events I should not be inclined to say that there was one which I had not yet discovered. I would reject the picture of regularity, which I could hardly ever apply,[227] as futile. One may say that the point of view which I take up towards one particular case depends upon the cases which surround it; depends upon whether, and to what degree, the phenomenon is the rule or the exception. The picture, the form of expression, the grammar which we are inclined to use, the language game which we are inclined to play, depends upon what it is that *usually* happens.[228]

But don't the words 'seeing red' mean to me[229] a certain private experience, or, a fact in the realm of primary experience, – which surely is utterly different from saying certain words? – How does he know that he has the visual experience red, I.E. how does he connect the word "red" with 'a particular colour'? In fact, what

---

[223]Only the word 'grocer' occurs by itself in MS 149 f.13v, no doubt as Wittgenstein's reminder to himself. In the present Edition the statement in the Wittgenstein Skinner MS parallels this word.

[224]After 'pan', 'somehow' is written on the main line in Skinner's ink hand and deleted with the same pen.

[225]The word 'has' has been written above 'had'; both words are in Skinner's hand.

[226]The words 'have never observed any regularity' are written above 'observe no regularity' in Skinner's pencil hand.

[227]This is an inserted pencil comma that seems to be in Wittgenstein's hand.

[228]This paragraph has some stylistic properties of a paragraph by Wittgenstein (1935c) in MS Notebook 149 ff. 13v-14r, which attracts one to consider the *possibility* that in this case Skinner might have been given this portion in written form by Wittgenstein; there are significant differences, however (for example, the absence here of a toothache theme, the change of ordering of themes with fragmentary or sketchy forms in MS 149, as well as a line deleting f.13v).

[229]The phrase 'to me' has a penciled wavy line under it that appears to be in Wittgenstein's hand.

does the expression 'a particular colour' here mean? What is the criterion for his[230] connecting the word[231] always to the same experience? –.

In fact, if he is to play a language game the possibility of this will depend upon his own and the other people's reactions. The game depends upon the agreement of these reactions. They, the players, must, in general, describe the same things as 'being red'. – "But if he speaks to himself surely this is different. For then he needn't pay regard to[232] other people's reactions[233]; he just gives the name 'red' to *the same colour* to which he gave it in previous cases.[234]" But how does he know that it is *the same colour*? Does he also recognize the sameness of colour as that which he used to call 'sameness of colour', and so on *ad infinitum*? It is quite true, he connects the words "red" and "the same colour" in accordance with ordinary usage, for example he would not say that he now saw the same colour which he had seen before, that that colour had been red and that the colour he now saw wasn't red – etc.

The words "'seeing red' means a particular experience" are senseless unless we can follow them up by "namely this" (pointing to[235] something). Or else they may stand for " 'seeing red' means an experience", as opposed to, say, a physical object; this is a proposition of grammar.

"What if I used language just for making entries in my diary? Without ever having learnt a language I could have invented a name for a colour sensation, say, the name "moo", and then used it to note down on what days I had this colour sensation." – That means, you play a private language game with yourself. But let's see what we are to say this game consists in. We mustn't say that you put down the word "moo" whenever you see red (or blue, etc.). For such colour words are used in our common language and are defined by means of samples which 'we all see'. Thus the expression suggests itself, that you put down "moo" whenever you see 'a particular colour'. But the former objection applies to the use of the words "to see" and "colour". For the use of these words obviously hangs together with objects of our common experience: samples, the human eyes, the act of looking, etc. So let us say that you write down "moo" whenever you have a particular experience. – Isn't it essential that it should always be the *same* experience? – You might say: "Surely we can change the meaning of a word during its use". That is, of course, true.

---

[230]'his' is written over the main line with an insertion sign, with the original word 'it' heavily deleted, all in Skinner's hand.

[231]Next to 'word' an expression is heavily deleted, which probably is 'and'.

[232]The words 'pay regard to' are written above 'consult', both in Skinner's hand.

[233]'reactions' has under it a wavy pencil line, which seems to be Wittgenstein's hand. Furthermore, we also given an alternative: "‖ the ways in which other people react ‖".

[234]The words 'in previous cases' are written above 'on a previous occasion', both in Skinner's hand.

[235]After 'to' the word 'this' is written and deleted in Skinner's hand.

33. I could, for example, in a language game like N°· 1 (Book I[236]) let the names of the building stones[237] change their names: say, in this way, that every morning A tells B what today he will call "plate", what "cube", etc.

34.[238] Or again one might play a game with a periodic change of names so that what on Mondays is called "plate" on Tuesdays is called "cube", on Wednesdays "column", etc. (Also A's silence may be a symbol: for example if A says nothing for some time B has to bring plates.) But suppose now I had described a language game (similar to N°· 1, Bk I[239]) in this form:

35.[240] "There is a builder and his mate. The builder, while working, sometimes produces various sounds (you may call them words), the mate brings him various building stones." Now would you call this a language game, or would you call the sounds which the builder produces a symbolism? What if I said: "It's a symbolism, the sounds change their meaning, but then you have admitted that symbols may change their meaning while we play a language game"? Would you be inclined to describe N°· 35 by saying "the words A ejaculates change their meanings"? We should[241] say: "We won't talk of a 'change of meaning' in 35, because no rule has been given which connects the sounds A produces with B's actions, and no regularity described in the change of such a connection". So instead of asking "Isn't it essential that it should always be the same experience?" I might have asked: "Isn't it essential that 'moo', if it changes its meaning, should change it according to a[242] rule?" For our purposes, however, the first question is as good as the second. – Then

---

236 "N°· 1 (Book I)' here refers to the *Brown Book*, Book I. The current manuscript comprises Books V and Book VI in the same series of volumes.

237 The phrase 'building stones' has a wavy pencil line under it, which appears to be in Wittgenstein's hand. In the Skinner manuscript of *Brown Book* I, the first occurrence of this phrase is on page 1, in §1, line 2. It reads in Skinner's ink handwriting: "B has to reach A building stones". On that page this is revised in Wittgenstein's pencil hand writing to read: "B has to hand building stones to A".

238 The margin number 34 introduces a manuscript line that begins with the last part of and completes the previous entry – i.e., "he will call 'plate', what 'cube', etc." The new sentence "Or again" introduces a fresh remark, whilst the usual line space left between remarks is not made. Accordingly, it is probable such an overlap indicates that Wittgenstein has decided part of the way through dictating this manuscript line that "Or again" should be taken as a separate remark, and so "34" is inserted in the margin to signal this.

239 This is a reference to the "Book I" that constitutes the *Brown Book's* first volume in the Wittgenstein-Skinner copy, which is without parallel in the structure or classification in the published Rhees version – the latter based on a carbon copy divided into two parts. As outlined in Chap. 2, the *Brown Book* manuscript contained in the Archive comprises six books, to which the first is being referred to above in the designation

240 In the manuscript page, N° 35, as with other numerals, is placed in the margin; yet the last part of the previous remark is placed at the beginning of the line, thus: "in this form". This phrase has been re-positioned in its place on the previous line. It seems clear that the line was written before the decision to commence a new proposition ("There is a builder...") was made or Skinner initially inadvertently omitted it.

241 The word 'should' is written above 'would' in Skinner's hand.

242 The word 'a' is written above 'some' in Skinner's hand.

in which case shall we say that you use "moo" at different times for the same experience? Is there a criterion for you that it is the same experience? You will say, that you recognise it to be the same. Then shall we say that you might recognise it to be the same without it really being the same; or does 'being recognised by you to be the same' here[243] mean the same as "being the same"? Does it make sense, in our example,[244] to say that two experiences are the same but are not recognised to be the same, or that they are not the same and are recognised to be the same? This question, as we shall later see, comes to this[245]: "Can you make[246] any use of the idea of a difference or identity of your experience if you admit of *no* possibility[247] of knowing whether these experiences are different or not? One might say[248]: "If you recognise this experience to be the same which you have had before it really plays no role whatever in what it is that you are recognising to be the same as long as only you recognise it to be the same." So let us say that when you have the experience called "recognising the present experience to be the same which before you called 'moo' " you make in your diary the entry "moo". But how are you to know that this is the case? (For remember no outward[249] criteria of which we might say they point to the fact that you have the same experience which you had before are here valid.) How can we say that an[250] experience justifies us in using for it either the word "moo" or the words "recognising the present experience to be the same which before I called 'moo' "? If we say that the latter experience must be recognised as the same which before we described in the same way we get no further. There are now two possibilities: Either in that private language of ours there is besides the word "moo" also such an expression as "the same impression". And in this case the description of our private language game will state in what kind of connections[251] the two expressions are used. Or, on the other hand, the only word of our language is "moo", then the description of the 'private language game' will say no more than that you are making the entry "moo" against some of the days in your diary.

When you speak about a private language game you are in fact thinking of a language game which a Robinson Crusoe[252] played with himself. But you would not say that he played a language game with himself unless he went through acts

[243]'here' is written in Skinner's hand over the line accompanied by an insertion instruction.

[244]The word 'example' is written above 'case' in Skinner's hand.

[245]The word 'this' is written above 'asking' in Skinner's hand.

[246]The words 'Can you make' are written above 'Are you making' in Skinner's hand.

[247]The word 'possibility' is written above 'way' in Skinner's hand, and the symbol '}' is written after them to group them, in what appears to be Wittgenstein's pencil hand.

[248]The main text offers a change from "say" to: "‖ make the queer remark ‖".

[249]'outward' has a wavy pencil underline in what seems to be Wittgenstein's hand.

[250]The word 'an' is written above 'certain' in Skinner's hand.

[251]The words 'in what kind of connections' are written above 'in what connections' in the same hand.

[252]'Robinson' occurs in MS 149 f.22v, though his game is 'lying to himself', with considerable differences of context.

analogous to those which make up a language game played between one man and another.

One could put our problem in this way: An ostensive definition could[253] be regarded as lying down, fixing, the meaning in which a word is to be used in our language. We might have, for example, laid down the ostensive definition for "red" consisting of a colour-sample with the word "red" written underneath it. Later on we refer back to this definition as one does to a statute.[254] Let us now consider the case of a 'private ostensive definition'. I have, for example, for my own use alone *and privately* christened an impression "red". (It's of course no use laying down this definition, say, in writing and by means of a sample. For the sample may change its colour and it can't be allowed any[255] voice in the decision[256] which impression it was that I called 'red'.) In what way does the act of having for myself named an impression "red" determine what impression I have to call red[257] in the future?[258] In what way can I be said to refer back to that[259] act of giving the impression a name when on a later occasion I use the word "red"? You say: "I just remember the decision I made then. Or I may look it up if I have laid it down, say, by means of a sample and the word[260] 'red'." Let us first consider the case in which you just remembered your decision. It makes no sense here to say that you remembered it 'rightly', or 'wrongly'. In fact if you say that you remember the decision to have been so-and-so, it comes to the same as saying the decision was so-and-so. But we can't here *explain* what "remembering the decision" means by saying that it is seeing what decision it was we had made[261]; we would be more inclined to give the explanation that "remembering the decision" meant that a certain decision came to us in a particular way. But in what way? Let us say that it is, for example[262]: A sentence comes into our mind of the form "I remember that I decided that 'red' should mean this impression", having a certain impression (you can substitute for this description of the phenomenon of remembering any other description of a mental phenomenon you please). Then 'referring back to the decision I have made about the use of the word "red" ' here means to say on a later occasion to myself "I remember that I have called this 'red' ", having while I say it a certain impression.

[253]The word 'could' is written above 'can' in the same hand.

[254]There is a large pencil written 'V' insertion mark here (placed after 'statute' and before the full stop), perhaps in Wittgenstein's hand, though there is no text present for insertion.

[255]The word 'any' is written above 'a' in Skinner's hand.

[256]'decision' has a pencil wavy line under it in what may be Wittgenstein's hand.

[257]Double quotes around "red" are deleted by Skinner's ink hand.

[258]The author here presents the milder alternative: "‖ an impression "red" help me to decide what impression to call "red" in the future ‖".

[259]The word 'that' is written above 'the' in Skinner's ink hand.

[260]'the red' is underlined in what appears to be Wittgenstein's pencil hand.

[261]The words 'what decision it was we had made' are written above 'what the decision was we had made', both in Skinner's ink hand.

[262]The words 'Let us say that it is, e.g.:' are written above 'Let us say, e.g. that it' in Skinner's ink hand.

But is this what we should call referring back to my decision? Doesn't 'referring back to a decision' mean that what I am doing now *agrees* with the decision I have made before? – But I only said, "having while I say it *a certain* impression"! I couldn't say 'the same impression' as, roughly speaking, the only criterion for it being the same was that I had said the words "I remember that I have called this 'red'". In fact the words "having . . . . . . a certain impression" were a meaningless flourish[263]; for not only could I not have specified the impression, but I had not even a right to use the word "impression". For this word has its well-known use only in our *common* language. – Supposing instead of just remembering the decision I looked it up, having laid it down in writing, etc. "Looking it up", I suppose, means that I look at what I have written (painted, etc.). But this again does not constitute 'referring back to the decision I have laid down'; unless I now act according to what I have looked up.[264] But the words "to act according to. . ." I have again taken from our ordinary language, and in playing your private language game you need not, in this sense, act according to your previous decision. "But what I do must seem to me to be in accordance with that decision". – Are you, in your opinion, *justified* in using[265] the words "in accordance with that decision"? Won't you be driven to say in the end[266]: "these words seem to me appropriate here", or something to this effect, and of these words again that they seem to you appropriate, and so on? It means nothing to say, "The words I use are *justified* by the impression I have", unless there is a rule for the justification of expressions of a certain form coordinating them[267] with descriptions of impressions by words and samples (that is, unless I can describe a game of justification of certain expressions played with words and samples).

Consider that when normally we use the word "red" to describe what colour it is we see, we never, or only in very exceptional cases, choose this word on the ground of remembering a previous occasion on which we had used it, or remembering the event of the word having been explained to us. We just use the word "red", without remembering anything. We use the word without any justification, – but that, of course, does not mean "we are wrong in using[268] it".

In[269] our private language game we had, it seemed, given a name to an impression, – in order, of course, to use the name for this impression in the future. – The

[263]'flourish' has a wavy underline in what seems to be Wittgenstein's pencil hand.

[264]The words 'what I have looked up' are written above 'this decision', both in Skinner's ink hand.

[265]The words 'in using' are written above 'to use', both in Skinner's ink hand.

[266]The words 'Won't you be driven to say in the end' are written above 'Must you not say', both in Skinner's ink hand.

[267]'coordinating them' has a wavy pencil underline, probably in Wittgenstein's hand.

[268]The words 'in using' are written above 'to use' in Skinner's ink hand.

[269]This paragraph has considerable agreement with a paragraph in Notebook MS 149 f.16r. A pencil line is above the first line in 16r, which is parallel with the opening of the above paragraph. Immediately prior to this pencil there is no parallel with the present MS. It is possible that the line was drawn by Wittgenstein to indicate that parts of this paragraph should be marked for transfer to the present and revision.

definition, that is, should have determined on future occasions for what impressions to use the name and for which not to use it.[270] Now we said that, after having given the definition, in certain cases you[271] used the word "red", in other cases you didn't; but we described these cases only by saying that you had 'a certain impression',[272] that is, we didn't describe them at all. The only thing that characterised them was that we used such-and-such words. Our definition didn't play the role of a definition at all, it did not justify the subsequent use of the word; and all that remains of this private language game is therefore, that you sometimes write the word "red" into your diary – without any justification whatever.[273]

The 'private ostensive definition' did not tie you down[274] to using the word in one way and not in another.[275]

"But[276] one *feels justified* when normally one says that something that one sees is red, although one doesn't think of a definition of 'red'." – Do you mean that when[277] you use the word 'red' to describe something you see you have, in general, a particular feeling which you call a feeling of justification? I wonder if this is so. But, anyhow, by 'justification' I didn't mean a feeling. But I think I know what makes you say that when, for example you tell someone that the book you're looking at is red you have a feeling of being justified in using the word "red". For you might ask: Isn't there an obvious difference between the case in which I use a word in its well-known meaning,[278] as when I tell someone: "The sky is blue

---

[270]Re. the parallel with this sentence and MS 149 f.16r. It looks as if Wittgenstein has dictated to Skinner the alternative, re-positioned between "‖ the definition should determine on future occasions ...... ‖ ‖ the definition should determine on future occasions ...... ‖", which does not occur in this position in 16r, while he has yet to decide whether or not to or how to insert it. Presumably the row of dots in the above main text stands for the remainder of the above sentence, i.e.: "for what impressions to use the name and for which not to use it". (Skinner appears to have begun inserting it above the line, then deleted the attempt above the first dot, due to lack of space.)

[271]The word 'you' written above 'we', both in Skinner's ink hand. This replacement also applies to the next use of 'you' in the sentence. The above term 'red' does not occur in the parallel in 16r, though it does later; and there are other variations.

[272]The words in single quotes 'a certain impression' are written above 'certain impressions' in Skinner's ink hand.

[273]The main text includes the alternative "‖that you sometimes write the word "red" into your diary‖, stripped of the ending – without any justification whatever.

[274]'not tie down' here is parallel with the use of the phrase 'not binding' in Notebook MS 149, f. 16r, though the order of this remark above is prior to the parallel paragraph in Notebook MS 149.

[275]The words 'in one way and not in another' are written above 'in one way rather than in another' in Skinner's hand; additionally, the main text poses an alternative: "‖doesn't tie you down to using the word *in this way*, and not in another‖".

[276]This paragraph is very similar to MS 149, f.16v-17r, though there are differences that include changes from MS 149's first person to the above 'one'; "use the word 'red'" to the above "sees is red", the addition of italics, etc.

[277]The word 'when' is written above 'whenever' in Skinner's hand.

[278]In the main text here is proposed the alternative: "‖ in an established meaning ‖", while the word 'an' is written above 'its' in Skinner's hand. Note that in Wittgenstein's composition in MS

today,[279] and the case in which I say any arbitrary word on such an occasion, for example, "The sky is moo". In this[280] latter case, you will say, I either know that I am just *assigning* a meaning to the word 'moo', or else I shall feel that I have no justification whatever for using the word.[281] "The word is just *any* word and not the appropriate word." – I quite agree that there is a difference between the experiences of 'using the name of the colour' – 'giving a new name to the colour', and 'using some arbitrary word in the place of the name of the colour'. But that doesn't mean, you have a feeling of appropriateness in the first case which is absent in the third. "But 'red', when I use it,[282] somehow seems to me to fit this colour". – We certainly may be inclined to say this sentence on certain occasions, but it would be wrong to say that therefore we had a feeling of fitting when ordinarily we say that something is[283] red.

36. But imagine this case. Someone[284] is looking along a row of patches of different colours and says: "Let me see if I still know which of these colours is 'red'? – Yes, I know". In such cases one sometimes says: "Something clicked in my mind when I came to *this* colour." (This, of course, is using a picture: When I have turned the wheel to a certain point, a catch clicks into place.) But did you conclude from the experience of clicking that what you saw was red? Supposing that looking at *this* colour your eyes suddenly widened and you gave a jerk,[285] – was it by the colour producing this reaction that you recognised it as being red? – Indeed such a phenomenon we call a phenomenon of recognition; but we call it that because, in other cases,[286] we use it as a symptom[287] that we have recognised a certain object *correctly*, there being ways of checking up on the recognition.

"But when you ask me: 'Do you know what 'red' means'? I answer, "Yes," after having brought before my mind a certain image."[288] But how is this *certain*

---

149 f. 17r, it uses the expression 'its well-known meaning', while in the above we see that the alternative 'in an established meaning' has been introduced.

[279]The manuscript here adds the alternative: "‖ as when, looking out of a window, I tell someone: "the sky is blue" ‖".

[280]After 'In', 'this' is written and deleted in Skinner's hand.

[281]'the word' is written above 'these sounds' in Skinner's hand.

[282]'when I use it' is not in MS 149 f. 17r; it is a typical instance of evidence of some of the differences between MS 149 and the present chapter.

[283]'was' in MS 149 17r, is here tuned up to 'is'. This parallel with MS 149, 17r's paragraph ends here, while the next paragraph in MS 149 has a parallel with remark 36 paragraph 3 ("'Can't a man. . .) in the present chapter, commencing with 'man. . .', though there are differences.

[284]This paragraph strikes a parallel with MS 149, f.20v.

[285]The words 'you gave a jerk' (the same as MS 149, f.20v) are written above 'you gave a little jerk' in Skinner's ink hand.

[286]The words 'in other cases' are written above 'under different circumstances' in Skinner's ink hand. MS 149f, 21r has 'under circumstances'; at this juncture the parallel starts to weaken and varying terms and themes concern the two manuscripts, though there is some continuity.

[287]The word 'symptom' is written above 'sign' in Skinner's hand.

[288]The word 'image' is written above 'sensation' in Skinner's hand.

sensation characterised? Only by that, that it comes when you say the word "red"? Or that it comes, and you feel satisfaction?

"Can't a man play a game of chess with himself, say in his mind, without showing to anyone else that he does so?" – What would you say he should do in order to be playing with himself a private game of chess[289]? Just anything? – I suppose you would say that, for example,[290] he imagines a chessboard with the pieces on it, that he imagines moves, etc. And if you were asked what he had to imagine to imagine a *chess-board*[291] you would explain it by pointing to a real chess-board or, say, to a picture of one, and analogously if you were asked what imagining the king of chess means[292] or imagining a knight's move.[293] – Or ought you to have said: He must go through certain private experiences which he has when – in the ordinary sense of the expression[294] – he plays a certain game of chess.[295] But what private experiences are there; and will any of them do in this case? For instance, feeling hot? "No! The private experiences must have the multiplicity of a game of chess".[296] But remember what we said about recognising a private experience as the same we had at another time. – If we go through with the idea of his private experience which we don't know, we can't talk of a 'certain private experience' either, because this expression is taken from the cases in which it alludes to a class of experiences we know, though we don't know which one of its members the other man[297] has. The private experiences which we imagined as unknowns X, Y, Z, ..., behind[298] our

[289]The words 'with himself a private game of chess' are written above 'a private game of chess with himself' in Skinner's (ink) hand. (The latter choice of expression parallels MS 149, f. 17v; on this verso there is a large insertion mark above this expression to indicate reversing the phrase order, which is implemented in the above text).

[290]In MS 149v. "e.g." is added over the line, whilst in the present edited MS it is moved to the main line.

[291]The words 'what he had to imagine to imagine a chess-board' that are written above 'what it means to imagine a chess-board'. Notebook MS 149, f.17v, uses the latter quoted expression. There are many more such examples. The above main text has the alternative: "‖ what it was he imagined if he imagined a chess-board ‖".

[292]The word 'is', as an alternative, is written above 'means' in Skinner's ink hand.

[293]The following option is added to the main text above: "‖ what it is to imagine the King of Chess or a knight's move ‖".

[294]The word 'expression' is written above 'word' in Skinner's hand.

[295]The words 'a certain game of chess' are written above 'a game of chess' in Skinner's hand.

[296]After this point in the paragraph, the parallel with MS 149 f, 18r begins to fragment, but is partially resumed below from 'If we go...' to '...mist'; but the intervening lines are not reproduced. Within the parallel here, as often with resemblances between the Notebooks, appear nouns, adjectives, phrases, quotation marks, plurals (not singular), dashes in the present MS that are absent from MS 149, f.183.

[297]The words 'the other man' are written above 'he', both in Skinner's ink hand.

[298]The word 'behind' is written above 'standing behind', both in Skinner's hand.

moves[299] lose[300] their identities and dissolve into a mist.[301] – Why[302] should you know better what 'having a sensation'[303] is like with the other man than what 'seeing red' is like? If you were very careful you would say, "having[304] a certain something". – And "having" is still too definite a word.[305]

So[306] what does giving to myself the ostensive definition of "red" consist in? – Now how am I to[307] describe it, shall I say, "It is seeing *this* colour (pointing to a red object) and saying to myself 'I see red'", – or is it: "Seeing a certain colour and saying 'I see red' "? – "The first version I don't like; it assumes that the other man has the same private impression which I am having". – But then how can I call it a 'colour'? Isn't it just as uncertain that I mean by 'colour' what they mean as that I mean by 'red' what they mean? The same, of course, applies to 'seeing'. (For I don't use this word here to mean an activity of the human eye.) The second version is justified, if I wish to say that it does not matter here to which of the colours (say, red, green, blue, yellow) I assign[308] the name "red".

Consider this case: If we tell someone to bring us a red object he makes sure that he knows what "red" means by calling up a colour-image before his mind. – So he

[299]The words 'our moves' are written above 'moves we make' both in Skinner's ink hand.

[300]Written 'loose' in MS.

[301]The expression 'dissolve into a mist' appears in Notebook MS 149, f. 18v. The mist theme is resumed later in Notebook MS 149, f.44r, where the sentence is used: "Die atmosphäre, die dieses Problem umgibst, ist schrecklich. Dichte Nebel der Sprache sind um den problematischen Punkt gelagert. Es ist beinahe unmöglich, zu ihm vorzudringen". Roughly: "The atmosphere surrounding this problem is terrible. Dense mists of [our?] language are situated around the problematic point. It is almost impossible to get through it." (There are considerable differences between MS 149; for example, a toothache theme emerges after this paragraph in MS 149 f. 18v, for which there is no parallel in the above.)

[302]This and the next sentence appear in 149, f.19r. They are outlined in the margin with a double arrow to link them to the previous folio 18v, para. 2, line 6 down (not noted in previous publications). This points at a locution that is not here reproduced in the above text – the core of which is: "The private experiences which we imagined as the background to the foreground of our actions". It evidently presents us with an opportunity to detect a thought process in Wittgenstein's revising of and selecting from the Notebook to craft a facet of a new family resemblance in his game.

[303]The words 'having a sensation', in single quotes, are written above 'having an experience' in Skinner's ink hand. Notebook MS 149, f.19r has a prior third earlier draft level with, 'what experiencing is like' (no quotation marks in MS).

[304]The word 'having' does not appear in MS 149, f.19r, and neither does the above 'And "having" is still too definite a word', nor do its two alternatives.

[305]Two alternatives are retained in the main text here: "‖ But "having" is still too definite! ‖" and: "‖ But, "having", – this is still too definite a word! ‖".

[306]The parallel with MS 149, f.19r now ceases, except for the occurrence of a colour theme, but it is there integrated with lying theme and a 'toothache' motif – both absent from the present edited MS.

[307]'to' is written over the line in Skinner's in hand, with an insertion instruction.

[308]The words 'I assign' are written above 'he assigns' in Skinner's ink hand.

can make sure, in this private way, of what 'red' means?![309] – But how does he know that the colour that[310] comes is the right one? (You are inclined to say that it is the right one if it comes in a certain way.) Suppose the man called up a colour-image and said "All right, I know what 'red' means" – he then goes and fetches a blue object, – shall we still say that he had made sure that he knew what "red" meant?

"He makes sure by calling up a colour image what the word red[311] means to *him*". Would you say that the word had meaning to him if it meant something else every time? And what is the criterion for the same colour coming twice? Which is the right colour, the one which comes now when I say red, or the one which I now remember came last time when I said red, supposing that these two are different?

Appendix: Untitled Archive Manuscript

Folio I, Pages 1–2

Note by the Editors

The following text is written on a page found in the Archive, and, if we can rely on its current position among the manuscripts, it should be placed at the end of Chap. 4.

The manuscript page is unlined paper, with a beautiful watermark – "Fine Indian Ivory", which is placed opposite an image of an elephant.

The Manuscript Text

Nov. 2ND Friday [1934].[312]

We think very largely in terms of mechanism and what it will do if it is of a certain kind. Our idea of mechanism is not that of its behaviour, but roughly its static aspect.

From its static aspect we go on to say what it will do

[309]The text offers the alternative: '‖ of knowing what "red' means"?! ‖'.

[310]'that' is written above 'which' in Skinner's hand; 'which', but not 'that', has a wavy line under it that appears to be in Wittgenstein's hand.

[311]There are quotation marks around "red", which are deleted in Skinner's ink hand.

[312]This day and date corresponds to 1934 (for which see Cheney (2000: 177).

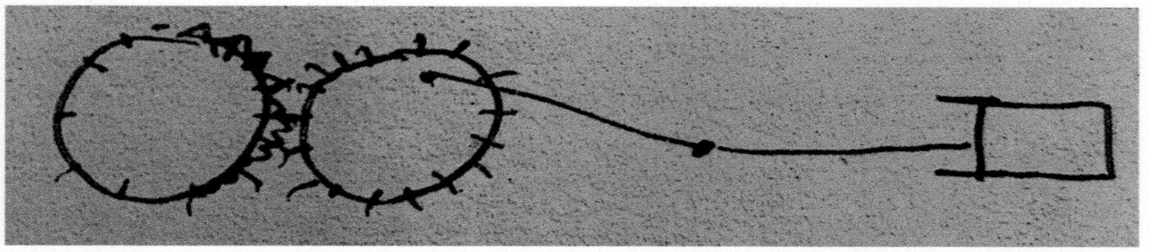

This picture stands for a certain form of behaviour

This picture is a disposition.

It is a hypothesis that this condition will be followed by a certain form of behaviour.

For us a certain mechanism often stands for a form of behaviour – I.E. it is our symbol for it.

We treat the picture in this way: by certain grammatical rules we deduce the behaviour. The picture is a geometrical drawing for us.

This is the reason why we think a certain state of mind 'understanding' can do certain things. We are mixing up the geometry and physics of a mechanism.

If we say that we can't prophesy[313] what a mechanism will do but we can say what it can do, then by "can" we mean we treat it as a rule for deducing certain forms of behaviour.

Question Could there be a transition case between these two cases?

The answer is that this can only be answered by an example. Why do I say this? You can't describe your transitional case now as you could describe a blue apple.

N.B. these notes are to be used sometime in the future.

"shld read?

Ex. 42 nor in the practice of it does a limitation of range plays 'a predominant' role etc."[314]

---

[313]'we can't prophesy' is written over the line with an instruction for insertion.

[314]"Ex. 42" may relate to remark 42 of vol. 1 of the *Brown Book*, which is in the Archive though not published in the present book.

Chapter 5
Philosophy

Wedn. Jan 17th[1] [1934]

We talked about the idea that we had to understand the order before obeying it.

The remark about thinking being the interpretation of signs referred to a discussion I had with Mrs. Knight.

If someone asked me what I did between 5 and 5.30 and I said I wrote a letter, then anyone of you could have described what I did – say I held a fountain pen, etc. If I had said "I expected Mr. Smith to come" what would correspond to the description you gave of the writing of the letter?

You can say "this is a peculiar activity of the mind: I can't say any more about it".

This is the way out when people say it is indefinable.

Let's see what it would be to expect Mr. Jones – is this another activity and what connection has it with[2] expecting Mr. Smith.

This chapter is comprised of what, in Skinner's handwriting on the cover of the original folder that may have contained it, is designated by him, in the following layout, as:

"Unedited Notes of Lectures by
Ludwig Wittgenstein
In Trinity College 1934
The Notes were taken by Sydney George Francis Skinner and are
In his handwriting".

This description "Unedited" is not accurate because some handwritten editing appears in the manuscript in Wittgenstein's hand. (See, for example, an entry for Ash Wednesday, 14th February – i.e., the second batch of notes with that date heading it). Consequently, we may take it Wittgenstein both read these notes, and they come with his knowledge of them. Wittgenstein's Moral Science' Tripos Lectures, at which the notes were taken, were advertised in the Cambridge University Reporter (13 January 1934, p. 522), with the heading 'Philosophy', and held in Trinity College.

---

[1] In the MS the date has an oblong box drawn around it – here reproduced as underlined, a practice that is used elsewhere in the Edition.

[2] 'the' is written and crossed out.

A. Gibson, N. O'Mahony, *Ludwig Wittgenstein: Dictating Philosophy*,
https://doi.org/10.1007/978-3-030-36087-0_5

If we say expecting Mr. Smith is walking about in the room – this is a rule of symbolism and is alright as long as we don't take the phrase to pieces. We define "expecting Mr. Smith" as *one* word. If I said "expecting Mr. Jones" is "looking towards the door" this is also all right as long as you don't take it to pieces.

This is not what you want – what are we to substitute for "expecting Mr. X". We want to apply 'expect' without previously being told. This might lead you to give up the idea that expecting Mr. Smith is a peculiar action of the mind, but then you might say 'expecting' is an action of the mind which can't be further explained.

What is[3] the connection with Mr. Smith?

Take[4] the example "expecting that Mr. Smith will come into the room".

If I ask you for the grammar of expecting Mr. So-and-so you are at liberty to give what you wish. But there are what you call awkward questions – awkward in that the answers given to such questions ordinarily won't do. If we say[5] expecting is a tension which will be relieved when Mr. Smith comes.

In this sense we could explain desire, wish, etc. but we must draw the consequences.

You must accept that you do not know what you wish, as you will not know whether the tension will be relieved when you get food.

You see, therefore, there are two cases. If a man says he knows what he wishes, then we say the object of his wishes can't be that[6] which he[7] supposes will relieve his tension as this may not be the case.

If someone says "Are you sure you are looking for a pencil" you will say "of course I know what I am looking for." This just tells us[8] the way you use the words 'looking for'. There may be a case where you use the word and do not know what you are looking for. Here is another use.

These are different cases.

If some says, "One does not know what one is wishing"[9] as Russell.[10]

There are two ways of using it.

This shows how complicated the use is. The way one knows one is expecting a man[11] is not generally the same as the way one knows somebody else is expecting so-and-so.

We have just described different usages.

---

[3] 'was' is written first and crossed out.

[4] 'What we want' is written before the beginning of the sentence and crossed out.

[5] 'the' is written and crossed out.

[6] 'the one' is in the main text, then crossed out and 'that' written over the line.

[7] 'he' is in the main text, then crossed out and 'he supposes' is written over the line.

[8] 'us' is supplied by Eds.

[9] After '…wishing"' "then I sa".

[10] I think the sentence should read 'If someone …', and the sentence seems unfinished.

[11] 'knows' occurs after 'man' and is crossed out.

If we go back to the question "You expected Mr. Smith to come into room between five and five thirty". Was it a process like whistling[12] a note or one like saying a sentence?

In general you would say this answer is given by *introspection.*

You would say: I prepared tea – I said something to my servant – looked out of the windows, etc. You would describe a series of actions in which Mr. Smith came in[13] in very different ways. These actions are what we call expecting Mr. Smith. You may say all these actions have some[14] characteristic but this is something very complicated.

If this is the question – what happens? Let's see by introspection.

What in a sense makes introspection futile is the idea that there must be some peculiar action "expect"; and where are we to find it.

'Expect' is a word which has never been defined. This does not mean it is indefinable. I will speak about this later.

If you[15] say, "But can't 'expect' be used in a more exact way", of course you can.

One question[16] that is always profitable: – "how did you learn to use the word?" – You heard your Mother use it in lots of different ways.

One used to measure in paces – but if you say "[17]this is different for different people. Can't we make it more exact?" – This has been done. What connection has the new with the old. The idea has been slightly altered on the one hand but on the other hand there is a connection between the two.

We can draw definite limits for the use[18] of the word 'expect' then you have altered the use. You cannot draw limits and leave it the same as before.

If we have a black circular patch on a white surface we could trace the outline with red ink.

It is like a dirty patch of which[19] someone says let's draw a sharp outline. This outline could not trace the exact limits as there were none, but this does not mean it has no use to draw it.

Suppose I said "expecting so-and-so" means saying I expect so-and-so. This is one of the standard cases. The criterion is that I say I expect so-and-so.

I could say I wish what I say I wish. This does not cover all cases I.E. the case of lies. If we say I think what I say I think. Saying is one of the activities we call thinking.

Sometimes we say saying and thinking are different; but thinking and saying are not two activities going along side by side.

---

[12]An allusion to Frank Ramsey's, "What we can't say we can't say, and we can't whistle it either".

[13]The manuscript first reads 'into', which is crossed out in favour of the above.

[14]'very' is written after 'some' and crossed out.

[15]'we' is written first and crossed out, with 'you' written over line.

[16]'is' is crossed out after 'question'.

[17]Open quote is inserted to match manuscript's close quote.

[18]'case' is written first and overwritten to read 'use'.

[19]'of which' is written over the line with a mark to indicate that it replaces a dash.

Suppose I said thinking and saying are the same thing. You would say "This is nonsense: you may think one thing and say another; or say something and not think at all."

How does it happen if I say "I shall be delighted to come," and think O heaven! I don't want to come. We ask the questions "are these two amorphous states?" If you ask yourself what happened all sorts of things happened. You may have said the thing to yourself, or imagined how boring it would be or only felt a feeling of discomfort.

Suppose you thought what you said; you may mean an amorphous state – or you may only say these words.

It is good to make a fiction of this kind. You could imagine the human language consisted of two things – you always sang your words. When you write a letter, you will[20] also give the melody.

We do distinguish between the cases – 'he meant what he said' and 'he did not mean what he said'.

You sometimes distinguish *strictly* between what he said and what he meant. What he said to himself was what he meant,[21] I.E. as asides in plays.

You can imagine beings who only thought aloud and if they said something they did not mean said an aside.

Using a verb like thinking makes us think there is one particular activity[22] thinking like saying. You might say thinking is saying to yourself. You might say one thing aloud and another thing to yourself. But this is not all we mean by thinking one thing and speaking another. Thinking is the interpretation of signs – what sort of thing is this interpretation and what do you mean by saying *we must* interpret[23] the sign.

If you said "Smoking forbidden" must be interpreted. Do we interpret both words? Yes! What is interpreting "smoking". Is it saying some words to yourself – in a different language.

If I read this notice in Spanish, and had to translate word for word – here you can say I interpreted each word separately.

I would not know a single thing that would *have* to happen. There is one case where I have to interpret words.

I am quite ready to acknowledge something as a fact of experience if it is one. Someone may say he has a picture for every word he reads.

Let us go back to the idea of people speaking and singing at the same time.

Could the singing replace what we mean by meaning.[24] (We say we know what a man means by the tone[25] in which he says it.)

---

[20] 'may' is written first, crossed out and 'will' placed over the line to replace it.

[21] 'and' is written after 'meant' and crossed out.

[22] 'like' is written after 'activity' and crossed out.

[23] 'this activity' is written after 'interpret' and crossed out.

[24] 'We say something' is written after the full stop and crossed out.

[25] A word is deleted after 'tone'.

Could you replace the meaning by the singing, or would one have[26] to say the singing must be meant one way or the other.

Supposing I say: I should like you to come to tea. These are mere words – they have got to be meant. Suppose 'meaning' them was some process, like singing say very low or to yourself. Will this sort of thing do for what you mean by meaning? You might say you could[27] there too.

Whatever you mean, you don't say to yourself that you mean it but this is what you mean. First of all you might say this – I said this but I meant this. Whatever process you mean has the multiplicity of your language because you can translate it into it.

You may say meaning is something internal and saying something external. We may find an external way of seeing what you mean.

You could imagine human beings who say one thing[28] and also write something at the same time. If they are the same, you say they mean[29] what they say, otherwise not. We may not be able to see what they write – but there are ways of finding out say by a mirror.

If meaning is some sort of process, it must be translatable.

We will go on to another point. What we have said applies to the words understanding and thinking.

I said I noted down the phrase "One can't obey[30] an order till one understands it." This phrase has a use and we must find it. At first sight it seems a general philosophic proposition.

I am not saying anything against this phrase, but you will see its use is a very specialised one.

I say to someone "go to so-and-so and fetch such-and-such a thing". When does the understanding take place, and must it go on the whole time.

If he understood it when I told him,[31] must he understand it again when he[32] does it?

What must he understand when he[33] does the action – how does the understanding accompany the action of fetching the chair.

He may just go and fetch. You may then say he does it then automatically. How do we distinguish between this and obeying it?

---

[26]'has' is changed to 'have' by Eds.

[27]We might suppose that the sentence should be left as it is, and Wittgenstein was positing the speaker ('you') as low-key personifying proxy for what is meant, as with singing in the previous illustration. If a word is omitted, the word could be 'be' or 'sing' however.

[28]'and s' is added and crossed out after 'thing'.

[29]'they mean' is conjecture for the phrase used in the manuscript, since it is: 'the may'.

[30]'understand' is written after 'obey' and crossed out.

[31]'did' is written after the comma and crossed out.

[32]'he' is supplied by Eds.

[33]'he' is supplied by Eds.

Why are you so keen on the person having an experience of seeing a chair between fetching it?

I suppose because you are anxious for some interpretation of understanding.

There is the possibility that when a person is asked to fetch a chair, he deliberates – but does this deliberation always go on?

You might say there is one definite difference between hearing an English order and a Chinese order.

If Mrs. Braithwaite[34] gives me a Chinese order, my first reaction may be have I heard aright. If she is Chinese, I may say what does she want?

If the order is in English – none of this happens. The words may be familiar. I may act on it (the order)[35] or not. If I do not, there may be an act of deliberation or there may not.

Another thing – hearing words of a language you know gives you certain definite sensations.

In some cases it calls up an image.[36]

There may be a certain specific reaction towards a word one has heard thousands of times in the same context. James[37] has said there is a particular sensation corresponding to the words "not", "in", "or".

If you concentrate[38] on these words, you may recognise different sensations for "not" and "with"[39] – but you don't mean you always have them. Suppose I had a list of words – English words and Chinese. I read it out – the subject says 'yes' if he understands the word, 'no' if he does not. All that happens is perhaps that I say 'yes'. He may see a lime-tree when I say 'tree'. Does he then think a lime-tree is a tree?

But doesn't this show something?

If I wanted to find out how much you knew about the English language and I read out the English[40] dictionary, doesn't this show something? – You may remember a sentence where the word is used, you may remember the corresponding German word. If you say 'no', when I say "compunction"; but this may be all that happened, etc.

There are two difficulties.

(1) You are looking for one process.
(2) You are looking for a momentary process instead of a series of processes.

[34]Married name of Margaret Masterman. See her contribution as joint author to "'Yellow Book' Notes" in Appendix [E].

[35]'(the order)' is written over the line.

[36]The manuscript has 'imagine'; amended to 'image'.

[37]William James (1890), *The Principles of Psychology*.

[38]Before this word 'correspon' is written and crossed out.

[39]Quotes for "with" are supplied by Eds.

[40]'wor' (and maybe the first part of the letter 'd' at the end) are written and crossed out.

The processes of understanding do not happen when I say the word tree necessarily[41] – Supposing someone said "I don't believe this – there is one particular process corresponding to the word understanding" – say imagining. Can I dispute that whenever the word tree is mentioned, Mrs. Braithwaite has an image of a tree?

No fact forces us to give a definition but certain facts tempt us to give one.

We have learnt the words of our language without a definition – such words as wishing, etc.

If now we want to give a definition, this definition will depend on what facts we have observed. If I feel a pressure whenever I use the word 'expect' you may now be tempted to give the definition of expecting as only applying to the case – when you feel the pressure.

The[42] cases – where explanations are needed are rare – suppose I said, "I expected him the whole day."

You will do the right thing if you are told 'to come to tea exactly at five o'clock.' You would say you didn't know how five o'clock exactly was to be determined and[43] needn't know.

This occurs in a restricted jargon.

The decorator says the room is twelve feet long.

The scientist says this is 25 cms long.

Totally different questions will have to be asked.

One does not ask in the first case at what temperature.

There is a doubt in the second case, but there is not in the first case.

The *situation* in general determines what it means, I.E. the 'dear Mr. So-and-so' at the beginning of a letter.

We try to give the word pace to a length which is roughly that of a pace. This is the real source of puzzlement.

Bright's[44] disease: in my book I talked[45] about an index as opposed[46]

---

[41] 'so happen after words' is written after the dash and crossed out.

[42] 'explan' is written and crossed out after 'The'.

[43] 'you' is written and crossed out in favour of 'and' over the line.

[44] A phrase is written and crossed out before 'Bright's'. It is not fully legible, but includes "If. . .I talk, it".

[45] 'called' is written before 'talked', then crossed out.

[46] This appears to refer to the *Blue Book's* examination of Bright's disease, using the above term 'index', so:

'Compare the grammar of this word, when it denotes a particular kind of disease, with that of the expression "Bright's disease" when it means the disease which Bright has. I will characterise the difference by saying that the word "Bright" in the first case is an index in the complex *name* "Bright's disease"; in the second case I shall call it an argument of the function "x's disease". One may say that an index *alludes* to something, and such an allusion may be justified in all sorts of ways. Thus, calling a sensation "the expectation that B will come" is giving it a complex name and "B" possibly alludes to the man whose coming had regularly been preceded by the sensation.' (Bright's disease is acute or chronic nephritis.)

Fri. Jan. 19th

If a poet[47] brings a battle on the stage, only a few points can appear but one must have a general impression of the battle.

I try to *give* you a few characteristic points which give you a general idea.

I want to give you a general survey of the grammar of thinking, intending, expecting, wishing, etc.

One of the difficulties is this: what am I trying to do. You are trying to describe how these words are used.

There are many different ways in which these words can be used in different ways.

Take the example: come to tea at "five o'clock exactly" or rather the words "five o'clock exactly". In the case of the tea, you know what to do. We can neglect the cases here of misunderstanding.

Suppose I say in astronomical calculations we use these words in a different way. In the case of the tea I may say ± 5 minutes or more exactly ± 1 minute. Suppose I talked about determining the position of the pointer, knocking, opening the door, you see now the way I am making it more exact is a way you couldn't have guessed. Going from ± 5 min. to ± 1 is what you already know to be more exact. Making arrangements about knocking on the door, say the first rap to be at five, you couldn't have predetermined what way I was to make it more exact. {If I were to describe the use of the word "come at five o'clock", there too would be all sorts of different cases.} I said it should be fairly easy to describe the use of the words above. Normally we use these words in different ways. There are also uses suggested to the person who philosophizes. We have to consider these. Take the example: wishing a person to come in from five to half-past five.

If I give the grammar, I have to give rules which are somewhere near the way you use it, and also near the way philosophy tempts you to use it.

Take the case of numbers: I might first talk of cardinal numbers etc. – but I also may have to give you[48] other uses. Nobody has probably made up an arithmetic of the cardinal numbers without five. – I don't mean our ordinary arithmetic not talking about five. It would be an arithmetic with different rules. (Or in the case of an arithmetic of the odd numbers, there would be no other numbers).

I say: "what was my difficulty in giving you a survey of the grammar?"

I must distort the way[49] they are used to one you are inclined to and also to other ways to show you this is not the only way it (the grammar) can be distorted.

I can only describe grammars which will impress you. I say that I do assume you use 'wish' to say you can 'wish from this time to this time'.

---

[47]This word has been partly removed by the effect of a rusted pin through the word. Thanks to Jonathan Smith for deciphering the reading.

[48]'uses' is written and crossed out after 'you' but restored.

[49]'you use them' is written first, crossed out and written over the line is the replacement "there are used"; presumably "they are used" was intended or dictated.

If I now ask what happens, you are at liberty to take what you like. I take cases which I assume come near to the way you do use it. If you do not agree I can only ask "how do *you* use it?"

I don't utter an opinion of mine. I hope to say something you would also say. All sorts of different things from 5 to 5.30. There is not one state which occurs the whole time, but there are lots of different states.

I suggest this: this is one possibility.

You may say – "I only say I am expecting so-and-so from 5 to 5.30 if there is one state which lasts the whole time." Then follows the awkward questions.

Suppose[50] you say it is like saying[51] you have a sleepless night when you did not sleep five hours. Then what did you do in the five minutes when you *did* expect him etc.

I have a feeling of[52] weakness and my body wishes food. Suppose I make these equivalent. If I then ask what it[53] is like if your body wishes whisky with a drop of sulphuric acid. You call one feeling your body needing food, another feeling needing water etc. If I substituted the whisky, you would not now understand this, but it would have to be explained separately.[54]

Take the word "eat": you know what eating a fountain pen means even if you have never done this. You understand 'eating a button' without a new explanation. If you know the feeling "the body needs food" do you know what "the body needs drink".

If I say it means "calling for drink" we now know what[55] 'the body needs *x*' means, i.e. "calling for *x*".

Mon. Jan. 22nd

What I tell you is only some good as long as you can use it. What I try to give you is philosophical experience.

Experience is the sort of thing when we say of a person "he has experience".

You get experience if what you see impresses you in some lasting way.

Later I knew I got nothing from all this travelling. This was an experience –[56] something I could do something with.

If a person does not easily get the right experience, you have the queer thing of the person saying, "what is the right experience?" You should get the hang of things, a certain *skill* in thinking yourself.

---

[50]Before 'Suppose' there is a partly legible expression, crossed out, which comments 'This is like x' with one other unreadable term marked 'x' here.

[51]The manuscript reads 'liking say' and is revised as above.

[52]'a' in MS changed to 'of' by Eds.

[53]'it' is inserted by Eds.

[54]'Separately' is spelled 'separatedly'.

[55]After 'what', 'needed x' is written and crossed out.

[56]'only I' is written after the dash and crossed out.

Partly I send you round journeys – you get nothing from them or are interested just now and then, which is no better. You should get some lasting experience from them.

You might say when the expectation is satisfied always means that a tension is relieved.

Russell's idea of expectation is this. - - – – –

You can say you got what you wished and yet get no satisfaction out of it at all.

In this sense you can say, "I wish or want to get an apple." The part of the sentence "to get an apple" is the object which I wish – (the object is ambiguous – I wish Mr. Smith did come – "Mr. Smith's not coming" is the object). Now there are lots of questions.

Does this phrase characterise a feeling of mine in an indirect sense?

As in Bright's disease.

Or does the phrase mean: We think that object will satisfy our hunger? In this case I would not know – it would be a supposition that my wish is for so-and-so. If, on the other hand, we say we know what we wish "I wished for so-and-so and got something else and was satisfied."

The process of wishing has a unique and direct expression of the form 'I wish that so-and-so will happen'.[57]

The process of wishing can be translated into this phrase "I wish etc."

This would be the case if wishing consisted in telling what one wished.

We are only considering the phrase "I wish that so-and-so will happen." What is the grammar of the phrase "that so-and-so will happen"?

In the case (as Russell) that what we wish we can't know. _____ If wishing is a feeling a hunger.

We know what will satisfy our appetite – our appetite consists in saying we want certain things. You wouldn't call hunger a thought – whereas wish and hope are called thoughts. This is because wishing and hoping can take the form of saying something. If we say by wish we mean some sort of hunger, then the object of our wish cannot be given in the hunger. When we say "our body wants salt" is this use.

Our criterion is future experience.

This is not the usual way we use it. As in "I wish to get a letter from so-and-so."

This is a description of what happens when I wish – it is a direct description. Whatever is happening can be translated directly into the expression "I wish for so-and-so."

Suppose wishing consists[58] in visualizing the event you wished for in a particular way. If we say this, then we can translate from the event of wishing directly into the language I wish for so-and-so.

---

[57] Single quotes are supplied by Eds.

[58] 'the' is written and crossed out.

I will make the still further supposition that wishing that so-and-so will happen means making a drawing of it. If someone asked me for the verbal expression, I could simply translate the drawing.

This is the case of looking at the wishing and translating it into so-and-so.

Another[59] – the simplest case – that the wish itself should be a linguistic expression. If wishing a thing is visualizing it or drawing it . . .

(Dreams have been said to be visualized events satisfying our wishes.)

(We are talking about the multiplicity which determines the event – not the multiplicity for hoping, fearing, etc. which will be given in any way as by gestures.)

Our process of wishing might be saying so-and-so with certain feelings – but these are amorphous and need not come in.

Divide the phrase:

I wish	so-and-so will happen
I fear	
I hope	

into parts. The first, the indication of our *attitude*; the second, the *object*.

The attitude I am not concerned with – as it consists of a word like "hope" etc.

The expression of the object I am concerned with.

We[60] said: the attitude might be a tension and the object that wish satisfies it.

"Our body wants salt."

Suppose the tension is something you see. Let's say wishing is a tension of the cheek muscles.

We say now the object of our wishes is what relieves the tension.

There is a different use of the word wishing in which the object does not express what will relieve the tension but is the direct expression of what is happening in the wish – as when you say it.

It makes the thing easy as an example if wishing is saying the thing you wish.

It may be anything else as long as it is translatable into this. (You could for instance be saying the thing you wish.)

Suppose I said I am afraid of so-and-so's getting ill. Suppose I say it is saying it. You will say this is absurd. It may consist of saying it with certain feelings. Whatever else you take must be capable of being translatable into this.

I am taking the class of cases where you say I must know what I am afraid of and *whatever* happens can be translated into "I am afraid of so-and-so".

[59]After 'Another', 'case' is written and crossed out in favor of '– the'.

[60]After 'We', 'satisfied' is written and crossed out.

Suppose we talked of hate: I said, “I hate Goodstein[61]”. Then you might ask, “Are you sure so-and-so is the cause of your feeling?” In this analysis you could not be sure of the object of your hate. (Hate can be transitive and intransitive.)

Suppose I had a fear – I can have a fear and know what I fear – and I can have a fear and not know what I fear. This is misleading. It is like saying two sentences. In one case **there is** no object.

You may say I know what you fear – it is so-and-so. You give a cause.

In the two cases:

I am afraid of so-and-so
I am afraid and do not know of what I am afraid

These two are not comparable – the objects are of a different kind.

– – – – – – – –

There is a case: You[62] say, “I am afraid of so-and-so.” I[63] say, “Isn’t it the case that you always fear somebody you knew when you were young.” Here you may change what you say you are afraid of.

These phenomena are terrifically complicated, I.E. as the phrase ‘the back of my mind’. Think of this.

It is enormously difficult to dictate something to you that will hold water.

Wednesday, Jan. 24th

It seems from the way I dictated that I meant it was wrong to say that in some cases we know what we wish and in others we don’t.

Suppose we say: If the dentist drills into my tooth under chloroform I have toothache but do not feel it.

If you ask, “Is this wrong” I will say it is a queer usage; but if I use this phrase in this case, why not. It may even be practical for some reasons

I.E. suppose when one is operated on with cocaine one screws up one’s face and holds it, then you might say you have unconscious toothache.

What is unclear is the word ‘know’. Find out what is the process of getting to know I.E. if there is someone talking in the next room you, get to know by going into the next room. If one’s tooth is drilled and one says one had toothache and this is the way you know it, this is a different thing.

You couldn’t make a man see what pains are by holding your hand to your face (you would have to give him a blow).

---

[61]Reuben Goodstein, who was in the audience, and received this manuscript after Skinner’s death.

[62]‘You say’ has been squeezed into a small space.

[63]‘If’ occurs before ‘I’ and is crossed out.

This is a totally different explanation from describing the operation on the tooth. These are of totally different kinds but they are each explanations.

Supposing one said: there is not much difference between[64] "a bad tooth and having toothache" and "a bad tooth and not having it". In one case you know it and in the other case you don't know it. This minimizes the differences between the cases.

Between the case[65]: You can use it in cases where it does not matter whether you see a person is in the room or not having other criteria that he is there. We can make such a use of 'know' work but it would just be a very misleading way of using it. We can perfectly use it in this way if only we go into and find out the use of the word "knowing" in this case.

One way of scrutinizing this use will be to ask yourself how do you "**get to know**" or "find out" in this case. We will see how we use the phrase "we know what we are afraid of".

Take this instance: I feel some sort of fear, oppression; and I ask myself, what is the matter? I have a certain sensation and into this sensation no object comes consciously. You may say, "It comes unconsciously."

I talk[66] about a fear which is just a sense of oppression. If one says "even if you *just* feel this fear, the fear always has an object only you don't know it."

We will meet this again when we talk about realism idealism solipsism – lots of people say the *same thing* in different ways. In one case the difference seems enormous and in another hardly anything at all.

Suppose we used "love" transitively and intransitively and talked about loving somebody and not knowing it.

Freud said he had widened the scope of psychology because he had discovered an unconscious realm of thoughts. This phrase suggests that the man has thoughts, which he[67] doesn't see.

The[68] analogy which comes into our mind when we hear unconscious thought and conscious thought is that of the unconscious thought being kept somewhere but not seen I.E. as the chair in this room and in the next room.

I have talked of "misleading". They do not mislead the man who uses these expressions practically. It is the philosophers whom it misleads.

I read a book the other day by a man called Spengler: there he uses the word Prussian which he uses it to signify a particular complex of virtues. He himself says not every Prussian has these – but why use this word? There is no harm in using this

---

[64]'between' is supplied by Eds.

[65]Here the manuscript begins a new page and repeats the last phrase 'between the cases' (but in the singular) from the previous page. This repetition can be taken (as above) as a means to dwell on the phrase, or might betray a break in dictation.

[66]Written in MS as 'talked', with 'ed' crossed out.

[67]After 'he', "doesn't know" is written and crossed out

[68]This sentence commenced with 'We' but is crossed out.

word for a complex of virtues; but we can't get rid of our ordinary use of the word. So that it is in a way cheating. This is exactly the same with the word 'knowing'.

Take the example of the word "infinite" in mathematics. If you remove the ordinary idea, the word dries up immediately.

There are cases where the word loses its power to mislead.

No one worries about this, I.E. as [with] imaginary numbers.

Compare multiplying by ¾ and multiplying by 3 and divide by 4. It doesn't matter which we say. In the first case we stress that we are multiplying by a number or by something that is roughly like a number. Saying "unconscious toothache" means we are drawing a parallel with something it revolts us to draw a parallel.

When people talked of imaginary numbers, what revolted was the analogy with numbers which they drew.

Suppose I said the equation $\boldsymbol{x^2 = -1}$ has no solution; but if we add one to the right-hand side the equation has solution 0. This is exactly the same as saying it had solution $\boldsymbol{\sqrt{-1}}$.

The expression: "$\boldsymbol{x^2 = -1}$ has no solution" has not the multiplicity of the expression: "it has the solution $\boldsymbol{\sqrt{-1}}$".

But I will add, how near it comes to having a solution.[69]

(Similarly with a line cutting a circle.)

If I express it in two ways,[70] it loses its misleading quality.

Take the case of the circle and straight line.

I said two opposite things – in one case we say: the straight line always cuts it, in the[71] other case the straight does not always cut.

In one case we say it does cut it but in an imaginary point

In other case we say it does not cut it but comes so-and-so near to cutting it.

The danger of using "Prussian" etc. is when we give over the reins to language.

Does the word 'unconscious' mislead Freud in his treatment of illnesses? This terminology is not responsible if he fails to cure a certain illness. **Where is he misled**?[72] He is misled in two places:

(1) Where he talks about it?[73]
(2) Where are mathematicians misled by a terminology like infinite?

Are they misled in their calculations? They are misled whenever they produce gas about their calculations – where they talk about their (the calculation's) importance.

---

[69]The manuscript's statement "how near it comes to having a solution" had double quotes that are crossed out, and after "near" there is a word written and deleted.

[70]'ways' is revision of 'says' in the manuscript.

[71]'say' and 'the' are supplied by Eds.

[72]This question is underlined twice, hence bold here.

[73]Notice the references to Freud a few pages prior to this use. There is no numbered successor for (1) in the manuscript; so '(2)' is supplied.

Another example comes to my mind:

People like Jeans[74] etc. said a few years ago that we have found out the floor on which we walk is not something solid but consisted of tiny particles at a great distance. This was supposed to be something terrifying. But we are not terrified – the floor is solid.

Another thing when we talk of a direction of time and then have the idea that we can reverse it.[75] Another example: the idea that when we buy nuts and we buy fifty; it is a round number, and when we buy fifty-one this is not quite a round number. In this case, one[76] is misled by the notation.

Friday, Jan. 26th

We said that even if we used a word in a rather queer way, we are not necessarily[77] misled I.E. a mathematician is not misled when he uses the word[78] "infinite" in his calculations.

The word infinite is not used arbitrarily. Can we say it is bad because it misleads us?

It is[79] not clear-cut between what words mislead us and what do not.

Someone looks very grave and wise and has a beard. Suppose I say, "shave off this beard". What will be left? In some cases it is obvious when a person has been misled and in others not at all obvious.

There is a definite case of someone being misled. It may be by the way the person tells me.

Someone else may tell me almost exactly the same thing so that what he said was quite true.

I.E. non-commutative algebra. If a man thinks he has got rid of old barriers, this is absurd.

If we talk of the 'sense of time', we may think so-and-so must have extraordinary brains to think about such a thing.

You say it always bothered me what it meant to reverse the sense of time. Suppose I say you meant so-and-so. Then you have been misled and I cure you.

I said Spengler has redefined Prussian to use a set of virtues. In what cases would a man be misled? Spengler honestly said, "Not everyone born in Prussia is Prussian". There will be many different cases where the misleading qualities of this notation will be seen. You will see it in how a man uses it.

---

[74]Sir James Jeans', who was also a Trinity man. Brian McGuinness (2008: 203 Letter 151 – to W. H. Watson) notes that this is the Sidgwick Memorial lecture, 1932.

[75]The manuscript has 'revert', is revised to read 'reverse'.

[76]The manuscript has 'says, which is crossed out.

[77]The manuscript has an unnecessary 'be' here.

[78]Skinner inserts 'the word' here.

[79]'It is' deleted, then restored to allow for indent and new line 'It is'.

There is another case of being misled where we are *worried* by the expression I.E. as in sense of time. Here we say let's point out the arbitrariness of this expression and then the worry will cease.

A man need not be misled by the expression "unconscious thought" as long as he remembers what it means. He will be misled as long as he isn't watchful.

A philosopher may be troubled and say how is unconscious thought possible: it is not possible. Then a scientist will say he has proved it and the philosopher won't know what to say.

This can be avoided if we look and see how we use the expression.

What is it like to despise someone and not know it? You meet A and someone B says he thinks you despise A. He may say, "would you like to dine with him or introduce him to your friends". This is quite a general case. You may take the criterion of what you would answer to certain questions if you ask yourself them.

What does this prove?

In[80] a sense you did not despise him as you would know it if you did. In a sense this certainly would be a discovery. Supposing I saw somebody and somebody said, "would you like to see this person in your parents' house?" This is a discovery and it may change my opinion but in a different sense I have not found out anything.

This is analogous to unconscious toothache.

This could be described in two ways – both perfectly correct.

One way – I did not despise him but I answered in this way to the question.

Second way – I despised the man and did not know it.

This you must remember says exactly the same. Otherwise one may say then how terrible I hate him without knowing it. Such things have happened in the expressions used in psychoanalysis.

When I talked of "knowing", I said one of the things to see was what[81] we called "getting to know". You may say this is getting on a sidetrack.

Suppose I wanted to explain a chair to you, I could show you by explaining, "sitting on a chair". This is part of the grammar.[82]

One thing I always recommended: "What's the criterion for such-and-such a thing? How do we know it? What's the proof?" Lots of people have asked, "How can we get to[83] know it?" People have answered, "We don't want this, we want the meaning." People thought it was a mere accidental thing how we got to know or found this out. In certain cases it doesn't matter, as in how do we know there is a man in the next room.

---

[80]Prior to the opening of this sentence commencing 'In', the manuscript has 'We would again in this' and it is crossed out.

[81]'what' is written twice, and the first is crossed out with 'was' written over it.

[82]'of the use' is written and crossed out, so maybe the previous full stop was added after the deletion.

[83]'to' is supplied by Eds.

Giving[84] the criterion or verification is giving part of the grammar of the proposition. Someone is murdered. A doctor comes and finds out when he died. Whatever we call the criterion of so-and-so shows what we accept as a criterion.

Suppose we said, "someone has toothache". Suppose we ask ourselves what do we call a criterion for the other person having toothache.

There may be lots of different cases as there may be lots of different answers to the question "Why do we accept this as a criterion".

You once learnt the meaning of the word toothache. You were told what the criterion of someone else having toothache is[85] I.E. as by his saying these words.

Suppose I notice a white spot comes on your hand when you have toothache. If someone says: "Why do you say this?"

I want to point out a difference. In one case I said I had the experience of seeing the white spot when you had toothache.

Can the same be said of your telling me you had toothache?

Can the same be said in every other case?

Take a different thing.

A doctor examines this man and says he's dead. First question might be "when do we call him dead". Shall we say, "when his heart has stopped"? Suppose for some reason we can't examine his heart but someone finds that the hair turns white when the heart stops.

Here we say we have a new method for determining death.

In this case asked why you say he is dead, you would answer you had found out. In the first case you might answer, "This is what we call death."

If we want to see how we use the word death, let's ask how do we know a man is dead. All sorts of things may be found out[86] about when a man is dead. You may say these do not help – what we want is what we call a man being dead.

Take the case "despise". I say if you want to know what this means, "getting to know" may mean describing what acts accompany despising, or what else would you use to describe somebody despising me. This is a contribution to the grammar of the word "despise".

In the case of the white hair it doesn't matter as you can say you have had such an experience. This is what the philosopher had in mind.

People muddled up two kinds of criteria. One kind is that through which you learnt the way to use the phrase,[87] the other that of experiences you have had.

When do we say a man has got gallstones? If he were operated on, we would say that was settled. These are not symptoms.

We talked about Moses and the different definitions. Actually asked, you would not know but could pick on a certain one. You might change your[88] definition later.

---

[84]'Showing out [the latter term is not entirely clear]' opens this sentence and is crossed out.

[85]'is' is supplied by Eds.

[86]'how' is written after 'out' and crossed out.

[87]Very probable reading of the term.

[88]'you' is written first and crossed out.

We can say, "We do not know definitions of words we use". We could have some talk about Moses – suppose someone says, "you have not got a definition, you do not know what you are talking about."

You might[89] take a definition, and then change it. But you would not change it to suit Napoleon or Bismarck; you would not change it arbitrarily but in a certain range.

Supposing we say this: –

Take the word chickenpox. This is an old word – it is much older than any knowledge of a bacillus. This clearly shows it was used in a different way.

Suppose there were several different rashes which *were* called chickenpox caused by several different bacillus.[90] Suppose doctors want to use different words. They may fix chickenpox for one particular one. But they do not change it outside a certain region.

Take the case of units of length – people a few thousand years ago measured length by the stride, ell,[91] etc. We[92] also have units of length of foot but here[93] we have changed[94] both the kind and the definition.

There is what one might call scientific definition which has a peculiar character.

We talk of one colour being lighter than another.

Suppose I said which is the lighter, this yellow or this green?

You may not know. Suppose science gives a definition. I define – take a photograph in a particular light, etc. If on the photo one is lighter than another then we say that one is the lighter.

There is a very characteristic thing – the scientific definition I.E. of chickenpox etc. One characteristic of these definitions is the use of numbers I.E. as in the talk that science reduces quality to quantity.

Mon. Jan. 29th

I gave you an example which was roughly this:

I said that one had always heard that people said that it was irrelevant to ask, "How do we get to know" to find what it meant 'to know'.

If you ask what is the movement of a body and I say let's find out how[95] we get to know a body moves, people say this is a[96] question of psychology and there might be beings who knew it without getting to know it.

---

[89] After 'might' it looks as if 'chan' for 'change' is started, then crossed out; and 'you' is written and crossed out after the word 'take'.

[90] *Sic*: bacilli.

[91] Old Germanic unit of measurement. Equals the length of a man's arm from elbow to tip of middle finger, i.e. cubit.

[92] 'have' is written and crossed out after 'We'.

[93] 'there' is written and crossed out after 'but'.

[94] 'changed' is written above the line and marked for insertion.

[95] After 'how' 'a body' is written and crossed out.

[96] Indefinite article is inserted by Eds.

To say how you can get to know somebody is in the other room is to give part of the grammar of knowing somebody is in the other room.

If I say after defining death as the stopping of the heart

"Is 'that the hair turns white' a criterion of death" then the answer is that here I give[97] an experience.

Take the two questions: How does one know?
How can one know?

One[98] asks, "How does one know?" meaning the symptoms.

When one asks, "How can one know?" this is asking what we call knowing. . . .

Suppose one says, "Do all[99] existential theorems construct that entity whose existence they prove?" I will say I can't prophesy anything: the word 'existential theorem' may be used[100] in another way.

There are two questions: "Do all etc.?"
And what philosophers ask: "Surely all etc.?"

The point was this: – is the question, "How can one know this?" a question of psychology or not. Or the question "How do you know":

That there is a chair here can be got to know in two ways:

I see it.

I shut my eyes and grope and find it.

That there is a chair here is a fact and one man gets to know it by seeing it, another by touching it, etc.

We say: "here is a fact: lots of people get to know it in different ways,[101] but this is irrelevant."

Why do we say these two experiences teach us the same fact? We can perfectly well say this, but that we are prepared to say this is a question we *fix*.

It seems as though the meaning of the words "there is a chair here" has been revealed to us by a revelation and then the question how can we get to know this by touch or sight.

<u>Example</u>
"There is a cylinder on the table."

---

[97]'here I give' is written over the line with marks to indicate to replace 'it' crossed out below this phrase.

[98]'One asks' is written after the first word 'If', which is crossed out.

[99]'all' is written above the line with instructions to insert it.

[100]'may be used' is written the same way twice, with first mention of it is crossed out.

[101]Singular 'way' in text is here pluralised by Eds.

I say: I get to know it by seeing it.
He gets to know it by touching it.

Am I compelled to use this form of expression?

These experiences have certain coordinations in experiences, I.E. I touch and see.

The form in which I presented it was "You and I got to know the same thing in two ways." Is this the form in which I have to present it? Could I not say?

"I have a visual experience"" "You a tactile experience".

You can imagine having the visual experience and no tactile experience, or have the tactile experience of a cylinder no-one can see.

What is said is alright; but it is just one of many forms of expression.

Let's examine the way you get to know the statement of the thing is a grammatical statement.

We will adopt this way of speaking:
"I specify two ways of getting to know tactile and visual".

I ask, "Could there be an experience of smell which told us the same thing".

Think of people only gifted with the sense of smell – we will suppose an object smells differently by the amount of air that can come to it.

Suppose I was a teacher – I would tell him this and then he would also get to know this.

We've made up this – we've made a statement and given a certain grammar to him for it.

We say, "That a cylinder stands on the table can be known in three ways" touch, sight, smell.

If I ask Mrs. Braithwaite how she got to know there was a cylinder on the table and she said sight, this would not be a grammatical question.

But if I ask, "How do you get to know," the answer to this question is a grammatical one.

Let's examine this case: there were experimental proofs of certain things happening if there was a cylinder on the Table I.E. smoke comes up.

I conclude there is a cylinder on the table. This is not part of the definition. When you who only had a sense of smell were taught this, you were taught directly, you did not get to know indirectly.

We said there were cases where there was a visual cylinder to which corresponded no tactile cylinder and vice versa.

Facts don't compel us to use Euclidean geometry in one case and non-Euclidean in another[102]: but they *suggest* it. They make it practical to use[103] one particular convention.

---

[102]'in another' is written over a deleted word, which seems to be 'not'.

[103]'to' is written after 'use' and crossed out.

Certain experiences may make me use certain conventions, but they do not compel me to use them I.E. I would not teach a man who smelt to call something a cylinder on the table unless I knew that this was coordinated by some experiences to seeing the cylinder.

Let's discard everything you might call direct criteria I.E. sight, touch. We only suppose it is there.

A[104] dog dies when there is a cylinder on the table. This is a hypothesis. A dog dying has nothing to do with the cylinder being on the table.

If you then say, "I never know it, I always suppose it," what does the word suppose mean? What is it opposed to? We use the words 'know' 'suppose',[105] and in certain cases we use them as opposites. Suppose I say, "I can never know this". What can this mean? I suppose you mean you never use the word 'know' in this case.

We can of course adopt this way of expressing: we never know a cylinder is on the table, we only suppose it.

Here we mean we are going to exclude the word 'know' and we are not going to oppose the word 'suppose' to anything.

If we wanted to make a convention about 'know the cylinder' we could make all sorts.

Suppose we say we know the cylinder is there if we see it. If we just see smoke we say we suppose[106] it to be there.

There is one characteristic in general about the use of the word knowing[107]: "you know it" is compatible with it not being the case I.E. my bed maker says she knows there are six[108] handkerchiefs in the laundry and there are five.

If we say this is a contradiction, then we mean we are not allowed to use the word know about future events but must use the word suppose. The word 'suppose' when used of an hypothesis is not opposed to knowing, and not opposed *by our convention*.

Suppose I say "we can never know, but only suppose an hypothesis", this is a statement of grammar.

Let[109] me use two words:

direct criterion

---

[104]'We sa' is written and crossed out before the start of this sentence.

[105]Single quotes (to match subsequent usage) are inserted by Eds.

[106]'know' was written first, then crossed out in favour of 'suppose'.

[107]'the way' is written after the colon and crossed out.

[108]After 'six' the abbreviation 'h. ch. s' is crossed out and 'handkerchiefs' is written over the line, with marks for insertion.

[109]Before the word 'Let', two lines are crossed out: the first is: 'We mean', and the second is written below the first, it is: 'the quest'.

indirect criterion

We call hearing a man in the next room and call this an indirect criterion and seeing him a direct criterion. We can use these words[110] in this way. If you go on and say we can only have indirect criteria, then you are not opposing indirect to anything any longer. You may point[111] out that there is nothing which categorically distinguishes direct and indirect criterion as we were using these expressions. It may be useful to do this.

There is an indirect criterion for another person having toothache. But you are clear there is no direct criterion. Therefore in a sense the word is too much.

The word "indirect" is *too much.* I haven't found anything that I call direct criterion.

Perhaps you will say "It is inconceivable I should directly experience your toothache" then we are making a grammatical statement.

Suppose you say "I[112] can have indirect criteria for my[113] toothache *surely*." But the one thing you probably would not allow would be an indirect criterion for your own toothache.

You can ask a question apparently: how do you know he has toothache when he holds his cheek? The answer could then be "Whenever I hold my cheek I have toothache." This answer would be coincidence of experience. If you then say, "Why do you say when he holds his cheek, he has toothache" then it comes to an end. We are getting only indirect criteria,[114] behaviouristic and psychological.

Wednesday, Jan. 31st[115]

Language as we use it is hardly ever used as such a calculus.

We don't think of the rules when we use it, but apart from that even when we are asked for rules, say a definition, we are unable to give it I.E. it has never been given. But in some cases we might give one which corresponds to the actual usage of the word. But this is an exception. Then you might ask why do we compare language to a calculus with fixed rules. The answer is that just those puzzles with which we have

---

[110]'these words' replaces the term 'them'.

[111]'mean' is written before 'point' and crossed out. May be this was Wittgenstein's oral slip self-corrected as he was speaking in view of *deuten/Bedeutung*.

[112]'I' is written over 'You' and the latter is crossed out.

[113]'you' is written first and crossed out in favour of 'my'.

[114]After 'criteria', 'such' is written and crossed out.

[115]The dates of entries are written and outlined by a boxed or half-boxed ink line to the upper edge of the page, on the right of the page next to the ringed page number. In the above case, the initial entry written is the "Friday, Jan. 26th"; it is crossed out and a separate box and date are placed to the left of it.

to do come from this point of view I.E. we are always referring to one particular kind of puzzle. The puzzlement which is the reason why we do this job is connected with this aspect *qua* a game proceeding according to fixed rules.

Take this case when we ask, "What is time?" (As with[116] St Augustine and others). One thing is clear that[117] what he wants is a definition (this is first approximation). This you will then say is unsatisfactory because it only brings him back to either one or several indefinable terms so even if he gets it, it will be no good.

The puzzle will be solved by something like a definition.[118] If we look at what he says, we will see he compares it to something like an ethereal substance and wants to know what it is made of. He compares it to measuring a length on a ribbon passing us.[119] You couldn't measure it as you do with a measuring rod if you only see a bit of it. The answer is that measuring time is quite a different thing I.E. a watch.

St Augustine knew precisely how it was measured, but at[120] the same time he was puzzled by it.

'Measuring time' is used in all sorts of ways, i.e. by feelings, etc. What is characteristic is this: St Augustine when he is troubled about time obviously treats it as though it had just one usage.

The puzzle is one due to presupposing that time is used in one strict way and then confusing it with other ways.

Take the case of Socrates saying, "What is knowledge". First he shows the pupils that there is a definition of surds; then he asks for a definition of knowledge. No definition is given.

Why does he ask for a definition of knowledge and not of meat? The answer seems to be that everyone knows what meat is, but not what knowledge is. This is very queer as we use the words 'time', 'knowledge' etc., every day and if we don't know what they are we would be talking through our hats.

He compares knowledge to the use of the word Surd. We would have to answer him by giving him several more or less exact uses which would[121] correspond to the actual uses of the word. This puzzle is connected with giving an exact grammar to knowledge and that is why we have to talk about exact grammars. We might express it: the man who is troubled about philosophical problems tries to give a law or *find the law* of the use of this word. Or the man who is puzzled about philosophy thinks he *sees* a law: either then finds where it won't do and is puzzled; or it won't do at all as in St Augustine's case where he could find no law being forced to look constantly on the analogous case of measuring space.

[116]'with' is inserted by Eds.

[117]After 'that', the manuscript has the incomplete: 'wants he w', which is crossed out.

[118]After 'definition', 'but this' is written and crossed out.

[119]'floating along on a band', which is replaced by: 'to measuring...passing us'. Cf. "ribbon of life", e.g. in Floyd (2016b: 68ff.)

[120]'at' is added by Eds.

[121]'act' is written and crossed out.

The way the puzzle works is this. First someone says, "What is time?" The answer is "the motion of the celestial body" as in St Augustine's time. This is a definition. St Augustine says, "This is not time". This should mean that the word time is not used synonymously to this phrase.

Secondly a wrong definition is given. Thirdly someone says this is not the right definition.

This is like the question "What's number?" Then we have the answer: it is "the scratch on the piece of paper." Then the third person says this is not three, meaning the word three is not used in the same way. Then he says, "What is three?"[122] Thinking another definition must now be given.

We do ask for the grammar of the word wherever we are puzzled by it. We are interested to know how the grammar of the word "three" *obviously* differs from the grammar of the sign. We do want to know this.

The philosophical puzzlement arises in this way. You think you see the law or way the word is used and are puzzled by it as soon as you state it. You may say if you philosophize you are preoccupied with language as a game with fixed rules.

What philosophy is has been expressed by Augustine when he says of time[123]

If you ask me I don't[124] know
If you don't ask me I know.

You couldn't say this of any problem of physics. From what he says it would seem as though philosophy were a making conscious of a thing: in a sense he says he knows but can't say it. What he means, he uses the word when he says he "knows".

You might say why not tabulate the rules.

You might write down the rules of the use of "measurement of time". But we are fascinated by the words.

There is an enormous fascination.

<u>Friday, Feb. 2nd</u>

We talked last time about philosophy more 'in general'.

There is one thing I wanted to say: –

I should like to make a comparison.

I should like to talk of the difference between the way Professor Moore does philosophy and the way I do it.

We constantly talk of meanings of words and give instances of different usages. So does Professor Moore.

[122] Quotations marks and question mark are supplied by Eds. on the analogy of usage in the previous sentence.

[123] 'says of time' is squeezed in under the line as an addition. There is a light mark after this phrase, which might indicate a full stop or hint of a dash written at speed.

[124] 'don't' is written over the line; after 'know' at the end of the first line a comma was written, then it is overwritten to insert a bold line break to separate the two lines as above.

The external difference is: –
though we constantly do this, we don't say the word is used in[125] three ways, say, but we *make up* our own ways. This is not a trivial difference. What I want you to remember is that a word when we use it, has the meaning we give it.

A word has not a meaning given to it from Heaven which we can investigate it. We look at the way we were taught the words I.E. as virtue, good. We have learnt it I.E. good is an extremely vague way, and that is the meaning. When we philosophise, we talk of exact meanings.

This can't possibly suit the complicated vague way we use it. The difference in the treatment I wanted to emphasize: –

We may investigate the way a word is used and find it has no exact meaning: to say all words have clear meanings is like saying the dirty patches have clear outlines if we knew them. This does not mean it is no use to give exact meanings.

I once said this was like saying the heat which the fire gives is no good because we don't know where it begins and ends.

If you make up an exact rule, I may make up another rule and show you that this would do just as well.

We are only dealing with troubles in the use of words, because you wrongly think you have got a rule, and then if you can't use it you are puzzled.

This is why we make up rules which are absolutely absurd – and deliberately in a way.[126]

We want to be clear it will not do to write so many different ways a word is used. There is a case we can do this as for the word "number" – here we make a list.

There is also a case where we can make a sort of ideal list – as in the case of games. We put down and say games *come near to* this or this etc.

If it is clear that we use some words in four or five strict uses as with[127] the number "one" and also some words explained by our parents in a hundred different ways.

Words like 'beautiful,' 'good' are like patches with extremely blurred outlines. You should be clear that it is possible to give the[128] different uses though it may take a hundred years. Other words we can describe the different uses in a very short time.

Suppose we give a definition that we say makes the meaning clearer. This will only be of use in very restricted case. We can leave out the question of vagueness altogether.

Take the case of the railway timetable. Here you will see vagueness does not come in at all. We were probably shown how to use it. We see when a train leaves London and arrives at Cambridge. Can the table only be used in parallel lines? Couldn't we use it in diagonal lines? Were we told that we mustn't? Mightn't there

[125] After 'in' the manuscript has 'a special', which is crossed out.

[126] Dash is inserted by Eds. to indicate a break for emphasis, since the phrase after it is written under end of the line ending 'absurd'.

[127] 'with', and the indefinite article later in this sentence, is added by Eds.

[128] 'all' is written in the main text, crossed out, and replaced by 'the' over the line.

be a case where we had to explain this? It doesn't make it more exact, but it *resolves a doubt*.

If you say the ostensive definition of red[129] is for normal eyes, then this is only adding something which resolves a doubt which might arise.

A clear-cut definition does[130] not make your use of the word clearer, but it may explain something about how it is used.

It teaches you the 'language'.

The definition is not meant to make the word clearer, but it gives you the same vague patch.

I want to talk about the difficulty in philosophy due to the fascination of certain words. If you say this to some people, they will say how queer! Philosophy must be a sort of psychoanalytic treatment. This is right in a way.

I have said before that philosophy is curing a disease. People will then say, "philosophy is not giving you anything new, it is just destroying a growth". I will say this is quite true: it really does do this.

Hardy proposed to write a book "mathematics for philosophers." This is like saying[131] the furniture to dust is for the chambermaid. A mathematician can "dust" his own theories.

He, the[132] philosopher, only needs examples. He doesn't need any bricks to build a house with.

It is quite different to say six feet is longer than five feet and to say this chair is longer than that. We shall see what an enormous amount of confusion this has caused. Twelve people are more than six people is a tautology. There are more people in the next room than in this room is a matter of experiment.

<u>Monday, Feb. 5th</u>

I gave you as an example of the misleading use of words the word 'philosophy' as I use it.

Whenever senses of a word are akin but widely different, there is an enormous difficulty. One case in which this is so is the word 'meaning'. If I use it for a strict grammatical explanation of the word, this is not the only use. But it is one definite

[129]What appears to be a comma has been crossed out with two short lines; a less probable view is that it is that it is a small quickly written '=' sign.

[130]'may' is written in the main text, crossed out, and replaced by 'does' above the line.

[131](After the definite article the manuscript has the incomplete 'furnishin', which is crossed out.) As Jonathan Smith has pointed out to me, the unusual word-order involving 'to dust' may be due partly to Wittgenstein's wish to parallel 'mathematics' with 'furniture' and 'philosophers' with 'chambermaids'.

[132]After the definite article, and before the comma, a word is written, though the last letter is not finished or not clear: if it is 'y', the word is 'only'.

use and it is healthy to stick to it. In some cases it is good to introduce, say, five different words. But this is misleading as it tempts us to say that a word has seventeen different meanings whereas this is not at all the case.

When you use a word in a sentence in a certain context, you get an idea of the word. You don't go by a definition say when you use the word plant. Biologists may make a definition, but you certainly don't have this in your mind nor will another 'definition' do.

Suppose Miss Ambrose is troubled by some philosophical problem, can I remove her trouble in spite of herself? Suppose someone says I have never removed his troubles. Is this something damning? (If you wanted an addition done, I could do your addition for you and remove your trouble.)

I don't think this is any argument against what we do.[133] It is only an argument against what I do as far as the particular person's trouble is concerned. I may not be able to remove his trouble. The trouble shows in that he can't ask a question – it consists in not being able to ask a significant question. If like a doctor who describes the symptoms of an illness, I would say he can't ask a significant question but *does* ask a question. When the trouble is removed, the question has not been answered but he ceases to ask it. What must you give a person that he no longer[134] has the temptation to ask such a question?[135]

I may not give the right analogy.

I may give the right analogy but the person may not be able to overlook or survey what I give him.

I haven't succeeded, but it does not mean what I have done is absurd.

I may give lots of descriptions of meaning, but the person may not have it ready.

How far does the thing I do compel you to remove your doubts? It just can't.

Sometimes the thing I do will work, and sometimes it won't work. Can I bring it about that you, whenever you are in a new trouble, can remove it yourself?

This depends partly on the imagination. What does for a trouble I have depended on for me to get rid of it?

It may be that the right picture comes into my mind.

If I can't answer it, it may happen that I shall never be able to answer.

The more pictures you have had, the less likely you are to be in a hopeless muddle.

I can only make it more likely.

The[136] sort of question which arises now may arise anytime again. It will arise as long as you are caught in your form of expression.

The puzzle is a sort of misunderstanding and the word which solves it may never be found.

---

[133]What appears to be 'But' is written to start the next sentence and deleted.

[134]After 'longer' there is written, 'asks these questions', which is crossed out.

[135]'to ask such a question?' is inserted below the preceding phrase at the end of the line.

[136]Before this word 'The', the manuscript has 'Could One say more', which is crossed out; 'One' is written with a capital 'O'.

Wednesday, Feb. 7th

We[137] reverted to the sense of "wish" "hope" "think" "expect" 'believe" where we are tempted to say, "I know what I wish" etc.

"How can you wish for anything to happen which may never happen?"

This is the way the difficulty may [be] expressed.

"Can I think a fact which does not happen?"

The idea is that I have a certain relation to a fact I.E. that he is dead. How can I wish or think a fact that isn't there?

It may be pointed out that there is not a single word after I wish but an[138] expression beginning "that".

People ask what is the object of the wish.

The answer given is that it is a "proposition".

The next question is "What is a proposition, I.E. King's College is on fire?"[139] Isn't this like hanging a thief who doesn't exist if we say we think a proposition?

Most people will talk about the shadow of a fact I.E. say "King's College is on fire" is a shadow of a fact.

Suppose[140] the shadow is a picture: I wish that Mrs. Braithwaite is dead. Suppose my object is a picture, then I can stab the picture.

How is this a picture of Mrs. Braithwaite?

Likeness is one way of comparing a picture with its object. Could I not project Mrs. Braithwaite on a sheet?

We might then say anything is a picture provided we know the method of projection.

You now say, "give both the picture and the method of projection".

Suppose the method of projection is given by words.

Take the old case of the railway timetable.

A man does not know how to read. We give an explanation by arrows. But how are we to explain this?

The map and the legend are a symbol and its explanation.

The legend says towns are like this, churches like this etc.

The legend is just an additional sign on the paper.[141] It has the form of an explanation.

The explanation always adds to the sign. It is just another sign.

We said the wish dealt with a picture, not Mrs. Braithwaite.

How do we know this is a picture?

---

[137]Before 'We', 'Who' is written and crossed out.

[138]Immediately after the ending of this word 'an', 'se' is written and crossed out.

[139]In a different context and distinct argument, Wittgenstein employs a variant form of this proposition in the *Blue Book* (p.39); it is instructive to assess the differences of context.

[140]Before 'Suppose', 'But a' is written and crossed out.

[141]'on the paper' is inserted above the line.

First you say	The wish has for object the fact.
Objection	How can it have the fact if the fact does not exist?
Way out	The object is a proposition which is a shadow of the fact.
Objection	How do we know it is the shadow of the fact?

What relation has the shadow to what it is a shadow of?

The shadow is a portrait. How do we know what it is a portrait of?

The usual way is similarity. A man may paint a portrait of me which looks more like my brother. Who is the portrait now of?

There are all sorts of things which make[142] a portrait mine. But there are standard cases.

If a painter paints a portrait of my brother from memory and he uses me as a model to assist his memory.[143] Then he may look at me while he paints.

If we put it in this way, we can say the wish and the thought are concerned with pictures. You could draw a picture of King's College on fire. How would you know it was King's College?

We can now say "I paint King's College on fire". We can ask all the same questions.

How do you know it is King's College? May not there be such a building somewhere else? etc.

Suppose I go a step further. I take another brother of, "I wish that so-and-so had happened". "I say that so-and-so had happened."

We take "I say King's College is on fire" which is equivalent to "King's College is on fire."

Can anything be more dissimilar than my saying certain words and the fact that King's College is on fire? What is the connection between the sounds and the building is on fire?

You ask me what is King's College and I point. You ask what is fire and I point again.

This is the method of projection – the method of projection is that we explain the words in this way.

If I call something a picture in the ordinary sense in which we talk of similarity, this is one use of the word picture.

If I show you a map, I don't say how similar this picture is to England.

'Similar' is a word which has one special use.

We call a good correct picture any amount of different things.

We call a good shadow all sorts of different things.

[142] After 'make', the phrase 'of it' occurs.

[143] 'to assist his memory' is inserted above the line.

You might say the painted picture of King's College on fire is more similar to King's College than the sentence.[144]

It you want a shadow, take the sentence. It may be all sorts of things, I.E. imagine King's College on fire. I suppose you see an image of K.C. on fire. Can't I ask you? "Isn't what you see another building? How do you know what it is you imagine?"

The answer is "I imagine *King's College* to be on fire."

If I ask you what is the relation, couldn't the relation just be that when I ask her[145] this, she said these words.

The[146] form of expression "I imagine so-and-so is the case" makes us think all these things must be present.

If I ask you "When was Napoleon[147] crowned?" you answer "1804". I say, "Did you mean the man who won the battle of Austerlitz". If you say "yes", do you mean it was present in your mind when you said it?

It is this using the past tense which is so queer.

Suppose[148] I ask you "Did you have biscuits at tea today". Here you remember it.

In the case of Napoleon, you do not remember it. Here you go and give a definition.

It is a good rule of thumb whenever you believe a thing etc. to replace this by saying the thing.

Accustom yourself to putting "saying" in place of thinking believing etc.

The troubles which we are concerned with are all in a sense trivial. One thing that led to these troubles is the expression in our language. I expect *him* and I expect that so-and-so will happen.

One thing that leads to the muddle is the word 'fact" another "object of fact' a third "I think *so-and-so*".

We could write a book on philosophy which was simply a collection of riddles. You would have to find the answers but the author would have to know them.

The confusion might arise "Can I imagine something which is not the case". "Can't I imagine[149] a centaur?" Yes, a centaur is something like a horse. You might

[144]The expression 'But the "words" are also on fire' occurs after the end of this sentence, at the beginning of the next line.

[145]'Her' rather than 'Him' may refer back to Mrs. Braithwaite, or Miss Ambrose, or Miss MacDonald. Note that after 'she' in the sentence there is a partially formed word deleted after it, perhaps 'sha'.

[146]Before 'I', a word is deleted, which seems to be 'We'.

[147]Note use of 'Napoleon' in the Rhees edition of the *Blue Book* p.39, yet different usage.

[148]'you' is written after 'Suppose' and crossed out.

[149]'imagine is written twice; one occurrence is crossed out.

be inclined to think that the first elements of what you imagine must exist.[150] You imagine the elements in a non-existent combination.

Here are two books – I can imagine them in a non-existent combination. Now take the elements. We say these must exist. We can't say of redness "it exists" because it has no sense to say it does not exist. Or "roundness", etc. We can say[151] a round object exists or does not exist.[152]

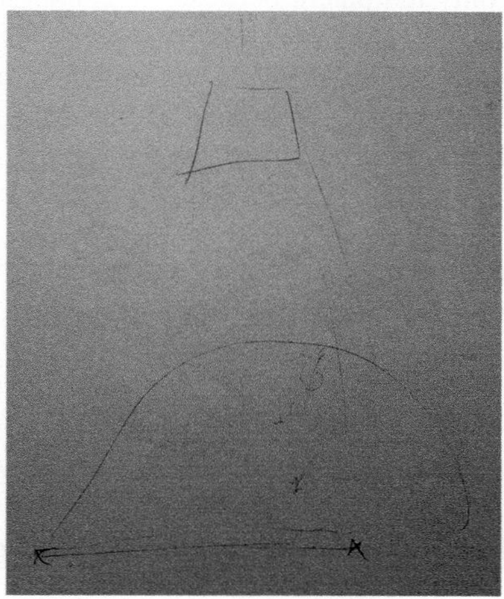

Suppose we said this: in order that anything should be a foot long, there should be another a foot long. I say, "This is quite alright. Mustn't there be a standard foot?" I seem to have said there must be two objects a foot long.

The objects are the measured object and the standard. But can I say the standard is a foot long? I cannot prove there are two objects of this length, as I could destroy one and the other would remain. One object was part of symbolism.

---

[150] Written 'exists' in the manuscript.

[151] 'say' is written over the line with marks for insertion, while before the beginning of the sentence 'We can say' is written out clearly and crossed out.

[152] On the back of the manuscript page there is a diagram, which is here reproduced. It appears to be a geometrical sketch of some relations of an object to its symbolism, presupposed as a standard, connected to a (mathematical notion of) a shadow, as extended over the next few paragraphs. This cluster seems intended as a puzzle to trigger exploration of these abstract mathematical relations in engagement with the language games here being explored. This theme is presented in the Pink Book (Chapter 3 above) and elsewhere in the present book. Compare and contrast the diagram above with the one on p. 233, under the date heading of Feb. 14th Wednesday. There the process of intentionally copying and projection is considered.

Supposing I said, there can't be only one red object in the world; there must be two red objects.[153]

Suppose I define red by pointing to a red patch.

Imagine[154] a Greenwich[155] red patch. I say there can't be only one red patch because the Greenwich patch is red. But the Greenwich patch is not red. This has the colour of the Greenwich patch.

If I define "red" by a[156] standard red patch, you[157] needn't[158] say what you imagine cannot be the only red patch, because we cannot say the Greenwich patch is red.

This does not apply unless you imply in our language there is a Greenwich red.

Can you use the word "one"[159] if it only applies to one case? Russell defines triplets by a Greenwich triplet.

Friday, Feb. 9th[160]

To say that the object of wish, expectation, thought is the shadow of the fact, there is the same problem. How can the shadow of the fact exist if the fact does not exist? We repeat the same question.

This is one of the many cases where an argument satisfies you because it goes one step.

We are satisfied with this motion of going one step back, or with doing something like it.

A reason may satisfy us. It is like stepping off the ladder on the ground or going one rung lower.

One of the most striking examples of this is using the phrase "it makes it probable."

Certain definite answers are possible, and then when the question "Why that?" the answer is, it makes it probable. The further question, "Why that?" seems somehow lost.

Whenever people have tried to justify induction, they have made this mistake.

---

[153]Objects' in the manuscript is written in the singular.

[154]Before 'Imagine', 'This is' is written and crossed out.

[155]Note that Wittgenstein discusses 'red patch' in the opening section of the *Blue Book*, though he does not use the 'Greenwich'. (N. O'Mahony notes 'Greenwich red' denotes the colour of the Time Ball on Flamsteed House, Greenwich, used since 1833.) He also uses 'red patch' in *On Certainty*, Section 53. This discussion appears to regress to his engagement with Russell's 1918 *The Philosophy of Logical Atomism*.

[156]'The' is written and crossed out in favour of 'a'.

[157]'if' is written above 'you' and yet is crossed out.

[158]'needn't' is written over the line above 'can't' and the latter is crossed out.

[159]'"one"' is inserted over the line with insertion marks.

[160]Note that this is the first occurrence of two dates that are the same – 'Feb.9th'. This set is paginated (1) to (6); the set after the present series, also with the same date, are paginated from (26) to (31). I think that this is evidence of re-drafting of the original notes, under Wittgenstein's guidance.

There is this tendency of providing an answer I.E. as saying the object of the thought is the shadow of the fact, this is the same thing: as the problem between shadow and fact remains the same as between thought and fact.

One of the facts that tempts us to say this is that a mental picture is really interpolated between the words and the fact. But we are where we were before. But I do not mean to say there is not such a shadow sometimes which mediates between the sentence and the outside object.

If in a chain I asked how is this link connected with that, there are two kinds of answer:

(1) These further links.
(2) But if I ask what connects two links, the answer is quite a different one – one goes through the other or is soldered to the other.

We are tempted when we are asked what is the connection to give a link, when this is not at all necessary.

The thing I ought to make clear to you is that all these questions do not help us at all. The whole trouble is that you think we are analysing this, I.E. "I wish that so-and-so would come". We will not get one hair-breadth nearer in the end, only this question will not shock us.

One could express this scruple about wishing a fact in a different way.

One could say the trouble arises through an ambiguous conception we have of a proposition being about something or containing something.

Suppose I said, "I saw the Judge and the Master of Trinity cross the Great Court today". This assertion contains the two assertions "I saw the J. etc."

If I say: "it isn't true I saw the Master of Trinity cross the Great Court today" does this contain the proposition.

Suppose I say the meaning of the proposition depends where I put the full stops.

Do we say the proposition contains the assertion?

We say the proposition "King's College is on fire" does not contain the assertion. But the two propositions connected by 'and' do contain the proposition.

Everything I have said could also be said about negation.

What does the negation negate? If it negates the fact, then the negation is false. People then say it negates the proposition.

Now on the other hand you might also assert that[161] the idea of the thought as having the object a shadow is alright if by shadow you mean[162] the picture of the thought. But then you could use the sentence.

Suppose you imagine a tribe which cannot speak and write but uses a pictorial writing.

[161] 'assert that' added by Eds.

[162] After 'mean' the incomplete expression 'the mea' occurs and is crossed out.

Suppose the language is used for leaders to make themselves understood to people whom they lead in battle. Why[163] shouldn't a plan of battle be transmitted[164] by a picture? Why shouldn't the order be given by a kind of hieroglyphics, I.E. a kind of picture writing? It would be much easier to call this the shadow and not suppose there is anything between. One looks at a picture of a man drawing a bow without supplementing the picture *qua* man by shapes and colour.

Suppose somebody shows you a photo and says that in order to appreciate the photo you must have a picture of the man given by the photo.

At the cinema, you don't miss the colours – you think in the medium of black and white, you don't imagine anything.

In reading a novel, you sometimes imagine things here and there but you don't imagine every word.

Someone once asked this question. A philosopher in the eighteenth century asked this question: could I ever[165] try to dance like this person. When he tried, it would be a totally different thing that happened.

Supposing we said that the fact this is the portrait of that consists in the picture having been derived from the face in a particular way.

Take this for the moment in the case of a calculation.

Supposing I said, multiply 25×25, write down the result.

If I say I portray Mrs. Braithwaite,[166] I copy this ellipse or I calculate $25 \times 25 = 625$. I could have said portray 25×25 in this particular way.

What it means to multiply 25×25 and to *mean* to multiply it.

Let's first say I write down the multiplication.

Then I say he made the portrait. But if he did not write it down and wrote it down wrongly, he did not make the portrait.

I write down 25×25. There is the normal way of multiplying it correctly

$$\begin{array}{l} \underline{25 \times 25} \\ 50 \\ \underline{125} \\ 625 \end{array}$$

Suppose we said, "So he intended to multiply it". I provide a case where he multiplied wrongly. He writes down the lines but the digits wrongly.

---

[163]'One' is written and crossed before 'Why'.

[164]'translated' is written before 'transmitted' and deleted in favour of the latter.

[165]'ever' is written over the line with marks for insertion.

[166]The MS read "Mrs. Br."

We will suppose he multiplies along $2 \times 5 = 60$ put down 1 and carry 9. This we call intending to multiply. If he said anything else we will say he intended something different. Supposing now I went a step nearer to common sense. We say if he just writes down, we don't know if he intended to do[167] so, he may have written it down by chance. We want now another criterion for intention.

I have replaced for the mental process of intention another process which goes on[168] paper, which doesn't matter a damn.

Whether this is just a matter of chance that he wrote it down depends on the process he went through. Whether he intended this or not depends on the process of deriving this from this. But why not call what he does with his hand and his mouth the process of deriving?

We can take *this case* where he did nothing else but spoke and scribbled.

You are inclined to say that[169] there is one mental process of deriving, which is the mental intention.

If you say he wrote this down but intended something else you will have to show how[170] this other intention shows itself. You might say it isn't enough that he should have said and written, then I say "just you say what is enough – what is enough in your opinion".

How do we know how[171] what he says hangs with what he wrote – then you will try to give an intermediate link.

He may point at what he says, say "2" when he says two, but how do you know he points at what he says?

Whatever process you give has the same trouble. You try to give a link that ends this with "intention".

This difficulty[172] could be interpreted by saying "every expression could be[173] a lie." But if we say "the wish" is the real thing, what is the wish? Could the wish itself be false? The difficulty is in what the process of lying is.

Suppose the process of lying consists in showing something on a piece of paper to a person and writing the opposite on the other side. This would then be the *process* of lying. There would be no further question. "Is this what he really meant?"

---

[167] 'do' is missing from the MS.

[168] 'on on' is written in MS.

[169] 'to say that' added by Eds.

[170] 'what' is written and crossed out on the line before 'how'.

[171] 'what' is written before 'how'; the latter term is written over the line.

[172] 'difficulty' is written over the line.

[173] 'is' is written in the main text and is crossed out in favour of 'could be' written over 's'.

Friday, Feb. 9th[174]

But this doesn't really remove our difficulty: for the question now is "how can something be the shadow of a fact which doesn't exist?"

I can express our trouble differently by saying, "How can[175] we know what the shadow is a shadow of?" The shadow would be some sort of portrait and therefore I can restate our problem by asking, "[176] What makes a portrait a portrait of Mr. M?" The answer which suggests itself first is "the similarity between the portrait and the man". And this answer in fact shows what we had in mind when we talked of the shadow of a fact. It is quite clear, however, that similarity does not constitute our idea of a portrait for it is in the essence of this idea that it should make sense to talk of a good and a bad portrait. In other words,[177] it is essential that the shadow should be capable of[178] representing things as in fact they are not. The obvious and correct answer to the question "What makes a portrait a portrait of so-and-so?" is that it is the *intention*. But if we wish to know what it means, "intending this to be a portrait of so-and-so", let's see what actually happens when we intend this. Remember what we said happened when 'we expect someone from five to five-thirty'. To[179] intend a picture to be the portrait of so-and-so[180] (on the part of the painter) is neither one particular state of mind nor a particular mental process, but there are a great many different combinations of actions and states of mind which we should call "intending . . .".[181]

It might have been that he was told to paint a portrait of 'N' and sat down before N going through certain actions which we call "copying N's face". One might object to

[174]As mentioned in a previous footnote under the heading of Feb 9th: 'Feb. 9th' on this page is the second occurrence of the same date. The previous set of pages is paginated (1) to (6). The present series of pages, also with the same date, are paginated in the manuscript from (26) to (31). This sequence includes the dates and writing for 'Monday, Feb. 12th' (pp. 28–31), and 'Feb 14th Wednesday' (p. 32). This is perhaps indicative of Skinner re-organising a previous copy under Wittgenstein's guidance. Wittgenstein's pencil hand: "Notes taken by Skinner (Beginning of lecture)". Alice Nelson (of the Wren Library, Trinity) pointed out to me that the positions of the rusted holes for the original pin marks in the manuscript pages are identical in these pages, which indicates assuming, as is most likely, that they were placed there by Skinner or Wittgenstein. For further reasons concerning the retention of this sequence of MS pages 26–32, see the first footnote under the lecture heading marked "Feb 9th" below.

[175]'put a' is written after 'can' and crossed out.

[176]'how do we know' is written' and crossed out before 'What'.

[177]'That is to say' is written and crossed out before 'In other words'. The deleted phrase also appears to have been written with a small initial and not capital 'T', with a comma originally giving pause to the previous sentence, which was subsequently changed to two separate sentences.

[178]'be capable of' is written above the line with marks for insertion, as is also the case with the following phrase 'in fact'.

[179]The original opening of this sentence – 'For the painter' is crossed out, and the 't' of 'To' capitalized.

[180]The expression 'is neither one particular' is written, crossed out and repositioned after the parenthesis in the main text.

[181]Close quotation mark supplied by Eds.

this explanation and say that the essence of copying is the intention of copying. I should answer that there are a great many different processes which we call copying something. Take an instance. I draw an ellipse on a sheet of paper and ask you to copy it. What characterises the process of copying? For it is clear that it isn't the fact that you draw a similar ellipse. You might have tried to copy it and not have succeeded, whereas someone else drew an ellipse with a totally different intention and it happened to be like the one you should have copied. So what do you do when you *try* to copy the ellipse? Well, you look at it, draw something on a piece of paper, perhaps measure what you have drawn and curse if you find it doesn't agree with the model. Or perhaps you say, "I am going to copy this ellipse and just draw an ellipse like it". There are an endless variety of[182] actions and words which have a family likeness and which we call "trying to copy". Suppose we said, "that a picture is a portrait of a particular object consists in it being *derived* from that object[183] in a particular way." Now it is easy to describe what we should call "processes of deriving a picture from an object" (roughly speaking, processes of projection) but there is a peculiar difficulty about admitting that any such process is what we call intentional representation. For describe whatever process, activity[184] of projection we may, there is[185] a way[186] of reinterpreting this activity.[187]

<u>Monday, Feb. 12th</u>[188]

– There is a way of reinterpreting this activity.[189] Therefore, one is tempted to say: this can never be the intention itself. The intention could always have been the opposite by reinterpreting the process of projection which we originally described. Imagine this case: suppose we gave someone an order to walk in a certain direction by[190] drawing an arrow pointing in this direction. Suppose drawing arrows is the language in which generally we give such an order. Couldn't this order be interpreted to mean that the man who gets it is to walk in the[191] direction opposite to that of the arrow? This would obviously be done by adding to our arrow some symbols which we might call an "*interpretation*". It is easy to imagine a case in which, say, to deceive someone you might make an arrangement that an order should

[182]'cases' is written and crossed out after 'of'.

[183]'thing' is written and crossed out before 'object'.

[184]'activity is written above the line with marks for insertion.

[185]In the main text 'seems to be' is written, crossed out, and replaced by 'is' above the line. Furthermore, 'an interpretation of them' is written next after 'is' but in the main text and crossed out.

[186]'activity' is written in the main text before 'way' and crossed out.

[187]After 'activity', written in the main text is 'should not call intentional representation', which is crossed out.

[188]Note that this lecture, which commences on MS page 28, continues the unusual pagination – pp. 26–32.

[189]This opening sentence is repeated from the previous entry date's last sentence, with the dash.

[190]'pointing' is written and crossed out after 'by'.

[191]'opposite' is written and crossed out after 'the'.

be carried out in the sense opposite to its normal one. The symbol which adds the interpretation to our original arrow could, e.g., be another arrow.[192] Whenever we interpret a symbol in one way or another, the interpretation adds a new symbol to the old one. Now[193] we may say that whenever we[194] give someone an order by showing him an arrow, when we don't do it "automatically" we *mean* the arrow in one way or another, and this process of meaning of whatever kind it may be could be represented by another arrow (in the same or opposite sense to the first), in fact it would be an arrow of some kind or other (being capable of being translated into an arrow). In this picture which we make of "meaning and saying" it is essential that we should imagine two spheres in which the two processes of giving signs and meaning respectively take place. Is it then correct to say that no arrow could be the meaning as every arrow could be meant the opposite way? Suppose we write down the scheme of saying and meaning by a column of arrows one below the other

$$\begin{array}{c}\rightarrow\\ \leftarrow\\ \rightarrow\end{array}$$

Then if this scheme is to serve our purpose at all, it must be clear from it which level is the level of meaning. I can for example make a schema with three levels, the bottom level always being the level of meaning. As a matter of fact one can talk of two levels even only in rather exceptional cases. But adopt whatever model or schema you may, it will have a bottom level and there will be no such thing as an interpretation of the bottom arrow. To say in this case that every arrow can still be interpreted will only be to say that I *could* always make a model of[195] saying and meaning which has one more level than the last schema. What one wishes to say is "every sign is capable of interpretation but the meaning should not be capable of interpretation. It is the bottom interpretation". Now I assume that you take the meaning to be a process accompanying the saying and translatable, and therefore equivalent to a further sign. You have therefore further to tell me what is the distinguishing mark between an outward sign and the meaning. If you do so for example by saying that the meaning is the arrow which you imagine as opposed to any which you may draw or produce by a gesture of your hand, you thereby say that you will call no further arrow; however it may be produced as[196] an interpretation of the one which you have imagined. All this will get much clearer if we consider what it is that really happens if we say a thing and mean what we say. Let us ask ourselves

[192] After the full stop 'in fact' is written and crossed out.

[193] 'what' is written and crossed out after "now'.

[194] 'we' is originally written as 'he', and the verb form to match this was 'gives'.

[195] 'wha' is written and crossed out after 'of.

[196] Appears to have omitted to include 'as'.

"If we say to someone, 'I should be delighted to see you' and mean it,[197] does a conscious process run alongside these words, a process which itself could be translated into spoken words?"[198] This will hardly ever be the case but to make my meaning clear I must give you an instance in which it would happen. Suppose I had a habit of accompanying every English sentence I said aloud by inwardly saying[199] a German sentence to myself. If then for some reason you call the inwardly spoken sentence the meaning of the one spoken aloud, the process of meaning accompanying the process of saying would be one which could itself be translated into outward signs. Another example would be that *before* any sentence which we say aloud we said the meaning of it (whether it be the same or opposite) to ourselves in a kind of aside. An example at least similar to the case we want would be saying one thing and at the same time seeing a picture before our mind's eye which says the same or the opposite and is the meaning. Such cases and similar ones exist but[200] they are not at all what happens as a rule, when we say something[201] and mean it or something else. There are of course a few real cases in which what we call 'meaning' is a definite conscious process accompanying or preceding the verbal expression, and itself a verbal expression of sorts of translatable into one. A typical example of this is[202] the aside on the stage. But that which tempts us to think of the meaning of what we say, as being in all cases a process of the kind we have just described, is our normal form of expression:

{To *say* something.
{To *mean* something.

which seems to refer to two parallel processes. A process which accompanies our words and which could be called the process of meaning them is the tone of voice in which we speak the words or one of the many processes similar to the rise and fall of voice. These accompany the spoken words not like a German sentence could accompany an English sentence, or writing a sentence could accompany speaking a sentence but in the way a tune accompanies words. This tune corresponds to the "feeling" with which we say the sentence. And I wish to point out that this feeling is the expression with which the sentence is said or something similar to this expression.

[197]Close double quotations marks are written after 'it', though they appear to be have been crossed out, and re-introduced after the closing question mark.

[198]'In almost ca' is written and crossed out here.

[199]'inwardly saying' is written over the line with instructions for insertion, whilst the original choice in the main line, which is crossed out, is 'visualizing'.

[200]After 'but', the expression 'they only form a very small part' is written and crossed out.

[201]'something' is written over the line, with marks for insertion.

[202]After 'is', a short illegible word is written and deleted.

Feb 14th Wednesday[203]

Our question was "Is it correct to say that that which distinguishes the process of intentionally copying, projecting,[204] portraying an object from that of unintentionally[205] producing a picture which may look like the object? Is it correct to say that[206] what distinguishes these two processes is the way in which the picture was derived from the object? Is it then correct then to say that what[207] characterises[208] intentional representation are certain processes of looking, measuring, content and discontent,[209] etc.[210]

Monday, Feb. 12th

Whether something *is* a picture[211] of something else, or what is the same *is intended to be*, depends on how the picture was derived.

We saw there was a difficulty in this.

Example: – I give someone an order to go somewhere.

---

[203]This entry for Feb. 14th is crossed out, with widely spaced diagonal lines. It is retained here as main text in this Edition, along with the rest of the manuscript, partly because it was pinned and retained in the overall dated lectures' order (that is to say: "Friday, Feb. 9th" MS page 26, "Monday, Feb. 12th" MS page 28, and its own date of "Feb 14th Wednesday" MS page 32). This choice of order prevailed, even though the next lecture date, after this sequence of pages 26 to 32, returns to "Friday, Feb 9th" – also replete with the usual pagination, whereby each lecture commences with and each lecture uses the sequence of 'page 1 to *n*', as with the pattern of lectures prior to this sequence of pages 26 to 32. Common and linked themes occur in and so connect all these lectures before, during, and after this current sequence (paged 26 to 32). For example, one theme concerns 'the shadow of the fact' (for example, Feb. 7th: 'shadow of a fact' (MS page 1); Feb. 9th (MS page 26); Feb. 9th (i.e. the second sequence with this date, which is reproduced in the next lecture in this Edition, which has the normal pagination of page 1 to *n*'): 'the shadow of the fact' (MS page 1).

[204]After the word 'projecting' an extended clause is added. Prior to this addition, there are what seem to be two deleted attempts to write it. First, "as it is do", which in view of the following "do" would have been "done"; this is deleted by a double line. Secondly, and separately on the next line is the word "as", which is crossed out separately (i.e., double lines – continuous through all the thus-crossed out words are used with the other deletions, whilst "as" has separate double lines through it). Thirdly, the resulting clause subsequent to these emerges as: "for instance it is done when we portray a face from".

[205]'intentionally' is written over the line with instructions for insertion.

[206]After 'that', the phrase 'this consists' is written and crossed out – each word by separate lines. This contrasts with the continuous line that deletes a longer sentence-fragment that is written after this phrase, namely; "the distinction between these two processes".

[207]'that which' is on the main line, whereas it is crossed out and 'what' has been written over the line with marks for insertion.

[208]After characterises, 'the process of' is written and crossed out.

[209]'discontent' is previously written after the word 'measuring', crossed out, and added here.

[210]This line ends the section of the unusually paginated sequence, MS pages 26 to 32.

[211]See picture motif in Lecture Feb. 7th.

I can give it by saying a few words or by pointing and saying a few words. I can give the word say just by pointing. This is the sign. How is to be interpreted?

Couldn't it be interpreted to mean, "go in the opposite direction"?

We can easily imagine a case where we try to do this.

Say if we wished to deceive. But how will I interpret this after giving an explanation?

If I say, "Please go this way ←,[212]" what would actually happen? You may go the other way saying I want to go the opposite way.

If the order ← means go the opposite way. I can put ←
→.

The second arrow is thought to be a mental arrow which is *the* interpretation. But do you have any such arrow?[213]

If the meaning is an expression, it must be expressible.

It does not then matter whether it is in our[214] mind or on paper.

Suppose you say, "Every arrow can line.

This could be called the sign arrow, and you say every sign arrow can have a different meaning arrow underneath.

It is only in very exceptional cases where there are two arrows. If I am asked what is in a letter I have received, I may read out a false sentence. Here is one of the few cases where there are two signs present.

The primitive picture is this: –

You say something	(1) You either mean it
	(2) You mean something else, which is buried away somewhere else.
	(3) You mean nothing. Here your mind is a blank.

It was said of Bismarck that he had an extraordinarily subtle way of lying by always saying the absolute truth when the other person was sure he was going to lie.

---

[212] Arrow is written and deleted – pointing the opposite way from the arrow that remains in the text.

[213] After the question mark 'in your' is written and crossed out, though there would have been space for the question mark to be written subsequently to this phrase.

[214] 'in our' is written over the line.

In the symbolic process there is nothing which by its kind is the bottom.

What does it mean that the arrow is capable of interpretation? That you can add an explanation, which either goes against the arrow so to speak.

Take two squares

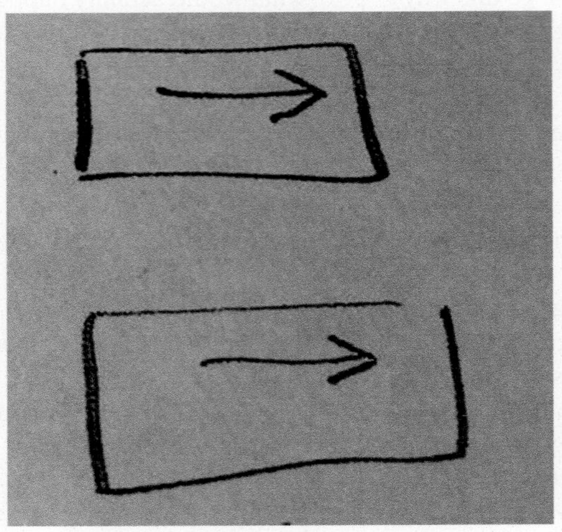

The top one contains the sign, the lower one the meaning. Here we have two containers. The bottom one could have the same arrow, the opposite or none at all. There could be three squares, but in this system there is no more interpretation. We can let it consist of ten boxes, or any. Whatever *you* decide, you must be ready to accept as the last thing.

The word "meaning" corresponds to a box, not to an arrow. You have to say that by "meaning" arrows you mean a 'kind' of arrow.

The experience of saying a thing in a particular *tone* of voice may be just what we call 'meaning a thing'.

If you mean by "meaning" something like a tone of voice – it may be the rise and fall. This would be something quite different. This would be the criterion for something being meant, and would not correspond to another spoken sentence.

Feb. 14th Wednesday[215]

[Wittgenstein writes:]

"Notes taken by Skinner (beginning of lecture)"[216]

---

[215]Namely, Ash Wednesday, 1934.

[216]This sentence in pencil is written in Wittgenstein's handwriting.

We said can one say that copying intentionally, making a portrait is deriving from a thing in a particular way? Can one say the characteristic of a portrait as opposed to a picture is this?

I said one could say this: – the difference between the case where I draw something which looks like so-and-so, and copy so-and-so's face but my drawing does not look like this, is characterised by the way it is derived. This characterised the intention, as the copy.

The objection to this was the intention could always be another thing and interpret whatever I did as a lie.

Could I not make someone believe I copied him by doing all these things and still did not copy him?

If the intention is anything at all, what is to prevent you from imagining something lying further back, unless you postulate it as something which lies furthest back.

When I cheated the man, doesn't it seem I have gone through all the things which I described as the characteristic of intending? But[217] in this particular case there is something further back.

Please describe what in the particular case your intention consisted in.

Could there be a third case where this was only done to cheat?

What does it consist in "to be angry" and "to pretend to be angry"?

Supposing, "to be angry" consisted in "frowning," etc.

But you say you could do all these things and 'pretend' to be angry.

I suppose by being angry you mean particular experiences.

Supposing someone said[218] being angry consists in frowning, etc.

One[219] answer might be "Oh this can't be anger, anger is something different."

Another better answer could be "these are part of the experiences of anger." You can add, "These are not the main experiences of anger".

It is easier for instance to make yourself[220] angry by knitting my brow[221] and doing other things. You work yourself up into anger. These are part of the experiences of anger and others follow.

At first you say these are quite different and anger is in a box with love, etc.

Suppose you read a piece of prose and read it with the right expression and feeling. Doesn't the right expression consist in just reading it as you do?

But suppose someone says you can cheat. But if you cheat, do you really[222] have just this experience.

---

[217] 'I say' is written and crossed out after 'But'.

[218] Note use of this expression in Lecture for January 17th.

[219] 'One' is capitalized.

[220] I.e. 'oneself' or 'myself'.

[221] 'knitting my brow' replaces the deleted 'frowning' in the manuscript.

[222] After 'really', 'read it in exactly the' is written and crossed out.

But there is not one primary experience – the understanding,[223] and a secondary experience the expression.

The expression is an experience. Why should this cheat?

Suppose you say I can intentionally knit my brows but I can't intentionally be angry.

This is something like saying I can't intentionally be flushed, but I can etc.

This is like saying, "the outward sign can cheat, but the inward sign can't".

But suppose you say, "can the man who reads the poetry cheat".

Take the case: – suppose someone said "I shall be delighted" in a particular tone, and I say he couldn't be cheating. I may have lots of experiences for saying this, but it is not anything further back which makes me say this. And it may be the same if it is I who say it.

The man who hears and the man who reads have one experience in common – hearing. Suppose you say the man who hears cannot know[224] the experiences of the man who reads and so can be cheated.

If all the experiences of the reader who reads with the right expression were put down, three quarters of them would be the experiences of saying and hearing.

Suppose we take a score to represent the experiences.

One part is the poem, another what the heart does, another a pedal note etc. The man who reads hears all these but the man who listens can only hear a part.

If I saw three parts of the score and expected the others to be something which they weren't, then I could say I was cheated.

But why shouldn't the three parts be all that matters.

Suppose what you mean by anger is characterised by the base note and if you do not hear this you may be cheated.

I will distinguish between certain parts which whether I see them or not might be changed and I would still call him angry.

There is the possibility that all the ones I see constitute anger and the parts I don't see are irrelevant.

I don't say this is always the case or even very often. You might think that all the parts which the other person sees are not the important thing and [the parts] the man feels are the important thing.

I will try to show[225] you this is not necessarily the case.

Imagine you were anaesthetized so that you had no feeling in your larynx but could hear.

Imagine all the inner organic feelings removed[226] and the only sense left that of seeing and hearing. So that you would be on the same level as the other man.

---

[223]Dash is inserted by Eds.

[224]After 'know', 'all' is written and crossed out.

[225]The text reads 'do'. Replaced with 'show' by Eds.

[226]Wittgenstein may have had his quick-tempered father in mind, who, after some six operations, which were not a success, died from cancer of the tongue.

Imagine people only have experiences of seeing and hearing, so that their anger would only be visual and audible.

I don't say our anger only consists in this, but these are part.

In such beings there would only be pretending to be angry where there was hearing without seeing, but none when there was hearing and seeing.

You might say a different thing: alright! Let's imagine beings only with auditory and visual experiences.

Then we have a case of cheating. But you may make the objection, "How do you know he sees and hears?" This is another objection but I shall not speak about it.

The man who has toothache holds his hand to his face.

Isn't this an experience of yours? This is an additional experience of toothache. Suppose I had a mirror constantly before me – here I would have the visual experiences of seeing myself very dejected, etc.

There are also the tactile experiences etc.

Couldn't there be such a thing as visual misery?

The idea of cheating always implies something is hidden and something else not hidden. Secondly,[227] if we say an expression always *can* cheat, this implies that the real thing is what is hidden (the parts audible are not the main thing.) The real thing is inward.

Now comes the difficulty for me to explain.

This may mean two different things.

It either may mean drawing a distinction between *his* experiences part of which are audible and visible and part in his stomach – if this is the distinction (we draw a line through the scores of his experiences, auditory experience, those[228] in his larynx muscles, etc.).

[227]This use of 'Secondly' is not preceded by 'First". Presumably the first point is the previous proposition commencing: 'The idea. . .'

[228]After 'those', 'who' is written and crossed out.

I say visual and auditory I will not call private, but tactile I will call private – I.E. I can't feel his heart beating.

Here we can't say those parts matter which we don't say.

We call the two parts public and private.

Suppose someone[229] thought only the private matter. But why should a feeling in my stomach matter more than what I hear.

Another thing you might say, "all the parts are private – I can't know what he hears, sees, feels in his stomach". I am going to leave this now. I am postponing a discussion of this, but it could have been brought in straightaway at the beginning.

We said the intention can't be the expression of intention but something inward.

Couldn't we have said it in this way, the intention is something private, the expression is something public? This is like saying – you can never know my intention – never know[230] what I hear, etc.

In the sense in which it is logically impossible to know, the knowing has no counterpart. Is this a matter of[231] experience that you can't know? Or a logical impossibility? Do we mean it makes no sense to talk of . . .? Then it is a question of the grammar of 'know', 'toothache'. Then it has no sense to say, "I can't know . . .".

Feb. 14th Wednesday[232]

We copy say a square or a face. The question was "does intentionally copying differ from unintentionally copying it by the process of deriving the picture from the object?"[233]

Suppose that by the process of derivation I just meant say what I draw if I draw a projection and draw projecting rays.

---

[229]'-one' is added over the line, with insertion mark, to 'some'.

[230]After 'know', 'my' is written and crossed out.

[231]'of' is inserted here by Eds.

[232]As noticed above, this the third use of the date (14th Feb. 1934) in this Lecture series. This perhaps arises from there being more than one lecture on the Wednesday, as well as matters discussed in the foregoing opening footnotes for each of the previous occurrences of this date.

The page numeral '1' is written to the right of this date, but it is crossed out, and replaced by '7'; and each page's numeral in this sequence of likewise corrected. This is unusual, since paginations typically commence with one for each lecture. This may have to do with the crossing out of the Lecture notes, reproduced above, which comprises the short entry for 'February 14th, Wednesday'. It could also relate to a very long lecture in which these different sequences were dictated. See the above opening footnote in this Edition for that Lecture, which shows that editing activity was in progress, concerning the notes beginning with Friday, February 9th – pages 26 through to 32. Skinner's correction from page '7' to '1' does not configure with this 26–32 page sequence, though it does show revision of ordering.

[233]This resumes and extends discussion of the theme that began (ninth of Feb discourse) 'shadow of a fact' linked to the earlier consideration of "copying is the intention of copying", etc.

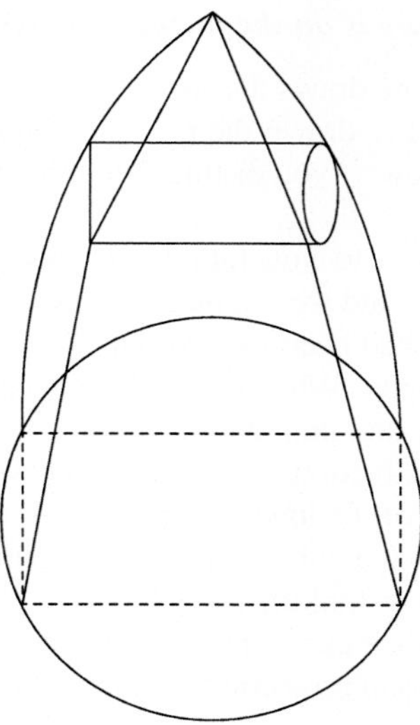

This is a simple case of 'projection' and 'object' (as we could call them).

There would be the case where he had just drawn the "dotted figure".

I am now proposing to distinguish between intentional and unintentional by saying the unintentional projection is when he draws the right shape without going thro' the process of drawing the rays.[234]

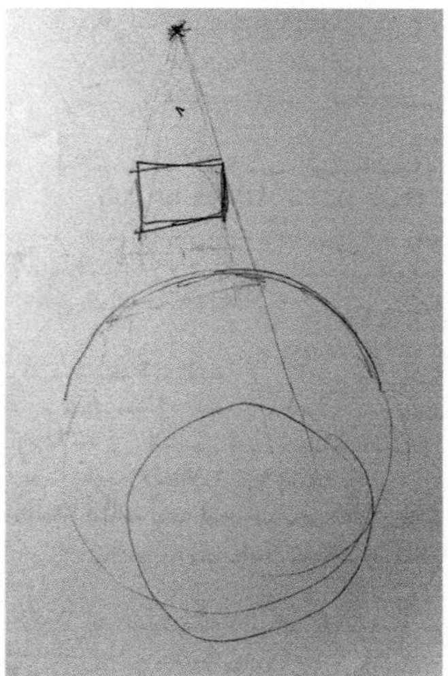

[234]A crossed-out incomplete sentence occurs after the end of this sentence. It reads: "This is a simple case for unintentional (similarly f".

In the manuscript this figure is on the facing page opposite the previous[235]

The intentional is when he draws the rays.

You say "Couldn't he have drawn the rays unintentionally?"[236]

Here you must tell me how[237] you distinguish unintentionally and intentionally in this case.

This example might seem to you fatuous. Is this true? Suppose someone had drawn the dotted[238] trapeze, and we doubted whether this was intentional or not; we might point to lines which had been rubbed out.[239]

Example. Take two points. What is the difference between drawing a[240] line between them intentionally and unintentionally?

I don't deny the distinction between intentional and unintentional — unintentional is all sorts of feelings – intentional means something else, say, concentrating yourself on the job. What is concentrating? There is something 'concentrating' as opposed to looking around.

We have reached one distinction. The one part of what we call intentional was what we could show process of projection, lines we could show. The other thing was we talked of feelings.

We distinguished between two main bits of intentionally representing – we drew one distinction between two different uses[241] of this.

One part is what is comparable to drawing lines of projection (this is the prototype).

The second part is when we ask what distinguishes drawing a line intentionally and unintentionally.

---

[235]Compare and contrast his diagram (in pencil) with the one on the facing previous page of the manuscript (MS p.6). Here Wittgenstein is employing a notion or projection from a point in geometry as an analogy; the idea of a 'shadow' in relation to such projection emerged in 1930s algebraic topology.

[236]This paragraph from "I am now proposing" to "draws the rays" is placed in parentheses that are crossed out.

[237]'how you' is written over the line; this replaces 'what' in the main text but is crossed out ('distinguish/es' is declined to accord with this alteration).

[238]Extra 'the' is omitted here by Eds.

[239]After the end of this sentence, another one is added, yet it is crossed out by a continuous line running through it plus diagonal lines. This reads: "It would not matter if I get into jams if you could profit by then seeing the way." After 'point', the word 'they' is written and crossed out; also, 'them' is written over the line, with marks indicating insertion.

[240]Here the indefinite article is written over the line, which replaces the deleted definite article in the main text.

[241]'uses' is inserted here by Eds.

Here we talked about concentrating, not being shoved about, etc. This is of a totally different kind.

Friday, Feb. 16th

We are talking about "wish" "expectation" "belief". I am going to use the word "thought".

The thought, we said, was that King's College was on fire.

This thought we said had an object.

The mere fact that we talk of the object of thought indicates a confusion.

This has one source[242] – our language says I expect so-and-so but also it says, "I expect *him*."

I expect a man seems like saying I shoot a man.

I don't expect a man in the sense in which I shoot him.

I expect a man is the same as I expect so-and-so will happen.

How can we expect something that it is not the case?

The way out. What we expect is something[243] intervening the fact.

I called it the shadow of the fact – the next station.

You may say there is no next station. That is one thing.

You may also say, in a psychological sense, that we can come nearer to it.

Could I not make a project: his face into something that it did not look like him at all? Could I not project it back into a painted picture of his face? This is nearer if by near we mean it makes it more *like*.

The idea of coming nearer – we would be tempted to say we come near if for instance this happened.

Someone asked me – what does so-and-so look like?

I give a weird projection from which he can by a long process get a picture of the man. Then we might say he came nearer.

This is the process of understanding a language as opposed to translating into another language.

If I gave you a telegram and a cipher, translating it into English would bring it nearer, but you already have everything.

You do not bring it nearer in any sense logically important, but it may be of great psychological importance.

When we think of the shadow or the sentence as the object of our wish, we said this brings us no further as the same question we asked about the fact can be asked about the shadow.

How can the shadow be a shadow of something which doesn't exist?

[242] In the manuscript 'sources' is in the plural. There was a slip in the spelling of the first half of the word, corrected by Skinner, which may have distracted him from correcting the plural.

[243] After 'something' an incomplete word is written and crossed out: 'bet'.

You might say we must imagine the shadow to be a picture the intention of which can't be mistaken, because the shadow is no good if we can ask the question "How do you know what it is a shadow of?"

The origin of the idea of this shadow is this: that we know pictures the intention of which we should say we can't mistake – it [is] obvious what they mean.

If I show you a photograph of King's College, if it is good enough, you say you can't mistake it.

As soon as you see it isn't obvious, i.e. what we generally call a picture is not the only thing that can be a picture the idea of the shadow intervening loses its charm.

We are tempted to think the idea of such a shadow will help us because we know certain types of pictures whose interpretation we don't doubt for a moment.

One class are pictures '*like*'.

Another are pictures with some words written underneath. We could make this step – certain the shadow is of no use at all to use for the simple reason that if we want a picture, the sentence will do for us.

That the sentence "King's College is on fire" is a picture is clear – the method of projection is when we give explanations of the words I.E. point to King's College, etc.

I went back from the idea which we have of the shadow to the sentence which I said could be the shadow. But I could[244] also say, "whatever the shadow is will do just as well." If the shadow is a painted picture, this will do.

Our next step would be to say how can the expression be the thought, surely the expression can always lie?

When we say the thought has an object, or are tempted to say the fact has an object, this is due partly because it seems to have sense to ask a question.

Suppose, "I say I wish, expect so-and-so will come." Somehow it seems in the wish that so-and-so will come seems already to have happened.

If it was something else, how could it have been the wish that *he will come*?

How can the wish determine what will satisfy it, because the wish is the wish and is something different from what will satisfy it.

We would be inclined to say, in an absurd way, if we wish something to happen, in the realm of wish it must already have happened.

I am explaining a disease of thought – only understandable to those who have had it.

If we say, "I wish so-and-so will happen", we think nothing short of describing what we wish will satisfy us – and this is true, as we said what we wished in the sentence. While we are wishing, we don't feel any difficulty – remember this.

We might say that the object of our wish is that which the explanation of this object explains. That which we explain it to be when we are asked what the object of our wish is. How would we explain what the object of our wish is – I.E. when we

[244] 'will' is in text and is crossed out in favour of 'could' written over the line.

wish the object King's College is on fire? This explanation will tell us what we mean by object. We may translate into another language, into pictures, into gestures – then we might say that this was what the object means. We might say the different signs used, gestures, pictures, etc. were shadows of the fact only: this is misleading as it then sounds that these pictures were the next best thing, the best thing being the fact itself. The comparison which here misleads us is this – we compare our case with that in which we say, "I can't show you the man himself, but I can show you a photo of him."

The confusion which you are in could roughly be put this way: that the fact you wish for seems as though it was a man that entered the room. Suppose this was so, and I asked you what you wished for, and you say I can't show you Mr. Smith now, but I can show you a photo of him. But I can show Mr. Smith. It seems as though[245] the fulfilment of the wish couldn't be shown to us while all the objects can.

Wednesday, Feb. 21st

I dictated last time that "thought" "wish" etc. must be about the actual things.

I.E. *Mr. Smith* and just Mr. Smith must come in.

The difficulty is that no substitutes are possible for explaining what we wish. I.E. nobody else entering the room will explain, "I wish Mr. Smith to come into my room".

What one feels is correct, but what produces the difficulty[246] is this.

I say, "I wish Mr. Smith to come in". In what sense does Mr. Smith come in? I say certain words, have certain thoughts etc.

But you say these do not bring Mr. Smith in.

How could you bring Mr. Smith in?

If you gave an explanation who Mr. Smith is, you would do this by pointing to Mr. Smith. A photograph of Mr. Smith would not do here.

But you do not point to Mr. Smith while you say the words. (You do not point to a man who is in America while you are thinking of him.)

The connection is not made while you wish, but before or after.

Mr. Smith must come in if you wish to explain "Mr. Smith" – a picture won't do.

But the connection is made before or after you say, "I wish Mr. Smith to come in".

The idea is that when you make the wish or have the wish you must be able to make a connection.

The connection is there in our calculus

---

[245]Extra 'as' here omitted by Eds.

[246]A word is crossed out here.

If I multiply 18×25, $18 = 1 + 1 + 1+$ etc. Does[247] this come in in the multiplication? But it *is* present in our arithmetical calculus.

Or in chess. A game of chess may end up in a mate. But does the mate come in in every move? When you make a move you *may* say to yourself that this is done to mate him. The fault you make is in thinking that the mate must be present while you make every move.

This is the same[248] as where you say meaning is a mental act. It would be a very peculiar act if it could make a connection between a word and its meaning.[249]

This is partly/part of[250] why we are revolted by the idea of replacing meaning by saying.

I once gave as an example of the use of words the man going with a written order to the grocer. You feel that in the order "bring me so many bananas" the bananas must be present the whole time. The bananas may come in at the beginning when the order is given and[251] the word "bananas" is explained to him, or at the grocers.

The connection which is made between thought and reality is not made by an obscure act of the mind – we do a certain calculation and the connection is made here.

This muddle pervades all philosophy.

It is based on a deep-seated expression, "I meant so-and-so" or "I thought *he* would come in."

Friday, Feb. 23rd

From the way our language describes things we are inclined to imagine that certain trains, mental states happen – don't happen in time but there is something corresponding to them which is like a state and is there all at once.

We are tempted to imagine that if we follow a particular train of pictures, this train must have been[252] somewhere before in a lump and we just pull it out.

This is an extremely deep-seated picture.

You can't now understand the disposition to use a particular picture – I am not going into this.

We are inclined to make particular comparisons – it is extremely difficult to get rid of them, or make others.

One of these is our use of the word "idea".

---

[247]Two words are written and deleted in the main text. What they are is almost entirely unclear. Possibly they were 'that class', whilst only the final letter 't' of the first word and the second letter of the last word may be 'l'.

[248]After 'same' 'case' is written and crossed out.

[249]The expression 'a word and its meaning' is written in the main text but crossed out.

[250]The word 'partly' is written in the main text with 'part of' written above the line, while both are ranked by '{' as joint choices of readings, and neither is deleted.

[251]After 'when' the expression 'the order is given and' is written over the line with instruction for insertion into the main text.

[252]The word 'been' is inserted by Eds.

What is our idea of the number three? To talk about idea may not be wrong.

If we think of three nuts, we can call this the idea. We think the idea is expressed[253] in the explanation we give.

If a particular train of thoughts comes whenever we think of three, we may call it the idea but it is no better than the sign.

I once gave an example. We write a letter and try to find an expression at some point for some idea in our mind.

I talk in a way deprecating about the word idea. I do not mean you shouldn't use it.

We use certain ways in odd ways – I mean in odd cases. Take the word idea. We say we have an idea in our mind and can't express it.

We may have words of another language or an image or a feeling we want to get rid of.

We have the idea that the idea is already and has only to be translated into words.

All sorts of things are the case, and this only very rarely.

We talk of analysing our idea of number.

We have roughly something before our mind[254] which guides us but how it guides may be any number of different things. Sometimes what guides me is[255] a gesture which I repeat.

Compare with remembering last night's dream when you lie in bed next evening.

If you want to bring back a particular expression, you imagine yourself in some position.

Take the case – let's analyze our idea of cardinal number. I say, "Have you an idea of cardinal number?" "Or of the cardinal number three?" You may say you have an idea of the cardinal number three.

You may think of certain things and reject them. I.E. ask, "Is three my three fingers?" "No, because it is also three apples." What you mean is the word "three" is not used synonymously with the words 'three fingers".

When you say anything like clearing up the idea of three – when you talk of analysing something before your mind, this is only a metaphor. You mean something like a particular[256] picture. You may have all the time a picture of three fingers.

If you say you were guided, you must see what it means. A line on a piece of paper may guide me to take my pencil along it. But also to draw a parallel line.

Or to make a projection or to draw something which looks pretty with the other line.

[253]After the word 'idea', 'is expressed' is written above the line with signs to indicate insertion in the line.

[254]After the term 'before', 'our mind' is written above the line with instruction for insertion.

[255]'is' is inserted by Eds.

[256]The word 'particular' has been crossed out and then, I think, written over again.

This is like Miss Ambrose being guided[257] in buying a hat by her dress.

Guiding in this case means all sorts of things, and in fact does mean all sorts of things.

There is one piece of advice I can give you.[258] If you think about philosophical problems and you are up against expressing a particular idea in your mind, be honest and say what your particular idea really consists in.

If you wanted to express your idea of the number three, and you have often before your mind's eye three fingers,[259] then don't be ashamed of admitting this. These crude things are enormously important.

Why shouldn't one be honest?[260]

It wants a particular technique to really definitely get hold of that crude picture which is really wanted. As soon as you have got it, things seem alright.

People who talk of fundamental intuition in mathematics – the word 'intuition' refers obviously to some psychological event which is utterly uninteresting. (Suppose Mr. Smith has a tendency to add one whenever he sees a number.)

'Plus one' is an expression.

This idea is akin to an idea which is much older, I.E. self-evidence.[261] People said logical laws were self-evident. What they meant was that if they asked people whether a law was true or false, they would "Yes" and not doubt.

If[262] you say by self-evidence you mean you would not have a doubt about it, as I would not have a doubt about a bed being in my bedroom.

You say "self-evident" brings to your mind the proposition "there is a bed in the next room" which you do not doubt.

This shows it is not what you want to say about logical laws.

If for instance we give the proof that you can't trisect the angle with the ruler and compasses.

This proof appears to give something which clarifies your idea of trisecting with ruler and compasses.

You might ask, "Is this the idea you had of trisection before you started."

[257] After 'Ambrose', 'in' is written and crossed out in the main text with 'being' written over it. The next word is 'guiding' in the main text, with the case ending crossed out with the alternative ending above the line resulting in 'guided'.

[258] 'piece of' is inserted here by Eds.

[259] After 'three', 'things' is written in the main text, crossed out and replaced by 'fingers'.

[260] After the question mark there is a deleted opening of a sentence, which commences 'One is' – a further word-fragment is illegible. There is a paragraph break, and the paragraph begins 'Someone says', which is crossed out, and re-commences with 'It. . .' as in this Edition.

[261] The subject of self-evidence is addressed in Wittgenstein's Lecture VII, reproduced in Chapter 5 of the present Edition.

[262] This sentence was originally started by 'If l' (perhaps the 'l' was to begin the word 'logical') and is crossed out and replaced in the main text by the opening in the present Edition.

Is it to look or measure so that the parts are equal? You may say you knew what divided an angle[263] into two parts or four parts etc.

The proof does not change your idea anyhow – it leads you a road you are inclined.

You could also say that here[264] you are eminently guided by an idea – you were guided by some idea of the trisection of the angle.[265] You recognise, when you accept the proof, certain similarities to things you are used to.

You would be inclined to say you were guided by your idea of the trisection of the angle – it depends what your idea was.

An idea is no damn good to us unless it is commensurable with what we say.

I.E. if it is not commensurable with an expression. It might be a picture on canvas. It is commensurable insofar as it is translatable into language.

You say, "I have idea." I say, "what is your idea", then I want you to translate it into language.

A person says he has an idea of the trisection of an angle. I say, "What?" The answer is expressed in sentences – he says you know what ruler is etc.

I say I am interested in the explanation or expression of the idea.

A man says he has an idea of the trisection of the angle, he says it is the idea of the trisection of the angle. Then this is no use to me. But if he tells me something else, then this is interesting to me. I will only be satisfied when he expresses his idea – gives me a further explanation.

The idea of trisecting in the proof was gradually remoulded. The remoulding in the proof went a way the man was inclined to go but needn't go.

We said if we wanted a picture for what we wished for the sentence will do. I want to talk about the objection to this that we can put an expression for what we wished as every expression can lie.

---

[263]After 'angle', 'and' is written in the main text and crossed out.

[264]For 'here' the manuscript first had 'her', which is crossed out and replaced over the line by 'here'.

[265]The next sentence began 'The land marks', but it is crossed out in favour of 'You...'.

Chapter 6
Visual Image in His Brain

What are we to say to a man who says he feels his visual image in his brain?

This is important because people often try to give a meaning to a proposition by saying they feel it is so, etc.

Hardy said that he *believed* Goldbach's theorem[1] was true.

All objections seem to be turned down by saying you feel it is there, and then you silence the man.

The question is not that you can't believe it, but what does it mean.

I gave the case of the diviner saying the water is three feet underground.

This has not at all been explained even if the words feel and 3 ft underground have been explained.

What makes us doubtful about the water-diviner is that he doesn't say he has learnt it. – Learning where he coordinated it.

He could have learnt it by saying that he had coordinated[2] feelings by experiment with different mixtures of gold and copper.

Our doubts could all be removed if he said he had learnt it.

This business of learning may mean all sorts of different things.

There is not one way of estimating length by eye but lots of different ways (this is of colossal importance to us).

I will enumerate.

In this chapter, the statement occurs "Then we might say the time at which a dream occurs is the time at which these processes take place in his brain". This is a component in Wittgenstein's investigation that is complementary to the one comprising the present chapter. Since the text in Chap. 4 is a new extension to the *Brown Book*, one finds evidence of a triangular link between these two chapters and the *Brown Book* concerning the subject that warrants further attention.

---

[1]Goldbach's Conjecture or method is titled 'theorem' in Hardy's own use: 'Goldbach's Theorem', in *Mind* (Vol. 38, 1929: 4); also, Wittgenstein develops his discussion in *LFM*, Lecture XIII: 123.

[2]After 'coordinated, 'the' is written and crossed out.

A. Gibson, N. O'Mahony, *Ludwig Wittgenstein: Dictating Philosophy*,
https://doi.org/10.1007/978-3-030-36087-0_6

(1) We estimate a length, and someone asks us how did we do it, and we can do it.
(2) How high is this building – it has four stories each about the height of a room.
(3) How did you estimate the breadth of the table – it looks like one foot.
(4) I imagined a foot rule put on it – it looks something like it.

These are all different processes.

You might say if I estimate four foot I must know what a foot looks like. This is not at all true.

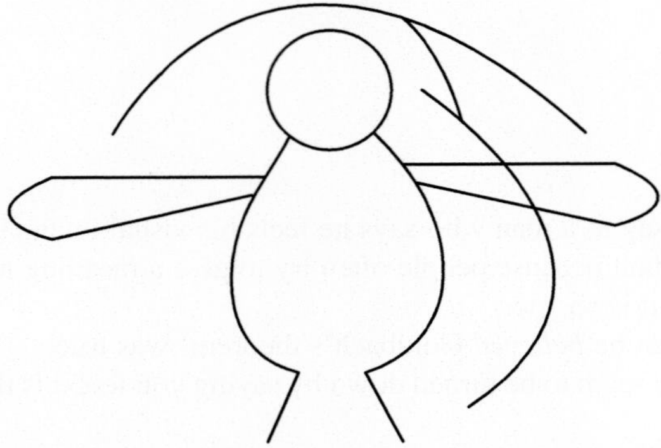

The third way is probably what is most interesting to us (one of the most striking things).

How long is this table? It looks like a foot.

He might remember another occasion when he had looked at something at this distance and then measured it.

But he may just look at it and say something.

I know what is difficult for you here.

We said before that we should not be surprised at the water diviner's statement if he told us he had learnt this estimation.[3]

Learning may have two meanings (characteristic).[4]

[(1)] Learning to estimate, etc. may either be the cause of your future mental and bodily acts.

I teach Mrs. Knight – this is a yellow patch.

Another time I say "imagine a yellow patch". (Making use of the teaching.)

I say paint me a yellow patch.

[3]Spelled 'estimations'.

[4]The text is somewhat incomplete regarding a few details here: (a) 'meaning' is in the singular; yet two types appear to be required in the following remarks. (b) Although an item has '(2)' affixed to it, there is no '(1)' itemized. So '[(1)]' is inserted at what may be the correct point, though it is also plausible to place it before the next sentence. The most likely position is the one chosen because when, elsewhere in the manuscript, there is a list to follow the sentence that stipulates the identity of the list, Skinner often does not offer punctuation after the last word of that sentence and there follows a paragraph break, as with the current example.

(2)Imagine a yellow patch.

How did she make use of my teaching?

What is the relation between my teaching[5] and the thing she does now?

There are two possible things you can say now.

I might have presented her[6] a little table of colours (you might not call this teaching).

If I call this teaching, teaching has two functions.

She made use of my teaching[7] in two ways.

My teaching was a sort of drill – one thing I drilled her to do something and give psychological explanation. When I say yellow she could bring up a yellow image.

It was like building up a mental mechanism.

It was like installing an electric connection.

– I can press a button – the connection may go wrong like forgetting.

In this sense the teaching was the cause of the understanding later and in this sense does not interest.

You might make hypotheses – she was born so that she could know the word yellow etc.

In[8] this sense teaching builds a mechanism which may work or not. (Once it is built.)

I said there were two ways of looking at teaching –

I gave her a table. If the table is the tuition she makes use of the table.

In one case the rule which I gave her was used when she did the thing – a little calculation.

In the other case there was no mechanism.

In the case where we estimated a length, in some cases there was a sort of calculation.

There was another thing which seemed to be entirely automatic.

You remember I said there was one case where you could ask how did you do it.

I will say you can always ask this, but they mean two utterly different things.

If you estimate a length, I may say: "*why do* you think that tower is 40 yds high".

One way of answering[9] is by four stories etc. – an answer giving a process by which I have arrived at the thing.

Supposing you said how long is this table – about a foot.

I may give an answer – I have so often measured such lengths I can do it.

---

[5]In place of the words 'my teaching' are two ditto marks to indicate the repeat of that phrase from the previous line.

[6]It is conceivable that 'her' was intended to be 'here', not least since there is an occurrence of 'her' that is crossed out and corrected to 'here' in Chap. 5's manuscript (Friday, Feb. 23rd lecture MS page 4).

[7]In place of the word 'teaching' is a ditto mark to mark the word 'teaching' in the line above.

[8]Prior to the current beginning of the sentence, the words 'On the other hand' are written in the main text and crossed out.

[9]The manuscript has 'answered'.

One gives a reason the other a[10] cause.

One gives the history of how I was able to do it.

The other a process which actually takes place when I do it.

You were at first astonished that an estimation could take place looking and saying.

There are two cases:

(1) Automatic
(2) A calculation.

I gave the examples of an order – imagine a yellow patch – and paint a yellow patch.

If I ask Mrs. Knight, how is it possible to say you could paint a yellow – she would say perhaps[11] I imagined a yellow patch and then painted it.

In the other order, what happens is *automatic*. This second is an utterly different business.

If we talk of estimating a length and obeying an order, we must realise there were utterly different cases.

In some there is a calculation and a transition – you then think this is the essence and are surprised when I say there is an automatic case.

There are cases of the calculation.

I gave Mrs. Knight a table – did I tell her how to use it, did I tell her how to use the words I gave her?

If the man says, “this is one foot” this is possibly all that happens or more may happen.

We call all these estimating a length – this is a family.

The importance of this will only become clear when you consider the use of a rule I give you.

A rule may serve to build up a mental mechanism or may be a part of your calculation.

Suppose the rules for moving the figures in chess – I know these rules and play chess – do I use these rules in chess. Had I not been told them I could not have played chess. They are not present to me. They are the history perhaps of my playing.

This does not concern a dispute.

The teaching of the rules and the rules themselves are history which builds up a mechanism which now functions without them.

If we play this way: the rules are put down on a table – for a Knight the Knights’ move is drawn

[10] MS reads ‘of’ here.

[11] After ‘perhaps’, the manuscript has ‘say’, which is crossed out.

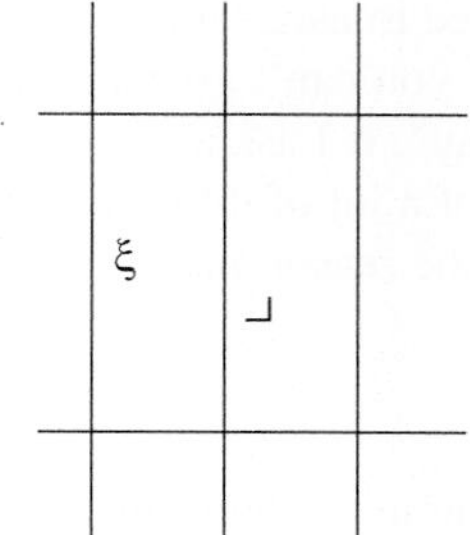

We copy these when we play.

This is not part of history. We use the rules when we play. If there is a dispute the rules come in.

(What we call teaching or giving a rule.)[12]

You may say the rules are at the back of your mind, I ask you what is the[13] criterion for this.

I may play a game without the rule coming in at all, but when asked I give the rule.

What I call cause and reason I am going to apply to the game of chess.

Suppose a man said why do you do this. You can give the calculation you went through, or you may say I have always done it that way.[14]

The reason is a calculation – why do you say $25 \times 25 = 625$: you give the calculation.

The calculation may be that's how I arrived at it or that's[15] how you could have arrived at it.

(The chain of reasons end – how did you do this – I use a table, etc.)

I meant to say the word why asks for two utterly different things which call for utterly different things.

So many people thought that the reason is something conscious at the end of a chain of clauses.

How did you get to Grantchester – either by foot – or you give a way.

Reason is a way of thought by which you arrived there, cause is a hypothesis.

The answer to the why of reason is giving a chain of thoughts you have gone through[16]

Cause is given from a series of coincidence.

---

[12]After 'rule' the expression 'may not come in or may come in' is written and crossed out. It looks as though the writer introduced the parentheses after writing the full sentence that ended with this quoted expression and omitted to alter the comma to a full stop. (Between this ending of a paragraph and the commencement of the next one, 'It' is written and deleted, with the rest of the line left blank.)

[13]'the' is supplied by Eds.

[14]After 'way', the words 'The question may be' are written and crossed out.

[15]'s is supplied by Eds.

[16]Written 'thro'.

We need only one experience to ascertain the reason.
You needn't give a cause if you can't give a reason.
You just say, "I went no way", if I ask a reason.
Giving a reason[17] consists of a lot of different cases.
What does questioning for the reason mean?

How did you arrive there?
How can you arrive there?

(or it may mean how did you learn to do this.)
[2 × 3 = 6 – there is the case "*this is a definition*"
Or I can give a "*calculation*" – or on the other hand[18] I can say of[19] the cause that I have learnt it]

If you say 2 × 2 = 4 is a definition, this is a reason.

This may be the end.

A man may say "I might have estimated it by imagining[20] a foot and applying it four times".

You can ask, "Might he have gone away".

A man may imagine an inch and count a foot or say it is a foot.

Examine yourself and see how you use the word reason.

I[21] say this is a foot. This is my reason A.

[If I said 4 foot, you would[22] think of feet, but if I say one foot, you needn't think of 12 inches.]

Suppose I say, "I see that (Reason B), a foot rule would roughly cover it" and then say, "Because it is a foot."

You must mean that you yourself must have imagined it – it must be the way you have gone otherwise it would have no means.

I estimated this is a foot.

Someone asks me – I say an inch would go into it twelve times. It might be a reason A – I arrived at it in this way. Otherwise it is a definition,[23] by one foot I mean twelve inches long.

Definitions have puzzled people enormously. It is difficult to distinguish between it being an arbitrary rule of signs.[24]

[17]After 'reason', 'are' is written and crossed out.

[18]'on the other hand' is written over the line without marks for insertion.

[19]'of' is supplied by Eds.

[20]'foot length' is written and crossed out in the main text and replaced by 'a foot' over the line.

[21]Before 'I', 'What about' is written and crossed out.

[22]After 'would', 'just' is written and crossed out.

[23]After the comma, a word is written and deleted, as is a short word following 'one'; both are illegible.

[24]The first batch of sheets comprising one sequence of pagination (pp.1–8) ends at the bottom of this page with the word 'signs'.

We[25] talked about the relation of the process of[26] teaching to the actual application.

We may someday find someone to whom teaching is unnecessary.

I tell someone, "Fetch me a chair from my room".

He has learnt the word chair by a process of drill in his youth.

It is imaginable a man may say he fetches the chair without ever having[27] been taught.

Supposing we have electric connection for a ticket. Is it imaginable we press the knob and the yellow thing comes out?

These are not quite analogous.

I said in what sense is the process of having learnt present in my obeying the order.

How is the mechanism involved in pressing the knob?

If we talk of the mechanism in you, this does not involve the teaching of your mother.

The electric connection does not involve[28] the plumber's action.

But what is this mechanism? Is it the mechanism in the brain? You will probably say, 'no'. I suppose you mean a mechanism[29] of thinking.

How is the[30] teaching involved in the mechanism of thinking? There is this case when your mother gave you a table. The table does not drop out like your mother.

The plumber does drop out.

The picture does not presuppose a painter.

Suppose you have seen a red patch and now talk about seeing something different.

There is the case where you remember your mother having taught you the chair.

But this is not essential.

The point is this that we eliminate the history of our action but not what takes place in our[31] action.

A man may quote a rule about the King of Chess.

The rules come into chess in two ways. They come in as the history of the game – *But* also come in as part of our *role* of thinking.

---

[25]Pagination in the manuscript starts again, from page 1.

[26]'the process of' is written over the line with a mark for inserting it.

[27]After 'having', 'once' is written and crossed out.

[28]'that' is written and crossed out after 'involve'.

[29]After 'mechanism', 'involve' is written and deleted.

[30]The word fragment 'mechanis' is written and crossed out.

[31]An illegible short word is deleted after 'our'.

It is perfectly clear that when[32]:

It is a shock for you to hear that you could have ever done this without being taught it.

What he does now is perfectly possible without having been taught the use of the word chair.

If Skinner has never been taught the use of the word chair and obeys my orders, you will say he is not really obeying my order because he has not understood.

There is no[33] reason in this case – so as to reason connection,[34] there is none.

You will say part of the use of the word chair is that Skinner can explain the word by pointing to a chair – here we have a transition like the table.

Skinner's action takes a certain time two minutes.

Although what[35] may happen is that Skinner hears the order and brings the chair.

You are tempted to say he understands it.

The understanding does not come in as an accompaniment to the game of chess. You can call[36] something the understanding.

Is what is done in those two minutes obeying the order?

In nothing that happens while he does it does the teaching come in.

What is interesting is that though you might not call this obeying if he had not learnt it –.

You would not call this obeying if he did not sometimes make use of a particular way of thought.

Suppose someone said, "A has never learnt[37] the English language", to B who has[38] known A for years.

What would B have to say?

He might say "I don't believe it" but surely it is imaginable or he might say, "A is an automaton."

It is a matter of experience that if there is one half of a sausage there was the other half.

If we know the latter part of the sausage –.

Involving the railway timetable is looking at the railway – or imagining it, etc.

Using the rule may have antecedents but it is just as thinkable that it did not have them.

If we eliminate the history entirely there are two cases.

(1) No rule is involved
(2) A rule is involved because it is now used, but not because of its being taught.

[32]The sentence ends here; but it seems to be a stylistic device to introduce the subsequent sentence.

[33]'no' is written over the line with a mark for inserting it.

[34]The word 'connection' finishes at the edge of the page, with the last part of the 'n' trailing down. This trailing may possibly gesture in the direction of a comma, which is accordingly inserted above.

[35]'what' is written as 'that' in the manuscript.

[36]The words 'what you like' are written and crossed out, with 'something' written over the line.

[37]'learnt' is written over the line.

[38]After 'has', 'never' is written and deleted.

If one separates[39] two things you are accustomed to associate together, etc. this is revolting.

We said in one sense the teaching[40] was the actual history of the understanding, obeying, etc. and in so far it may not be necessary.

In the other sense the teaching was involved in the application.

In talking about estimating we saw there was the possibility of looking and saying something.

There are cases where there is the question "how did you estimate the length" or "why did you say four feet".

There was[41] one answer where we say this seems one foot and there seem four of them.

There is one case where it seems[42] no answer of this sort[43] can be given.

It strikes you as puzzling that the teaching of estimating does not come in in the process.

This is because you think of the teaching as supplying you with a process for arriving there. The teaching supplies you with reason.

The teaching also drills you and then you do whatever you do.

You may refer to the teaching as the cause why you react in such and such a way.

Or as the table you have with you.

I could put it in a slightly different way:

When you do something following an order: fetch a red thing and do it.

When you think[44] of the relation of this to the teaching, you want to justify your action by the teaching.

The teaching is only the justification if it shows you the way to go.

As a cause it is not a justification.

I *taught* someone the meaning of red[45] by an ostensive definition. Then I give him an order. Then I ask him why he does it (such as paint a red). He says this is red and this is the same colour.

(In the case of red one never does this, but a more special colour.)

He justifies his action by the way I taught.

He goes through[46] a certain process in which my teaching comes.

---

[39]The text has 'separate'.

[40]The word 'history' is crossed out.

[41]'was' is written over the line.

[42]'it seems' is written over the line.

[43]'of' is written next to 'sort' and is crossed out.

[44]'thing' is corrected by Eds. to 'think'.

[45]The words 'and' and 'by pointing to' are written and crossed out – the former and latter quoted expression separately.

[46]Written 'thro'.

I say, "point to something red" and she points to the strand and I say why – she says "this is red".

What does this mean?

Is this a justification of the first sort?

When she pointed to this, there was no process she could point to as arriving at this.

Take the case where I gave her a pattern, she says you called this so-and-so and I painted this. If I say "why". She may say don't you want people to paint what you have called red.

If she says I pointed to this as I have learnt that this is[47] red. This is a causal connection.

What reason that I can give you ought to satisfy.

Is it stupidity on your part that you are satisfied at all?

Take the case of justifying going to the train by having looked at the timetable.[48]

One case I give a process I have gone through.

Another I give something which ought[49] to satisfy an intelligent person I.E. a reason.

What ought to satisfy an intelligent person.

Take the[50] example of red patch. I say, "Why if you were shown this did you do it automatically. You ought to have had a reason". Could you reproach him? Clearly not, as he would only have had another reason.

Supposing one said, "Do you suppose this jacket will last another fortnight". I say, "Yes." You ask "Why."

I say, "The jacket has lasted a long time, but not so long as to fall to pieces".

Or I think 'this canvas will last another term as it has lasted[51] two terms and ought to last as long as they usually do'.

I could then say these "canvases don't change much".

If you are philosophically inclined and still say "why," you will[52] at the end say "all these things make it probable". If I go on and say 'why' you won't be able to say anything.

At first sight it appears you must know all the reasons.

When you look into it, you see there are one or two sensible reasons and then you come to an end.

If there is a sensible question to ask, there will be an answer.

The chain of reasons has a beginning.

---

[47]The word 'is' is missing and is inserted here by Eds.

[48]'time-table' replaces the actual text, which has 'timetime'.

[49]'ought' was originally written as 'out', and the 'gh' was squeezed in afterwards.

[50]'the' is inserted by Eds.

[51]The form 'last' appears in the manuscript; altered to 'lasted' above.

[52]The letters 'pr' are written and crossed out.

If someone says "surely everything must rest on a reason". This is like saying "the house rests on the earth, surely the earth must rest on something." I say there is no occasion for a thing to rest on anything.

I want to formulate the other case.

It is something like this – suppose I asked for the reason for making this move in chess – I might have to say I haven't gone thro' any process, but you might give a legitimate reason laid down in the rules of chess. ________________

4

I wanted to say[53] there is an idea (1) for anything you do there must be a reason (2) the chain of reasoning has no beginning.

Supposing one did say this that wherever you are you must have got there from a previous place.

If you added, "you could" it would not seem so absurd.

Wherever you are you could have got to this from ten yards and to that from another ten yards away, etc.

This is like saying that "space is infinite" or "time is infinite".

This is what one is aware of when one says, there is an infinite chain of reasons.

This does not mean there can be an infinite chain of reasons; *there can be any number of reasons.*

This is like saying the series of cardinal numbers is infinite.

You sometimes hear the remark "the causal chain never ends[54]" "there is always a 'why'."

Supposing I said, "however far[55] you go back, you can always add another reason." This is like saying you can always add[56] another one.

Suppose I said, "However rich a man is, he can always have one more shilling".[57] Suppose this was understood that riches have no end and that there is someone who can have an infinite number of riches.

You can have any number of riches.

Supposing a fairy said, "Wish for anything from 1/- to 1000/- or wish for any number of shillings.

Here she did not limit the possibility, but[58] I still had to tell her a number of shillings.

---

[53]The words 'one is inclined' appear after 'say' and are crossed out.

[54]'ends' inserted by Eds., though it is 'end' in the manuscript.

[55]The word 'far' is inserted after 'you' in the MS.

[56]'add' is written over the line.

[57]'shillings' in the MS.

[58]The words 'she did not limit' appear after 'but' and are crossed out.

If one said "Why did you do this" ___[59] "so-and-so always repeated this to me when it came to mind." This is a hypothesis based on many different experiences. The reason is based on one experience.

Motive – I must know why I did it.

In one case I say, "surely I must know why I did it".

This idea is that this is a cause which I know from the inside.

The truth is "there is no cause".

You are hungry and go to the room – you say to yourself you are hungry. This is the motive.

He knows the motive.

I want to talk about thought reading, from the Monday lecture. Meet at eleven Friday morning.

[59]There is a comma in the middle of this line. It could be that the line was intended as deletion of the comma; yet the line is long enough not to be regarded only as a sign for removing a punctuation mark.

Chapter 7
The Norwegian Notebook

I see a towel hanging on the wall from a nail with folds, curves, shades of light and dark. I[1] draw a very exact picture of this towel. With this picture I[2] show other people how the towel looked when it was hanging on the wall. I[3] use this picture to hang a towel of the same size and shape, as indicated by the picture, and under the same lighting conditions, again in the same position[4] on the wall. Here I will have[5] reproduced the same set of circumstances as indicated in the picture. The picture enabled me to keep a record of a certain set of circumstances involved in the towel hanging on the wall. It is important also to remember the circumstances under which the picture was painted; I sat before the towel constantly looking at it, drawing with a pencil on a piece of paper before me, rubbing out lines, painting, mixing colours, a process taking time and involving lots of different activities. Also when I start to draw the towel, I don't already have in mind all the different things I will do: I may have a general idea of what I will do, but most of the things I do will be brought about by what happens while I am doing the drawing.

What then do I refer to[6] by a certain set of circumstances involved in the towel hanging on the wall and in what way is the picture a record of it? One thing I might say would be the visual spatial shape of the towel. But by saying this I have so far said very little.

---

[1]Illegible word is crossed out after 'I'.

[2]After 'I', 'can' is written and crossed out.

[3]After 'I', 'could' is written and crossed out.

[4]'in the same position' is written over the line, with instruction for insertion.

[5]After 'have', two expressions were written and crossed out, one above the other.

[6]'refer to' is written over the line, to replace the crossed out 'means'.

A. Gibson, N. O'Mahony, *Ludwig Wittgenstein: Dictating Philosophy*,
https://doi.org/10.1007/978-3-030-36087-0_7

How will I explain to another man what[7] the visual spatial shape is? I suppose, say, by some such process of drawing as I have described (as it is not sufficient just to point).[8]

So already I am introducing[9] something very complicated into[10] my explanation. While[11] drawing I will be *constantly* looking at the towel. It seems therefore that in talking of the visual spatial shape I don't necessarily mean something seen at a glance, taken in at a glance. If[12] I make a drawing of[13] the shape of the towel, it is not like saying,[14] "the colour of the towel is white", although if I think of the shape of the towel I think of it as something which is seen[15] at one glance.[16]

Have[17] I[18] if I look once at the towel[19] seen all that is necessary to be seen to draw its shape? – No. Even if I stare hard at it, it will[20] not be enough to draw its shape by, I will need to look at it,[21] and this is absolutely necessary.

---

[7]After 'what', 'what I mean by' is written and crossed out. Given that the last word of the sentence – 'is' – is in its normal position, it seems that this expression was deleted before the end of the sentence was written.

[8]This parenthesis and contents are written above the line.

[9]'introducing' is written, crossed out, replaced by 'doing' (which is also deleted) above the line, then dashes under 'introducing' indicate restoration of this first choice.

[10]After 'complicated', 'with' was originally written and crossed out in favour of 'for' above the line, which was then crossed, and 'into' was written after it above the line.

[11]The word 'Also' opened this sentence but was crossed out.

[12]'but' originally opened this sentence and is deleted.

[13]Originally in the main text 'describe the circumstances' was written and crossed out in favour of 'make a drawing of' above the line. ('circumstances, say', using the word a second time, is also written and deleted).

[14]'making a description of the same form or saying' is written first, then deleted and replaced over the line by 'such a statement as saying'. With the exception of the word 'saying', this expression is also deleted.)

[15]'is seen' is written over the line, and the original reading, now crossed out, is 'can be seen'. Also, above the deleted 'seen', the fragment 'I see' is written and deleted.

[16]After the end of this sentence, a parenthetical sentence is written and crossed out, which reads: "(In certain cases it can; if the towel is laid out flat I will say the towel is square'.)" Prior to 'I will', 'I can' was written, and deleted in favour of the former.

[17]'Have' is written over the line, with the original 'Can' crossed out.

[18]Illegible word is written after the first-person pronoun.

[19]After 'towel', a partly recognizable phrase is written and deleted. It reads 'that I ‖'; here '‖' may indicate the illegible word, which might be 'have'.

[20]After 'will', 'probably' is written and crossed out.

[21]After 'it', there are some revisions that are deleted: 'over and over again and they' is written over the line, and crossed out, with marks for insertion immediately after 'it'. 'it' on the main line is succeeded by the sentence-fragment: 'looking at the towel plays a very great part in discovering it'. (Here 'discovering' is probably the word, but somewhat uncertain due to heavy deletion marks.)

But isn't[22] it right to say I see the spatial[23] shape of the towel when I look at it? Is[24] it something I can[25] then draw? Is it something I can remember?[26] If I look at it ten minutes later, will I see the same shape again? Will I notice that it is the same shape?[27] If I say, "No, it now has a different shape", what have I noticed[28]? I may be able to see[29] certain definite changes, as one corner is now to the left or it hangs down lower.[30] Or while looking at the towel I may say,[31] "the towel[32] is changing its shape".[33] What connection have[34] "Seeing the towel" and "drawing the towel"?[35] Sometimes "seeing the towel" means looking at it very closely. "Keeping it before my eyes"[36] sometimes only means[37] saying "there is a towel there". In this second case I just look at the towel without any thought as to its shape, colour, position, etc.

Are[38] we to say, saying "there is a towel there" involves seeing the towel[39]? (Is seeing to be like taking a photograph of it?) I[40] say to a man, "Is there a towel hanging on the wall in the next room?" and he goes and looks, and says, "Yes, there is a towel there",[41] I say then, "Did you see it?" He may give lots of different answers

[22]These two opening words are written over the line with instructions to replace the original first word "I".

[23]'visual' is written first and crossed out in favour of 'spatial'.

[24]Parenthesis is written and crossed out before 'Is'.

[25]'can' is written over the line above a word, which is deleted and illegible.

[26]This sentence is written above the line, with instructions for insertion.

[27]This sentence is written above the line with instructions for insertion.

[28]'something' is added over the line with instructions for insertion, and then deleted.

[29]'say' is first written in the main text, then crossed out, and 'see' is inserted above

[30]A sentence is crossed out here in the main text, which reads: '(Given several different pictures of the towel, I look at the towel and point to one of them and say "the towel has this shape" or I say "the towel has none of these shapes.")' The expression 'look at the towel' is added above the line; the final parenthesis was written, then deleted, and another parenthesis written after this.

[31]'see' is written above 'say', and neither is deleted.

[32]'its shape' is written in the main text, then replaced by 'the towel'.

[33]After 'shape', the main text has expression within double quotation marks: '"the wind is blowing the towel about"', which is crossed out.

[34]'What connection have' is written over the line, with instruction for insertion.

[35]After the question mark a sentence-fragment occurs in the main text and is crossed out: 'have certain resemblances. When.'. Although the sentence-fragment does not have a replacement, the deleted 'when' is replaced by the word 'Sometimes', which comprises the first word of the next sentence, as above.

[36]This quoted expression is written over the line with instructions for its insertion.

[37]'means is written above the line with instruction for insertion, and 'only' is squeezed into the main line in smaller script than the main line's writing.

[38]A sentence is written above the top line of writing, as if added later, and it is crossed out, namely: '(There are of course occasions when it will have sense to talk of seeing the shape of the towel at one glance.)'

[39]After 'towel', a word is written and deleted, now illegible except for the final 't'.

[40]'I' was preceded by 'If', now crossed out.

[41]Before 'I' there was a heavily deleted word, which might be 'Supposing'.

to this question, he may wonder why I asked it, he may not remember if he saw it, he may say “Yes I saw it, it was white and hung from such and such a nail”, he may say, “Yes”. Need I ask him this question, does an answer to[42] it necessarily give me any more information, does it make his information more reliable? If the answer to this question is necessarily yes, why should[43] I need to ask it? (I might have asked him: “Do you see a towel hanging on the wall?”[44] Does this mean anything different from, “Is there a towel hanging on the wall?”

I might mean by it[45] “do you see it or do you only[46] feel it?” as I might have asked if the room was dark, e.g. I ask, “Is there a towel hanging in the next room?” The man says, “Yes.” I say, “Did you see it?” He says, “No, I felt it.”)

But if the next room is well lighted, will the man answer the question by saying “Yes, I see it, but haven’t felt it”. In[47] some circumstances I would say the two questions were different,[48] in[49] others I might not trouble to make a distinction between them.

In a well-lighted room I will not want to ask[50] if the man sees a particular object when he looks at it. There might be a great many different circumstances when I might[51] ask the question, “Did you see it?” and a great many when I would not ask it. It seems then whether I ask the question or not depends on (definite[52] outward) circumstances, and also whether it has sense or not depends on (definite[53] outward) circumstances[54]: “Do you see the towel?” might be asked[55] in some cases[56] if I want to find out if a man has good eye sight or not, if he sees. But when I ask the question in such a case as that above, do I require such information as I get in this case, or are

[42]The phrase ‘an answer to’ is written over the line, with an instruction to insert it.

[43]‘Should’ is written over the line, and the original ‘do’ it replaces is crossed out.

[44]Here a close parenthesis was written and is crossed out. The parenthesis is closed a few lines later, after ‘. . .felt it”.)’

[45]‘by it’ is written over the line, with instruction to insert it.

[46]‘only’ is written over the line, with instruction to insert it.

[47]‘In’ is written over the line and replaces deleted original ‘Under’.

[48]‘were different’ is written over the line with instruction for insertion.

[49]‘in’ is written over the line and replaces the deleted ‘under’.

[50]The words ‘think of asking’ are written above ‘want to ask’.

[51]‘might’ is written over the line and replaces the original ‘would’.

[52]The word ‘definite’ is written above ‘(outward)’ within the parentheses with no deletion indicated.

[53]The word ‘definite’ is written above ‘(outward)’ within the parentheses with no deletion indicated.

[54]After ‘circumstances’, ‘asking’ is written and crossed out.

[55]‘asked’ is written over the line with an insertion mark to place it after ‘be’.

[56]After ‘cases’ the sentence continues, but is now crossed out, as follows: ‘like asking “do you see?”’. This now-deleted expression also has a replacement for ‘like asking’, which is written over this expression as ‘to find out’ – this is deleted.

we at all trying to do the same sort of thing. Suppose I am in doubt whether another man is seeing what I am seeing, and I ask him and he tells me he is seeing quite different things, what am I to do about it? I don't yet[57] know. Am I to say he is lying; am I to think I am seeing an hallucination, or that he is? (In a great many cases seeing the towel is having one's eyes directed on the towel.)

Seeing is associated with all sorts of specialized kinds of behaviour[58] (and with all sorts of specialized kind of external conditions)[59] – in some cases I say a man sees because his eyes are[60] open, or because he walks, acts, moves about, in a particular way, says certain things, refers to certain objects about him – as blindness is associated with a quite different kind of behaviour. Or again because he tells me he sees.

Suppose someone says to me "Do you see?" and I reply, "Yes, I[61] see", I may then follow this up by pointing to[62] a table and saying "I see this table" and such like. (Making an experiment is only one[63] kind of behaviour in[64] certain circumstances, there are lots of other kinds of behaviour; or again[65] the experiment may be made more or less thoroughly.)

Under[66] what circumstances is[67] what I call the experience of seeing to be something which I am aware of. There are clearly cases as when sitting in a room[68] where I look about me being aware of my surroundings,[69] telling myself that I am seeing such and such things, seeing things,[70] looking at things closely.

---

[57]'yet' is inserted above the line with parentheses around it, which are deleted, and there is an instruction for insertion in the main text after 'don't.

[58]'all sorts of kinds of behaviour' is written over the line, together with the word 'specialized' written above this addition, all for insertion as they appear above. The expression 'all sorts of specialized kinds' replaces the now-deleted phrase, 'of a particular kind'.

[59]These parentheses and all their contents are written above the line for insertion after 'behaviour'.

[60]'eyes are' is written above the line, with a mark for insertion.

[61]After 'I', a heavily deleted word is written. Possibly, but not probably, it is 'can'.

[62]After 'to' there is a deleted, illegible word.

[63]After 'one', the word 'particular' is written and crossed out.

[64]'in' is written above the line to replace the crossed-out 'under'.

[65]'again' is written above the line with an instruction for insertion.

[66]The next five manuscript pages have large crossed lines through them. This does not obscure what is written. In view of their qualitative and thematic continuity with the previous part of the manuscript, and with the subsequent continuity with the subsequent part of the manuscript that is not crossed out, these manuscript-deleted pages are reproduced as part of the main published text. The whole sequence of such retained yet-deleted paragraphs commences with the word 'Under' and terminates with the term 'red'. Such retained paragraphs are identified by a footnote placed at its ending.

[67]'Under what circumstances is' is written above the line with instructions for insertion.

[68]'as when sitting in a room' is written above the line with instruction for insertion.

[69]'of my surroundings' is written above the line with instructions for insertion; this expression replaces the deleted original entry 'that I am seeing'.

[70]'seeing things' is written over the line.

There are other cases where I cannot in any sense talk of being aware of seeing, but where I still wish to talk of seeing. What part do these cases of being aware that I see play in the whole (scheme of)[71] activities and thoughts and actions bound up[72] with the idea of[73] "seeing". It is clear that on the one hand they play a very important part, on the other hand they are not the only important cases, there are many other equally important classes of cases.[74] (There are cases of telling myself[75] what I am seeing, as[76] when looking at a landscape.) In what sense can I then talk[77] of seeing being a private, personal activity? Where I am looking round at[78] examining different objects,[79] the special colours, shades of colours, the impressions which these objects give me. But am I then troubled by the thought of not being able to communicate these impressions to other people, am I thinking of these impressions as something incommunicable or rather won't[80] I think that these can be easily communicated to other people, just by telling them.[81] Or again do I think that there is some exact impression which cannot be communicated. If I am trying to think of the right words to express my impression, what sort of activity is this? Certain words come into my mind which do not satisfy me. If I look up to the sky and try to think of the right words to[82] describe its colour, I think of blue, then light blue, then a pale light blue, then perhaps an ethereal transparent light blue. If I compare this with what happens when I look at the sky on[83] another occasion, one thing I can say is that I will think of words in somewhat the same qualitative order (red, dark red, burning dark crimson red, looking at the Northern light). Also these words seem to have quite different[84] kinds of exactness, I.E. I can quite easily explain what I mean by blue[85] to someone who didn't know the language but what I mean by light, aethereal,[86] transparent are

[71] 'whole' and '(scheme of)' are written above the line, with an instruction mark for insertion under 'whole' but not the rest, while the latter is written over the deleted word 'general' in the main text. The next word in the sentence, 'activities' also has a heavily deleted illegible word prior to it.

[72] After 'bound' and connected to it is a deleted word-fragment, probably 'up' or 'in'.

[73] 'the idea of' is written in the main text, crossed out and then re-written above the deletion.

[74] This phrase is written over to make the original form clearer, as with some other parts of words on this page (e.g. 'other', 'cases').

[75] 'telling myself' is written over the line and replaces 'being aware of', which is deleted.

[76] There was a parenthesis before 'as', which is deleted.

[77] 'talk' is written 'talking' with the ending deleted.

[78] 'looking around at examining' is written over the line and replaces the deleted expressions 'pointing out to myself, telling myself the'.

[79] 'I am seeing' is written after 'objects' and is deleted.

[80] 'won't' is written over the line and replaces the deleted 'don't'.

[81] This latter clause is added over the line.

[82] After 'to', 'sky' (a slip for try'?) is written and crossed out.

[83] 'on' was changed from 'of' by Eds.

[84] After 'different', the following expression is written and crossed out: 'fxxxtxxxis' kinds of exactness', and the term 'kinds' is written to replace the crossed-out word 'degrees'.

[85] 'red' is written and crossed out in the main line in favour of 'blue'.

[86] I retain 'aethereal' with the ligature (obsolete), which varies from the above spelling 'ethereal'.

more difficult, and still further more difficult to explain why I connect them together in a sentence. It is as though I had learnt[87] to make certain calculations with words. (To find the product 25 × 25 and to find the right words to describe the colour of the sky.) When I first looked at the case of having an impression and trying to find the right words to fit it, it looked as though I had some private impression which was[88] only communicable if I found the right words to express it, and this might seem to make it correct to say I have private experiences which cannot be communicated. But it is clear I can try to communicate,[89] I can learn to communicate more elaborately, I have certain rough ideas and techniques how to communicate.

(Trying to find the right words is like painting the sky or drawing some object.)[90]

My impressions are only personal in so far as I give personal expression to them in such statements as "I see a light blue ..." "I feel ...", otherwise my impressions only refer to outward[91] circumstances which I can refer to by pointing. But if I have an hallucination, this surely is not an outward circumstance, something I can point to, or if I do point to it other people will not know what I am pointing to. Of course I wish to talk of cases of seeing hallucinations, and perhaps of giving no indication that I am seeing them.

What[92] would it mean to say actively "examining[93] the object I am seeing" it?[94]

What are personal experiences? Would I call looking at a red object and saying this is red a personal experience? Another man looking at the object[95] also says: "this is red". Is it right as to say: "O then it is only my personal response, I see it is red,[96] it is a fact that it is red".[97]

Suppose he says: "But there's no table there, you are pointing at empty space". Here is obviously a case where the man sees something[98] which I do not see. Am I then to say that this shows that what I see "the table" and point to, is not something common to us both and is therefore something private to me. I can say this. But this is a very rare case. And I can only say this because the other man has *told* me that he does not see the Table. In ordinary cases such a state of affairs does not arise, he will

---

[87]'learnt' is written over the line and replaces the deleted 'been taught'.

[88]After 'was', 'dependent' is written and crossed out.

[89]'them' is written in the main text and crossed out.

[90]This sentence and its parentheses are crossed out with horizontal lines through the writing, in addition to the large diagonal lines that are used on this sentence.

[91]After 'to' a word occurs, now deleted, which may be 'external'; then 'outward and' is written, with 'and' deleted.

[92]This paragraph commencing here is more heavily crossed out than are the previous pages.

[93]'examining' is written over the crossed-out expression 'looking at'.

[94]After 'seeing' there is a deleted, only partly readable, fragment, which is: 'it clearly staring'

[95]After 'object', 'will' is written and crossed out.

[96]'I see it is red' is written over line.

[97]This ends the crossed-out section mentioned above (whose commencement is marked by a line and a page break), i.e. from 'Under what circumstances'.

[98]After 'something', there is a deleted clause added: 'which I do not see different.'

not want to know whether I see the table, the chair, the bed which he sees. (When thinking of personal experiences like "I see" think in terms of the circumstances involved, not in terms of mental phenomena.[99] Behaviour is only *one* of the circumstances, others are external[100] conditions.[101]).

Another case when I talk of what I see being private to me is when nobody else is present. It may very often happen[102] when I look at a table[103] in a room by myself that I think to myself[104] "Nobody else is seeing this table, therefore what I see is my own private picture." But certainly one of the things that make me call it[105] private[106] is that no one else is present.

Again another thing I might think is[107] that nobody else is standing in just the same position as I and therefore will see this table in just the way I am seeing it. But it isn't that I[108] don't think this is like the case of the hallucination. If another man comes to where I am standing I expect him to agree with me if I tell him or draw a picture for him of what I see.[109]

When I think of seeing a table, I sometimes look upon it as an experience, happening at the moment and independent of past and future actions and memories. If I say later, "I saw so-and-so".

When I think of seeing a table, I sometimes refer to it as an event[110] happening (at a particular moment). [But one must not mix up 'seeing an event happening' and "I am seeing the event is happening."] In what way does it happen? If it is happening. Is it an event which other people are seeing? Can I say I saw the table at the time when I did see it and whether I remember it or not is irrelevant?

---

[99]This later clause is written over the line with instructions for insertion.

[100]'external' is written over the line, which replaces the deleted 'the lighting'.

[101]An end parenthesis is written here, as is a full stop, and 'my', as well as the expression 'my knowledge of the man's character'; all these are deleted, while the end parenthesis is restored.

[102]The words 'may' and 'happen' are placed above 'very often happens', though the latter is not deleted.

[103]'say' is written after 'table' but deleted.

[104]After 'myself' the expression occurs 'that what I see must be something', and it is crossed out.

[105]From 'certainly' to 'it' is written above the line and replaces the deleted 'it is only'.

[106]After 'private' there are some deleted words in the main text: 'because' appears to be written in the usual soft pencil line, and is deleted, whereas the next few words that are in a sharp pencil form – 'from the fact that'. Above this expression 'is that because' is written, with 'because' deleted.

[107]'I' and 'think is' are written above the line, and the latter expression replaces 'be' in the main text.

[108]'it isn't that I' is written over the line and replaced the crossed-out 'this', whilst the thus redundant

'I' before 'don't' is not deleted.

[109]The following passages are crossed out in the manuscript, though their quality and relevance to the theme attracted AG to include them in the main text of this Edition. They are separated from the main undeleted text by a single line across the page at their opening and ending.

[110]The word 'something' is written above 'an event'.

Do I say when I am seeing the table some event is happening? I certainly don't mean that something is going on before my eyes.

Is all that I mean by seeing a table having my eyes directed on a table? Clearly not. For I wish to talk hallucinations after images – nor do I generally know when I see a table that my eyes are directed on it.[111]

(Do I wish[112] "I see the table" to refer to an event happening. Yes, I do, but what sort of an event it is I may[113] not yet know. I may at the time think I am seeing a table and afterwards say it was a hallucination.[114]

But by saying seeing the table is an event I may wish to say that it is not an action[115] done by me as carrying a chair from one room to another. Seeing a table or other objects, is, I would like to say, one class of events happening in the world.)[116]

Can I say the picture I have when I see the table in this room is something more simple than the table regarded as an object? Suppose a man asks me what a table is and I point to a table and say "this is a table", this would be *one* way in which I could answer his question – by letting him see the[117] table. Another answer would be to give a description of a table.

Again in what way do I refer to the table which I see as a picture (or sense-datum)? (Sometimes[118] I regard it as[119] an object that I will use for eating on, writing on, and[120] think of it in this way, and use it in this way. The[121] appearance of the table only comes in[122] when another man wishes to know[123] what a table is, and when I wish to tell another man something about a table.) I am accustomed to talk of[124] seeing different views of the table. I may be in a room in which there is a table

---

[111]This ends the passages that are deleted in the manuscript.

[112]After 'wish', a heavily deleted expression occurs, which probably reads 'to use'.

[113]Before 'may', 'don't' occurs and is crossed out.

[114]The following section, which is separated by Eds. by two continuous lines, is deleted in the manuscript.

[115]'an action' is written above the line with an instruction for insertion.

[116]These last two sentences are crossed out.

[117]'the' is written above an undeleted 'a'; note that the use of a dash in this sentence is also attended by an undeleted comma.

[118]This opening of the sentence is written over the line, and it has two precursor versions that are crossed out, so: the main text has 'On ordinary occasions don't'; above this is written: 'how can I come to have this'.

[119]'an' is written above the line.

[120]This and the subsequent use of 'and' are written over the line as an ampersand, with insertion mark for insertion.

[121]'The question of' opens the succeeding sentence, and is deleted, with capitalizing '[T]he' in the new opening.

[122]'in' is written over the line, replacing the crossed out 'up'.

[123]'to know' replaces the deleted 'for an explanation of'.

[124]'am. . .of' is written above the line, replacing the deleted 'talk about'.

in the centre. I look at it first from one side of the room, then from the other and then from one corner of the room.[125]

Suppose these different views are represented by photographs. What do these photographs enable me to do? If I want to know how the table looks from different parts of the room I can look at the photographs and see, without having to walk to different parts of the room. I have a sense-datum of the surface of my hand.

[1.][126] "It looked green but it is red". A and B are looking at a chair. A says, "The green of this chair goes well with the green curtains". B says: "O, but this chair isn't green, it's red". A looks again at the chair and says: "Of course it's red, but it looked green just now".

[2.] Compare this with another case: A has gone into a shop with a red sample in his hand which he wishes to match. He selects that cloth which matches with this sample when put side by side with it. When he goes out into the street, he finds the two colours do not match. He says: "They don't match now, but they looked the same in the shop". B then tells A that colours which look alike in artificial light very often do not look alike in daylight. A can test this by experiment.

On the other hand in case N°. [1] B would probably not offer A any explanation. It would not be an experience with which he was familiar.

Should A say in case N°. [2] that the colour of the cloth in the shop[127] was like the sample[128] and was not like the sample in daylight, or should he say that the colour of the cloth was not like the sample but only appeared like it in the shop. It is clear that whichever he says[129] no confusion is likely to arise, as in this case A is quite clear[130] about what he is saying. If another man C is in doubt what A is saying, A will take him into the shop and show him that the colours of the cloth sample are[131] alike. But suppose when A takes C into the shop, they find that the colours are not alike,

[125]The continuous lines from 'Suppose' to 'hand' are crossed out.

[126]The numbers '1' and '2' are used later on for these cases. So these numerals have been introduced in [] here by the editors for clarity; also the next page the manuscript continues the sequence '3.' to '6'.

[127]There is a lightly drawn line to suggest that part of the sentence after 'was' should be transposed; yet there is not further instruction, and it seems the instruction is not completed, and perhaps may be ignored. This line may be connected with the lightly drawn three large crosses drawn over the remaining and the next two paragraphs. This light crossing out continues until, and including, the sentence "...my eye sight is not good".

[128]His phrase is written over the line and replaces a deleted and illegible word – as with another word after 'was'.

[129]This opening to the sentence replaces the deleted version prior to it on the main line, which is: 'It is clear that it does not matter which he says'.

[130]After 'clear' a deleted sentence-fragment occurs in the main text, which is: "how he is using these expression[s]'.

[131]The expression 'colours...are' is written above the line, with the replaced 'two colours are' crossed out.

should[132] A then say that they were alike when he[133] was in the shop half-an-hour ago, or should he only say they appeared alike.

A[134] says[135] they were alike then; but after[136] asking the shop assistant who says they were not alike, A then[137] says: "They appeared alike to me then."

[3.] The same[138] things happen as in case N$^{o.}$ [2]. C[139] is surprised that A made such a mistake. A says: "They were alike in the shop". He takes C into the shop and shows him that the two colours are[140] alike.

[4.] As in [3.] But A and C when they go into the shop find the two colours are not alike. A says they were alike half-an-hour ago when he was in the shop. He asks the shop assistant who backs up what he says.

[5.] As in [4.], but the shop assistant says they were not alike. A asks the shop assistant who wears glasses whether his eyesight is good. The shop assistant says it is not. A says, "They were alike then, but they did not appear alike to the shop assistant,[141] who has bad eye-sight.

[6.] As in [5,] but the shop assistant says he has good eyesight, while A has bad eyesight. A says: "They were not alike then but appeared to me alike. My eye-sight is not good".

****[142]

A is sitting and leaning on a desk with his elbow on it. B comes into the room. A tells him he has been doing this for the last two years. B tries to reach the wall behind the desk but stumbles and complains of being knocked badly on his knees.[143] B does this several times, each time the same things happening.

---

[132]The word 'will' is written above 'should'.

[133]'he' written above the line replaces the manuscript's deleted 'I'.

[134]'If' appears before 'A' and is deleted.

[135]After 'A', 'says' is written in the main line, then deleted, with 'insists' written over it and this in turn is crossed out, with 'says' being restored over the line.

[136]'but after' is written above the line over three words – two are illegible; the last reads: 'and'.

[137]'then' is written over the line to replace the deleted 'will probably'; following this stylism, the above term say[s] is misspelled, here corrected by Eds.

[138]'same' is written over the line.

[139]Before 'C', 'But' is written and deleted.

[140]'are' is inserted over the line.

[141]This term is written over the line.

[142]Asterisks supplied to mark the beginning of a fresh page, which also seems to mark a break or pause in the sequence.

[143]Before 'B' the sentence-fragment: 'After doing this [a] second time' is crossed out.

Is it right to say now that it becomes (more and more)[144] pointless for B to say there is not a desk there (Wisdom[145]).

But suppose lots of people[146] said that A and B went about Trinity drunk and doctors testified that they were drunk, would it then become pointless to say there wasn't a desk there (Redpath[147]).

But if B says there is a desk there, this entails that this does not happen (Paul[148]). 'But I suppose this would mean that he would, if it did happen'.[149] We can say it is pointless for B to say, "there is not a desk there[150]" *provided* this doesn't happen, I.E. that doctors testify that A and B were drunk. But if this does happen, we might still find that all the doctors who had testified had been mad.

Is this also entailed (Redpath)?

If one says it is pointless for B to say there is no desk there, does one mean that *no* future events might make him say there wasn't a desk there? On the other hand B may say: I didn't assume at once there was a desk; I did various things, asked various questions. Surely after doing all this I can definitely say that there is a desk there.[151]

If something does make me say there wasn't a desk there, then I shall be saying something wrong (Paul).

The proposition "There is a desk there" is a proposition about a physical object and we are accustomed to[152] verify it in a particular way, in some such ways as B did in the case above. But we are also accustomed to the fact that physical objects behave in a particular way. Physical objects in general don't suddenly disappear; don't change their appearances. Suppose that the phenomenon of physical objects

---

[144]'(more and more)' is written above the line.

[145]I.e., A.J.T.D. Wisdom, later to become a Cambridge Professor of Philosophy. We have marked the introduction of a question by members of Wittgenstein's audience by indenting the start of such contributions, though it is not so marked in the manuscript. The beginning of such a question is not quotation-marked, which thus involves a decision as to its start; the style and manner of these questions, and also Wittgenstein's contrasting style are strong indications of the commencement of such questions.

[146]'went abou[t]' is written and crossed out.

[147]Theodore Redpath (who later became a Fellow in English at Trinity) recalled to Arthur Gibson the experience recorded in this manuscript.

[148]George Andrew Paul (Trinity: Affiliated Student, 1934, Exhibitioner 1935, Senior scholar, 1936, Sen. Scholar, Fellow, from 1939). In (Paul et al. 1936: 1–2, 70), Paul may be reflecting his grasp of some of the above debate, though he did not acknowledge it. Paul there cites Wisdom (1936, 156–57, etc.), relevant to the above context where Wisdom is also mentioned.

[149]This sentence is crossed out in the MS..

[150]'not' and 'there' are written above the line, as is the next use of 'doctors' which replaces 'people'.

[151]There is an alternative reading in the MS, so: "‖ I can say nothing will ever make me say there wasn't a desk there ‖".

[152]Here Wittgenstein's use of the expression 'we are accustomed to' recurs elsewhere and is worthy of comparative analysis; cf. Klagge and Nordmann (1993a: 293). There are a few other similar terms in both contexts, whilst the arguments and points made are different.

disappearing became more and more frequent, would we then any longer be inclined to accept the criterion of B's knocking[153] the desk and asking A questions. The sentence "There is a desk there" would lose its point in the sense that if we use it now, we expect and are accustomed to find that the desk is in the same place ½ hour later[154] or if not somebody has removed it or burnt it, but if the desk might explode or burst like a soap bubble at any instant, or if we were constantly liable to delusions of such experiences as B had, the sentence "there is a desk there" might become practically useless, at least the present use which we make of it would break down. What would make a sentence "there is a desk there" lose its point would be changes in the experiences we are accustomed to have[155]; B's experiences don't make it more pointless to say "there isn't a desk there", it is rather[156] all the experiences and ways depending on these experiences in which[157] we are accustomed to talk about physical objects which make it pointless.

We might say that if we said after such an experience that there wasn't a desk there it would mean that we had changed our use of "there is a desk there", given it a new verification. One thing that strikes when we think of what B did is that he really needn't have done so much.[158]

We think[159]: Whether B says "there is a desk there" or "there is not a desk there" *generally* depends on the experiences which he has at the time he says it. In general B will[160] say one or the other, and will not wait till he has seen a doctor who will tell him whether he is drunk or not, or whether he is subject to[161] delusions.

Suppose I ask myself or ask other people: How do I know that all my present experiences aren't a dream or a delusion. Can't I imagine that I go up to London tomorrow and return in the evening and find there is no Trinity College here, and that everyone tells me there has never been any Trinity College. But if this happens need I say my present experience I am having now was a dream or never happened. I am at the moment assuming that when I find Trinity College[162] has disappeared I still find I have the old body of mine with which I am familiar, the same face, hands, etc. I shall

---

[153]This word may be 'kicking'.

[154]'the' and 'is' either side of 'desk', and '½ hour later' are written above the line, the latter with an instruction mark for insertion.

[155]After the semicolon 'and' is written and crossed out.

[156]After 'rather', 'changes' is written above the line and deleted.

[157]'depending on these experiences in which' is written over the line, with an instruction for insertion. An abbreviation ('exp.s)' is used for 'experiences', which itself is an insertion above this inserted sentence-fragment. Skinner's way of writing 'ex' is so unusual that it is evident that this abbreviation is for 'experiences', which occurs in the previous line.

[158]The sentence, replete with parentheses is crossed out; it is indented and occupies a separate paragraph in the text.

[159]The word 'think' is inserted above the first choice 'say' on the line.

[160]'make' is first written on the line, crossed out and replaced by 'say', placed next to it, on the main line.

[161]A word, written on the line, is crossed out and illegible.

[162]'College' here is written over the line, with a mark instructing insertion.

have to separate my past experiences (as I remember them)[163] and memories into some which I call delusory and some which I call real, and my basis for doing this will be comparing these memories with what other people tell me about my past.

What puzzles me in contemplating my present experiences is that I can't say definitely which experiences[164] I shall always call real and which ones I could call delusory. The point is that I can imagine circumstances which would make me call some of them delusions, but I think it is essential that I should not imagine circumstances which would make me call all of them delusions – because if I do that, how am I to say that it was *I* who was having these delusions?

One[165] of the sources of Paul's[166] trouble seems to me to be the way we are misled by the uses of the past, present and future tenses in our ordinary language. I am inclined to think that a proposition "Gasking[167] is sitting there", "There is a desk there", entails another proposition in the future, namely "Gasking was sitting there", "There was a desk there".

[163]This parenthetical clause is written over the line.

[164]'experiences' is written over the line.

[165]This paragraph is written on a separate page, while there are five blank lines on the preceding page.

[166]I.e. George Paul, mentioned above.

[167]See Gasking and Jackson (1967).

Chapter 8
Self-Evidence and Logic

VII[1]

I want to go on now to a chapter[2] by itself.

By this chapter I have to interrupt what I have been doing – talking about generality.

This is logic I must talk about.

This chapter title does not head the original manuscript, though the two expressions have been chosen because of their uses in the manuscript.

---

[1]There are unpublished notes by M. MacDonald in Rhees's Papers in Trinity College, based on the same lecture series as that of Skinner's manuscript (for which see various footnotes below); VII has the date of 1.2.35 at its head. There are extensive and significant differences between these versions; there is additional material in the present reproduced in Skinner's manuscript. For example, Lecture Seven in MacDonald's notes is about 680 words long, whereas Skinner's is over 2200 words long, as well as having diagrams that are absent from MacDonald's. [The MacDonald notes state that they derive from the lecture series that commenced at the beginning of the Michaelmas Term 1934, which are resumed in the Lent Term on 21.1.35. If we follow the date of MacDonald's LECTURE SEVEN, Skinner's VII would derive from material delivered on 11.2.35. The first six lectures are covered in the MacDonald notes. The previous Lecture – Six – in MacDonald (dated by her to 6.2.35), which is absent from Skinner's manuscript, tackles a subject that is entirely different from the seventh lecture, i.e. rule following and adding numbers to series ("add 10"). Her notes of Lecture Six end arbitrarily in mid-sentence, with "Whenever a rule is used there" (p.2). Skinner's opening for Lecture Seven is presented as a fresh start, and dictation of a chapter for a book. (So, we should not conclude that in Skinner we have an arbitrarily cut manuscript without its previous chapters because, for example, its opening differs from MacDonald's and Ambrose's. The context matches Ambrose's (1979: 133, para. 3), though neither agrees in specific phrasing.]

[2]The references to 'chapter' are absent from Ambrose and MacDonald, whilst present in Skinner's version.

A. Gibson, N. O'Mahony, *Ludwig Wittgenstein: Dictating Philosophy*,
https://doi.org/10.1007/978-3-030-36087-0_8

One of the great difficulties which is in[3] logic was this – what is the criterion for a proposition being a proposition of logic.[4]

By some it was taken to be self-evidence.[5] This was what both Russell and Frege said. At certain points where this is a difficulty one tends to bring in psychology.

The word self-evident is taken from psychology.

It is what everybody is in his right senses would say is true.

It is the highest degree of plausibility. But it is clear that what is self-evident may not be to another.

People did not mean to say these propositions were self-evident to Europeans.

The self-evidence seemed to be an objective one, not a subjective one. It seemed to lie in the proposition.

It had been said also that logical propositions were tautologies. They were redundant, they were in a sense such that they needn't be said.

But if they needn't be said, what is their use?

This whole investigation was made much easier by something Frege had done.

Frege explained such notions as '*v*' '.' '~' by means of the words 'true' and 'false'. He did it in a very queer way.

Frege had a most extraordinary[6] theory – this is an example of greatness in a mistake.[7]

I will give it in a few words, but I have no hope that any - one of you will be able to see.

Frege had this idea that every sign had a sense and a meaning.

'The morning star' and 'the evening star' had the same meaning but different senses.

He went on to say of propositions: the sense was what it said, and it could have two meanings, 'the true' and 'the false', $\sim p$ means 'the true' if p means 'the false'. His idea was to define a function of p by means of a list. The idea of defining a function by a list.

---

[3]'is in' added by Eds.

[4]Such a question requires many undeveloped researches. A case of deep progress from one angle is Ganesalingam (2013), which shows the way forward.

[5]In MacDonald's (11.2.35.) version the lecture notes begin as follows: "What [sic] the criterion for a proposition being a proposition of logic? Criterion used to be self-evidence, a word taken from psychology." (There is no use of "Russell, and Frege said" in MacDonald's version, though she mentions Frege in a related connection.)

[6]The abbreviation 'exam.' (example) is written after 'extraordinary', then deleted presumably in favour of 'theory'.

[7]This sentence is absent from MacDonald (1935) Lecture SEVEN, p.4. It is typical of those that are absent from her typescript and present in Skinner: i.e., an evaluative overview by Wittgenstein of a position or philosopher. This appears to indicate that MacDonald was listening for, or only wished to note, a narrower scope of data than Wittgenstein offers.

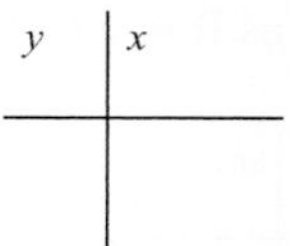

For the argument p 'the true', the value of $\sim p$ 'the false' and for p "the false" the value of $\sim p$ "the true".

p	$fp = \sim p$
T	F
F	T

What Frege did not seem to see was this: that if you can say this sort of thing, this table itself can be taken as a symbol of the function.[8]

It seemed as though what was said in this table was saying something about negation.

What was written down was only another symbolism for $\sim p$: it was just a translation.

I write.

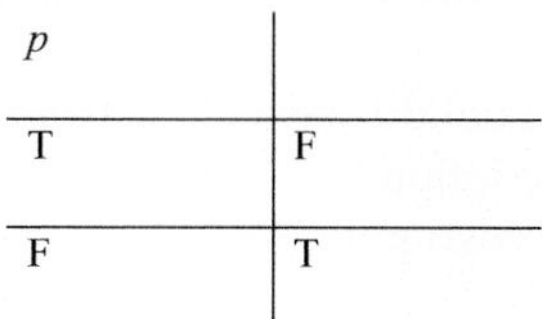

p	
T	F
F	T

$=\ \sim p$; this is another symbolism for $\sim p$.[9]

The importance of this was that this schema was another way of writing $\sim p$.

[8]Depending on how we compute the re-ordering from this above equation back to the beginning of the manuscript, there are 831 or 377 words before this equation to the opening of the manuscript. This contrasts with MacDonald's 200 words back the beginning of the chapter. The difference between 831 and 377 is because Skinner's manuscript re-orders the chapter and adds new chunks of discussion.

[9]This line, which commences with '=' is parallel with the table and its first line in the MS.

Suppose I wanted to write *pvq* as Russell uses it, I write.[10,11]

P	*q*	
T	T	T[1419]
T	F	T
F	T	T
F[1420]	F	F

These are the truth possibilities. In the third column, I put variations of T and F. One can write the third column by translating from ordinary English.

I said here "this is only false if both are false." This is not an experiential proposition; it is a rule which I embody in the notation.

One can do away with[12] the first two columns by writing them in a particular order.

I write it say

(*p.q*) [F.T T T].

Now[13] if once you see that this says nothing about *p v q* but is another way of writing it, then it is found[14] what the characteristic of a logical proposition is.

When I first wrote down these, I had a different notation.

The reasons I had for writing in this way you will hardly understand at the moment.

I wished to show the essential difference[15] between the symbol for a proposition and another thing, say a description.

I wrote a proposition with two poles.[16]

---

[10] An obvious but important point: the pronoun "I" presumably identifies Wittgenstein, as elsewhere in this chapter, not least since Skinner would hardly pitch his authority against Russell's.

[11] The letters in this column are written very lightly, in contrast with the bold emphasis of the letters in the other columns.

[12] 'with' is added here.

[13] 'the' is written after 'Now' and crossed out.

[14] Written in MS as 'find'.

[15] Written 'different' in MS.

[16] The symbols are: "W *p* F W *q* F W." Note the vertical mark contiguous to F, drawn on the lower connecting line.

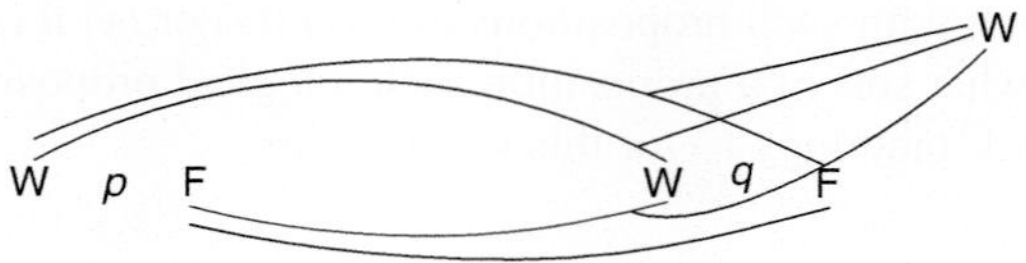

Suppose I write $p \vee \mathrm{F}\, q$

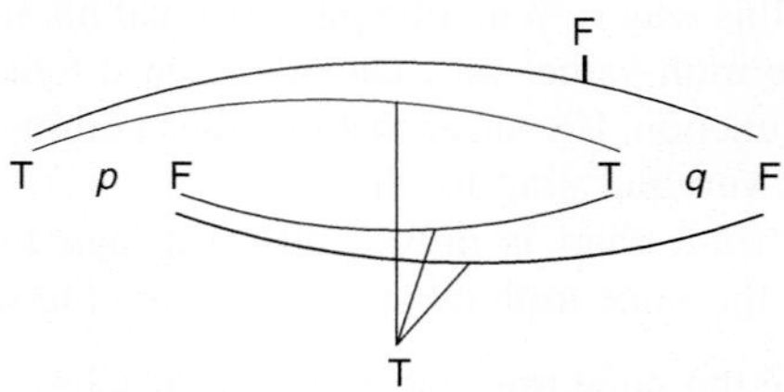

I will write here $p \supset q$.

$p \supset q$ was meant to be false only when p was true and q false.

If now you make a logical proposition by putting p equal to q i.e. $p \supset p$.

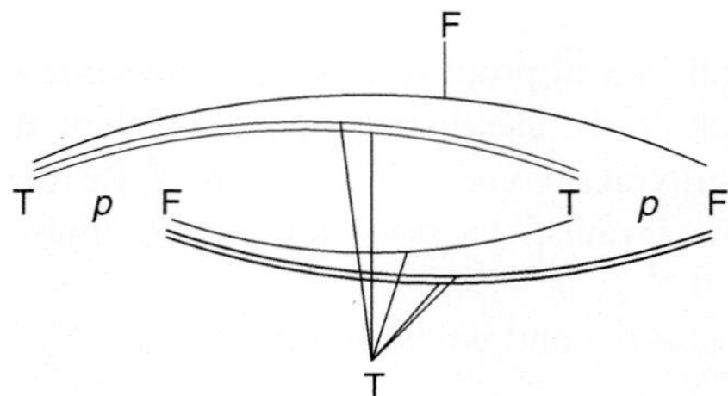

Here you have one argument p and this has two arguments T and F.

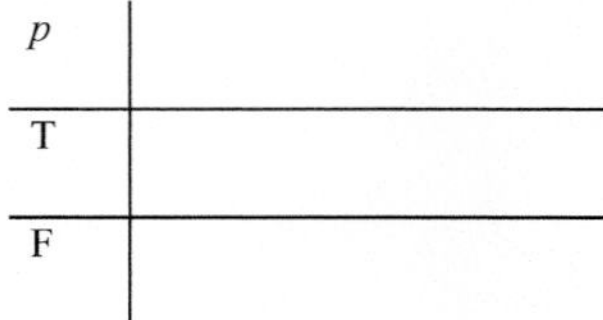

p	
T	
F	

Here the proposition seems to have only one pole.

Here you can give an extremely easy rule to describe what is a logical proposition.

The merit of this is that it gives a quite easily overlookable rule to see what is a logical proposition.

This shows a logical proposition has nothing to do with self-evidence.

[I have not dealt with such propositions as $(\exists x)$ f(x) or (x) f(x).]
Let's look at what sort of a proposition such a logical proposition as $p \supset p$ is.
I used the word 'tautology'[17] for this sort of thing.
First puzzle:

(1) What is the use of writing this down if it is a tautology?[18]
(2) What sort of generality is this when we say this about all propositions.

The first point, "what is the use of such propositions?" becomes much more puzzling if I put it in this way.[19] When Frege explained his functions by means of a list when he gives the truth-values on each side, then it looked as though this said something about the function. If you see that this says nothing about 'negation', etc., the business becomes very puzzling for this reason.

If what I call the truth chart is only a different sign for the function, if two functional signs have the same truth table then they are the same signs.

$\sim\sim p = p$ – they have the same truth table. This is alright in ordinary props.

What about tautologies? Here then all tautologies mean the same, that is: nothing.
$\sim p. \sim p = \sim p$ made no difficulty at all.
But if this is so, then I can write

$$p \supset p = p v{\sim}p = {\sim}(p.{\sim}p)$$

You can say then all logical propositions have the same sense, namely none.

Why on[20] earth then do we use dozens of them if they all have the same sense? The answer is obvious if you ask yourself what sort of use do we make of them. If we tell someone "If it rains, it rains," we don't tell him anything. Therefore it is not this use at all we make of it.[21]

If we have a truth of $p\ v\ q$ and we write this[22]:

[17] First written in the plural, then changed to singular.

[18] The word 'taught' is written before the term 'tautology', then crossed out.

[19] After the end of this sentence, 'If you' is written then deleted.

[20] 'on' is written as 'one'.

[21] The following incomplete sentence is written, broken off, then crossed out: "(The use will have nothing to do with giving us inf)".

[22] This equation is an example of those that are not present in MacDonald's notes.

p	q	
T	T	T
T	F	T
F	T	T
F	F	T

This I call a tautology. Should we call this a proposition or not? Should we say it is nonsense or sense?

This is at first sight a mere question of words.

A good case, I will say, can be made out for not calling it a proposition.

'Nonsense' we call such structures as 'yellow chairs grows' or 'slithy toes'.

These I called sense-less. This stressed the quantity of sense is zero. It stressed this is a degenerate case of a proposition. Again that it is senseless could be the verbal translation of

$p \supset p$ is senseless could be translated into[23]

$$q.p \supset .p. = q$$

A logical proposition adds nothing to a proposition.

There are such disputes as, "Is $p \supset p$ senseless or is it nonsensical?"

All I did was to make out a case for this.

I said that in some ways a tautology plays the same role in this calculus as 0 in arithmetic.

Again whether you allow in your language such a structure as $q.\ p \supset p$ or not (just allowing q) is as you please.

Why do we develop these, then?

I will give one case in which it may become obvious.

Let us example one tautology in particular,

$$p \supset q : p. \supset .q.$$

If p implies q is true and p is true, then q is true.

Adding 'true' is like making the arrow $\rightarrow$ very clear by writing »»$\rightarrow$.[24]

This is supposed to be a logical proposition which allows us to draw the conclusion from p and $p \supset q$ to q.

[23]'p ⊃ q' is written after 'into', then heavily deleted in favour of 'p ⊃ p'.

[24]'To that' begins the next line, yet is deleted.

This which I have written down is a tautology. You can see this by the notation I have given.

This therefore says nothing.

On the other hand if we regard it as the logical proposition by which we go from p and $p \supset q$ to q then this obviously seems to say something.

The other extraordinary difficulty seems to be this. Let us call this a proposition. This I can't say in this I have inferred it, rather I say I can infer it.

$$\frac{\begin{array}{c} p \supset q. \\ p \end{array}}{q}$$

Did I need another rule to derive this by means of this?

This seems to be a link but why shouldn't there be other links to be connecting this.[25] What has this to do with the inference? Is it another proposition or is it a rule?[26]

It isn't a rule as a rule should say something and this says nothing.

The solution of this problem is this.

We can make up rules by, say:

$$q.(p \vee q) = q$$

Here $p \vee q$ adds nothing to q I.E. $p \vee q$ follows from q.

Notice in the tautology we had twice[27] the sign of implication. This sign seemed to be connected with the word "follows". You remember in our symbolism with the two poles, we let all the truth functions coalesce.

In such a picture it is clear what it means to say it has no sense.

The sign ' $\supset$ ' seems to have something to do with the word 'follows'.

What is the connection?

[28]I said I could give a rule of inference by writing down an equality.

I said I could give it in a totally different way. I shall say if I have a proposition

$$\mathrm{P} \supset \mathrm{Q}.$$

[25]The next sentence was commenced by 'This seems to', but has been deleted.

[26]'Is' is spelled 'If' in MS. (Also, as with mathematicians of the period, Wittgenstein and his amanuensis sometimes end a conditional antecedent with a full stop, and (presumably for emphasis) commence the consequent with a capital letter.)

[27]Note that the original word order has been retained here, rather than the more usual 'twice had'.

[28]This sentence originally commences with a phrase, now crossed out; the first word is 'The', while it is possible that the second term is 'same'.

If then instead of P and Q I put such functions that this becomes a tautology then we say Q follows from P. I.E. a rule of inference.

If I read $p \supset q : p. \supset . q$ off, is this a rule of inference?

These are different.

If you write $p \supset q : p \supset q =$ Taut.

then this is an inference. That it is a tautology is that it allows the inference.

You *might say whenever such a thing is*[29] a tautology then we accept the inference[30,31,32].

$(p \supset q.p).q \;=\; p \supset q.p.$

You will have noticed, I had two equations. In the first on the right-hand side I had the letter T. I must justify this. I explained what I meant by tautology by the true and false schema.

I want to explain how Russell could have written it in his notation and what part in it the word tautology would play.

Russell could have written his primitive proposition[33] in this way.

p ⊃ pis a tautology according to our true and false schema.

I could have written

$$p \supset p = \text{Taut.}$$

[N.B. the assertion sign ⊢ plays in Russell's case no other rule than the full stop. The function of it is to show where the sentence begins. The beginning of the sentence is marked by the assertion sign. In Frege the assertion sign was meant to correspond to a mental sign as opposed to a possible assumption. I don't usually say to you "It rains," leaving it to you to find out whether it is or not.

We could imagine a language in which every statement consisted of a question and a yes or no.]

In Russell's notation there would be different kinds of equations.

$$I.E. {\sim}(p.{\sim}p) = Taut.$$

Another kind would be this: $\sim\sim p = p$.

---

[29]These italics correspond to broken underlining under these words, which is not matched word for word; so it may indicate a pause for the reader and or the author to reflect on the phrasing.

[30]A symbol was written after the '=' and before 'p ⊃ q', then deleted; it seems to be 'q'.

[31]Notice how the right-hand side of the equation has a parallel with "$\vdash p \supset q.p$" in Wittgenstein's MS 222, p.151 remark §212, which manuscript is drawn upon for the earlier part of his *Remarks on the Foundations of Mathematics*.

[32]This underlining is reproduced from the MS.

[33]'as' is written then deleted in favour of 'in this way'.

$$Or \quad q.(p \vee q) = q.$$

The primitive propositions will have to be written with 'Taut'.

[34]We might use the word 'Taut' (or some other sign -∞-) only once in my primitive proposition. I could form the logical product of Russell's primitive propositions and put it equal to Taut.

I could give this rule of transformation:

$$\textit{From} \quad p \supset q = \text{Taut} \quad \textit{I transform to} \quad p.q = p$$

If I have the case.

$$p \supset q : p. \supset .q = \text{Taut}. \tag{1}$$

I could transform this to.

$$(p. \supset q.p).q = p \supset q.p \tag{2}$$

This would be my primitive rule in the calculus.

This becomes important when we talk about Russell's propositions about generality.

Next time I shall show clearly the connection between (1) and (2).

Then I will talk about the generality of tautologies.

Then I will talk about in what sense Russell's propositions about generality are tautologies.

Then I will talk about what has been said: that I showed that mathematics consisted of tautologies. This is nonsense. The truth is extremely simple: that this which I am doing is a bit of mathematics.

VIII[35]

We talked the other day about inference, implication and tautology.[36]

One of the things I should like to explain is[37] the exact relation of that *method* which uses the T – F symbolism and that method which Russell uses.

---

[34]'What' commences this sentence, but is deleted.

[35]MacDonald's (1935, p.5) date for this lecture is 13.2.35.

[36]Most of this lecture is absent from Ambrose's account, and MacDonald's (1935, pp. 5–7) typed copy is very different from Ambrose's. MacDonald's opening 'sentence' is: "Inference and implication", which among other things misses off 'tautology'.

[37]The word 'is', which is inserted over the line, could be inserted by Wittgenstein's hand since he sometimes (but not typically) writes a looped 's' such as is the case in the manuscript here. Skinner typically writes 's' very much like the printed 's' (as with Skinner's writing in the present manuscript, for example VII, pp.4–5, VIII, p.4, etc.). So the attribution is uncertain here.

Russell sets out from a certain number of primitive propositions, to develop from these he sets out further (other) propositions.[38]

That which produces a queer muddle is the relation between such a system as Russell's and its application.[39]

What I am going to say will lead to a comparison of such a system as Russell's (and its application) and such a system as Classical mechanics, say as given by[40] Newton's three laws (and its application of law).[41]

The translation into the T–F symbolism showed the difference between it[42] and its application and that of another system.

Like[43] Russell's primitive propositions:

Examples – $p \vee p . \supset . p$

$q . \supset . p \vee q .$

We might ask a question or refrain from asking it – namely "Why did he pick these?"

First answer: "These are always[44] true for all ps."

This would only satisfy primitive minds. We discard this asking whether he tried them for all propositions.

The next answer: They are tautologies; that's why he chose them.[45]

---

[38]Wittgenstein's pencil hand (which is a different pencil line style and width from Skinner's) adds the word 'further', with his parentheses around '(other)', and he placed square brackets around '[From these he sets out]' with an arrow to indicate that he re-orders and adds to Skinner's hand to indicate, as far as one can see, the result above.

[39]The word 'to' is inserted above 'and'; the word 'between' is added above 'application' in Wittgenstein's hand; also, there is an underline.

[40]The phrase 'as given by' is inserted in ink over the line in Skinner's hand, together with new ink parentheses (those which open a parenthesis after Russell's name, and close after the first use of 'application', as well as the final two parentheses for the last phrase of the sentence). A pencil parenthesis around 'or Newton's Laws' has been deleted. Wittgenstein's hand lightly adds a final phrase that appears to be 'of law'.

[41]Behind this reference to Newton lies the possibility that Newton's 2nd Law may have a family of subsets within it (cf. J. C. Baez (2011)).

[42]'itself' is first written then deleted in favour of 'it'.

[43]"like" appears to be in Wittgenstein's hand.

[44]The word 'always' is diagonally crossed through in pencil with one long stroke, which seems to be in Wittgenstein's hand as elsewhere on the same page.

[45]These three lines are in Wittgenstein's handwriting, though written on the reverse of the sheet for this context. The last line is encircled in the same pencil as the symbols, whereas the main text is in ink in Skinner's hand. These details are not present in MacDonald (1935); nor in Ambrose (1979: 138–39).

$$p : p \vee p . \supset p$$

$$p \vee p . \supset p$$

$$p \vee p . \supset .p : \vee : p \vee p . \supset .p . : \supset : .p . \vee p . \supset .p .$$

Anyone of us if we wished to see if these were correct will say something about them.

e.g.[46]

$$q \supset r . \supset : p \vee q \supset p \vee r$$

If you don't read this with a clear punctuation in your language it will seem cumbrous.

You[47] may say to yourself: "In what cases is it true in what false?" This is in fact what I have embodied[48] and given in the T – F notation. But remember this is[49] nothing but another notation.

When I[50] say they are tautologies I mean that *by some process of thinking*[51] he has arrived at the view[52] that they are logical propositions.

But this is only a translation And if we only say it in a different way,[53] why do we need to say it all? Couldn't you imagine people who didn't do this? You would have to say, "Yes."

I wouldn't in all cases go thro' this process and the process in any case is a process of translation.

I said we could begin with these

$$p \vee p \supset p$$

$$q \supset p \vee q$$

$$q \supset r . \supset : p \vee q \supset .p \vee r$$

There was[54] something about the T – F notation[55] which avoided the abyss which is not avoided here.

---

[46]Changed from "I.E." to "e.g.". This detail is mentioned because Wittgenstein's hand alters Skinner's handwriting, for which see Chap. 2.

[47]Before 'You' there is a vertical oblong, which may be there to indicate a new paragraph, since it is towards the end of a continuous line and after a full stop.

[48]The words 'embodied in' are added above 'given'.

[49]Here in the MS the incomplete expression 'anothe' is written, then deleted.

[50]'you' seems to have been written first, then deleted, with 'I' (presumably Wittgenstein's first person) is squeezed into the available space.

[51]This italicized expression is underlined.

[52]'at the view' is added by Eds.

[53]'way' is not in MS.

[54]'are' is written first then deleted.

[55]'there' is written then deleted.

If you say you can begin here, then you may ask, "Why?" If it is not answered by saying they are tautologies, you may be tempted to say they are true.

When we talk of the application, we are talking of the application of logical propositions to ordinary propositions e.g.[56] "I am leaving the room," etc.

The T – F notation shows the relation between the tautology and ordinary propositions.

Suppose we said it is all a question of axioms.

If all you mean was that Russell started out with these and did certain things with them, etc.

If you call them axioms,[57] remember you don't know anything about their application.

A mathematician will say this doesn't matter.

This is alright as long as you are then not tempted to connect this with propositions.

If you look at it in this way nobody would have troubled about this.

But all this did assume that it was in some way comparable with ordinary props, I. E.

"if *p* implies *q* or *r*."

Let's try to translate the language we used last time into Russell's notation.

What in Russell's notation corresponds to our word 'Tautology'?

Russell does not use the word at all nor apparently anything corresponding to it.

At first sight it looks like this.

I am going to explain now what appears to be the relation between Russell's and our view and how they seem to contradict.

We are inclined[58] to say Russell assumed certain axioms and deduced certain propositions.

We on the other hand seem to say something about these axioms, namely they are tautologies, hence we seem to go one step further than he.

We could put the apparent difference as: "Russell assumed them, while we justified them."

We needn't say Russell assumed them: we can say what corresponded to our form of tautologies is his taking them as primitive propositions.

The calculation is in fact a group developed out of these six primitive propositions by means of certain operations.

And if you now asked me what would correspond to my sentence, "____ is a tautology" in Russell's primitive propositions would be answered by saying, "It is

[56]Note above in VIII.2. that Wittgenstein has changed I.e. to e.g. in line with present context.

[57]What appears to be Wittgenstein's hand draws a line from 'axioms back to 'ordinary propositions' above.

[58]MacDonald (1935, p.5) has a different attribution: "Russell was inclined". Prior to this sentence she has two schematically sketched points not present in Skinner: "You do not explain the connection with pns. By calling them the prim. Pns. Axioms. It is a pure game."

saying that it is[59] a proposition derived from one of Russell's primitive propositions."

It has not been said at all how these are applied say to "It rains."

Where Russell was so terribly misleading was that these expressions look and were taken to be propositions in the ordinary sense."

If we wish to put Russell right, we could say Russell took these expressions and made other expressions.

Then comes the task of explaining what is the relation between these and such propositions as

$$p \vee q \supset r \text{ etc}$$

Russell, beside these six primitive propositions, used a primitive proposition in words

I.E. "Anything implied in a proposition which is true is true."

If we translate this into Russell's notation, it comes to saying.

"If $p \supset q$ and p is a true proposition then q is a true proposition."

This comes to $p$enable us to draw the inference $\supset q. p : \supset . q$ which he does not[60] wish it to. For he says this is true if p is false.

What is this which[61] enables me to draw this inference? It must obviously be a rule of inference.

In what sense do the words, "Anything implied by a true proposition is true" enable us to draw the inference that $p \supset q. p : \supset . q$ does not.

Is this the same as to say, "Anything implied by a proposition if it is true is true"? In all other cases he would say this was the same.

The point that he wouldn't say it was the same was that $p \supset q. p : \supset . q$ is not a rule. The point really is that he ought not to give this proposition at all. He said too much.

He did not need to say it in this way at all. The rule he wanted to give was

$$p \supset q.p : \supset .q = \text{Taut.}$$

Now I ask, "Should he have given this rule at all?" Need he have given this rule in the form at all? The answer is that he never applied it except to his primitive propositions and their consequences. He could have written this as a rule, but given as a *general* rule it was no good as it was only applied to his primitive propositions and their consequences.

---

[59] 'is' is inserted by Eds.

[60] 'not' is inserted above line.

[61] The word 'which' is written in Wittgenstein's hand, over the line with an instruction to insert it.

As he began with this general rule of inference couldn't he have proved this (as he did in fact). If he gives this $p \supset q . p : \supset . q$, this is not at all a rule he is allowed to use.

We shall put it: every logical proposition is a rule of inference from one normal proposition to another normal proposition.

Therefore his "$q \supset p \vee q$" etc. has as its application the use allowing the inference from q to $p \vee q$.

This also $p \supset q . p : \supset . q$ is as all others as a rule of inference (between normal propositions).[62]

We shall have to distinguish between two

Kinds of inference:

(1) The inference from one proposition to another normal proposition.
(2) "The rule of inference" in his use of[63] logical.

The rule of inference which he should have given for his system we could call "Rules of logical inference" as opposed to "rules of material inference" by which I infer from your coming into the room to "It rains."

The rules of material inference are the logical propositions.

The rule of logical inference is not a logical proposition.

It is a[64] rule how to handle a certain 'group' of symbols.

Logical is defined by this group – the operation which enables one[65] to go from one member of the group to another must be specified for this group.

[If anyone of you has troubled to look closely at mathematical proofs and where they seem paradoxical and interesting – it is a tiny point, just a tiny distortion of the truth.[66]]

I will go on to talk about the point I left unsettled when I said one could write the rule of inference both in this way

$$p \supset q : p . \supset . q = \tau$$

and in this way

$$(p \supset q : p).q = p \supset q : p.$$

I will put it slightly differently.

[62]'Therefore' is written on a separate line, deleted.

[63]'in his use of' is added by Eds.

[64]'grou[p]' is written and deleted.

[65]'one' is inserted by Eds.

[66]The manuscript reads oddly: "has seemed where they/seem/paradoxical and interesting is a tiny point". The word "/seem/" is marked for insertion, and it is retained above. This squares oddly with "has seemed", which we delete, and judge that Skinner intended this. We add '– it' to mark the proposition here deemed "tiny".

If we have an implication $p \supset q = \tau$
this comes to $p.\, q = p$
I could[67] translate these two into my writing and they would become clear.
But[68] I want to do it in a different way

$$p \vee q = p.q \vee q = q$$

In Russell's symbolism $p\,.\, q \vee q = q$ would not be proved.
(Russell never wrote $p \vee p = p$ – he did not start out with equalities.)

$$\sim p \vee q = \sim p \vee (p \vee q) = p. \supset .p \vee q$$

$p \supset p \vee q$ is one of Russell's propositions.

$$p \supset q = p. \supset .p \vee q$$

Our transformation has led us to this. We have shown $p \supset q$ is a logical proposition.

I want first of all to talk of a mistake I made. We said we could write down all truth formulas[69] in this way

p	Q	r	s	t	
T	T	T	T	T	
F					
.	. —	—		— .	

In the last row we write some combinations of Ts and Fs.
When the question arises: what about $(\exists x)\,.\,\phi x$? –A general proposition.

We have said it is most tempting[70] to take this as a logical sum (and $(x)\,\phi x$ a logical product).

I should like to go into a false idea.

There are two tendencies if you try to analyze such expressions. They have something with a logical sum or product.

There is an objection to this.

[67] Skinner starts writing 'sh', deletes it, then writes 'could'.
[68] 'it' is first written and then deleted it in favour of 'I'.
[69] Written 'Frs' in text.
[70] Written in MS as 'tempted'.

If I say "all men in this room have hats", I can't replace this by "Mr. Wisdom has a hat and . . ." because then there must be another proposition "And these are all the men in this room".

There is a way of getting out of this.

Suppose you talked of 'things' in this room instead of "men".

Here I could enumerate the individuals and not need to say they are all.

If I enumerate the men in this room then man is a predicate to the things that are men.

What does it mean if there is such a thing as enumerating all things. To say that these are all is not a proposition about the world but a grammatical proposition.

This I said was alright because in $(x)\boldsymbol{x}$ I talked of all individuals, not all men. This would apply to such things as permutations or the roots of an equation.

I saw later I was mistaken in supposing that there could be an enumeration of individuals. In such a case this $(\exists x)\ \phi x$ would really be a logical sum.

Here the individuals would not be given by enumeration but by saying they are all individuals.

Suppose I wished to say in some language all permutations of the letter a, b, c, d were used.

I could do it in this way or I could do it by enumeration.

The atoms were given in the same way as we say "all combinations of a, b, c, d".

I made up another function $(\mathrm{N}x)\ fx$. This was a logical product of negations, I.E.

$$\sim f(a).\sim f(b).\sim f(c)$$

and so on until we have gone through all the individuals in the word.

This is true only in the case when all are false

p	q	r	s	t	
F	F	F	F	F	T
F					F

. . —— ——

I wrote this [T . . . .].

The mistake here if you ask if this is a logical product the answer should be 'No' simply because there is no number of Fs written.

All this arose from a puzzlement about the expressions $(\exists x)\ f(x)$ and $(x)\ \phi x$.

This might seem to be done because I was keen on reducing these to this notation. The idea was not to reduce the number of primitives to one.

It had a very important reason or rather two.

One thing that troubled me for years; we talk of propositions, all sorts.

I said there must be something which is the general form of all propositions.

The other thing was this: – If from $(x)\, f(x)$ there followed $f(a)$, if this was really the relation of following, then it had to be the same relation as $p \vee q$ follows from p.[71] What all these latter relations had in common was clear from the T – F scheme.

I said to myself: It had to be a relation between the truth of $(x)\, f(x)$ and the truth of $f(a)$.

It had to be related in the way one truth function was contained in another.

There is another difficulty which I have not mentioned which must be understood to understand the business.

Whenever we tried to unify things, there were[72] people who would say, "Why couldn't they be different things?" This was a scientific question – I was clear it wasn't that, it wasn't that we could go on and discover five or six different things.

This again has something to do with the idea of logic which Frege and Russell had – that logic was a science about certain objects.

From this point of view there seemed to be no trouble in discovering a new sign. (I.E. Keynes[73] discovered a new sign.)

This seemed to be[74] utterly wrong ideas.

I only saw the mistake I had made much later when I saw that the words "this says something" or "this is a proposition" does not refer to one calculus.

In Russell's ideas, there seemed to be *one fundamental* calculus and every other calculus could be based on this.

If there was one fundamental calculus, one had to show what was characteristic of this calculus.

After discarding the idea of logical constants as a science, there was a word 'sense,' a word 'proposition', a word 'generality', a word 'logic'.

To "proposition" there corresponded the word "logic", to the word "logic" corresponded "mathematics".

Mathematics was based on logic which treated of propositions.

It became imperative to discover the general form of 'proposition', 'logic' etc.

If you have one idea of logic, you must be able to give the formula of logic, the formula of proposition.

That's what I thought I had done – the T – F notation. This replaced the word "proposition".

It was a question of showing one formula for logic; otherwise it may be anything.

This joined up with saying "one could give the general form of a mathematical state" as one could give the general form of logic, etc.

They were all equivalents in a row.

---

[71] Skinner wrote '*p* follows from *p v q*'. Wittgenstein revises and reverses this order – as above, having encircled the original "v q", so that only 'p' is left, and he wrote 'v q' to follow after Skinner's writing of 'p'.

[72] 'were' is inserted here by Eds.

[73] 'did' is written and deleted.

[74] A word is written, perhaps it is 'become', and is deleted and 'be' written over the deletion in Skinner's hand.

This breaks down when you see there is not one idea of proposition, that one calls lots of different things propositions, a family and not bounded.

It is therefore a natural science, there are logical calculuses.

You can make inventions in it but not discoveries.

There have been people recently who introduced new logics. *This is not what I am doing*.

At the end of the eighteenth century people began to talk of non-Euclidean geometry.[75] This form 'non– ...' made a great impression. This led to the idea of a 'Non-Aristotelian logic.'

One, I think consists in this.

It has three possibilities, 'True,' 'False,' 'Possible,' say T, F, P.

You would make a table of *p, q* and vary the T, F, P underneath.

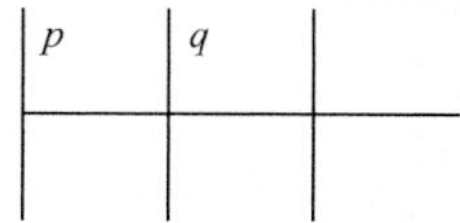

There is a huge danger in making up such games. Taken as a game there is no reason I shouldn't do anything.

The idea of such a game is that it destroyed prejudices. You say: You think it looks always in this way, well, it needn't.

The question is: "Why do you make up this game"?

Unless he says it amuses him, he must show us its application.

This makes it appear the man has found a new country, has extended logic, has made a discovery.

You might ask a question: If I criticize this man's calculus and say it is no use, why did I give you the T – F notation?

I will answer this and you can see whether it applies to the other.

Taken as a piece of calculus the T – F calculus is as dull and useless as any other.

The point of it was to make clear the relation between the calculus and its application.

It has only a value as a medicine if you have a particular disease in your brain.

The 'T – F – P' notation does not serve the purpose – it seems to make this if anything more muddled.

[75]MacDonald (1935, p.7) does not have this sentence, while she does have: "Logic is a calculus and you can make inventions in it but not discoveries. But I am not taking the view of Lewis and the Warsaw school that there are many different 'logics'."

IX[76]

We must talk about what it means to talk of a discovery in mathematics.

A false appearance is created: three valued logic.

The points is what do[77] you use it for?[78]

There is another aspect, which I must talk about straight away.

This is the connection between the idea of a proposition and the so-called laws of logic. At first sight it appears the laws of logics, for instance $\sim (p. \sim p)$, are propositions about propositions.

We have seen they are tautologies and say the same thing, that it[79] is nothing.

The temptation to say these are laws which hold for every proposition is comparatively small.

We have not gone into the point: what does it mean to say

$$p \vee \sim p$$
$$\sim (p \sim p)$$

hold for all propositions.

Ask such questions as: Have you tried it for all propositions? Do you know it?

What is misleading is the way we talk about it.

If I said, "all apples are sweet", you would immediately ask how I know it, etc. It is quite obvious that our case here is not the same.

What is it that holds?

I suppose it holds for one particular proposition if it holds for all.

Take "the black-board is shabby."

Then we have seen that this says nothing about blackboards.

If $p \vee \sim p$ has any use, it might be some sort of inference. We might on some occasion pass from p to $\sim p$.

[Is he at home? No. Then he is not at home.]

Why does one look at $p \vee \sim p$ and $\sim (p . \sim p)$ as two fundamental laws.

They have a very impressive form. *This* or *that*.

It is called the law of excluded middle.

It is like saying, "This is red or green, there is no third."

It has this impressive form (it is an important thing in life to say either this or that).

---

[76]Dated by MacDonald to 18.2.35.

[77]'do' is squeezed in space after 'what' and 'you' written.

[78]'for' is inserted by Eds.

[79]'it' is written after 'that' and deleted; restored here by Eds.

Suppose in human life $p \vee \sim p$ had not been said for many ages and then a man said this $p \vee \sim p$, this would have produced probably a great impression. It would be like a riddle for the whole human race.

We have said $p \vee \sim p$ could be used as a rule of inference.

It could be used a part of a rule,

I.E. $\rightarrow p \vee \sim p =$ Taut. $\rightarrow$ I.E. a rule.

What about this rule holding for all propositions?

Brouwer[80] has said there are propositions for which the law of excluded middle does not hold.

This sounds like a *discovery*.

I could show you there was a definite temptation to say this.

What would be the answer to this?

If I said I had found propositions for which this didn't hold, I suppose you would say you wouldn't call this a proposition.

You may say the propositional game is not played like this.

Here you see a justification for the[81] definition: "A proposition is everything that can be either true or false."

What does it mean?

Iron is everything that combined with sulphuric acid has such and such properties (those of iron sulphate).

It seems as though you can *try to deny* it and if you can it is a proposition.

What am I to try, what is to be the result?

There is an angle from which this makes some sense.

As you give rules to children: if you want to know whether this is nominative or accusative ask who or whom.

Or tell a child to say the not form.

Or if you are doubtful about the spellings, write down two versions side by side.

You thereby rely on a part reaction.

This rule is really of this sort.

I have put the thing slightly wrong; the whole point[82] is that the words "true" or "false" sound like "red" and "green", what we usually call properties.

What does it mean to say: "a table can either be true or false"?

To say a proposition is everything that can be true or false comes partly down[83] to saying that a proposition is something played with this rule

---

[80]Note that Goodstein refers to Brouwer in his letter, reproduced in Appendix [J]. Also, from 1929, Wittgenstein's mathematical colleague Max Newman, as early as 1923 (p.6, footnote 1), had taken on board a similar use of Brouwer to the above, which is reflected in Newman's teaching of undergraduates such as Goodstein and Skinner. Newman (1926) combined the foundations of mathematics with theoretical physics.

[81]After 'the' 'sta' is written and deleted.

[82]'point' is written over the line in Skinner's hand.

[83]'down' is inserted by Eds.

$$p \vee \sim p = \text{taut.}$$

There are two versions:

(1) A proposition is everything that is true or false.
(2) A proposition is everything that can be true or false.

(1) is straightforward nonsense.

The 'can' in (2) introduces a new element which is dangerous.

If you say, "Only a proposition can be negated" this sounds as though there was negation and something else which was a proposition or not. I.E. negation is a shape and a proposition is what fits the shape.

You say 'apple' and 'negation' don't fit while 'the apple is red' and negation do fit.[84]

But suppose I say 'apple is not true' means 'apple is sweet'. Is there such a thing as keeping the idea of negation constant and seeing whether it fits an apple?

~ Apple: what is it to mean negation here? Aren't we deluded into thinking we have tried to put the two together?

Of course this delusion comes from the use of the word 'meaning' and that idea that it is an idea: here, we say, we have two ideas which don't fit.

If you have ideas I.E. say '~' and 'apple' to yourself what is not fitting? Is it that you get sick?[85]

Further we do use such expressions as "Not an apple". You say this means: "I don't want an apple." But what we say is "Not an apple." That is, we do have a game in which we use "not an apple."

Perhaps you have the idea of two games not fitting: the game which we usually play with "not"; and the game we play with "apple".

We have talked about something similar before: about the rule we would have to give identity in order that it should express negation.

What we showed was that this was a rule connecting the word "not" and the word "negation".

We use Russell's calculus in such a way that for p and q we don't put propositions but such words as "apple", "man", etc.

What do you call 'to negate apple'? I can say it, but you will say I don't mean it. I would say I don't make any use of it, but it's quite possible a use can be made of it.

When I said one did as a matter of fact make a use of "not an apple," you might object that here as a matter of fact apple is a proposition.

Hence the example is not quite fair.

Let's see in what sense it has a new meaning, in what sense not.

The relation between the two meanings we might express by saying: In one case it means 'apple,' in the other 'bring me an apple.'

---

[84]See Wittgenstein's small notebook MS 157b, f.28r for the 'T' negation sign, and f.26r for 'Apfel'.

[85]Quotation marks for '~' and 'apple' are supplied by Eds.

In the primitive language this is 'apple' and the way it is used. You say, "If we translate it into our ordinary language it is 'not bring me an apple'." But this in the first case isn't true.

Rather shouldn't you say you regard it[86] as one of the thousand things you do call proposition.

Let's say it now in a revolting way: "You are saying now I can negate an apple."

What can you do with an apple? Eat it. All I have done is to put ~ and apple side by side.

How can one negate such a thing? – We aren't asking to do something with it.

You might say why should we be[87] tempted to call either the word 'apple' or 'not an apple' not a proposition.

You should now show us the transitions from what we ordinarily call propositions and this.

What is usually called the meaning when the meaning is something accompanying the word is something like the gesture.

We look at the word as something more artificial and the gesture as something more natural.

If you ask me if one can make[88] use of such a game in which there are 'or,' 'not,' 'and' and words like 'chair,' 'table,' etc.

I have given one use.

I say now 'We use words and *by these words we mean orders*.'

This last [remark][89] is suspicious.

I say by the word 'chair' we mean 'bring me a chair'.

When I say I describe the use by saying I am going to use 'book' as I generally use 'bring me a book'. This refers to the practice. But the practice does not come into my language; it is not a rule of my language.

If you ask, "Is there any use for this game?", I will say, "Yes, I will describe you one."

If you ask, Is the 'or' 'not' in this game the same as our 'or' 'not'?

What is the criterion for this?

Is it the use? This is shown in the games.

Is it the associations? The gestures may be the same – very likely.

Question: 'Is the word "not" in "It does not rain" the same word as in 2 × 2 does not equal five.'[90]

---

[86]'it' is supplied by Eds.

[87]'be' is supplied by Eds.

[88]'make' is supplied by Eds.

[89]'remark' is supplied by Eds.

[90]The varying uses of single and double, and no quotation marks, in direct speech is worthy of attention here; they do not seem ad hoc – i.e. not a result of arbitrary transcription by Skinner. Re. the latter, note that the sentence that commences *'Is the word for "not"*, was initially opened with a double quotation mark, which was then changed to a single. See Heal's (1995) analysis of dialogue and argument and its use of punctuation in the *Investigations*.

He lays down a game of v or $\sim$ and says whatever one puts in $p \ v \sim p =$ taut is a proposition. This is one way in which you can use the word proposition.

We said we could do away with[91] the words 'true' and 'false'.

We could do the same with proposition.

We could say, "The proposition it rains is true."

Then we could leave out both the words "proposition" and "true" and just say, "It rains."

Suppose you say: Suppose it is not just played with words like table, but say words 'blue,' 'table,' 'long,' etc.

I can either describe the game or its application.

When I talked about the words 'apple,' 'table' used as orders, that was immediately alright for you as it was of obvious use.

But what is a practical thing for us depends on the kind of life we lead.

I explained the use by giving an example which immediately seemed a good way of using it.

But I could have chosen an[92] example where you would not at all have seen the use.

We could imagine an absurd game in which a person was taught to react to words in all sorts of way[s], I.E. if I say 'lift' he has to smash something etc.

But you say: What's the use of this?

I say I don't know[93] but why shouldn't it have a[94] use?

This has something to do with what is called pragmatism, I.E. saying a proposition is true if it is useful.

Here he has a particular idea of use, I.E. of a man in laboratory[95]

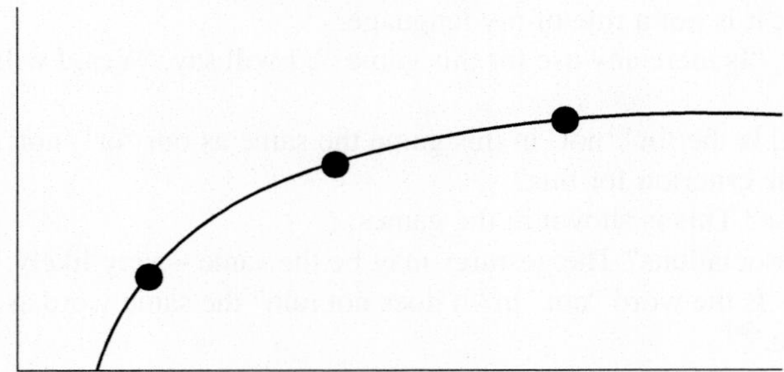

[91] 'with' is added by Eds.

[92] 'an' is added by Eds.

[93] 'who' is written, then crossed out.

[94] MS has 'an' not 'a'.

[95] It is instructive to compare and contrast this graph, and reference to "lab", etc., with Ambrose 1979: 142. MacDonald's lecture notes have no diagrams at all for this lecture in contrast with Skinner's; this not untypical.

Here the chart must predict something: this is what is meant here by useful.

But if the scientist is a lunatic and pays men a great deal to get the wrong chart, then here the wrong chart is very useful in another sense.

All that I want to say rests in this notion of fitting (i.e. the idea that negation does not fit an apple).

I will try to talk about something different but which is closely connected.

Russell introduced symbols of "three- or four-term relation".

Russell said jealousy was as three-term relation and love was a two-term relation.

In what sense is love a two-term relation?

What has a man discovered who discovers that love is a two-term relation?[96]

Question: It does seem in this case as tho' something was being said as to how 'love' was used.

Suppose someone said: Is there such a thing as a two-term relation? You say: "Yes, there is one, love is one."[97]

The first impression is this.

Suppose Russell had been in[98] doubt if there were 3-term relations and had then thought of jealousy.

It seems as though we found a natural phenomenon which fitted a certain schema, and that without it the schema would have been empty.

It seems that in a way the usage depends on certain natural facts. If there are these facts, we can use them: otherwise we can't.

I can say, "What have you discovered?"

Have you discovered two people in love with each other or two people not in love?

I could imagine there was no love in[99] our[100] tribe and then came to another and discovered nobody was in love.

If I could discover no two people were in love, I could also use now my 2-term relation as I would also have discovered love.

Suppose you said: He might have discovered a use for the word "love"; and for the phrase "A loves B".

The symbol itself would, I would say, already give me a perfect justification for using the word.

We[101] might make a discovery, I.E.[102] that such an expression can be useful but this usefulness depends on our actual way of living. I may use a thirteen-term relation and then discover that it is useful.

---

[96]See an earlier, brief, use of a similar point in Wittgenstein (2005: 384).

[97]It is worth comparing and contrasting this form of expression and its context with Ambrose (1979: 142).

[98]'in' is supplied by Eds.

[99]'with' is written first, deleted and 'in' placed over it.

[100]It looks as if 'as' was first written, and then changed to 'our'.

[101]'These are' is written before 'We' and deleted.

[102]We have moved back this occurrence of 'I.E' to this point from its position in the MS immediately after 'useful', since it appears to have been intended to mark the clause that it introduces.

But there is no such thing as discovering a use which gave it content and meaning. I can discover something which gives it use in the sense of practical use.

The calculus which I make now with the words does not receive any contents with what is found – this is what makes it absurd to say. . . .

Imagine we had the words 'red' 'green' 'yellow' but used them without ever pointing to coloured patches.[103]

If we take language, we might make two distinctions

(1) The[104] words, their transformations etc.
(2) In our use of the word 'blue' there is reference to a sample blue[105]:

Is this patch part of the application or of the language? The patch is part of the language. It is a paradigm.

Let's see what role the phenomenon of love play[s] in the use of the word "love".

What was wrong with our example of the man from one tribe going to another?

Suppose we have one couple which served as our explanation of the word 'love'. This couple would be like our patch and serve as a paradigm.

You understand that this paradigm seems to have the word 'love' as[106] content. Apart from this sample nobody need be in love.

What it seems is that we had to have a paradigm. But this belongs to language. The paradigm enlarged the game: it was no application of the game.

Can we say we have found a phenomenon which gives 'A loves B' sense? We have added it to the game with the word 'loves': we enlarged it by putting[107] in a paradigm.

It means nothing to say:

[103]Term in singular in MS.

[104]Lower case opening in MS, here capitalized to match (2).

[105]The original manuscript is written in black including the sample patch. Wittgenstein could have instructed a blue colour to make up the patch but did not. He is inviting us to think of the notion of the colour blue rather than a patch of a particular shade of blue.

[106]'as' in the MS is written 'a'.

[107]'in' is written after 'putting' then deleted.

fits the word 'blue'.[108] We have made up a game with 2-term relations and a paradigm.

We thought we had made up a hollow form into which we found something accidentally fitted.

X

Last time I first of all talked about what it meant[109] to say it was not possible to negate a thing only a proposition.

We left this and asked what was added to a three-term relation if a certain phenomenon in nature was found.

This links up with all the different cases, of trying to build up a logic which was prepared for all eventualities.

We don't leave anything open at all, we make a calculus with 13-term relations. It[110] will probably look like $R(a, a_2, a_3 \ldots a_{13})$.

The sample belongs to your language, not to the application of your language.

You need the sample to say these 13 things are not in this relation.

To discover the relation of love would be a genuine discovery (in natural history).

To take a sample and say this is being in love.

Let's see if we can find anyone like this, this is not a discovery.

We are really here[111] talking about the application of mathematics.

I will go back to that business about negating an apple.

I will first ask you what's your criterion for negating an apple.

If your criterion is that the word 'not' should be accompanied by a feeling or gesture, I should say 'why not'.

You may say this hasn't been to the point.

What you really mean is that it should be used in a particular way – that makes the negation.

If you say this, how can you ask whether you '*can*' use this with the word 'apple.'

I suppose by putting it with the word 'apple'[112] you are presupposing a use.

We gave a use of ~ apple. Here I did make the ordinary use of ~ apple. What seemed wrong to you was that I didn't use it as it would be expected from the word 'apple' standing here. But it is no use asking 'can you use it' if you have already decided not to use it.

What does the use of apple include? Does it include the use of negating it?

Where does the fitting come in?

---

[108] This shape is the same as that in Wittgenstein's 1935 C Notebook (MS 148), p.39.

[109] 'it meant' is written twice, second occurrence deleted.

[110] 'We' is written before 'It' then deleted.

[111] The word written is 'hear', a mistake for 'here'.

[112] After 'apple' the pronoun 'I' is written and part of another word, which are crossed out.

We are wavering between two different aspects of it.
One aspect is this:
Apple is one thing, idea ready-made.
It is comparable to a shape, say[113]

Negation is some one definite shape.
say[114]:

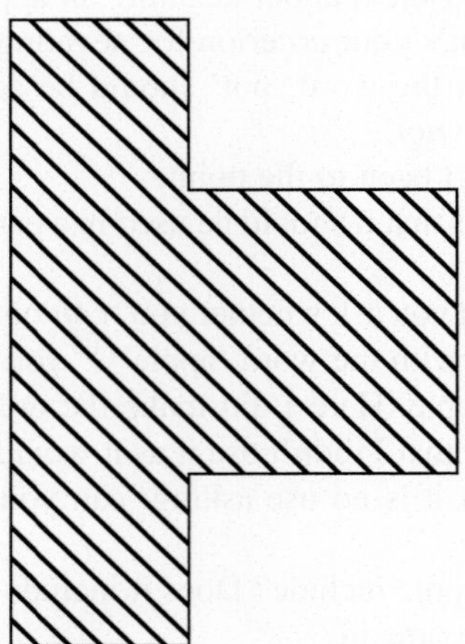

[113]Notice that while Skinner's notes for Lecture 10 are replete with illustrations, MacDonald (1935: 10–13) has none.

[114]This shape and its complementary form are the same as that used in Wittgenstein's unpublished 1935 Notebook MS 148 (pp.37 and 39). The discussion is different, though some underlying concerns aim at the same end.

The point is now: Do they fit?

The second aspect is this:

The negation is not completed till its use is[115] completed.

If you treat them as characterised by their use, you can't say their use doesn't fit, as ~ Apple must then have its use given.

The two ideas between which you are wavering are two ideas about meaning:

(1) It is something in your mind present while you are using it.
(2) It is the use.

The thing which you must understand is that as soon as we talk of use; we cannot talk of fitting.[116] Because of the use they have, they must have together.

It has something to do with what I once wrote, "You can give a sign a wrong meaning."[117]

When we say we cannot negate apple, then it seems to us:

(a) This is not an experiential statement.
(b) That we can describe what we can't do.

Suppose we have used the word "not" as we usually use it: also the word "apple".

Suppose someone says: What you can't do is to use the expression "~ Apple" as you used "~ Apples are sweet".

I could now say this.

When you say I can't use this like that, how am I to[118] describe this?

I could ask this: Does the use of the word "apple" itself exclude the negation or doesn't it yet?

If you don't say this, I don't see why we shouldn't use it, at least as far as apple is concerned.

The same is true of "not".

How do we fix the use of the word "apple"? I suppose by the rest of the usage and by general rules.

Either we have already given the rule that the word 'apple' can or cannot be negated. The same applies to not.

If we have not forbidden such a combination, there is no reason why we shouldn't have it.

You might say that I have been giving a use of '~ apple' and this contradicts me.

For ~ was used roughly as ~ is used but apple was not used as apple is used.

If I say 'apple' when I want an apple, no one would normally say it was being used in a different way.

---

[115]'The word 'its' is written in the MS; changed to 'is'. The first 'it's' is written with the apostrophe in it; 'not' is inserted above the line.

[116]For analysis of "fitting", see Luntley (2015: 148–50); versus tracking, cf. Summerfield (1996: 100–38).

[117]He states the negation of this in the *Tractatus* §5.4732: "we cannot give a sign the wrong sense".

[118]'to' is added by Eds.

We have here two bodies which fit or don't fit.[119]

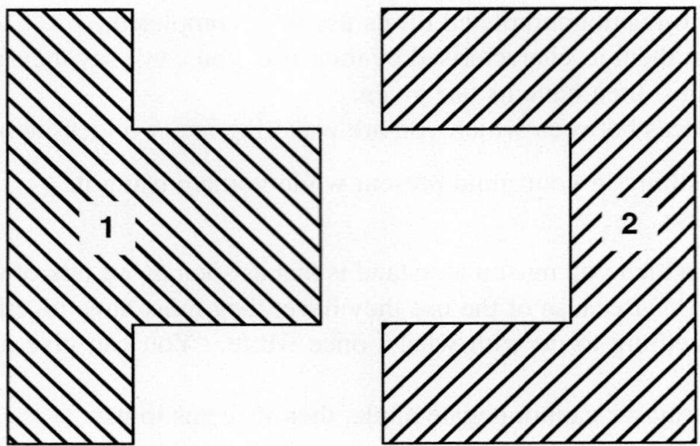

That this fits this may mean two different things. If I give you the measurements of the hollow shape, you know what fits it.

I.E.[120] you give at the same time the measurements of what fits. This is in the geometrical sense.

But suppose (1) was capable of contracting and expanding.

In one case to say they fit only one set of measurements is required, in the other two sets of measurements and a physical statement. [Consider (1) in a certain temperature.]

Let's consider now the fitting of '~' and 'apple'.

If I say this:

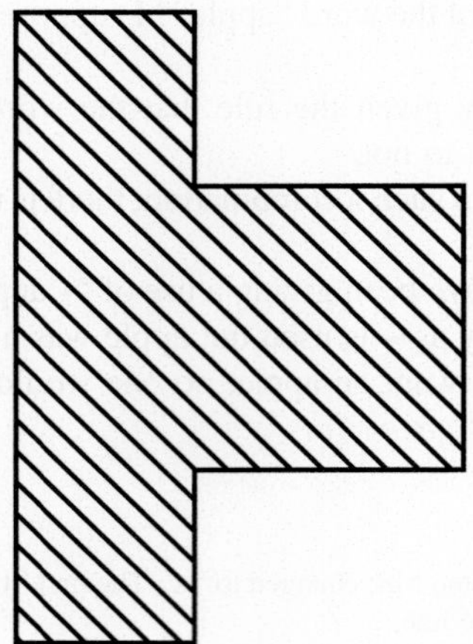

[119]The numerals '1' and '2' are written in each 'body' respectively.

[120]After 'I.E.' 'to' is written and crossed out.

fits this:

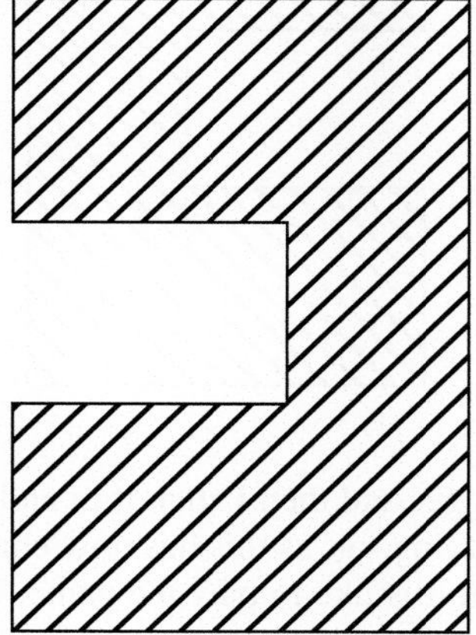

This may be an experiential statement. Here the measures need not be given.
Let's examine the proposition:
"You can't fit this

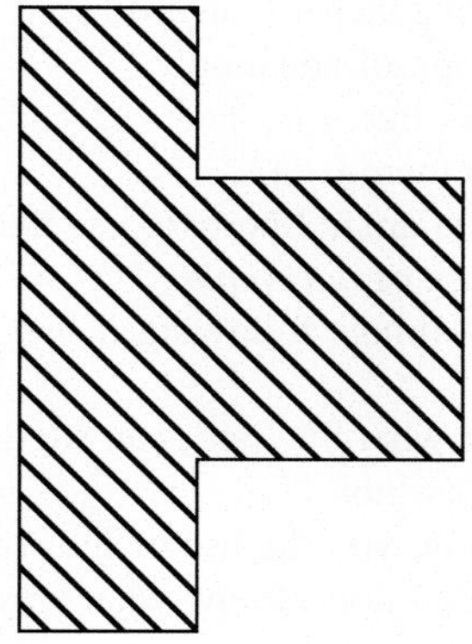

and

together."

Suppose one looked bigger and then I stick it in and it does fit.

You say: This is not playing fair as this has now changed size.

You must describe what it is like to play the game. Then I will know what I am allowed to do. Because just now fitting means they have the same diameter.

We might talk about the uses of "~" and "Apple"[121] being pliable.

I am able to do whatever you can describe to me.

This has something to do with the question of the application of mathematics.

We say this piece of chalk is longer than this and also this piece of[122] chalk is 2 inches, that piece is three inches.

To say 3 inches does not fit 2 inches comes down[123] to saying $3 \neq 2$; and is a rule.

Let us fix the use of the sign of negation.

Could I say negation looks like ~ ( ) or ~ .?

Does it say partly what is there ( )?

If it says it partly, then you are at liberty to use the rest as you please.

I say to you: Give me a use of *not Apple*.

You can insist on certain things following it, i.e. you may say, "I don't allow names of fruit to follow it."

Could one say: By saying what fits 'not', can we describe the way 'not' is used?

Suppose we say, "apple fits 'not'."

Suppose my task is to show you the use of the word 'to'. I say: "to fits apple". Have I told you anything? You don't know what way of fitting I mean. You know how 'sons' fits apple.

What would you do to describe the way you mean?

Why was this procedure humbug?

---

[121]Note that 'Apple' is spelled with a capital A in the MS.

[122]'the' written before 'chalk' is crossed out.

[123]'down' is inserted by Eds.

In order to explain something, I should have to[124] explain the way 'to Apple' is used.

Suppose I had explained the word not to you and I then said "'to' takes apple as 'not' takes apple." To say 'to' takes[125] the argument apple does not say anything except that 'to apple' is one configuration of the game. (Chess.)

Instead of 'apple' take the name of a colour 'red'.

I have 'is red' and write 'go is red'. Here you could ask me whether the use of 'go' is described by what fits it.

Here you could make a guess. You might think it was some object or an after image.

Whereas it might be something quite different, I.E. nothing is red.

I might say: "go" fits 'is red' as "this strip of material" fits "is red."

You might then say that it does make some sense to say it gives the use.

It suggests something, but it could suggest all sorts of different things.

In the end we have to explain it.

I can give you examples of describing the use.

One[126] can't lay down the use of the word is given by what other words fit it.

The question I will go back to is:

Are mathematical equations tautologies and if not, what is the relation of mathematical equations to tautologies?

If I have such a proposition as 2 + 2 = 4 is this a tautology?

There are two things to say for it being a tautology?

(1) You say 2 + 2 = 4 is self-understood.
(2) There is a tautology which very often has[127] been mistaken for this equation.

The tautology I mean is:

"If there are two things there and two things here then there are 4 things here or there."

You know Russell could have written

"*f* is satisfied by *x* and *y* and not by more things"

by[128]:

[124] 'it' is first written and then deleted with 'to' replacing it.

[125] 'app' written after 'takes', then deleted.

[126] Before 'One' 'There' is written then deleted.

[127] The MS reads 'have', changed here to 'has'.

[128] Above the equation retained in the main text, the following draft equation has been written, and then crossed out: "$\exists(x\,y).\,(f(x).f(y)) : \mathrm{f}\,(\mathrm{z}).\ \supset\ z = x\ v.\ z = y$". Written above this equation is: "$\exists\ (x)\ y\ z$". All this is deleted; and below it the above equation in the main text is written and retained.

$$(\exists x\, y)(\, f(x).f(y)) : \exists(x\, y\, z)f(x).f(y).f(z)$$
$$\supset z = x.v.z = y.$$

In this proposition of Russell's there is a sign of equality: this can be eliminated altogether.

I will introduce this now.

Russell used in his symbolism a sign of equality (as in the one I have).

I.E. in

$$\exists(x\, y\, z)\phi x.\phi y.\phi z \;\; \supset x = y.v.x = z$$

This misled us to think there was such a thing as a proposition x = y or $x = x$.

The use we make in arithmetic of the sign of equality we can always look at as part of a rule. (For this we write that.)

In Russell's notation, it looked as though there could be an experiential proposition of two things being equal.

If in Russell we write:

$$(\exists x, y)fx.\, f\, y.x = y$$

This comes to writing $(\exists x)\, fx$.

We make a general rule: we talk of one sign if there is one; two[129] signs if there are two different signs.

Russell wrote $(\exists x\, .\, y)\, f\, x.\, fy$ and left it open whether they were equal or not.[130]

I write

$$(\exists x\, y)fx.\text{fy v } (\exists x).fx.$$

Suppose I wish to say in Russell's notation: "There is only one x such that fx"

He writes $(\exists x)fx : (\exists x, y)fx.\;\, f\, y \;\supset$ x. y

I write: $(\exists x)\, fx\, .\sim(\exists\, x,\, y)\, f\, x\, .\, f\, y$

The merit of this writing is that it makes it impossible to write $x = y$ or $x = x$.

In Russell there was a proposition

$$(\exists x).x = x.$$

It seems as though one could say, "There is a thing which exists."

No such proposition is allowed in my notation.

---

[129]'different' is written after 'two' and crossed out.

[130]In a typical variation and in a shortened form, (MacDonald 1935, p.13), this sentence reads differently as: "Russell leaves open the question whether x = y but I say we do not want to leave this open but only if they are not equal will we use two letters." There is no text in MacDonald for the rest of the lecture, whilst here Skinner has some fourteen lines of extra detail.

I will write the above proposition (E 1 x) fx.
If I wish to write

$$(\exists x\ y) f\ x . f\ y . \sim (\exists x\ y\ z) \ldots$$

I will write (E z x) $f\,x$[131] [This is the phrase: there are two things which. . . .]

XI

There seemed to be a great difficulty in bringing in[132] the idea of identity into logic as it seemed to bring in a relation between objects.[133]

$x = y$ seemed to involve[134] some relation between the two things about which we talked.

If I use such an expression[135] as:

$$(\exists x\ y)\phi(x) . \phi y \supset x = y$$

I seem to have to say something about x and y.

If we write this, we must be allowed to write:

$$(\exists x, y) x = y$$

What, then, does this mean?

I say this to make it clear what was the point of that notation which did away with the equality sign.

We also had such a prop as $(\exists x)\ x = x$.

To many people, we said, this seemed to say there was a thing.

The temptation to write $a = a$[136] is removed altogether by[137] this notation. This was regarded once as fundamental in logic and was called the law of identity.

---

[131] 'as' is written after this equation, then crossed out.

[132] The word 'in' is written twice here – last word of the line and first word of the next line.

[133] Although the full stop is raised and lighter than others, it seems certain that this is the meaning of the mark.

[134] Insofar as there is a parallel between Skinner and MacDonald (1935, p.13) here – and the latter's notes are much shorter and lack Skinner's technical detail, it is plausible to suppose that her use of 'to involve' is the missing expression that Skinner has omitted here. A sentence later MacDonald has the statement: "Russell's notation is about the thing referred to. But I never say, this chair is identical to itself."

[135] 'exampl' is written before 'expression', then crossed out.

[136] This equation is mentioned in MacDonald's (1935: 25) unpublished "Lectures on Logic, etc.", though the contexts are very different.

[137] 'with' written, crossed out, and 'by' inserted over it.

Russell's notation suggested as strongly as anything the proposition $a = a$.

I will go on to the use of the identity sign in:

$$1 + 1 = 2.$$

I want to discuss the idea that 1 + 1 = 2 is a statement or abbreviated statement of the form, "if I had one apple in one hand, and if I had one apple in the other, then I had two apples in both hands".

We have a sign (E 1 x) ϕx

I want to write in this[138] symbolism the above sentence.

"I have only one apple in one hand"[139] will be written

(E 1 x) ϕx . x is a thing ϕ is have apple in hand[140,141,142]

$$(\mathrm{E}\ 1x)\,\phi x\,(\mathrm{E}\ 1x)\ \ \psi x.\sim(\exists x)\ \phi x.\psi x.\ \supset$$
$$.\supset\ (\mathrm{E}\ 2x)\,\phi x\,v\ \ \psi x.$$
$$(\mathrm{E}\ 2x)\Phi x = (\exists x, y)\phi x.\phi y.\sim(\exists x\ y\ z).\phi x.\phi y.\phi z.$$

I can now instead of 1 + 1 = 2 take

$$15 + 27 = 42$$

$$(\mathrm{E}\ 15x)\phi x.(\mathrm{E}\ 27)\psi x.\sim(\exists x)\phi x.\psi x.\ \supset .(\mathrm{E}\ 42x)\ \phi x.v\ \psi x$$

This[143] was not the addition, as we didn't know what to put after the last E till we had added 15 and 27.

Imagine this: I have here 15 and 27. I will put 56 instead of 42.

Is this correct?

I will write the whole like this:

$$(\mathrm{E}\ 15\ x)\ \phi x.(\mathrm{E}\ 27\ x)\psi x.\mathrm{Ind}.\ \supset .(\mathrm{E}\ 56x)\phi x.v\,\psi x \qquad (1)$$

Suppose I wrote down 2 + 2 = 5.[144]

---

[138]'Russell's' written before 'this' and crossed out.

[139]Quotation marks are added by Eds.

[140]Opening parenthesis is written and crossed out before the negation sign.

[141]Closing parenthesis is written and crossed out after 'x'.

[142] ⊃ written twice contiguously – once at the end of the line and once at the beginning of the next line.

[143]'This' is written twice, one crossed out, with extra space between – perhaps to enable indenting the sentence.

[144]This sentence is crossed out.

In my rules of addition $a + b = c$. There are some which are proved by means of others.

For instance $1 + 1 = 2$ can be treated as a definition of 2.

I can treat $3 = 1 + 1 + 1$

5[145] $= 1 + 1 + 1 + 1 + 1$

I can then prove $5 = 2 + 3$[146]

If I wrote $2 + 3 = 6$, you needn't say this is wrong, but that '6' only means what we mean by 5.

If you say $2 + 3 = 5$, then you are assuming a particular calculus.

In our calculus as it is $2 + 3 = 6$ is wrong.

(1) is a contradiction. It may or may not.

How are we to find out?

Suppose I did it by means of the Truth–False notation.

I could write (a) in the long way.

We would have in the 1ST bracket 15 letters then 16 letters, then 27 and 28 letters of the alphabet, and lastly 56 and 57 signs.

How should I know whether all this was a tautology or not?[147]

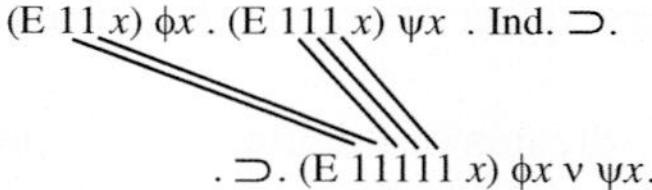

This is now a tautology or continuation.

In order to know what I have [I have] to go thro' a kind of calculus which I could write down by the lines,

I have calculated $2 + 3 = 5$.

The result of the calculus is that the whole thing is a tautology.

$$= \text{Taut.}$$

Whether it is or not I only know by adding, by a process which[148] comes to adding.

---

145 '3' is written underneath and changed to '5'.

146 '3' is written oddly.

147 This symbolism reads:

$$(\text{E } 11x)\phi x.(\text{E } 111x)\psi x.\text{Ind.} \supset .$$
$$. \supset .(\text{E } 11111x)\phi x \; v\psi x.$$

Skinner's drawing of the lines is much more informative and technically communicative than MacDonald's (1935: 14) Lecture Eleven.

148 'we' is written then crossed out.

We thought this might come to[149] 2 + 3 = 5.

It is in one case a tautology in the other a continuation, but we don't know which.

2 + 3 = 5 corresponds to the implication being a tautology, I.E. corresponds to the equation which puts this equal to tautology.

I could instead of a simple calculus of addition surround it unnecessarily by[150] symbols like E and taut, etc.

I could then instead of 2 + 3 = 5 write

$$(\mathrm{E}\ 11x)\ldots \qquad = \qquad \mathrm{Taut}$$

This would have to have the same multiplicity as the calculus with which[151] a child calculates 15 + 27 = 42.

If you want to know whether 15 + 27 = 42 or not write down[152]

$$(\mathrm{E}\ 15x)\ldots\ldots$$

and examine whether it is taut or not.

But to do this you have to go through a calculus of the same multiplicity as that which you show 15 + 27 = 42.

There is besides a very queer thing in the idea that this is the same as[153] 15 + 27 = 42.

Nobody if he adds ever dreams of thinking of any propositions.

This joins on to Russell's theory of cardinal number.

Russell tried to define the cardinal number in this way: he said it was a class of classes.

The first idea he introduced was the idea of correlation.

The idea was this. They[154] first introduced the idea of numerical similarity, of 'being equal in number'.

He said two classes were equal in number if they were correlated one–one.

---

[149] 'to' is added by Eds.

[150] Word order reversed; it is written: 'by unnecessarily'.

[151] The MS reads 'which with' and 'with' is deleted. I suggest the above reverse order, rather than deletion, was intended by Wittgenstein.

[152] As with the previous use of the form four lines above, to the far right of this '=' and what might be 'Tau' is written, and then crossed out.

[153] 'the' is written and crossed out.

[154] May be Frege and Russell, or Russell and Whitehead in relation to their *Principia Mathematica*.

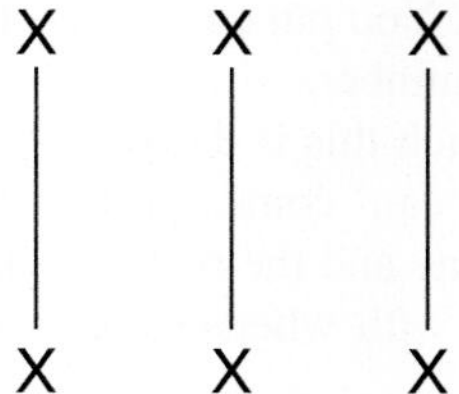

If they are correlated this way, he and Frege said they were equal in number.

They said the number 3 was the class of all triads, this would mean the class of all classes correlated to one of them, (0 0 0).

We have, say, one triad which serves as prototype and the others are correlated one–one.

In fact then you can then take any.

Russell took the number of '0's.

Then he took 0, 1 and took them as the prototype of couples.

We could perfectly well take as the prototype of couples this

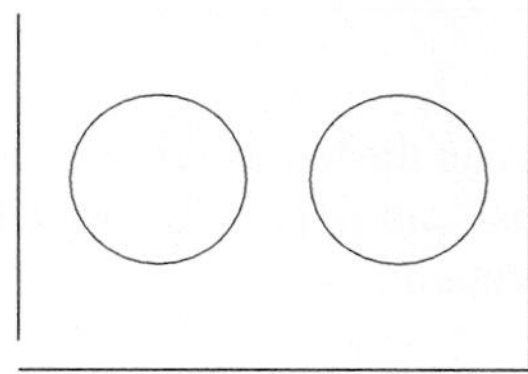

Russell and Frege then said this, that the no. 2 is the class of class one–one correlated to

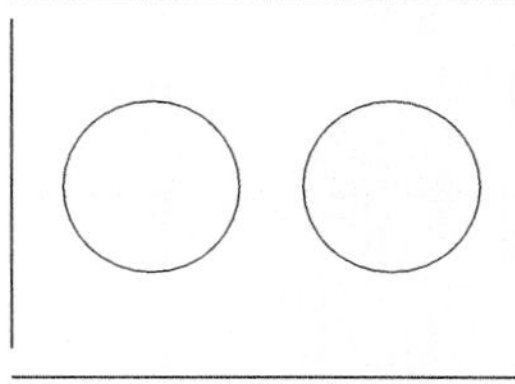

Amendment to this: Say 'can be correlated'.

This suggestion won't work.

What does it mean to say[155] that they can be correlated one–one?

I suppose I can try.

[155] 'to say' is written above the sentence.

You have saucers and cups. You put cups on each saucer. If a cup or saucer is left over, you have not the same number.

This is the source from which this is derived.

This problem of the word ‘can’ comes in in[156] the difference between what we call the geometrical straight line and the real straight line.

The question is connected with whether the class is given in extension or by a function.

I write

$$(\mathrm{E}x, y)\phi x.\phi y \quad (\text{this meaning `there is only'}).$$

Suppose I had two lists A and B.

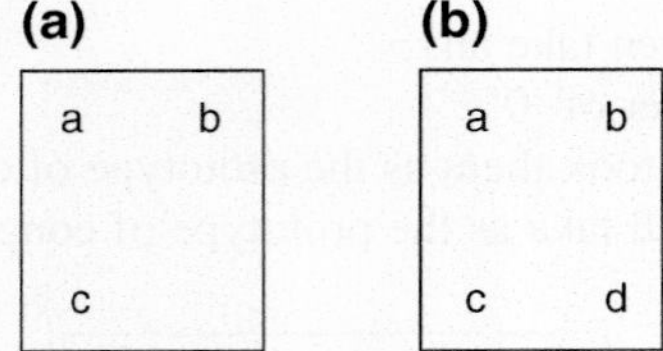

Suppose I asked someone: can the letters on A be correlated with the letters on B? Is this the same question as “can *a, b, c* be correlated with *a b c d*?”

This is a question of experiment.

XI 8

If I have this kind of case in which I determine names, equality by correlated

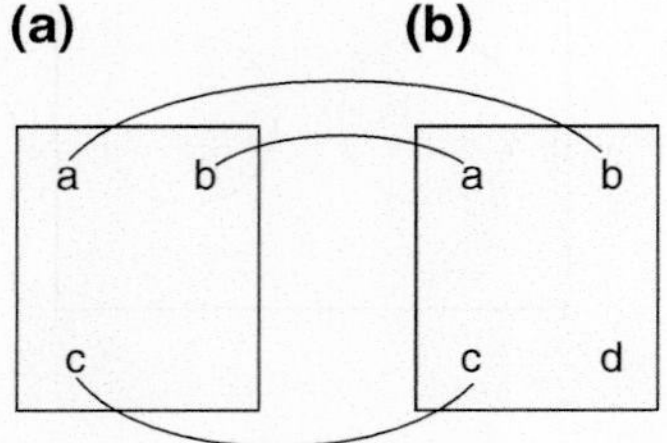

This is clear that this is like some sort of measurement of number. This is comparable to a method of measuring number.

156‘in’ is written twice.

We are now talking of the measurement of number.
One way is a one–one correlation.
For instance I give each man in the room one of the nuts in my hand.
I give another way:

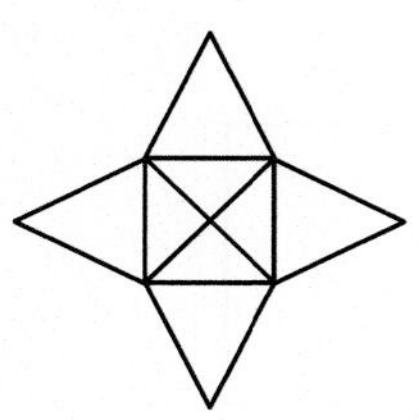
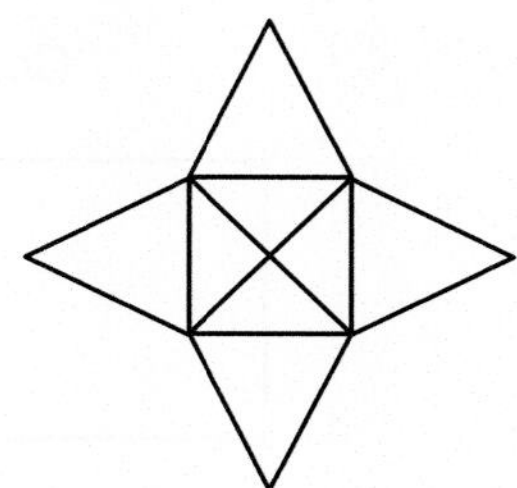

I say: "there are the same number of intersections." Here you do by seeing the figures are the same.
Another way,

	x x x	x x x
	x x x	x x x
	x x x	x x x
Or another	x x	0 0
Or	0	1

There are all sorts of different ways of determining whether two classes have the same.
One method applies to 'visual' number.
These are 1 2 3 4 5
Here if we see 5 points here and 5 there, you say same number.
Or again the rows are visual numbers

x x x	x x x
x x x	x x x
x x x	x x x
x x x	x x x

Here 12 is not a visual number.
Here we are talking about numbers given in this way.

x x x x x
0 0 0 0

What is meant by the possibility of these being correlated:

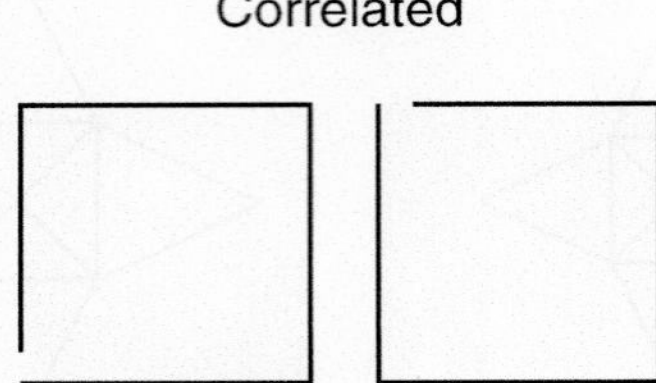

We say these two have the same length if they can be superimposed.

I suppose it means our criterion for their having the same length is there, being superimposed.

If we have superimposed them and take them apart, are they of the same length?

We say under certain circumstances they have the same length, under others not.

It[157] is in no way a straightforward thing to say whether they have the same length or not.

One possibility is that once they have[158] been superimposed, you always say they have the same number.

You see in different cases I will say all sorts of things.

If I put thirty dots you know by just looking at them I wouldn't know their number.

But I would still be able to see if one appeared or one[159] disappeared.

Russell did not say they had the same number if they could be correlated (as this makes of it all something empirical):

0 0 x x

Can these be[160] correlated? This is a hypothesis.

If we talk in this way, we might never know if they had the same number or not.

Russell and Frege said, however, that two classes were of the same number if they **were**[161] correlated.

Russell said, we need not bother about a real material correlation.

---

[157] 'On the other' opens the sentence and is crossed out.

[158] 'the same' is written and crossed out.

[159] 'didn't appear' is written, and two dots undeleted on line as if the remaining expression was intended as another instance.

[160] The word 'be' is inserted by Eds.

[161] 'were' has a thick double underline in the manuscript.

If I have for instance two things *a, b* and two things *c, d*.

Russell said there was always one correlation existing between these two.

He said there was one function only satisfied by *a, b* and one function only satisfied by *c, d*.

$$\text{The first is} \qquad x = a.v.x = b \qquad (1)$$

$$\text{and the other is} \qquad x = c.v.x = d \qquad (2)$$

If I want to say there are 2 and only 2 things which satisfy (1)

I say $(\exists x\ y)\ x = a\ .\ v\ x = b : y = a\ v\ y = b$

The correlation is this[162]

$$x = a\ v\ y = d : x = b\ v\ y = c$$

Why does this look as if it were a correlation between *a, b* and *c, d*

It[163] is because of the identity sign.

This formula is a way of correlating these four letters.[164] But it doesn't mean there is, unless you mean this does correlate the 4 letters.

XII [165]

I talked about Russell's theory of number.

Russell defined number as a class of classes similar to a given class.

He also defined similarity or numerically equal. ...

This idea of two classes being numerically equal if they can be correlated one–one is taken from the sort of test we know.

---

[162]The five lines below in this footnote are written in the main text above after "this", and indented in the following manner; yet they are deleted in favour of the equation above in the main text:

$$(\mathrm{E}x)x = a$$
$$(\mathrm{E}\xi)\xi = x\ v\ \xi y$$

This proposition is a relation between *x* and *y*:

$$x = a\ v\ x = c :$$
$$: y = b\ v\ y = d$$

[163]A capital 'W' is written before "It" and deleted.

[164]The letters 'ur' are written in pen over the pencil writing of "four", and 's' is added in ink to "letter".

[165]Although Skinner includes diagrams in this and the next two lectures, MacDonald's (1935: 15–20) copy of notes does not have any.

We also noticed there [are] many ways in which we saw numerical equality. Similarity or 'having the same number' refer to two different things.[166]

(1) If we say one has the number 3 then the other has the number 3.
(2) We may have a tribe, which does not make use of numerals but only of numerical equality.

For instance they distribute food, say nuts and they want each man to have the same.

One tribe does this[167]

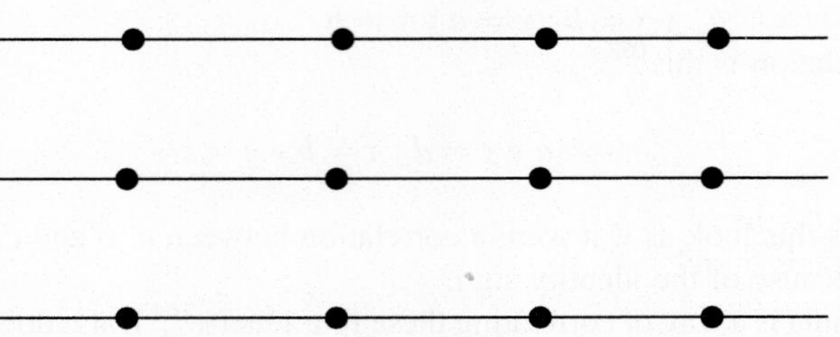

Another tribe does this. They put 3 like this

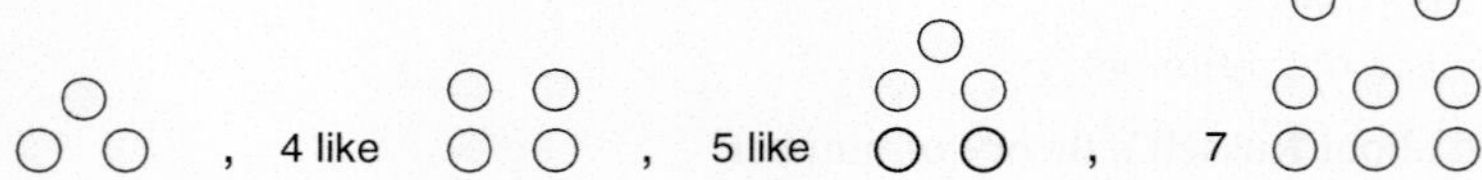

They learn by heart particular figures
Or

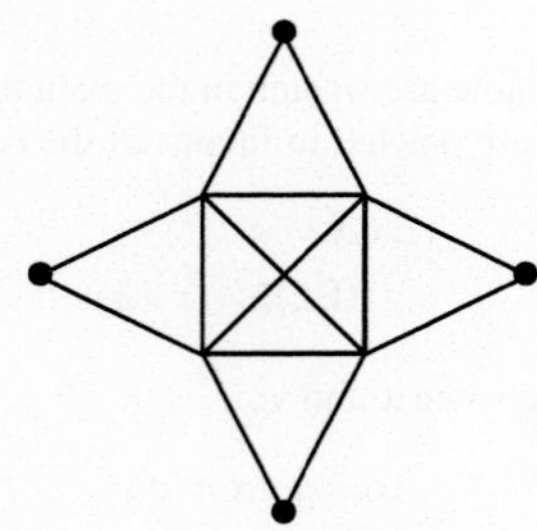

(what we call 8).
They lay out certain stars of nuts and deal them out.

[166]'test' is written first, and then crossed out.
[167]These lines are crossed through.

Another way of knowing numerical equality is this

They go *a, b, c, d, e, g*[168] → and stop at *g*.

Our alphabet has a particular order.

Imagine a tribe which used letters, but had never put them in a particular order.

(Suppose all words were made up of *a, b*. Then this is easily imaginable.)

I said they counted the nuts *a, b, c, e, f, g*[169] as we count.

But they may have fixed a certain group of letters A, D, R, E.

Here counting means just saying them all.

Let us consider again the version of our definition of similarity when we say 'can' be correlated.

The use of correlation of counting is exactly comparable to measurement (there are different kinds of measurement and they needn't agree).

We shall first talk of correlation such things as holding hands, joined by pieces of string, sitting to the right, sitting on top of each other.

If we call this correlation, then we can decide what we want to do.

I can first say they have the[170] same number if they are correlated.

Let's say I fix the meaning of numerical equality if they are joined one–one with strings.

'Numerically equal' is a short phrase for 'being joined one–one by string'.

Then comes the other: two classes are numerically equal if they can be joined by strings one–one.

This is a definition which can be used.

I will talk now about equality of colour.

I say two colours are equal if in a certain light at 6 inches apart they look the same.

Again I give another definition. I call two patches of the same colour if when they are brought together I see no dividing line.

You are first given the phrase 'two colours are equal'. If someone says this, how am I to verify it? My definition tells me how to do this.

In this case you can't say are they still equal[171] when they are taken apart.

All we have is the test.

There can be also the case of 'conjecture' –

I will talk of two patches of colour projected by a lantern on the wall.

We see the two patches not together. We have the definition they are equal if when they are together there is no dividing line.

Someone says, "They are equal." This is a conjecture.

Or he may have seen them together. We have then to fix whether we will say once they have been together they are equal.

---

[168] 'f' is missing from this series.

[169] 'd' is missing from this series.

[170] 'the' is inserted by Eds.

[171] "if" written and crossed out.

Let's have a practical case in which this definition of numerical equality would be used.

A tribe[172] writes down three letters A, B, C and they count by laying their nuts on the letter.

They could say: "We call two heaps equal if they both go on the scheme A B C."

I will introduce[173] a name of their number.

We can either call A B C the number or we give the letters an order and the name of the number is the last letter.

Then they could say.

We are going to say a man has C nuts if they can be put one–one on the letter C.

This defines the expression, "Mr. Smith has C nuts."

We have given the use of the phrase, "He has C nuts."

We must see what would become of Russell's definition.

'Class' of all classes. I will talk[174] about what this 'class' means.

A class is generally represented by a list.

Russell thought of a class as defined by a common property.

Russell introduced 'class' to be able to talk about it as a list.

He seemed at the same time to be able to define it by a property.

The question which both Frege and Russell tackled[175] was this:

Frege had said the number was the property of a class.

This seemed to have two meanings[176]

Take the property

(1) having blue eyes
(2) being in this room
(3) being a man.

There are five blue-eyed men in this room.

Five is the property of being a blue-eyed man in this room.

The property 'being a man in this room' has the property 6.

Frege wanted to say, "Gans[177] and Paul are 2"; but this would imply Gans and Paul have a property which only they have: and there doesn't seem any reason there should be.

Frege said he had found one.

---

172 'A tribes' is written, altered to 'tribe' above. The expression "They can" precedes this term and is crossed out.

173 'of' appears before "a", which is crossed out.

174 'talk' is inserted by Eds.

175 'tackled' is inserted by Eds.

176 The MS reads 'meaning'.

177 I.e. Abraham Gans, an American student attending Wittgenstein's class in 1935, along with Skinner, Goodstein and others. (Cf. McGuiness (2008: 283) notes that Gans was with Moore on occasions when Wittgenstein was in dialogue with both; sometimes spelled 'Ganz' – cf. as noticed here by McGuinness; also, Ground and Flowers (2015: 565); Klagge and Nordmann, (2003: 346).)

$$X = \text{G}.v\, x = \text{P}.$$

Both Russell and Frege said they could formulate[178] classes intentionally because they had found a function which could define them.

How are we now to define '3' in the way Russell did it assuming our tribe which puts nuts on A B C for 3.[179]

The number 3 is the property of a class, namely the property of being able[180] to be laid out one–one on these three letters.

I can give this definition but have I gained anything by it?

You must see why Russell and Frege were so keen on being able to define the number 3, on being able to say: 3 = ... so-and-so.

You could ask why on earth we should define the number 3, if we define it we define it by other things we have not defined.

Why should he bother to define '3' and not 'have'?

This[181] is a most important problem.

You will notice that philosophers don't try to define everything, but there are a few things they have a thousand times tried to define.

We want to see what is in common to those things for which they feel a craving for a definition.

This arises from a question which seems in a way unanswerable.

If I say, "What is a chair" I can give some sort of answer.

If I say, "What is the number 3", I don't seem to be able to give an answer.

This question arises from a misunderstanding: in fact a jumble of misunderstandings.[182]

There is the word 'meaning' in our language. ....[183]

There[184] is another mess added to this. If I say, "Give me seven apples," seven isn't meaningless, it seems to have a function.

The best way is to replace "meaning of" by "function of" to get clear.

---

[178]'formulate' is inserted by Eds.

[179]After the question mark, the following sentence is written and crossed out: "The number 3 is class of all collections of nuts which can be put on A B C." "collection of" is also written above the line over 'nuts'.

[180]'able' has had 'cap' squeezed in to as prefix to 'able'. Since 'to be' has not been changed to 'of being', I have left the above first version.

[181]'W' starts the sentence and is crossed out.

[182]'People' is written after the end of this sentence and crossed out.

[183]'I give' is written and crossed out on the next line, after this line of dots.

[184]'There' is written twice, and the first crossed out, which perhaps indicates that the beginning of the sentence is to be indented, as above.

Supposing I bought something for 3 shillings. If you say "did you get something for 3 shillings?" I[185] say "yes". But I may have[186] got something –[187] permission to sit in a chair in a theater.[188]

Or if I say to a cab driver: "Here's three shillings, drive as q uickly as you can", what do I get – velocity you might say.

I said before I was going to tell you about a particular mess.

"Seven" seems to occur in two contexts

(1) There are seven men.[189]
(2) 2 + 7 = 9.

One, sentences of ordinary life.
Two, mathematical propositions.

You may see now how the two difficulties join.

If I write 1 + 3 = 4 and ask, "Has '3' a meaning?"

it seems as though I play about with signs, it has no longer this 'function' in an ordinary proposition.

Such a question is confused by giving certain alternatives.

First they ask "What is the number 3."

Then they look about: they say it is the scratch.

Others say: this can't be the case as we don't use 3 for the scratch,[190] and they look about for something else to be the number 3.

"The number 4 is nothing but the sign 4."

One answer: You might have used the shape 5 instead of the shape 4.

What you could say is:

I have two expressions the sign[191] "4[192]" and[193] also the sign "the sign '4' ".

The man[194] says "4" can be replaced by "the sign '4' ".

We don't say the sign 2 and the sign 2 is the sign 4.

Or I can say, "the sign 4 is painted red" but not "4 is painted red."

Let me give the sign 4 a name.

Certain letters have names, i.e. *y* or *ε*.

I will have then two signs. One sign is '4' and the other 'dash'. The word 'dash' is.

---

[185]'you' is written before 'I', and crossed out in favour of 'I'.

[186]'have' is inserted by Eds.

[187]Dash is inserted by Eds.

[188]This US spelling of 'theater' is in the manuscript.

[189]MS says 'mean', which is possibly short for meanings but more likely 'men'.

[190]The comma interprets a very light dot in the manuscript.

[191]'word' is first written, then deleted, with 'sign' written over it.

[192]The word 'four' is written, then crossed out in favour of the numeral.

[193]'the words' is written, then crossed out.

[194]'who' is written after 'man' and crossed out.

the name of the sign 4.

If I say to Paul 'Write down a 4', I would have say: "Write down a dash".

The formalists' question comes to saying, "Have 'dash' and '4' the same meaning".

I will first of all say the words 4 and dash have different uses.

There is a further difference.

If you say, "What is dash?", I can point to 4.

But then you say, "What is four?" and there is nothing to point to.

(Similar to this is the need felt for a definition of time. Is time the motion of the celestial bodies?)

This was not the only reason people craved for a definition.

When people say, "I have 7 apples," 7 is not a substantive and people don't look for any substance.

But when we say "Mathematics treats of numbers, the number 7, etc." there it is a substantival phrase.

Suppose I say: 1 + 1 + 1 = 3 Df.[195]

You say, What about '+' and '1'?

There is the use of '1' which I can give you.

Is '1' definable? Yes, any word can be defined, replaced by other words.

Is it '1' defined in our calculus?

If we gave a definition that also would be one of the grammatical rules (grammatical rules by which the use is given).

We said the question had been asked, "Has four a meaning or is it a meaningless mark?"

The difficulty arose through two uses.

(1) There are 4 men: here it seems to have a meaning.
(2) 2 + 2 = 4: here it just seems to belong to a game.

There then arises the question, "Does '4' have a meaning in such sentences[196] as, "There[197] are 4 apples in my hand and not in 2 + 2 = 4."

What is the relation of the use of 4 in a sentence and a mathematical proposition?

The mathematical proposition is a rule about the usage of the word 4.

The relation between a mathematical proposition about 4 and a material proposition about 4 is the relation between a rule of grammar and a sentence in which it can be used.

Suppose I told you a story and said,

"I gave Mr. Smith 2 apples, and then another 2 apples: ... later he ate these 6 apples ...".

Is the story a false one?

---

[195] 'Df.' for 'Definition'.

[196] 'questions' written first, then it is overwritten as 'sentences'.

[197] 'They' is the form in the manuscript, taken to be an error for 'There'.

Suppose you say, "he must have had one before," or "they must have reproduced themselves".

Or again: We got an elliptical cake and divided it into eight parts

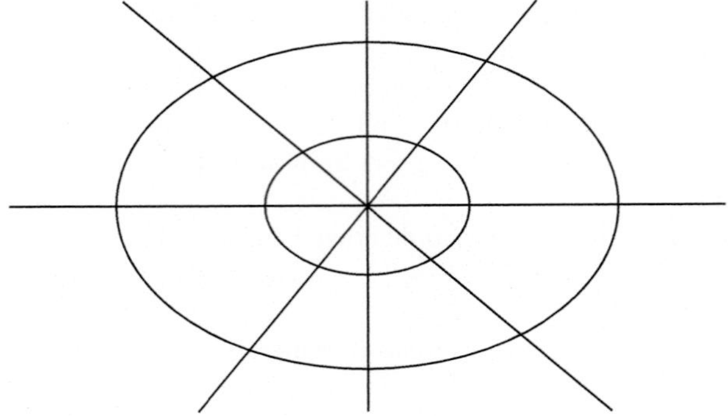

such that the angles were equal and then we ate the eight equal bits.

Compare a circle[198]

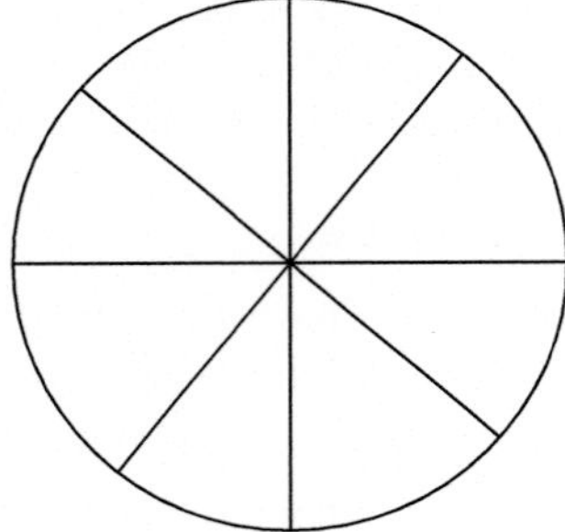

The geometrical proposition applied to a story says, "It does make sense" or "it does not make sense." It doesn't say whether it is true or false.

Let's ask again whether 4 has a meaning or is a meaningless mark

Take the sign '~'.

It is used in all sorts of propositions. "I was not at the theatre."

Also in $\sim\sim p = p$.

It is not a meaningless mark, you would say in "I was not at the theatre."

There is an obvious temptation to say it has no meaning in $\sim\sim p = p$.

A[199] man says in $\sim\sim p = p$ there is a temptation because it is about the mark 'not'.

---

[198] 'They' is written before 'Compare' and crossed out.

[199] 'One' is written first and crossed out in favour of 'A'.

$\sim$ I go to the theatre:
in this sentence it has meaning, while you might say $\sim\sim p = p$ gives it meaning.

The normal sentence about the mark $\sim$ is it is[200] 2 inches long, it is white,[201] etc.

Are the rules of chess about the pieces of wood? If you operate with words 'be about' you get into a mess.

$\sim\sim p = p$ don't give any property about '$\sim$' they give something about the use.

If you say $\sim\sim p = p$, you don't say you are now at liberty to use '$\sim$'. What does a rule 'allow' you?

The question is "What is a rule?" Is it a sentence saying you are allowed to do something, has it a date?"

You can't answer this. This is certainly not a sentence. It is not either also of what we have just said. I can give you another answer "I can show you[202] how to use this sign." "I'll give the use of it, that's all."[203] This use is not that of a sentence referring to human beings or to time.

Is 2 + 2 = 4 about 4 or not? Or is it about the scratch 4?

If the sentence "I have 4 apples" is about 4, then 2 + 2 = 4 is not about 4 in that sense.[204]

If you say the word '$\sim$' is used amongst us in such and such a way, then this is a statement that is,[205] say ethnological. This is not expressed by $\sim\sim p = p$.

Let's get clear about 'rules' and 'statements'.

If you talk about a rule $\sim\sim p = p$ and do not call it a statement or command.

You might say the statement or command[206] you gave contains the rule, it is a radical (as in chemistry).

I could say: "Do according [to] this rule $\sim\sim p = p$," using the rule as part of the command.

If you want to know the connection between the use of a rule and the command, I say it occurs in all, they all make use of this radical.

Take the command:

"Use p for $\sim\sim p$.

Are these anything but words? What makes it a command? I suppose it is its use.

You may translate it by "in this book $\sim\sim p = p$."

This is done sometimes.

$\sim\sim p = p$ is very like a statement "in this book I shall use $\sim\sim p$ for p."

---

[200] It seems that 'it' is written first, then overwritten as 'is'.

[201] Perhaps 'white' because Wittgenstein drew the negation sign in chalk on a blackboard.

[202] 'you' is inserted over the line.

[203] Opening quotation mark is supplied by Eds.

[204] There follows the sentence, which is crossed out:
'If you say it is about the scratch 4, then I say, "Be careful, it is about the use of 4."'

[205] 'that is,' is added by Eds.

[206] 'common' written first; then overwritten as 'command' with 'you' added over line. The word 'gave' is preceded by 'can' and crossed out.

You say, "I know how to use the word 'not'."

(A person may know how to play the violin, but be quite incapable to say how to hold it[207] ....)

Suppose you say 2 + 2 = 4 is a statement about what we do.

Suppose I write:

We do this 2 + 2 = 4.

There will be two cases.

(1) This which uses the words "we do" as opposed say to the Chinese.

We might use it this way.

The question is What's the use of these words?

All there was to it you made[208] under certain circumstances a use.

(2) Is there any wrong if we carry these words "we do" along with us? I could write.

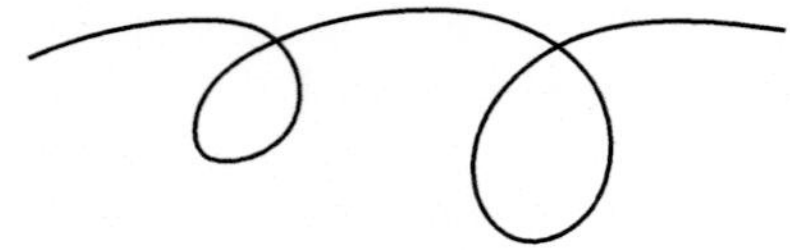

underneath it.

Suppose a tribe has a natural tendency to write 2 + 2 = 5, then you may call 2 + 2 = 4 a law if it is laid down.

We don't do this.

How are children taught it? They are taught to say "two + two equal four" and get an apple if they do it, say.

I made this once clear to myself by saying "It is like a garden path." A path is not a command to go this way, or a statement to go this way. A path is like a rule, I might say.

---

[207] 'it' is added by Eds.

[208] 'made' written first; then crossed out and 'did' written over it and crossed out. Furthermore, 'made' then written over the line, to restore the original choice. 'certain' written over the line and inserted in main text.

XIII

Wisdom.[209]

I began by putting[210] my difficulty in the form "What is a rule?"

I think one can do something to indicate the sort of thing people want done when they ask this question.

When you talked of 2 + 2 = 4, I should say in old-fashioned language you were making an analysis of arithmetical proposition.

I want you to indicate an answer to this question by giving an analysis.

It seems to me one can ask this question and give an answer to it.

The fact to me that there are a great number of different kinds of rules does not seem to me to prevent this.

____.

Suppose[211] we took statements made in this class – these could be[212] classified explanations, questions, instructions, etc.

This[213] may be very useful.

We could also make a stupid classification. I.E. if we put recipes alongside instructions.

In a factory where they produce machinery there are blueprints. Such a blueprint is some sort of communication given to the man at the lathe. Is the blueprint an order, a statement, suggestion, etc.?

There are different cases.

Mr. Wisdom said the question "What is a rule?" though antiquated by our language still had sense.

I say "*In certain cases* it does[214]; in others you are misled when you ask this question."

In[215] children's first reader, things are classified under about a dozen headings – furniture, drink, food, garment.

---

[209]Margaret MacDonald's lecture notes for Wittgenstein's lecture on 4.3.1935 (MacDonald 1934–35: 18) ascribe third person speech to Wisdom ("Mr. Wisdom asks"), whereas here we have Skinner's preservation of Wisdom speaking in his own first person – which if accurate, is the work of the amanuensis taking down actual speech, rather than MacDonald's and Ambrose's transposition and reduction to handy notes. Ambrose and MacDonald's notes sometimes switch or disagree about first and third person ascriptions of theirs and Wittgenstein's 1st person. A.J.T.D. (John) Wisdom was appointed as a lecturer at Cambridge in 1934, which led to a Fellowship at Trinity. Wittgenstein comments on his engagement with him in his letter to W.H. Watson (cf. McGuinness 2008: Letter 180).

[210]'by putting' is inserted above line.

[211]MacDonald's LECTURE THIRTEEN (given on 4.3.1935), presenting notes that are roughly similar to the above sentence, prefaces them with "Wittgenstein replies that".

[212]'be' is missed out here.

[213]A word is deleted after 'This'.

[214]Here ';' is oddly written: it might be a comma.

[215]Three words are written, then crossed out, and placed before the start of this sentence. The first word is 'Wisdom'; the other two cannot be read.

Suppose you say, "What about spinthariscope?" This isn't included.

Talking in a primitive language I should probably say $2 \times 2 = 4$ was a rule.

I should say I wouldn't call it a command or suggestion etc.

What follows from this?

If it follows for you that therefore I put rules separate, then it does follow.

________.

When you ask "What is..?" the answer is generally "It is . . . ..," a definition.[216]

Mr. Wisdom's classification is not any one, it is bound up with one particular phenomenon, namely "tone of voice."

What tone of voice is bound up with $2 \times 2 = 4$, you would say at once "that of a statement."

That this particular division Mr. Wisdom referred to sticks so hard is due to its being bound up with the tone of voice, and therefore with the action of the whole[217] body.

If a child did ask, "What is a spectrometer?" in this primitive classification, I should say "piece of furniture" – it is at the outside of the limit of furniture.

<u>Wisdom.</u>

You have a rule when it is a statement, command, suggestion of a certain kind.

You could do this. But if you now say this is what a rule is, then I say I can give you things we call rules which are none of these or if you say they are one of these you don't feel comfortable about it.

Suppose I had[218] said: This is a statement about apples (I have 4 apples in my hand) and this is a statement about the use of signs ($2 + 2 = 4$).

This is very confusing indeed.

If you make a statement then the question is[219] first whether it is true or false.

I might say something different: I am talking about $2 + 2 = 4$ as a radical.

If I use the word rule, it will always be in particular cases where I wish to oppose 'this' to 'that'.[220]

In our language there are verbal forms of statements, commands, etc. questions, etc.

Let's assume a language in which these are only distinguished by the tone of voice (I.E. "Mr. Smith leaves the room").

Could I classify these by the written sentence?

As far as the written sentence goes I couldn't give such a classification.

---

[216]On the next line a sentence commences and is crossed out. It reads: "Or it may be".

[217]A word is written and heavily deleted here.

[218]'had' is written over the sentence, marked for insertion – as is the first indefinite article in the next sentence.

[219]A word is heavily crossed out after 'is'.

[220]The parallel sentence by MacDonald (4.3.1935), p.19, varies from the above, as follows: "If I use the word rule it will always be in a particular case where I wish to oppose a rule to something else."

Could you in this case classify $2 \times 2 = 4$ with "I have 4 apples"? Not as far as the use goes. But with the other classification I couldn't do anything at all.

Take the case of the blueprint given to the workman.

We could call something a command or a statement according to what games they occur in.

Here if we do this, we see there will be[221] many more distinctions.

As there are so many cases, it ceases to be interesting.

As long as we talk of tone of voice, the distinctions are fairly clear and are interesting.

But in the other they become obsolete and uninteresting.

Take the case of a rule in a game.

Direct a man by giving him sentences consisting of four letters.[222]

a	↑
b	↓
c	→
d	←

For the order.

a b b d c c c

the man is to move steps as.

[221]Inadvertent indefinite article after 'be' was removed by Eds.

[222]Note a similar chart, and parallels in discussion, in the manuscript MS 116 (*Philosophische Bemerkungen*, XIII), pp.143–44, but with a, b, c, d; but matched with arrow order →, ←, ↑, ↓ respectively.

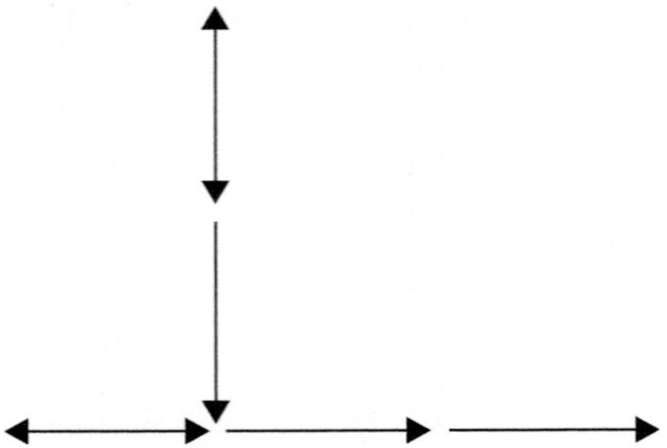

The table would be called a rule (if I was say accustomed to change it each month).

The command would not be a rule, here.

Suppose the command was to make a certain ornament

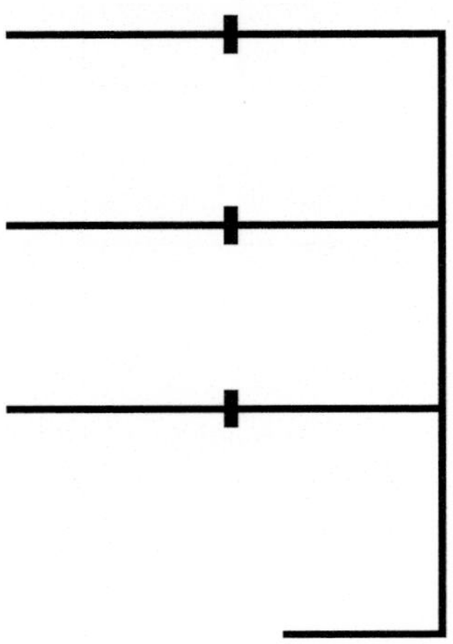

a b c c a

Here we would be inclined to call this a rule for this[223] ornament. At the same time it is a command.

If you say why is one a rule and not the other, I might say a rule is something generally used in many cases.

[223]'this' is 'his' in MS.

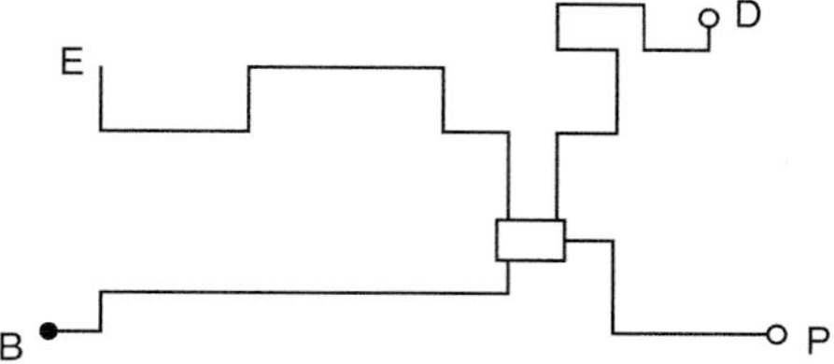

If I said how does one go to the[224] 'Butcher' what you said would be a description.

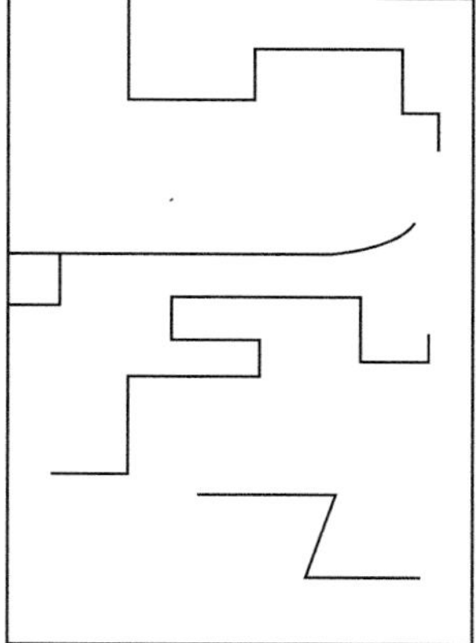

If I gave this as what you had to do every day, then you would probably call this a rule.

To call something a rule, we request a certain kind of generality.

In a particular case it might be helpful to say this.

If I am Professor of Logic and say this, it means nothing at all. – In the particular case this phrase has use.

Sometimes it is business to ask is this a rule or not, sometimes it is not.

<u>Example given by Wisdom.</u>

We could classify relations into.

(1) Those who were tall or short, male, female, blonde, etc.
(2) Into brothers, mother, father, etc.

______________________________.

He[225] said one was *a priori*, the other not.

This was because if I gave all the relations, he could have.

[224] After 'the', 'other' is written and crossed out.

[225] Presumably Wittgenstein has resumed speaking, and this 'he' indicates John Wisdom.

This is misleading unless you have already given your scheme of relations (and then filled it in).

Suppose you've made the tree without putting things in and then put them in.

Wisdom → I think this classification is given in the way (2) is given.

____.

The[226] form of the tree is a statement of natural history.

I must point to the fact that ordinary relations are very simple. The usefulness of the schema lies in the great simplicity.[227]

This is a fact due to the particular natural history.

In the case of utterances we did also give a case where there was a simple scheme, namely tone of voice.

XIV[228]

Look at the sentence: "There are no other men in this room but Jack and John."

What's this about? The first answer is Jack and John and perhaps then about the room.

If you say, "There are no other fair-haired men . . ." then you may go on and say it is about the quality of being fair-haired.

If the word 'two' is in it, you can go on and say it is about two.

Then again you can say it is about negation.

Everyone would say $\sim\sim p = p$ was about negation.

If you say, "there are 2 men in this room" is about 2,[229] $2+2=4$ is not in this sense about 2.

The relation of $2+2=4$ to its application, is the relation of a grammatical rule to its application.

Last time we talked about the statement

"$2+2=4$ is a rule for the usage of the signs 2, +, 4, =".

There is something wrong about this.

A mathematician to whom I said this might feel uncomfortable.

It has been said that $2+2=4$ is not so much a rule of the game as a position of the game.

We can have a rule.

[226]Wittgenstein resumes speaking here.

[227]It seems that four words have been written and crossed out; only the first can be read – 'of'.

[228]The MacDonald notes mention 6.3.1935 to be the date of the lecture.

[229]After the comma, 'then' is written and crossed out.

If we write	$2 + 2 = 4$
Then	$2 = 4 - 2$
	$0 = 4 - (2 + 2)$

This has often been compared to[230] moving pieces on a chess-board.

We might have a game in which we tried to transform one expression into as few moves as possible into another. In this game we would have rules, namely say the commutative etc. laws.

$$25 \times 24 = 600$$

It has often been said this equation tells us a truth.

Suppose we had a tribe which never used large multiplications, but had a game in which they had large multiplications.

I want you to see why 2 + 2 = 4 or such lose utterly their characteristics of human utterances.

a b b c We might call this an **instrument**:

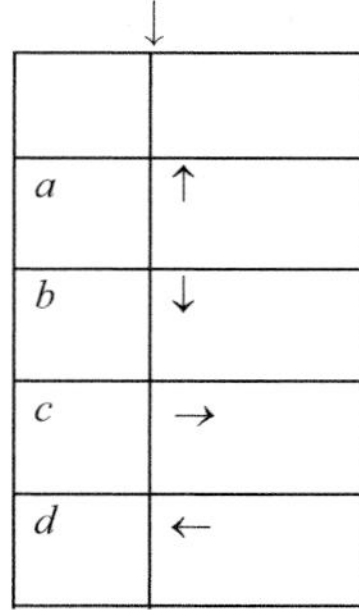

The chair is there (1)

I believe the chair is there (2)

(1) → (2) may be quite different kind[s] of statements. Sometimes they are just the same statement.

(1)→seems to be about the chair while (2) is about my state of mind.

[230]Two words are written and heavily crossed out, with 'moving' added over the line.

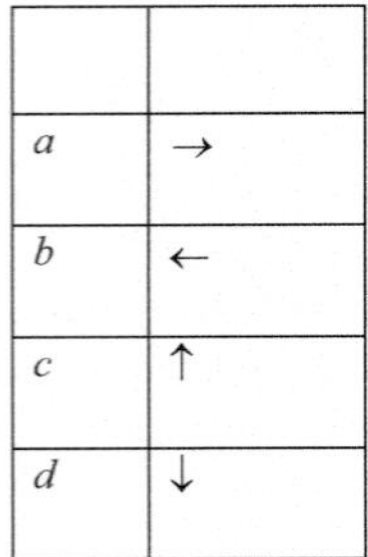

a	→
b	←
c	↑
d	↓

This could be used as a position in a game.

A school has suggested that $2 + 2 = 4$ is nothing but a position in a game.

We gave two definitions of numbers:

(1) the property of a class that can be correlated . . .
(2) the property of a class to be correlated . . .

I gave the example of tying strings to.

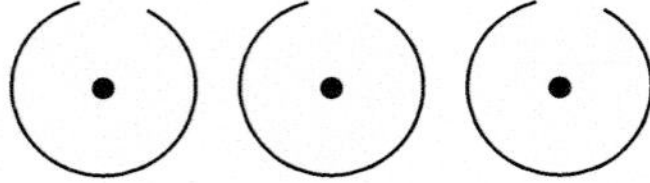

. . . the chairs in this room.

If you ask me if it is possible to define 3 as the property of being capable to be correlated one–one to a paradigm.

This sounds very nice as it sounds like a definition.

For it says "*it is* the property . . . and so on."

Suppose I have a sentence "I have 3 apples." The definition of 3 is the property of being correlated one–one to:

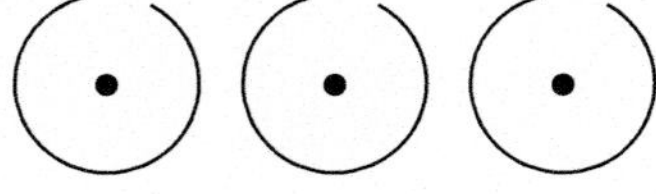

This has shown me a use of the number 3 in the form of a linguistic definition.

Russell did not say this.

The idea which both Frege and Russell had was this.

I have

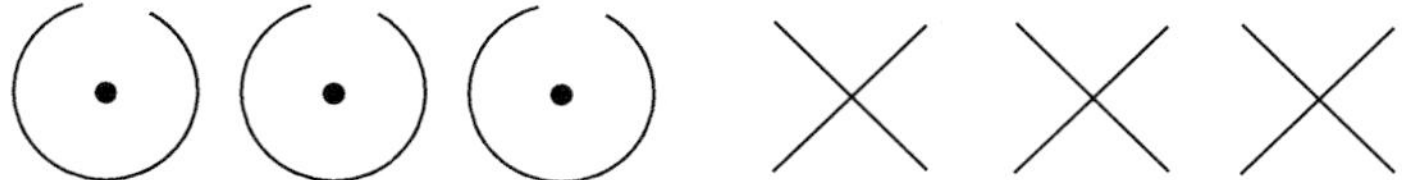

There is a definite tendency to say if they can be correlated they are correlated.

The idea is that the 'can be' was a very faint line joining them while being correlated was having this line very thick.

In Euclid there is an axiom. A straight line[231] can be drawn between any two points.

This is often put in the form: "A straight line joins any two points."

If a man says "There is a straight line joining any two points".

If someone says, "Surely not", then he answers, "It can't be seen: it is a geometrical line."

Suppose I called the geometrical line the possibility of a line:

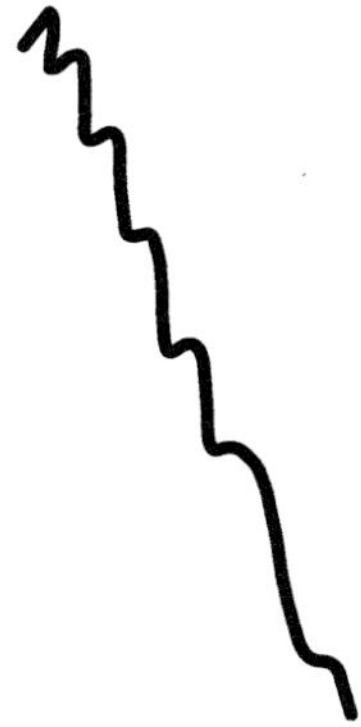

If we had *a b c* and *d e f*, Russell would say these are *logically* correlated.

$x = a.\ y = d.$ This only holds for x being a and y being d.

$$x = a.y = d.v.x = b.y = c.v.x = f.y = g$$

This correlates *a b f* to *d c g*.

If we have any class of things, how are we to know whether they are correlated or not?

How do you know the things in these ovals have the same number?

[231] 'line' is added here by Eds.

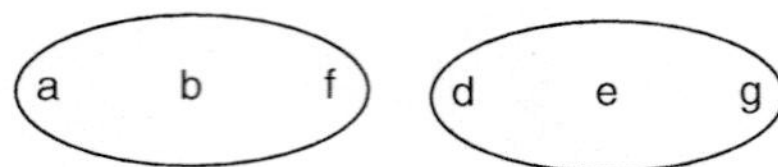

Suppose you say they have the same number if they can be correlated by this.

How does this help me?

I can't know whether they have the same number unless I have made this correlation, written it down.

Can I try and see if this chair equals this chair?

This sign I will say gave us the result of a calculation.

We said Russell tried to define number by numerical equality.[232]

There is a difference between saying "There are as many combinations as roots of a certain equation"[233] and "there are as many nuts as chair."

One corresponds to a measurement and the other to a calculation of number.

$2 + 2 =$ number of solutions of equations of the four degree.

The number of chairs = number of people in this room. This is not a statement of mathematics while the other is.

Look at Russell's correlation and try to use it in this case.

There are some people in this room and some in the next.

Russell tells me there are the same number if there is a one–one correlation between each man in this room and each man in the next.

Suppose I had found there were the same number, I made out two lists:

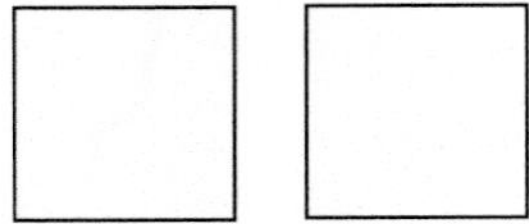

Russell would come along and say: You say there are the same number and he would write out his expression.

It depends on how the classes of the people in this room and that.

Does it make sense to say A, B have a different number from *a, b*?[234]

What about an experiment whether

[232]Written 'numeral' in the manuscript.

[233]Close quotation mark is added by Eds.

[234]After the question mark, the next line commences with 'In order"; yet is crossed out.

A	and	D
B		E
C		F

have the number. You could regard a process of correlation[235] as an experiment. But if asked whether they had the same number you would say, yes.

But an objection might be here made that I can overlook both.[236]

1 *a b c*
2 *a c b*
3 *b a c*
4 *b c a*
5 *c a b*
6 *c b a*

The whole of this is one calculation.[237]

Suppose I say I have given[238] my children seven names the permutations of *a b c*: I would say this is impossible.

[235]After 'correlation', 'of' is written and crossed out.

[236]'Suppose' introduces the box and is crossed out.

[237]Although this lecture in Skinner's manuscript has very many diagrams, yet MacDonald's has none.

[238]'given' is squeezed into a small space at the edge of the page, though legible.

XV[239]

Russell's theory of number.[240]

Remark.[241] → Russell's idea of correlating members of class[242] by his relation using identity.[243]

Difference in the use of "correlation" as applied[244] to Russell's[245] case and any other case.

If a cup stands on every saucer, they are correlated. But if the cups and saucers were on different tables, can they not be correlated in some other way? What about a correlation by strings?

In Russell's case if his correlation does not hold, then no other can hold, I.E. if his doesn't hold, it doesn't make sense to say any other holds.

Consider the word 'correspondence' instead of 'correlation' (Frege).

The[246] idea of corresponding[247] does not seem to imply any idea of sitting on, or something else rough and material but something of a different nature.

If here are three men and here three chairs, there is then a correspondence between them whatever the physical facts may be (i.e. if the chairs were in Africa, etc.).

Do you see where this idea of correspondence is taken from?

It is apparently not an idea taken from physical correlation.

I want to hint it is an idea taken from maths.

I will now make a distinction between an experiment and calculation.

We will take a very simple calculation.

Take the abacus.

Suppose I had ten beads o o o o o o o o o o

One separates 4 from them o o o o o o o o o o

---

[239]As with lectures I and XVI, the above Lecture text in the manuscript is also written in ink. MacDonald's Lecture Fifteen notes as delivered on 11.3.1935. Skinner's word-count for the present manuscript of the lecture is approx. 1300, whereas MacDonald's is around 790. Skinner's MS displays many more symbolic illustrations and diagrams than MacDonald's.

[240]This opening and the term 'Remark" do not occur in MacDonald's (1935) notes, nor in Ambrose's (1979: 159).

[241]This underline is in pencil, while the text is ink. It is possible that this term 'Remark', with its underlining, is a custom deriving from the currency in German and the formal style of 1930s Cambridge mathematics lecturing – the latter which are laid out in the style of numbered "Chapters", typically with divisions into axioms, lemmas, premises, proofs, theorems, with numbered "Remarks" in such a sequence. This also obtains for contemporary Cambridge research publications (see H. Baker 1937: 259).

[242]Wittgenstein appears to be mentioning Russell's use in his *Principles of Mathematics* (1903: 267–69, §256).

[243]The opening remarks up to this point are absent from MacDonald's (1935, p.23); nor are they in Ambrose's 1979: 159 publication).

[244]The spelling in the manuscript is 'apply'.

[245]See Russell's use in the latter's *Principles of Mathematics* (1903: 260–67, §251–55).

[246]'If these' commences this sentence, and the expression is crossed out.

[247]This form of 'corresponding' also appears in Wittgenstein's MS151 (1936, p.2) where a geometrical source for the term is explored.

We can work with the abacus alone or both with the abacus and numerals.

The beads may be our numerals, or we may use the beads for counting and have spoken numerals.

We have an abacus each. A sells B 1 cow and B puts one o[248] across then 2 cows, etc. and A makes B pay so many shillings. This would be using the beads as numerals.

Suppose with the abacus we make the calculation two and three is five.

We could use an abacus in such a way that there would be no arithmetic proposition demonstrated by[249] it.

We might use the beads of the abacus because[250] we can use them easily while we can't move the cows.

Suppose the abacus had a thousand beads and I had the numerals 461 and 527.

These wouldn't do as we couldn't distinguish them.

So we will use the abacus with spoken numerals.

000 00 00000

$$3 + 2 = 5$$

Is this a calculation or an experiment?

We took a certain number of beads, put a certain number by them and counted them.

Need we find 5?

Suppose we found 6.

Suppose you say "We would have to get 5."

Would you say the same in an experiment?

I.E. the specific weight of iron is 7.5.

Could the specific weight of an element change?

You say the specific weight will be 7.5 if you take the proper precautions.

Suppose you did take[251] what we call the proper precautions and got 7.8.

We should not, I suppose, be prepared to say the specific weight has changed.

I suppose then a big investigation would start.

We might then say:

We could say Mr. Hardy[252] has found it 8 so it in this case is 8.

We could also say: the specific weight of iron is 7.5 whatever anybody finds.

If 1000 people find a different specific weight, we shall have a number of hypotheses.

There is one way here – that of holding this specific weight 7.5 constant.

---

[248] 'one 0' is inserted over the line after 'puts'.

[249] 'by' is inserted by Eds.

[250] 'because' is written above the line with signs for insertion.

[251] 'take' is written above the line with marks to insert.

[252] Since Hardy was a founding father of early twentieth century pure mathematics in Cambridge, and was averse to recognising the relevance of pure to applied mathematics, then this use of his name about 'experiment' may well be ironic.

Consider this mechanism – Three wheels:

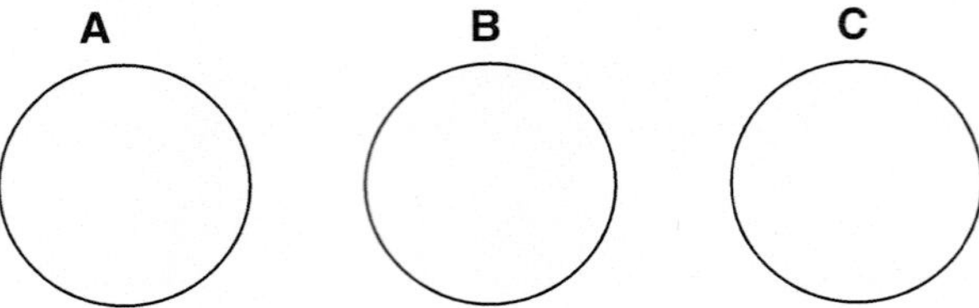

A is to turn in a particular way. B is the wheel I turn. C is a parameter. Whatever I want C to do, I must vary what I do with B to give A what it should do.

I am regarding the case where we keep C still.

If I say that I shall stick[253] to 7.5 as the specific weight of iron, what does the sentence "the specific weight of iron is 7.5" become?[254]

It becomes a rule of our language.

Suppose iron had always been found to be 7.5.

By[255] resolving not to change this, whatever we may find in the future, it is easy to describe what we have at the present.

Take a calculation 571 + 824.

Then take nuts: put 571 nuts and add 824 nuts and then count them. Need we get the same result?

Suppose we get a different result. What do we say?

We say we have made a mistake.

If we put 3 apples and then add 2 apples and we get 4 apples. Suppose we say, "One *must* have vanished."

What is the criterion for vanishing? It may mean that I see it vanish. But we had a box even if we didn't see anything we call vanishing, we would say "so-and-so must have vanished."

This always shows we are dealing with a calculation.

Suppose we make the calculation 275 × 899.

———

———

2789113

and write it. Haven't we made an experiment? We have put the chalk on the board.

---

253 Stick' is written over the line.

254 Quotation marks are added by Eds.

255 'Go back' commences the sentence, then is crossed out.

We say 275 × 899 is 2789113: and this serves us to describe experiments: and if I make experiments with nuts,[256] this may show some of the nuts have disintegrated.

I have particles of chalk which in a very short time rearrange themselves[257]

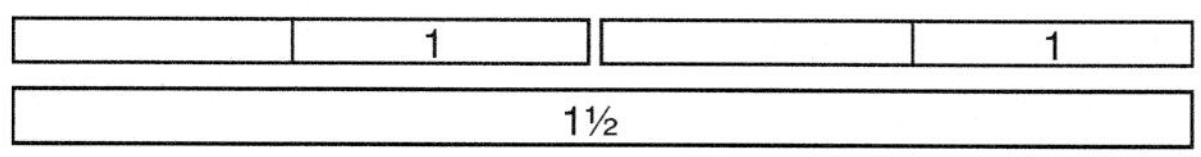

as measured by a rod separately.

Here we might find it convenient to say 1 + 1 = 1½.

We say Russell's correlation seemed to have the queer property[258] that if it didn't hold, no other could hold.

Take a – very simple – correlation: that of lying on parallel vertical lines.

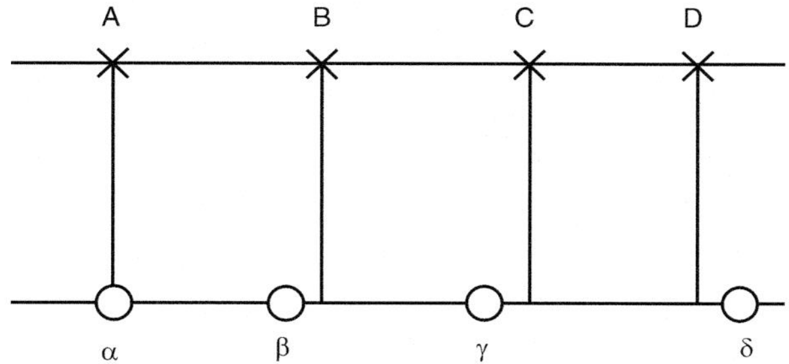

One could imagine a geometrical condition that they could be correlated in this way. These three points haven't this.

I.E. if the three points were equidistant, then they could be correlated.

We say if these don't lie on geometrical straight lines, then they can't lie on real straight lines.

Russell says if we have two sets of material entities[259]:

[256]'with' is written over the line with instructions to insert.

[257]The following expression and punctuation is written after the above sentence, then crossed out: 'If I say: "With normal chalk 275 x 899 gives so-and-so'.

[258]'property' is written over the line with marks for insertion.

[259]This illustration, and some others like it, are absent from MacDonald's MS 1935 (pp.23–24); also absent in Ambrose's (1979: 159–61) partial copies of this lecture.

If there is a one–one correspondence then they can be correlated in a material way, if there isn't then there can't be.

Suppose we asked how are we to find if there is a one–one correspondence between them.

Here Russell has given us a method of doing so

$$x = A.y = \alpha.v.x = B.y = \beta.v.\ldots.$$

What I want to show you that what he has given you here is really a way of calculating that A, B, C, D and α β γ δ are the same number.

Would you say the Latin letters and Greek could have a different number? Wouldn't you say they couldn't?

Mathematics could use as a mathematical method correlation.

Many tribes count 1, 2, 3, 4, 5, 6, 7, 8, 9, 10 many after a certain number.

1 1 1 1 1 1 1 1 1 1 1 1 1 1 The Difference or the Number Can no Long Be Recognised

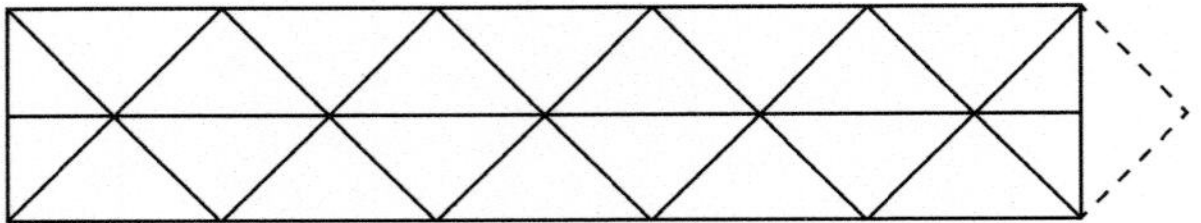

This might represent 19 and 20 which could be recognised. Such a notation would actually be used by the mathematics of the tribe.

We might have propositions.

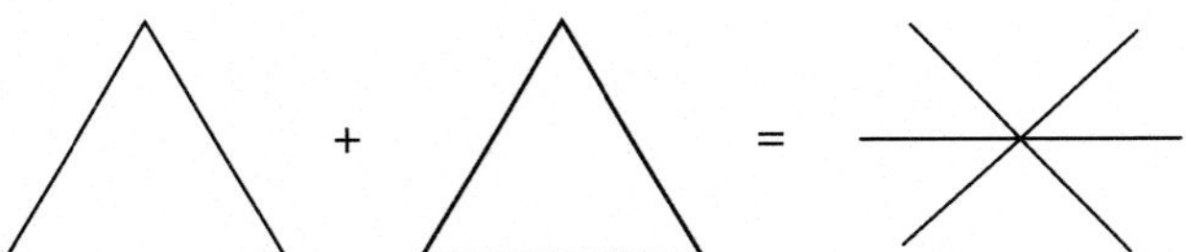

(This might be very interesting indeed.)

We have a way of finding whether the nuts of one box and the nuts of another can be connected one–one by strings.

We should say if they can be connected one–one by strings then they can also be connected in the Russellian way.

XVI[260]

I[261] wanted to give a survey of what we had said of Russell's theory of cardinal numbers.

Let's consider a theory ‖ to Russell's (a sort of parody). –.

Suppose that the things we are talking of are points in a plane.

By points I mean the intersection of two colour boundaries.

Take then two classes of such points.

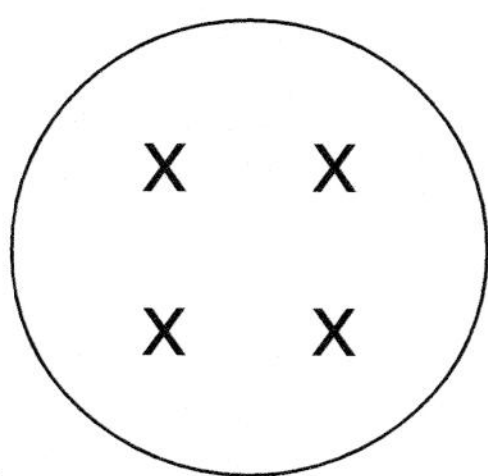

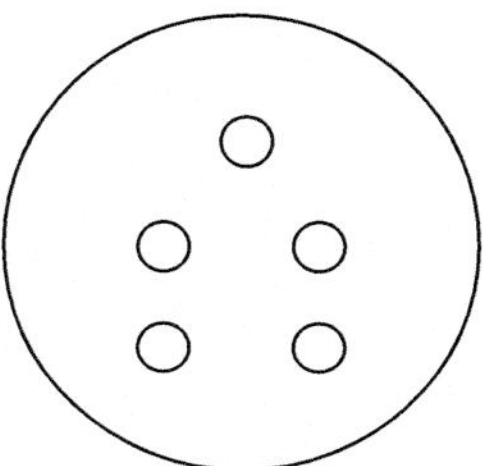

Russell's idea is that the two classes have the same number if they are 1–1 corr.[262]

1st we might say two points are correlated if a line is *actually* drawn.

I shall note[263] the number of points is equal if real straight lines[264] are drawn, but I will say if geometrical straight lines exist.

Definition: Two classes have the same number if geometrical straight lines which connect them (1–1)[265] exist.

If I say this, this seems to solve my trouble, as now I am utterly independent of what is actually done with them.

First I could ask what is your criterion for there existing these geometrical lines (11) correlating them.

1st you may say I can take a ruler and draw them. You say if I can draw them actually, then the geometrical lines exist.

Suppose I had counted each say 5, wouldn't you then say geometrical straight lines existed connecting them (1–1)?

---

[260]If we follow the parallel with MacDonald (1935), LECTURE SIXTEEN, p. 24, the lecture on which this is based was given on 13.3.35. Skinner's version is about 2500 words long, whereas Ambrose's is approximately 700 and MacDonald's text is roughly 600 words long.

[261]It is typical here of Skinner that he takes down dictation from Wittgenstein in the first person, whereas MacDonald (1935) LECTURE SIXTEEN, p. 24, typically has fewer or no first-person narratives.

[262]This sentence is similar to MacDonald's Lecture Sixteen, p.26.

[263]The manuscript has: 'not'.

[264]The manuscript has: 'line'.

[265]Dash is inserted by Eds.

The criteria are just as before manifold.[266]

I can say it means just the same to say there exists the same number of points in the two ovals as to say there exist geometrical straight lines connecting them (1–1).

When the elements we are counting are points in ovals, we can give a theory[267] to Russell's if we say there exists a system of geometrical straight lines connecting them (1–1). Here since[268] admittedly this doesn't say anything about reality, it says no more or less than that they have the same number.

We said to say that[269] a geometrical straight-line joins two points says it makes sense to say a straight-line joins two points.

Thus we can say "two sets of[270] points have the same number if it makes sense to say lines join them one–one":

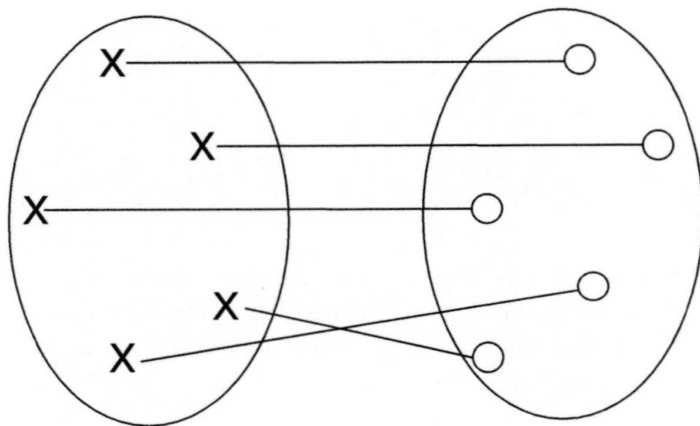

This proposition which says that it makes sense to .... will be a proposition of grammar.[271]

<u>Case.</u>

[266]The text has 'manyfold'.

[267]'theory' is written over line with instruction for insertion.

[268]'since' is written above 'though' with the latter crossed out.

[269]After 'that', the expression 'to say theory' is written and crossed out.

[270]'sets of' is written over the line.

[271]This illustration is similar to mapping from one domain to another in graph theory, which the author perhaps had in mind.

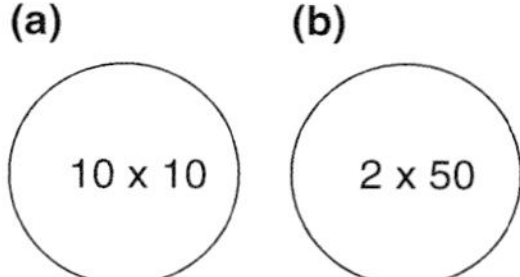

Suppose I said in the boundary A there are ten rows each ten dots, in B there are two rows of each 50. Then you will say there are the same number. But this is a proposition of mathematics $10 \times 10 = 2 \times 50$.

You can say $10 \times 10 = 2 \times 50$ if the units are geometrically correlated.

When we defined equal numbers of points by saying a system of geometrical lines connected them (1–1).[272]

What was the charm of this?

When we said there existed geometrical straight lines it sounded as if we said there was a correlation, whereas what it means is that it makes sense to say there was a correlation.

This is in fact the feeling which Russell's theory was meant to give you.

Take the[273] class of all Englishman and of German.

Nobody thinks there is a correlation connecting them. But it seemed as though in a sense there was an ideal correlation.

. . . ∧ ∧ ∧ ∧

Let's once more consider the "can" in the "number of Englishmen and German are equal if they 'can' be correlated."

If in this connection of 'being able to'[274] correlate, we talked of 'being able to,' we make an assumption in that certain possibilities interest us, others don't.

When[275] we talk of correlating two people, you want to know what is the condition which interests us.

x o

Two points. I say I assume I have the strength, I won't die, I have chalk,[276] etc. What was that condition in which you were interested?

"The condition," you have to answer, "is that there are two."

---

[272] Dash is inserted by Eds.

[273] 'the' is inserted by Eds.

[274] 'be able to' is written over the line.

[275] 'You' written, deleted; then 'When' is written.

[276] 'I have chalk' is written over the line and marked for insertion.

It comes to saying there must be[277] one point here one point there. Here the statement loses all its meaning.

Example (this is a very helpful example).

I will introduce a ‖ notion to that of logically possible. I will make up a notion of chemically possible. I will give two criteria for it.

Suppose I have a flask of oxygen and of hydrogen I could make water by making them combine.

Again, with[278] oxygen and nitrogen.

You said the first was chemically possible, the second was not chemically possible.

Couldn't oxygen and nitrogen combine to give water.

Couldn't we call something nitrogen burning in oxygen? Yes.

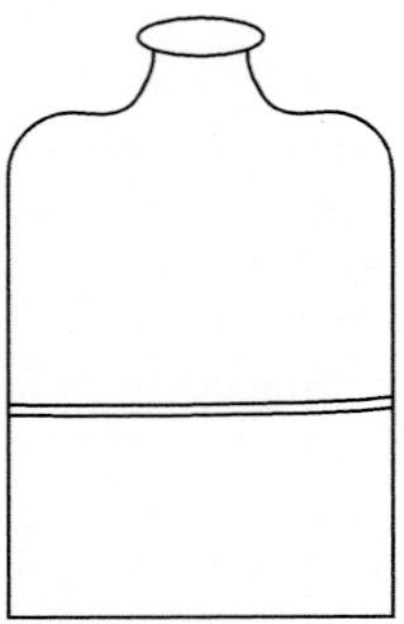

I have a flask sealed. I put in it oxygen and nitrogen (while before there was mercury in the flask).

Suddenly there is a condensation going on which by tests is water.

We say this is impossible, meaning that it has never happened.

I will give another meaning of chemically possible.

You know chemical formula can be written structurally.

Take another compound

[277] 'be' is inserted by Eds

[278] 'with' is supplied by Eds.

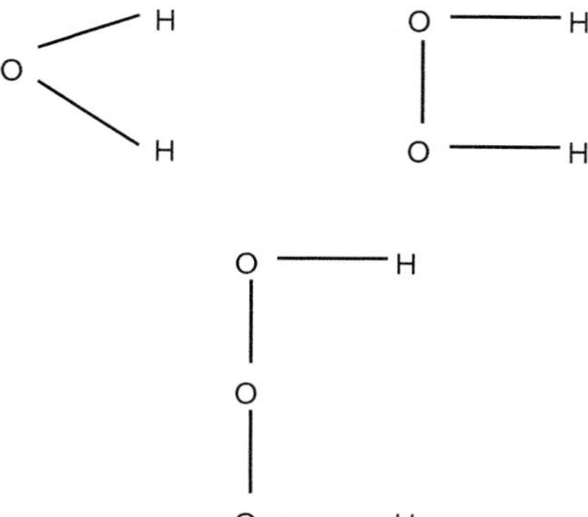

Each[279] O has two valences and H one.
Every chemist will say there is such a compound O_3H_2.
By the rule of valences we can write down O_3H_2.
On the other hand O – H we might say was 'chemically impossible.'
Suppose a chemist found under very high temperatures:

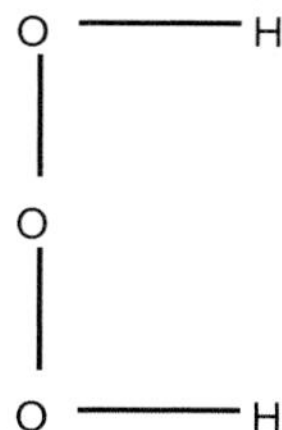

existed, then he would say now he has found this which was chemically possible.
Do you see there is something we talk of 'possible'?
Neither O – H or.

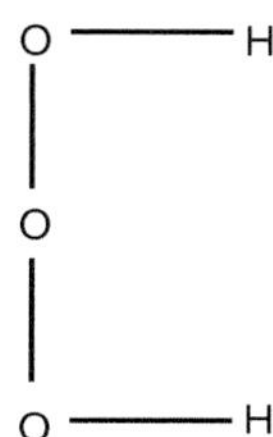

[279] After 'Each', 'other' is written and crossed out.

exist.

Does O_3H_2 exist more than OH? We might say this is possible, that's not.

What does it mean to say that this is possible when neither of them exists[280]?

Suppose[281] I said, "Where is the phenomenon of possibility" to be looked for.

I think we could say in the symbolism we use. And the symbolism, I might say, is justified by experience – justified as useful, practical.

The essence of logical possibility is what is allowed in our language where what is allowed is[282] justified not by being true or false but by being judicious or injudicious.

I will go back to the idea of the possibility of (1–1) correlation.

By correlation I will mean correlation by a piece of string.

I will show what I mean by possibility of correlation.

I have several dots or pegs:

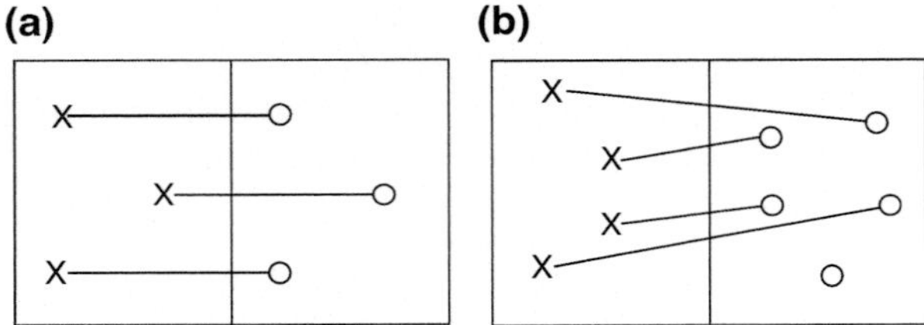

In order to see whether I can correlate these with strings I draw lines.

In A I draw lines and say, "Yes, it is possible." In B I draw lines and say, "No, it is not possible."

Why do I say it is possible in A?

We will take another example first.

You know it has[283] very often been said that what is logically possible is thinkable (or imaginable as others have said).

One might have answered what is[284] imaginable for Mr. Smith will not be for Mr. Jones who is dull.

This is not what they meant. What they meant by 'imaginable' was[285] 'belong to a[286] system'.

---

[280]'exist' is written in manuscript.

[281]'One might sa' is written and crossed out before 'Suppose'.

[282]'determine' is written and crossed out after 'is'.

[283]'has' is replaced by 'is' in MS by Eds.

[284]'more' is written and crossed after 'is'.

[285]'was' is written over the line.

[286]Indefinite article is inserted by Eds.

Suppose you said, "Could you imagine a man sitting on a chair." Suppose you draw a man sitting on a chair. We could call this the possibility of sitting on the chair, meaning here that there exists the picture sitting on a chair in our system of pictures.

You might say the possibility or impossibility can be proved.

Here is the proof of the possibility:

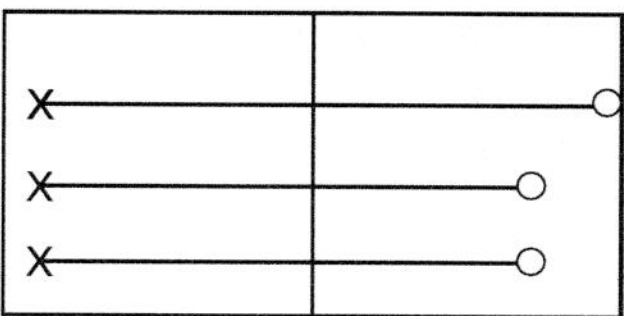

and here of impossibility:

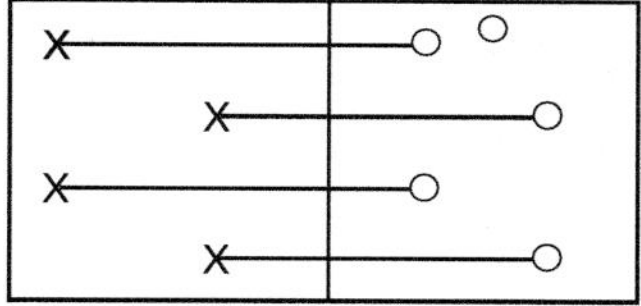

If I call these proofs I wonder whether they are experiments in this sense or not. You say I made an experiment and was able to draw the lines in one case but did not succeed in the other.

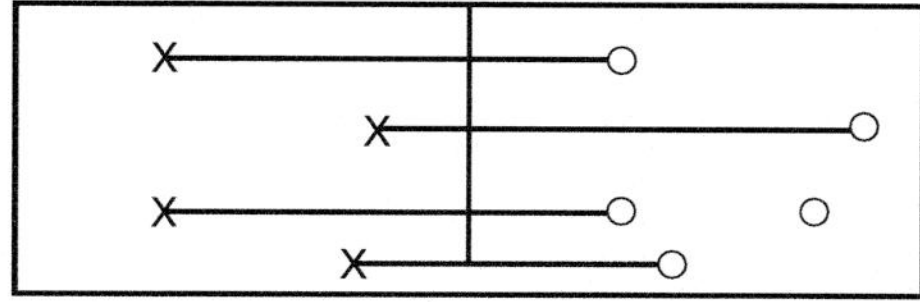

I have to stop because I have toothache.

This shows what we mean by one–one correlation; it is part of our language.

This is on the same level as saying I can show you that it is possible for a man to sit on a chair by drawing him sitting on a chair.

I have two rows of three points and six points as on an hexagon.

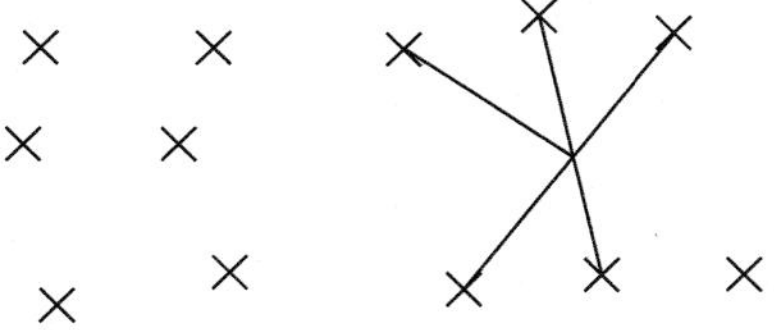

Seeing these two Figs. I say it is possible to connect them one–one by strings. I could even say it is possible if they have the same number.

We could take it to be the criterion for being connected by counting them

1, 2, 3, 4, 5, 6.

Can I conclude anything?

This is just taken as what we call it is possible to connect them one–one: this means[287] count them and see if they have the same number.

There seems at first sight to be a possibility independent of any physical possibility. We could then ask, "What is your criterion for this[288] possibility?"

I would say this possibility is that there should be a certain picture in the system of pictures we use.

When we look at Russell's correlation, we get the feeling there is pre-eminently one criterion for there being correlated one–one.

If this is so, it is in fact that they have the same number. Only he expressed this in the form that there seems to be a correlation.[289]

Russell[290] said the criterion that they are correlated in the material way was that they are correlated in the immaterial way by means of identity.

We could on the other hand make an agreement on one particular way, say, of drawing lines which we would call trying out the possibility.

I want to say that what we did here

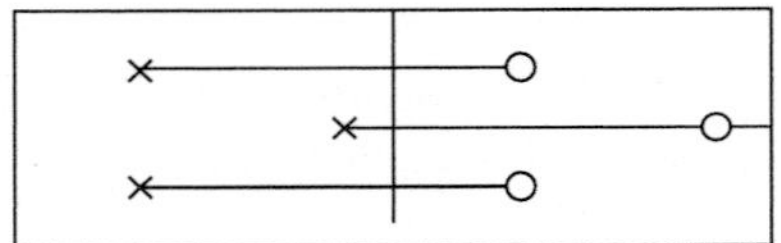

was really a geometrical construction (though there are certain things absent which make geometrical construction in Euclid).

There seems to be a trying and not succeeding which is relevant. If I have connected a set of red pegs to a set of blue and some of the blue are left over.

There was in fact no trying here. Part of the blue are left over.

I could say I had shown there was a greater[291] number of blue ones than red ones.

If our joining them with lines[292] was an experiment, then it did not interest us as it could have[293] failed with string.

It only interested us as it defined the same number.

[287]'it' is written, then deleted after 'means'.

[288]After 'for', 'your' is written and crossed out in favour of 'this' written above the line.

[289]'correlated' in the manuscript is changed to correlation'.

[290]'Russell's' in the manuscript, changed to 'Russell'.

[291]In the manuscript 'great' occurs, changed here to 'greater'.

[292]'with lines' written over the line; also 'line' is written and crossed out after 'was'.

[293]'have' is missing in the manuscript and supplied by Eds.

Suppose I tried to connect them with lines and failed. Then I tried with string and succeeded.

I assume there are a great number which I can't overlook.

I try to connect them with strings and succeed. What will you say?

Some of the dots must have vanished. (Miss Macdonald).[294]

Suppose I asked[295] "Were there more red dots than blue dots or had they the same number?"

Two cases (1) After joining them with chalk, you may say it is very unlikely you will be able to join them with strings.

or (2) You say it is impossible.

(1)[296] can be experimentally tested and find out as to[297] whether you can or not join them with strings. Here joining them with chalk is an experimental test for the other.

Then there is the case where you say it is impossible to join them with strings.

Here we say "it makes no sense to say they are joined with strings and if we do join them with strings we say they are not these but different ones".[298]

To say you are capable of joining them with strings although you haven't done so with chalk, is on the same level as saying you can join with strings although they haven't the same number. These are on the same level.

Take such a case as this.

Imagine lots of dots here and there:

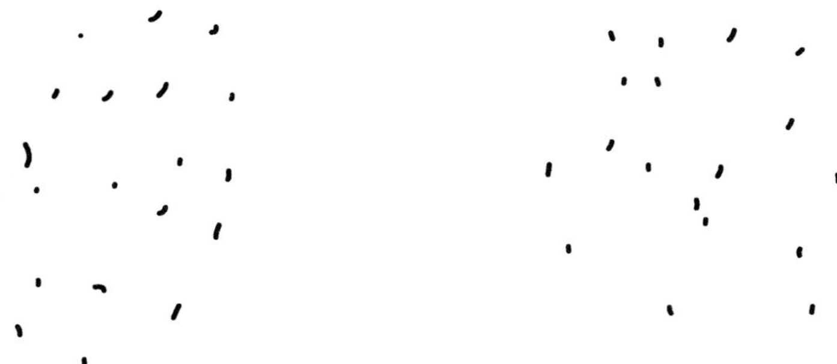

Let our criterion for their having the same no. be counting. We get 20 and 18.

I ask, "Can we correlate them one–one by line?"

There are two answers

[294]'(Miss Macdonald)' is squeezed into the margin, and on the next full line, is written: "If you say 'must'" is written, and crossed out.

[295]'Who' occurs and is crossed out after 'asked'.

[296]'Suppose' is written and crossed out before (1).

[297]'as to' is added by Eds.

[298]Close quotation marks are added by Eds.

(1) Possibly, let's try and see. This will show we said a series of words here and another set there and then two words which did not agree and then joined them.

(2) It is impossible. This means that if you now can join them, you make the rule that you will now say that two have appeared or two disappeared.

When we use the word "logical possibility" it means we are giving some rule about our expression. We say if we do correlate 18 points with 20 points then we say that some other points have appeared.

At first sight it has a clear and obvious meaning to say if two classes have different numbers I cannot correlate them (1–1). This seems as though I am prophesying a failure – like saying "Don't try, you won't succeed." Whereas in reality I am giving a rule which says that we are not two classes that[299] have different numbers and are correlated (1–1) and if two tests give us different results then at all costs I am to say there has been a mistake.

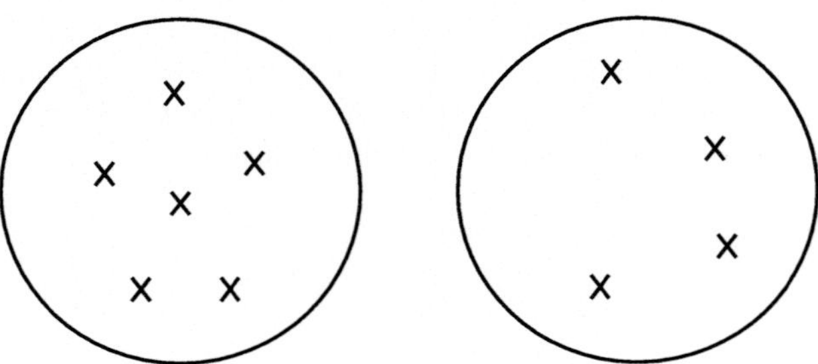

Let us go back to Russell's statement as we put it that the numbers of the two sets were equal if there existed geometrical straight lines joining them (1–1).

It seemed that if there did not exist geometrical straight lines joining them one–one, then there could not exist any other system joining them (1 1).

Our point was what was the criterion for geometrical lines existing.

One way would be to count them and say they had the same number – another *to actually draw* chalk lines.[300]

To say that unless they have the same, they cannot be joined one–one by a material correlation.

---

[299] 'that' is added by Eds.

[300] The italics indicate a line of dashes under these italicized words.

If we join them by chalk lines and in spite of some being left over, I am capable of joining them with strings. I have to say: this is not so that some points have appeared; others have disappeared.

Next Term[301] Note. Remember that we have no necessity for a definition of number. The whole charm was that number seemed to be a substantive, the generic name of things which mathematics dealt with. It seemed then we could analyze the notion of '3' namely "the class of class of triads". These triads seemed to be related by being connected by "very thin lines." As soon as you see Russell's correlation by identity is a chimera, for we cannot find out whether they are correlated in this way than in seeing if they can be connected by lines etc.[302]

Next Term[303]
April 24th 3 o'clock.

[301]The last two words are crossed out.

[302]This was the last lecture of the Lent Term. Lectures continued in the next – Easter – Term. We do not know why Skinner's manuscript of Wittgenstein's lectures does not include the other Terms. Even so, the manuscript here is self-contained, with a few detailed signs of Wittgenstein tuning it. It may be that Wittgenstein did not view the other Terms as something he wanted to place alongside this manuscript and wished this manuscript to be a composite member of the Archive that he sent to Goodstein.

Some unpublished lecture notes of the Easter Term Lecture, 1, p.1, typed by Margaret MacDonald (1935) commences with: "Recapitulation of Russell's theory of number. There is certainly something tempting in Russell's idea. But the whole idea of trying to define Number springs from a misunderstanding. We do not need a definition of Number. A definition of Number can only 'reduce' the idea of number to a set of indefinables. If this had been Russell's intention, this would have been very unimportant. But it was not his intention. The reason for giving a definition was an attempt to answer the question. What is a Number? It was an attempt to get rid of this puzzlement by getting clear about the grammar of the word 'Number' and of the numerals. We do not define Number any more than we define the King of Chess apart from the actions of the various kings of the sets. Any definition of number loses its meaning when we cease to ask the questions 'What is Number?' We try to get clear about how we use the word Number and do not think that here is the digit '3' and somewhere else intangible which is the number '3'."

[303]This is April 1935.

Part III
Calculations in the Hand of Francis Skinner

Chapter 9
Mathematical Calculations in Skinner's Hand

Introduction

Arthur Gibson and Niamh O'Mahony

The presence of this chapter's contents in this book requires some explanation. The manuscript from which the transcription is drawn is entirely different in style from the other manuscripts comprising the Archive, and it contrasts significantly with Wittgenstein's usual published work. It is also placed here rather than as an appendix because it was placed by Wittgenstein in with his own collection of dictated manuscripts that he sent to Reuben Goodstein soon after Skinner's death. Rationale for its appearance here amounts to the following: Wittgenstein posted them to Goodstein to be considered for publication. We would benefit from Goodstein's judgement about it; yet he did not express a view. If this were adopted as sufficient negative ground for resisting its publication, then such a position, *ceteris paribus*, opposes the publication of any of this Archive – including the *Brown Book*, since Goodstein was also silent about the whole Archive.

It is possible that Skinner composed the calculations under the direction or influence of Wittgenstein – perhaps he was interested in the way Skinner excavated areas for attention by virtue of his style in the calculations, not least with Wittgenstein's interest in Euler's work on primes.[1]

The transcription of the manuscript in this chapter is here also presented out of a duty of care to Francis Skinner's right to be recognized for his devotion to the emergence of Wittgenstein's philosophy and compositions in this Edition. Even with the probability that Skinner is the author, it is yet propitious to include the manuscript here for the above reasons, and allow that Skinner's role in its composition is as the recipient of a degree of Wittgenstein's influence.

[1]The case of Schlick's "Diktat für Schlick", as analysed by Oakes and Pichler (2013) is relevant for progress on this point.

A. Gibson, N. O'Mahony, *Ludwig Wittgenstein: Dictating Philosophy*,
https://doi.org/10.1007/978-3-030-36087-0_9

The manuscript papers reproduced here in this chapter contain calculations in the integers mod 257, which involve calculations concerning proof of Fermat's Little Theorem.[2] Basically the calculations are standard, though not in some ways in style and are not very advanced, which in contrast with typical research standards. Even so, significant mathematicians often researching for truths hidden in elementary questions, frequently make progress by reducing a complex case to a simple example. So, Wittgenstein's yen to employ simple examples is not in itself naïve nor relevantly limiting. In view of Littlewood's and Hardy's highly complimentary assessments of him in this regard, there is no reason to object to Wittgenstein aiming for original philosophy or a new logical angle, or a creative idea for future maths while probing calculations. (Recall that there were no computers in 1934.) Some of the style is oddly quirky: as if the mind is switched off in favour of brute number-crunching. But 'quirky' can lead to original mathematics and open-mindedness the manuscripts may be the remains of a device to trigger the unexpected. Furthermore, these Fermat manuscripts are incomplete – there might be more somewhere; they do contain a self-evaluation; our Edition offers some of this in the conclusion at the end of this chapter).

A number of mathematics professors[3] in Cambridge urged the publication of this manuscript as a historical document while a few philosophers saw no point in so doing. It appears that there is nothing in existence quite like this, surviving from Cambridge mathematics of the 1930s. The contents of the manuscript transcribed are to some degree unusual, even a little bizarre and unprecedented, though with familiar main traits. The calculations seem to have the job of exposing nuances or contrasts in carefully composed elegant mathematical procedures, driven by no order other than random exploration within a series of numbers and relations. Somewhat like unusual secondary twists in a language game using standard outcomes.

They do not amount to proofs, while deploying the results of Fermat's Little Theorem. The style of these calculations appear to manifest some of Wittgenstein's presuppositions: the calculations have an almost obsessive feel to them. They do not employ any elegant short-cuts, but are brute-force calculations, as if searching for unexpected patterns or moves away from standard routes. It is as though the writer has said to himself: 'I know what these results are, but I want to see them for myself, by switching my brain off and just doing the calculations.' These calculations allow for a difference between mechanical calculation and a driven mind composing calculation. Or perhaps these are numerical calculations to confirm some guesses the author has made – but this is unlikely, as the guesses are very easy to prove. Anyone who had been exposed to these things would surely also have been exposed to their proofs and the underlying explanations.

The specific contemporary context for the manuscript's concerns is not easily identified. Despite this, progress can be made. We can reconstruct parts of a picture

---

[2]Of course, this Fermat's Little Theorem is entirely different from Fermat's Last Great Theorem, whilst the Pink Book (Chap. 3) ironically refers to the latter as having been proved in the 1930s – proof only achieved a generation after Wittgenstein death, by Wiles and Taylor.

[3]E.g. I. Leader, M. Perry, M. Atiyah, M. Hyland. This is not due to any original discovery. Rather, for its striking character.

of pertinent research activity that frames the manuscript, including the earlier research of some of Francis Skinner's Cambridge teachers, within the circle of contacts that Wittgenstein esteemed. Twelve years before Francis arrived at Trinity, Hardy (with Ramanujan[4]) devised a theorem concerning the normal number of prime factors of a number *n*.[5] In 1925, four years before Francis came to study at Trinity, his future teachers Hardy and Littlewood[6] devised conjectures treating asymptotic (i.e. limiting) aspects of the distribution of primes.[7] In 1932 – Francis's second undergraduate year, his teacher at King's College[8] – A. E. Ingham (1932) published his book on the distribution of primes, while Skinner was engaged with Wittgenstein when being taught by Ingham.

The manuscript's calculations in the first sequence are concerned with some calculations in the integers mod 257. The number 257 is featured, but its significance is not explained. The calculations make sense because it is a Fermat prime, being of the form 2^(2^n) + 1; namely 2^(2^3) + 1.[9] So it starts with a calculation of the powers of 3 mod 257. There is no explanation of what is going on, but it is clear from the numbers that this really is a calculation mod 257. They are calculated right through to 3^{256}, which is 1. This is a confirmation of Fermat's Little Theorem (that a^{p-1} is always 1 mod p, for any prime p and any a that is not a multiple of p). Since no earlier power of 3 is 1, this also shows that 3 is a primitive root mod 257. So perhaps this was the aim of the exercise, though there are much simpler ways to see that 3 is a primitive root. Or perhaps the writer needs a table of all powers of 3 because of the calculations that come next.

Next we have some sums of powers of 3. For example, the sum $3^2 + 3^4 + 3^6 + \ldots + 3^{256}$ of all the even powers of 3 is calculated, as is the sum $3^1 + 3^3 + 35 \ldots + 3^{255}$ of all the odd powers of 3. Curiously, these sums are written down but not evaluated; each is zero, since the first is the sum of all powers of 3^2 (and the sum of the powers of any element that is not 1 must be zero) and the second is 3 times this. Then we also have the four sums $3^a + 3^{a+4} + 3^{a+8} + \ldots$, for $a = 1, 2, 3, 4$. Again, the expressions are not evaluated, even though each must be zero. There are many more of such calculations.

Some of the sums are roots of quadratic polynomials with coefficients. Again, these actually boil down to things like 'the roots of the polynomial $x^2 + 0x + 0$ are

---

[4]Hardy and Ramanujan (1917).

[5]That the normal order of the number $\omega(n)$ of number n's distinct prime factors is $\log(\log(n))$.

[6]Hardy and Littlewood (1925).

[7]Of course, this research moved on, while the pioneering work of Hardy-Littlewood still stands, for which see Ivić (2013: 14–18, 63, 71, 75), not least developed by Bombieri (cf. Ivić, 2013: 41, 43–44).

[8]As reported in the Introduction, some of the manuscript paper that comprises the Archive has a King's College watermark in it, which may derive from Skinner taking it from a King's room in which he was taught. It is worth exploring the possibility that Wittgenstein himself gathers the paper from King's, during periods of teaching there.

[9]Note that in Gibson's (2010b: 72) reference to this number, there is a printer's error, i.e. "2^23", which should be as above and read "2^(2^3) + 1".

0 and 0,' but this is not mentioned. Of course, nowhere is it stated that 'we are working mod 257,' so maybe the various sums do not collapse down to zero, but in that case it is hard to see where we *are* working. This is especially in view of the fact that the powers of 3 that are being added together in all these sums really *are* being calculated mod 257.

Maybe the manuscript is working part of the time in the integers mod 257 and part of the time in the integers – but in that case many of the results stated would not be true, so this is unlikely.

One sequence of calculations that is puzzling is the sequence that comes next. It is not joined to the above calculations; yet the sequence is written in the same style and identical manuscript paper to the calculations. It seems to involve some variables (the Z_i and the R_i) that could be related to the earlier calculations, but it is not clear what they are or how they are related. There is no evaluation of these contents in the rest of the manuscript; this might be because there are missing pages – though there is no evidence for this, or due to a wish to leave the problem in this form to puzzle some readers. However unlikely it may seem, this might be a purposively incomplete language game to highlight the effect of absence on assessing what language game it is.

There is also a page containing some relations between powers of roots of a polynomial and the roots themselves: for example, the expression in terms of some symmetric functions in the roots for the sum of a certain power of the roots.[10] This might be regarded as an example of thinking mathematically. It is somehow experimental, yet out of place among the other calculations. If something were added, it might become another language game of significance, which may be the point in its presence. Such examples as there are in the Archive might thus be taken to be the mathematical equivalent of natural language examples of use by which Wittgenstein explores terrain, its discontinuities, its various topologies and geometries. This could point the way as to how to embark on an investigation of arbitrary games within mathematics, whose rules are not arbitrary.[11]

[10] See Appendix [A].

[11] There is a page of geometry that contains the expressions of two standard triangle-geometry theorems about the radii of circles associated with triangles. There is also a sketch that complements this.

Transcription of Mathematical Calculations

MS Page 1[12]

$3 = 3$[13]
$3^2 = 9$
$3^3 = 27$
$3^4 = 81$
$3^5 = 243$
$3^6 = 215$
$3^7 = 131$
$3^8 = 136$
$3^9 = 151$
$3^{10} = 196$
$3^{11} = 74$
$3^{12} = 222$
$3^{13} = 152$
$3^{14} = 199$
$3^{15} = 83$
$3^{16} = 249$
$3^{17} = 233$
$3^{18} = 185$
$3^{19} = 41$
$3^{20} = 123$
$3^{21} = 112$
$3^{22} = 79$
$3^{23} = 237$
$3^{24} = 197$
$3^{25} = 77$
$3^{26} = 231$
$3^{27} = 179$
$3^{28} = 23$

---

[12]Since there are no indications of page numbers on the manuscripts that comprise these calculations, page numbers in the above form are added to this transcription to keep track of references made in the Editors' Conclusion at the end of the transcription.

[13]This sheet on which on the recto side in ink is written the number series from "$3 = 3$" to "$3^{187} = 100$", the verso has penciled geometrical shapes (intersecting triangles, some within circles, etc.). There are also two equations written; they are:

$$r = \frac{2\Delta}{a + b + c}$$

$$R = \underline{2\Delta}$$

The geometrical shapes on the verso sheet have parallels with similar, but not identical, drawings by Wittgenstein in some of his large lecture notebooks, for example MS 148 (1935a), pp. 9–21, 27.

$3^{29} = 69$
$3^{30} = 207$
$3^{31} = 107$
$3^{32} = 64$
$3^{33} = 192$
$3^{34} = 62$
$3^{35} = 186$
$3^{36} = 44$
$3^{37} = 132$
$3^{38} = 139$
$3^{39} = 160$
$3^{40} = 223$
$3^{41} = 155$
$3^{42} = 208$
$3^{43} = 110$
$3^{44} = 73$
$3^{45} = 219$
$3^{46} = 143$
$3^{47} = 172$
$3^{48} = 2$
$3^{49} = 6$
$3^{50} = 18$
$3^{51} = 54$
$3^{52} = 162$
$3^{53} = 229$
$3^{54} = 173$
$3^{55} = 5$
$3^{56} = 15$
$3^{57} = 45$
$3^{58} = 135$
$3^{59} = 148$
$3^{60} = 187$
$3^{61} = 47$
$3^{62} = 141$
$3^{63} = 166$
$3^{64} = 241$
$3^{65} = 209$
$3^{66} = 113$
$3^{67} = 82$
$3^{68} = 246$
$3^{69} = 224$
$3^{70} = 158$
$3^{71} = 217$
$3^{72} = 137$
$3^{73} = 154$

$3^{74} = 205$
$3^{75} = 101$
$3^{76} = 46$
$3^{77} = 138$
$3^{78} = 157$
$3^{79} = 214$
$3^{80} = 128$
$3^{81} = 127$
$3^{82} = 124$
$3^{83} = 115$
$3^{84} = 88$
$3^{85} = 7$
$3^{86} = 21$
$3^{87} = 63$
$3^{88} = 189$
$3^{89} = 53$
$3^{90} = 159$
$3^{91} = 220$
$3^{92} = 146$
$3^{93} = 181$
$3^{94} = 29$
$3^{95} = 87$
$3^{96} = 4$
$3^{97} = 12$
$3^{98} = 36$
$3^{99} = 108$
$3^{100} = 67$
$3^{101} = 201$
$3^{102} = 89$
$3^{103} = 10$
$3^{104} = 30$
$3^{105} = 90$
$3^{106} = 13$
$3^{107} = 39$
$3^{108} = 117$
$3^{109} = 94$
$3^{110} = 25$
$3^{111} = 75$
$3^{112} = 225$
$3^{113} = 161$
$3^{114} = 226$
$3^{115} = 164$
$3^{116} = 235$
$3^{117} = 191$
$3^{118} = 59$

$3^{119} = 177$
$3^{120} = 17$
$3^{121} = 51$
$3^{122} = 153$
$3^{123} = 202$
$3^{124} = 92$
$3^{125} = 19$
$3^{126} = 57$
$3^{127} = 171$
$3^{128} = 256$
$3^{129} = 254$
$3^{130} = 248$
$3^{131} = 230$
$3^{132} = 176$
$3^{133} = 14$
$3^{134} = 42$
$3^{135} = 126$
$3^{136} = 121$
$3^{137} = 106$
$3^{138} = 61$
$3^{139} = 183$
$3^{140} = 35$
$3^{141} = 105$
$3^{142} = 58$
$3^{143} = 174$
$3^{144} = 8$
$3^{145} = 24$
$3^{146} = 72$
$3^{147} = 216$
$3^{148} = 134$
$3^{149} = 145$
$3^{150} = 178$
$3^{151} = 20$
$3^{152} = 60$
$3^{153} = 180$
$3^{154} = 26$
$3^{155} = 78$
$3^{156} = 234$
$3^{157} = 188$
$3^{158} = 50$
$3^{159} = 150$
$3^{160} = 193$
$3^{161} = 65$
$3^{162} = 195$
$3^{163} = 71$

$3^{164} = 213$
$3^{165} = 125$
$3^{166} = 118$
$3^{167} = 97$
$3^{168} = 34$
$3^{169} = 102$
$3^{170} = 49$
$3^{171} = 147$
$3^{172} = 184$
$3^{173} = 38$
$3^{174} = 114$
$3^{175} = 85$
$3^{176} = 255$
$3^{177} = 251$
$3^{178} = 239$
$3^{179} = 203$
$3^{180} = 95$
$3^{181} = 28$
$3^{182} = 84$
$3^{183} = 252$
$3^{184} = 242$
$3^{185} = 212$
$3^{186} = 122$
$3^{187} = 100$
$3^{188} = 70$[14]
$3^{189} = 210$
$3^{190} = 116$
$3^{191} = 91$
$3^{192} = 16$
$3^{193} = 48$
$3^{194} = 144$
$3^{195} = 175$
$3^{196} = 11$
$3^{197} = 33$
$3^{198} = 99$
$3^{199} = 40$
$3^{200} = 120$
$3^{201} = 103$

[14]On the verso side of this sheet on which on the recto side in ink is written the series from "$3^{188} = 70$" to "$3^{256} = 1$", the verso has pencil written equations of the form:

$$= \quad (a_1 a_2 a_3 a_4)\left(a_1{}^3 a_2{}^2 a_3\right), \text{etc.}$$

$3^{202} = 52$
$3^{203} = 156$
$3^{204} = 211$
$3^{205} = 119$
$3^{206} = 100$
$3^{207} = 43$
$3^{208} = 129$
$3^{209} = 130$
$3^{210} = 133$
$3^{211} = 142$
$3^{212} = 169$
$3^{213} = 250$
$3^{214} = 236$
$3^{215} = 194$
$3^{216} = 68$
$3^{217} = 204$
$3^{218} = 98$
$3^{219} = 37$
$3^{220} = 111$
$3^{221} = 76$
$3^{222} = 228$
$3^{223} = 170$
$3^{224} = 253$
$3^{225} = 245$
$3^{226} = 221$
$3^{227} = 149$
$3^{228} = 190$
$3^{229} = 56$
$3^{230} = 168$
$3^{231} = 247$
$3^{232} = 227$
$3^{233} = 167$
$3^{234} = 244$
$3^{235} = 218$
$3^{236} = 140$
$3^{237} = 163$
$3^{238} = 232$
$3^{239} = 182$
$3^{240} = 32$
$3^{241} = 96$
$3^{242} = 31$
$3^{243} = 93$
$3^{244} = 22$
$3^{245} = 66$
$3^{246} = 198$

$3^{247} = 80$
$3^{248} = 240$
$3^{249} = 206$
$3^{250} = 104$
$3^{251} = 55$
$3^{252} = 165$
$3^{253} = 238$
$3^{254} = 200$
$3^{255} = 86$
$3^{256} = 1$

MS Page 2

$s_1 = 9 + 81 + 215 + 136 + 196 + 222 + 199 + 249 + 185 + 123 + 79 + 197 + 231 + 23 + 207 + 64 + 62 + 44 + 139 + 223 + 208 + 73 + 143 + 2 + 18 + 162 + 173 + 15 + *$
$+ 135 + 187 + 141 + 241 + 113 + 246 + 158 + 137 + 205 + 46 + 157 + 128 + 124 + 88 + 21 + 189 + 159 + 146 + 29 + 4 + 36 + 67 + 89 + 30 + 13 + 117 + 25 + *$
$+ 225 + 226 + 235 + 59 + 17 + 153 + 92 + 57 + 256 + 248 + 176 + 42 + 121 + 61 + 35 + 58 + 8 + 72 + 134 + 178 + 60 + 26 + 234 + 50 + 193 + 195 + *$
+ 213[15] $+ 118 + 34 + 49 + 184 + 114 + 255 + 239 + 95 + 84 + 242 + 122 + 70 + 116 + 16 + 144 + 11 + 99 + 120 + 52 + 211 + 100 + 129 + 133 + *$
$+ 169 + 236 + 68 + 98 + 111 + 228 + 253 + 221 + 190 + 168 + 227 + 244 + 140 + 232 + 32 + 31 + 22 + 198 + 240 + 104 + 165 + 200 + 1.$

$S_2 = 3 + 27 + 243 + 131 + 151 +$ 74[16] $+ 152 + 83 + 233 + 41 + 112 + 237 + 77 + 179 + 69 + 107 + 192 + 186 + 132 + 160 + 155 + 110 + 219 + 172 + 6 + 54 + 229 + *$
$+ 5 + 45 + 148 + 47 + 166 + 209 + 82 + 224 + 217 + 154 + 101 + 138 + 214 + 127 + 115 + 7 + 63 + 53 + 220 + 181 + 87 + 12 + 108 + 201 + 10 + 90 + 39 + 94 + 75 + *$
$+ 161 + 164 + 191 + 177 + 51 + 202 + 19 + 171 + 254 + 230 + 14 + 126 + 106 + 183 + 105 + 174 + 24 + 216 + 145 + 20 + 180 + 78 + 188 + 150 + 65 + 71 + 125 + *$
$+ 97 + 102 + 147 + 38 + 85 + 251 + 203 + 28 + 252 + 212 + 109 + 210 + 91 + 48 + 175 + 33 + 40 + 103 + 156 + 119 + 43 + 130 + 142 + 250 + 194 + *$
$+ 204 + 37 + 76 + 170 + 245 + 149 + 56 + 247 + 167 + 218 + 163 + 182 + 96 + 93 + 66 + 80 + 206 + 55 + 238 + 86.$

$t_1 = 81 + 136 + 222 + 249 + 123 + 197 + 23 + 64 + 44 + 223 + 73 + 2 + 162 + 15 + 187 + 241 + 246 + 137 + 46 + 128 + 88 + 189 + 146 + 4 + 67 + 30 + 117 + 225*$
$+ 235 + 17 + 92 + 256 + 176 + 121 + 35 + 8 + 134 + 60 + 234 + 193 + 213 + 34 + 184 + 255 + 95 + 242 + 70 + 16 + 11 + 120 + 211 + 129 + 169 + 68 + 111 + 253 + 190 + *$
$+ 227 + 140 + 32 + 22 + 240 + 165 + 1.$

---

[15] Eds think this is 213 rather than 215.

[16] Eds think this is 74 rather than 24.

t_2=9+215+196+199+185+79+231+207+62+139+208+143+18+173+135+141+-
113 + 158 + 205 + 157 + 124 + 21 + 159 + 29 + 36 + 89 + 13 + 25 + *
+226+59+153+57+248+42+61+58+72+178+26+50+195+118+49+114+239+84
+ 122 + 116[17] + 144 + 99 + 52 + 100 + 133 + 236 + 98 + *
+ 228 + 221 + 168 + 244 + 232 + 31 + 198 + 104 + 200

t_3 = 27 + 131 + 74 + 83 + 41 + 237 + 179 + 107 + 186 + 160 + 110 + 172 + 54 + 5 + 148 +
166 + 82 + 217 + 101 + 214 + 115 + 63 + 220 + 87 + 108 + 10 + 39 + 75 + *
+ 164 + 177 + 202 + 171 + 230 + 126 + 183 + 174 + 216 + 20 + 78 + 150 + 71 + 97 + 147 +
85 + 203 + 252 + 109 + 91 + 175 + 40 + 156 + 43 + 142 + 194 + 37 + 170 + *
+ 149 + 247 + 218 + 182 + 93 + 80 + 55 + 86

t_4=3+243+151+152+233+112+77+69+192+132+155+219+6+229+45+47+209+-
224 + 154 + 138 + 127 + 7 + 53 + 181 + 12 + 201 + 90 + 94*
+161+191+51+19+254+14+106+105+24+145+180+188+65+125+102+38+251+
28 + 212 + 210 + 48 + 33 + 103 + 119 + 130 + 250 + *
+ 204 + 76 + 245 + 56 + 167 + 163 + 96 + 66 + 206 + 238.

u_1=136+249+197+64+223+2+15+241+137+128+189+4+30+225+17+256+121+-
8 + 60 + 193 + 34 + 255 + 242 + 16 + 120 + 129 + 68 + 253 + 227 + 32 + 240 + 1.

u_2 = 81 + 222 + 123 + 23 + 44 + 73 + 162 + 187 + 246 + 46 + 88 + 146 + 67 +
117 + 235 + 92 + 176 + 35 + 134 + 234 + 213 + 184 + 95 + 70 + 11 + 211 +
169 + 111 + 190 + 140 + 22 + 165.

u_3 = 215 + 199 + 79 + 207 + 139 + 143 + 173 + 141 + 158 + 157 + 21 + 29[18] +
89 + 25 + 59 + 57 + 42 + 58 + 178 + 50 + 118 + 114 + 84 + 116 + 99 + 100 + 236 +
228 + 168 + 232 + 198 + 200.

u_4 = 9 + 196 + 185 + 231 + 62[19] + 208 + 18 + 135 + 113 + 205 + 124 +
159 + 36 + 13 + 226 + 153 + 248 + 61 + 72 + 26 + 195 + 49 + 239 + 122 + 144 + 52 +
133 + 98 + 221 + 244 + 31 + 104.

u_5 = 131 + 83 + 237 + 107 + 160 + 172 + 5 + 166 + 217 + 214 + 63 + 87 +
10 + 75 + 177 + 171 + 126 + 174 + 20 + 150 + 97 + 85 + 252 + 91 + 40 + 43 +
194 + 170 + 247 + 182 + 80 + 86.

u_6 = 27 + 74 + 41 + 179 + 186 + 110 + 54 + 148 + 82 + 101 + 115 + 220 + 108 + 39 + 164 +
202 + 230 + 183 + 216 + 78 + 71 + 147 + 203 + 109 + 175 + 156 + 142 + 37 + 149 +
218 + 93 + 55.

u_7 = 243 + 152 + 112 + 69 + 132 + 219 + 229 + 47 + 224 + 138 + 7 + 181 + 201 + 94 + 191 +
19 + 14 + 105 + 145 + 188 + 125 + 38 + 28 + 210 + 33 + 119 + 250 + 76 + 56 + 163 +
66 + 238.

[17] We think that this fuzzily-written number is 116.

[18] We think that this is 29.

[19] This number was originally 64, with the 4 crossed out and 2 written above it.

u_8=3+151+233+77+172[20]+155+6+45+209+154+127+53+12+90+161+51+254+-106 + 24 + 180 + 65 + 102 + 251 + 212 + 48 + 103 + 130 + 204 + 245 + 167 + 96 + 206.

MS Page 3

w_1 = 249 + 64 + 2 + 241 + 128 + 4 + 225 + 256 + 8 + 193 + 255 + 16 + 129 + 253 + 32 + 1[21]

w_2 = 136 + 197 + 223 + 15 + 137 + 189 + 30 + 17 + 121 + 60 + 34 + 242 + 120 + 68 + 227 + 240[22]

w_3 = 222 + 23 + 73 + 187 + 46 + 146 + 117 + 92 + 35 + 234 + 184 + 70 + 211 + 111 + 140 + 165

w_4 = 81 + 123 + 44 + 162 + 246 + 88 + 67[23] + 235 + 176 + 134 + 213 + 95 + 11 + 169 + 190 + 22

w_5 = 199 + 207 + 143 + 141 + 157 + 29 + 25 + 57 + 58 + 50 + 114 + 116 + 100 + 228 + 232 + 200

w_6 = 215 + 79 + 139 + 173 + 158 + 21 + 89 + 59 + 42 + 178 + 118 + 84 + 99 + 236 + 168 + 198

w_7 = 196 + 231 + 208 + 135 + 205 + 159 + 13 + 153 + 61 + 26 + 49 + 122 + 52 + 98 + 244 +104

w_8= 9 + 185 + 62 + 18 + 113 + 124 + 36 + 226 + 248 + 72[24] + 195 + 239 + 144 + 133 + 221 + 31

w_9 = 83 + 107 + 172 + 166 + 214 + 87 + 75 + 171 + 174 + 150 + 85 + 91 + 43 + 170 + 182 + 86

w_{10} = 131 + 237 + 160 + 5 + 217 + 63 + 10 + 177 + 126 + 20[25] + 97 + 252 + 40 + 194 + 247 + 80

w_{11} = 74 + 179 + 110 + 148 + 101 + 220 + 39 + 202 + 183 + 78 + 147 + 109 + 156 + 37 + 218 + 55

w_{12} = 27 + 41 + 186 + 54 + 82 + 115 + 108 + 164 + 230 + 216 + 71 + 203 + 175 + 142 + 149 + 93

w_{13} = 152 + 69 + 219 + 47 + 138 + 181 + 94 + 19 + 105 + 188 + 38 + 210 + 119 + 76 + 163 + 238

w_{14} = 243 + 112 + 132 + 229 + 224 + 7 + 201 + 191 + 14 + 145 + 125 + 28 + 33 + 250 + 56 + 66

w_{15} = 151 + 77 + 155 + 45 + 154 + 53 + 90 + 51 + 106 + 180 + 102 + 212 + 103 + 204 + 167 + 206

---

[20]This number might be 192.

[21]w_2, w_4 and so on are written alongside w_1, w_3.

[22]This line and the lines below it, are on the right-hand side of the page.

[23]Here the numbers from + 117 + 92 + 35 + 234 + 184 + 70 + 211 + 111 + 140 + 165 are crossed out and the numbers from + 235 + 176 to 22 added below them.

[24]The numbers from + 26 + 49 + 122 + 52 + 98 are crossed out and the numbers from + 195 + 239 to 31 added below them.

[25]This appears to be 20, not 70.

$w_{16} = 3 + 233 + 192 + 6 + 209 + 127 + 12 + 161 + 254 + 24 + 65 + 251 + 48 + 130 + 245 + 96$

$x_1 = 64 + 241 + 4 + 256 + 193 + 16 + 253 + 1$ ‖ $x_2 = 249 + 2 + 128 + 225 + 8 + 255 + 129 + 32$

$x_3 = 197 + 15 + 189 + 17 + 60 + 242 + 68 + 240$ ‖ $x_4 = 136 + 223 + 137 + 30 + 121 + 34 + 120 + 227$

$x_5 = 23 + 187 + 146 + 92 + 234 + 70 + 111 + 165$ ‖ $x_6 = 222 + 73 + 46 + 117 + 35 + 184 + 211 + 140$

$x_7 = 123 + 162 + 88 + 235 + 134 + 95 + 169 + 22$ ‖ $x_8 = 81 + 44 + 246 + 67 + 176 + 213 + 11 + 190$

$x_9 = 207 + 141 + 29 + 57 + 50 + 116 + 228 + 200$ ‖ $x_{10} = 199 + 143 + 157 + 25 + 58 + 114 + 100 + 232$

$x_{11} = 79 + 173 + 21 + 59 + 178 + 84 + 236 + 198$ ‖ $x_{12} = 215 + 139 + 158 + 89 + 42 + 118 + 99 + 168$

$x_{13} = 231 + 135 + 159 + 153 + 26 + 122 + 98 + 104$ ‖ $x_{14} = 196 + 208 + 205 + 13 + 61 + 49 + 52 + 244$

$x_{15} = 185 + 18 + 124 + 226 + 72 + 239 + 133 + 31$ ‖ $x_{16} = 9 + 62 + 113 + 36 + 248 + 195 + 144 + 221$

$x_{17} = 107 + 166 + 87 + 171 + 150 + 91 + 170 + 86$[26] ‖ $x_{18} = 83 + 172 + 214 + 75 + 174 + 85 + 43 + 182$

$x_{19} = 237 + 5 + 63 + 177 + 20 + 252 + 194 + 80$ ‖ $x_{20} = 131 + 160 + 217 + 10 + 126 + 97 + 40 + 247$

$x_{21} = 179 + 148 + 220 + 202 + 78 + 109 + 37 + 55$ ‖ $x_{22} = 74 + 110 + 101 + 39 + 183 + 147 + 156 + 218$

$x_{23} = 41 + 54 + 115 + 164 + 216 + 203 + 142 + 93$ ‖ $x_{24} = 27 + 186 + 82 + 108 + 230 + 71 + 175 + 149$

$x_{25} = 69 + 47 + 181 + 19 + 188 + 210 + 76 + 238$ ‖ $x_{26} = 152 + 219 + 138 + 94 + 105 + 38 + 119 + 163$

$x_{27} = 112 + 229 + 7 + 191 + 145 + 28 + 250 + 66$ ‖ $x_{28} = 243 + 132 + 224 + 201 + 14 + 125 + 33$[27] $+ 56$

$x_{29} = 77 + 45 + 53 + 51 + 180 + 212 + 204 + 206$ ‖ $x_{30} = 151 + 155 + 154 + 90 + 106 + 102 + 103 + 167$

$x_{31} = 233 + 6 + 127 + 161 + 24 + 251 + 130 + 96$ ‖ $x_{32} = 3 + 192 + 209 + 12 + 254 + 65 + 48 + 245$

MS Page 4

$y_1 = 241 + 256 + 16 + 1$ ‖ $y_2 = 64 + 4 + 193 + 253$ ‖ $y_3 = 2 + 225 + 255 + 32$ ‖ $y_4 = 249 + 128 + 8 + 129$ ‖ $y_5 = 15 + 17 + 242 + 240$ ‖ $y_6 = 197 + 189 + 60 + 68$ ‖ $y_7 = 223 + 30 + 34 + 227$ ‖ $y_8 = 136 + 137 + 121 + 120$

[26]This line "$x_{17} = 107 + 160 + 214 + 75 + 174 + 89$" from x_{17} is crossed out.

[27]Eds think that this is 133, but it could be 138.

$y_9 = 187 + 92 + 70 + 165$ || $y_{10} = 23 + 146 + 234 + 111$ || $y_{11} = 73 + 117 + 184 + 140$ || $y_{12} = 222 + 46 + 35 + 211$ || $y_{13} = 162 + 235 + 95 + 22$ || $y_{14} = 123 + 88$[28] $+ 134 + 169$ || $y_{15} = 44 + 67 + 213 + 190$ || $y_{16} = 81 + 246 + 176 + 11$

$y_{17} = 141 + 57 + 116 + 200$ || $y_{18} = 207 + 29 + 50 + 228$ || $y_{19} = 143 + 25 + 114 + 232$ || $y_{20} = 199 + 157 + 58 + 100$ || $y_{21} = 173 + 59 + 84 + 198$ || $y_{22} = 79 + 21 + 178 + 236$ || $y_{23} = 139 + 89 + 118 + 168$ || $y_{24} = 215 + 158 + 142 + 99$

$y_{25} = 135 + 153 + 122 + 104$ || $y_{26} = 231 + 159 + 26 + 98$ || $y_{27} = 208 + 13 + 49 + 244$ || $y_{28} = 196 + 205 + 61 + 52$ || $y_{29} = 18 + 226 + 239 + 31$ || $y_{30} = 185 + 124 + 72 + 133$ || $y_{31} = 62$[29] $+ 36 + 195 + 221$ || $y_{32} = 9 + 113 + 248 + 144$

$y_{33} = 166 + 171 + 91 + 86$ || $y_{34} = 107 + 87 + 150 + 170$ || $y_{35} = 172 + 75 + 85 + 182$ || $y_{36} = 83 + 214 + 174 + 43$ || $y_{37} = 5 + 177 + 252 + 80$ || $y_{38} = 237 + 63 + 20 + 194$ || $y_{39} = 160 + 10 + 97 + 247$ || $y_{40} = 131 + 217 + 126 + 40$

$y_{41} = 148 + 202 + 109 + 55$ || $y_{42} = 179 + 220 + 78 + 37$ || $y_{43} = 110 + 39 + 147 + 218$ || $y_{44} = 74 + 101 + 183 + 150$[30] || $y_{45} = 54 + 164 + 203 + 93$ || $y_{46} = 41 + 115 + 216 + 142$ || $y_{47} = 186 + 108 + 71 + 149$ || $y_{48} = 27 + 82 + 230 + 175$
$y_{49} = 47 + 19 + 210 + 238$ || $y_{50} = 69 + 181 + 188 + 76$ || $y_{51} = 219 + 94 + 38 + 163$ || $y_{52} = 152 + 138$[31] $+ 105 + 119$ || $y_{53} = 229 + 191 + 28 + 66$ || $y_{54} = 112 + 7 + 145 + 250$ || $y_{55} = 132 + 201 + 125 + 56$ || $243 + 224 + 14 + 33 = y_{56}$

$y_{57} = 45 + 51 + 212 + 206$ || $y_{58} = 77 + 53 + 180 + 204$ || $y_{59} = 155 + 90 + 102 + 167$ || $y_{60} = 151 + 154 + 106 + 103$ || $y_{61} = 6 + 161 + 251 + 96$ || $y_{62} = 233 + 127 + 24 + 130$ || $y_{63} = 192 + 12 + 65 + 245$ || $y_{64} = 3 + 209 + 254 + 48$.

$z_1 = 256 + 1$ || $z_2 = 241 + 16$ || $z_3 = 4 + 253$ || $z_4 = 64 + 193$ || $z_5 = 225 + 32$ || $z_6 = 2 + 255$ || $z_7 = 128 + 129$ || $z_8 = 249 + 8$ || $z_9 = 17 + 240$ || $z_{10} = 15 + 242$ || $z_{11} = 189 + 68$ || $z_{12} = 197 + 60$ || $z_{13} = 30 + 227$ || $z_{14} = 223 + 34$ || $z_{15} = 137 + 120$ || $z_{16} = 136 + 121$

$z_{17} = 92 + 165$ || $z_{18} = 187 + 70$ || $z_{19} = 146 + 111$ || $z_{20} = 23 + 234$ || $z_{21} = 117 + 140$ || $z_{22} = 73 + 184$ || $z_{23} = 46 + 211$ || $z_{24} = 222 + 35$ || $z_{25} = 235 + 22$ || $z_{26} = 162 + 95$ || $z_{27} = 88 + 169$ || $z_{28} = 123 + 134$ || $z_{29} = 47$[32] $+ 190$ || $z_{30} = 44 + 213$ || $z_{31} = 246 = 11$ || $z_{32} = 81 + 176$

$z_{33} = 57 + 200$ || $z_{34} = 141 + 116$ || $z_{35} = 29 + 228$ || $z_{36} = 207 + 50$ || $z_{37} = 25 + 232$ || $z_{38} = 143 + 114$ || $z_{39} = 157 + 100$ || $z_{40} = 199 + 58$ || $z_{41} = 59 + 198$ || $z_{42} = 173 + 84$ || $z_{43} = 21 + 236$ || $z_{44} = 79 + 178$ || $z_{45} = 89 + 168$ || $z_{46} = 139 + 118$ || $z_{47} = 158 + 99$ || $z_{48} = 215 + 42$

---

[28]The number 88 replaces a number that has been crossed out – which appears to be 162.

[29]This number seems to be 62, with the 2 over-writing a 4.

[30]This number may be 156.

[31]The number 138 has been written above the number that appears to be 90 – which has been crossed out.

[32]This number is not entirely clear but appears to be 47.

$z_{49} = 153 + 104$ ‖ $z_{50} = 135 + 122$ ‖ $z_{51} = 159 + 98$ ‖ $z_{52} = 231 + 26$ ‖ $z_{53} = 13 + 244$ ‖ $z_{54} = 208 + 49$ ‖ $z_{55} = 205 + 52$ ‖ $z_{56} = 196 + 61$ ‖ $z_{57} = 226 + 31$ ‖ $z_{58} = 18 + 239$ ‖ $z_{59} = 124 + 133$ ‖ $z_{60} = 185 + 72$ ‖ $z_{61} = 36 + 221$ ‖ $z_{62} = 62 + 195$ ‖ $z_{63} = 113 + 144$ ‖ $z_{64} = 9 + 248$

$z_{65} = 171 + 86$ ‖ $z_{66} = 166 + 91$ ‖ $z_{67} = 87 + 170$ ‖ $z_{68} = 107 + 150$ ‖ $z_{69} = 75 + 182$ ‖ $z_{70} = 172 + 85$ ‖ $z_{71} = 214 + 43$ ‖ $z_{72} = 83 + 174$ ‖ $z_{73} = 177 + 80$ ‖ $z_{74} = 5 + 252$ ‖ $z_{75} = 63 + 194$ ‖ $z_{76} = 237 + 20$ ‖ $z_{77} = 10 + 247$ ‖ $z_{78} = 160 + 97$ ‖ $z_{79} = 217 + 40$ ‖ $z_{80} = 131 + 126$

$z_{81} = 202 + 55$ ‖ $z_{82} = 148 + 109$ ‖ $z_{83} = 220 + 37$ ‖ $z_{84} = 179 + 78$ ‖ $z_{85} = 39 + 218$ ‖ $z_{86} = 110 + 147$ ‖ $z_{87} = 101 + 156$ ‖ $z_{88} = 74 + 183$ ‖ $z_{89} = 164 + 93$ ‖ $z_{90} = 54 + 203$ ‖ $z_{91} = 115 + 142$ ‖ $z_{92} = 41 + 216$ ‖ $z_{93} = 108 + 149$ ‖ $z_{94} = 186 + 71$ ‖ $z_{95} = 82 + 175$ ‖ $z_{96} = 27 + 230$

$z_{97} = 19 + 238$ ‖ $z_{98} = 47 + 210$ ‖ $z_{99} = 181 + 76$[33] ‖ $z_{100} = 69 + 188$ ‖ $z_{101} = 94 + 163$ ‖ $z_{102} = 219 + 38$ ‖ $z_{103} = 138 + 119$ ‖ $z_{104} = 152 + 105$ ‖ $z_{105} = 191 + 66$ ‖ $z_{106} = 229 + 28$ ‖ $z_{107} = 7 + 250$ ‖ $z_{108} = 112 + 145$ ‖ $z_{109} = 201 + 56$ ‖ $z_{110} = 132 + 125$ ‖ $z_{111} = 224 + 33$ ‖ $z_{112} = 243 + 14$

$z_{113} = 51 + 206$ ‖ $z_{114} = 45 + 212$ ‖ $z_{115} = 53 + 204$ ‖ $z_{116} = 77 + 180$ ‖ $z_{117} = 90 + 167$ ‖ $z_{118} = 155 + 102$ ‖ $z_{119} = 154 + 103$ ‖ $z_{120} = 151 + 106$ ‖ $z_{121} = 161 + 96$ ‖ $z_{122} = 6 + 251$ ‖ $z_{123} = 127 + 130$ ‖ $z_{124} = 233 + 24$ ‖ $z_{125} = 12 + 245$ ‖ $z_{126} = 192 + 65$ ‖ $z_{127} = 209 + 48$ ‖ $z_{128} = 3 + 254$

MS Page 5

$R_h = S_{257-4h}$

$z_1 = R_1 = S_{253}$
$z_2 = R_{16} = S_{193}$
$z_3 = R_4 = S_{241}$
$z_4 = R_{64} = S_1$
$z_5 = R_{32} = S_{129}$
$z_6 = R_2 = S_{249}$
$z_7 = R_{128} = -S_{255}$
$z_8 = R_8 = S_{225}$
$z_9 = R_{17} = S_{189}$
$z_{10} = R_{15} = S_{197}$
$z_{11} = R_{68} = -S_{15}$
$z_{12} = R_{60} = 17$
$z_{13} = R_{30} = 137$
$z_{14} = R_{34} = 121$
$z_{15} = R_{120} = -223$
$z_{16} = R_{121} = -227$
$z_{17} = R_{92} = -111$

[33] Eds consider this to be written 76 rather than 70.

$z_{18} = R_{70} = -23$
$z_{19} = R_{111} = -187$
$z_{20} = R_{23} = +165$
$z_{21} = R_{117} = -211$
$z_{22} = R_{73} = -35$
$z_{23} = R_{46} = 73$
$z_{24} = R_{35} = 117$
$z_{25} = R_{22} = 169$
$z_{26} = R_{95} = -123$
$z_{27} = R_{88} = -95$
$z_{28} = R_{123} = -235$
$z_{29} = R_{67} = -11$
$z_{30} = R_{44} = 81$
$z_{31} = R_{11} = 213$
$z_{32} = R_{81} = -67$
$z_{33} = R_{57} = 29$
$z_{34} = R_{116} = -207$
$z_{35} = 29 = 141$
$z_{36} = 50 = 57$
$z_{37} = 25 = 157$
$z_{38} = 114 = -199$
$z_{39} = 100 = -143$
$z_{40} = 58 = 25$
$z_{41} = 59 = 21$
$z_{42} = 84 = -79$
$z_{43} = 21 = +173$
$z_{44} = 79 = -59$
$z_{45} = 89 = -99$
$z_{46} = 118 = -215$
$z_{47} = 99 = -139$
$z_{48} = 42 = 89$
$z_{49} = 104 = -159$
$z_{50} = 122 = -231$
$z_{51} = 98 = -135$
$z_{52} = 26 = 153$
$z_{53} = 13 = 205$
$z_{54} = 49 = 61$
$z_{55} = 52 = 49$
$z_{56} = 61 = 13$
$z_{57} = 31 = 133$
$z_{58} = 18 = 185$
$z_{59} = 124 = -239$
$z_{60} = 72 = -31$
$z_{61} = 36 = 113$
$z_{62} = 62 = 9$

$z_{63} = 113 = -195$
$z_{64} = 9 = 221$
$z_{65} = 86 = -87$
$z_{66} = 91 = -107$
$z_{67} = 87 = -91$
$z_{68} = 107 = -171$
$z_{69} = 75 = -43$
$z_{70} = 85 = -83$
$z_{71} = 43 = 85$
$z_{72} = 83 = -75$
$z_{73} = 80 = -63$
$z_{74} = 5 = 237$
$z_{75} = 63 = 5$
$z_{76} = 20 = 177$
$z_{77} = 10 = 217$
$z_{78} = 97 = -131$
$z_{79} = 40 = 97$
$z_{80} = 126 = -247$
$z_{81} = 55 = 37$
$z_{82} = 109 = -179$
$z_{83} = 37 = 109$
$z_{84} = 78 = -55$
$z_{85} = 39 = -101$
$z_{86} = 110 = -183$
$z_{87} = 101 = -147$
$z_{88} = 74 = -39$
$z_{89} = 93 = -115$
$z_{90} = 54 = 41$
$z_{91} = 115 = -203$
$z_{92} = 41 = 93$
$z_{93} = 108 = -175$
$z_{94} = 71 = -27$
$z_{95} = 82 = -71$
$z_{96} = 27 = 149$
$z_{97} = 19 = 181$
$z_{98} = 47 = 69$
$z_{99} = 76 = -47$
$z_{100} = 69 = -19$
$z_{101} = 94 = -119$
$z_{102} = 38 = 105$
$z_{103} = 119 = -219$
$z_{104} = 105 = -163$
$z_{105} = 66 = -7$

$z_{106} = 28 = 145$
$z_{107} = 7 = 229$
$z_{108} = 112 = -191$
$z_{109} = 56 = 33$
$z_{110} = 125 = -243$
$z_{111} = 33 = 125$
$z_{112} = 14 = 201$
$z_{113} = 51 = 53$
$z_{114} = 45 = 77$
$z_{115} = 53 = 45$
$z_{116} = 77 = -51$
$z_{117} = 90 = -103$
$z_{118} = 102 = -151$
$z_{119} = 103 = -155$
$z_{120} = 106 = -161$
$z_{121} = 96 = -127$
$z_{122} = 6 = +233$
$z_{123} = 127 = -251$
$z_{124} = 24 = +161$
$z_{125} = 12 = 209$
$z_{126} = 65 = -3$
$z_{127} = 48 = 65$
$z_{128} = 3 = 245$

MS Page 6

z_1 and z_2 are the roots of $x^2 - y_1x - y_5 = 0$
y_1 and y_2 are the roots of $x^2 - x_1x + (x_{32} + x_{19}) = 0$
y_5 and y_6 are the roots of $x^2 - x_3x + (x_{29} + x_{18}) = 0$
x_1 and x_2 are the roots of $x^2 - w_1x + (w_{14} + w_{16} + w_1 + w_8) = 0$
x_3 and x_4 are the roots of $x^2 - w_2x + (w_{15} + w_7 + w_2 + w_{13}) = 0$
x_{31} and x_{32} are the roots of $x^2 - w_{16}x + (w_{16} + w_6 + w_{12} + w_8) = 0$
x_{19} and x_{20} are the roots of $x^2 - w_{10}x + (w_{10} + w_{15} + w_2 + w_3) = 0$
x_{29} and x_{30} are the roots of $x^2 - w_{15}x + (w_{15} + w_5 + w_{11} + w_7) = 0$
x_{18} and x_{17} are the roots of $x^2 - w_9x + (w_{16} + w_4 + w_9 + w_1) = 0$
w_1 and w_2 are the roots of $x^2 - u_1x - 1 = 0$
w_3 and w_4 are the roots of $x^2 - u_2x - 1 = 0$
w_5 and w_6 are the roots of $x^2 - u_3x - 1 = 0$
w_7 and w_8 are the roots of $x^2 - u_4x - 1 = 0$
w_9 and w_{10} are the roots of $x^2 - u_5x - 1 = 0$
w_{11} and w_{12} are the roots of $x^2 - u_6x - 1 = 0$
w_{13} and w_{14} are the roots of $x^2 - u_7x - 1 = 0$
w_{15} and w_{19} are the roots of $x^2 - u_8x - 1 = 0$
u_1 and u_2 are the roots of $x^2 - t_1x - 4 = 0$

u_3 and u_4 are the roots of $x^2 - t_2x - 4 = 0$
u_5 and u_6 are the roots of $x^2 - t_3x - 4 = 0$
u_7 and u_8 are the roots of $x^2 - t_4x - 4 = 0$
t_1 and t_2 are the roots of $x^2 - s_1x - 16 = 0$
t_3 and t_4 are the roots of $x^2 - s_2x - 16 = 0$
s_1 and s_2 are the roots of $x^2 + x - 64 = 0$

Editorial Conclusion for this chapter

Here is a preliminary assessment and evaluation of some of the above details in Francis Skinner's manuscript, insofar as it is judged in its present extant and (maybe) incomplete form, for which there is currently too little historical context to facilitate decisive inferences.

The first page of the manuscript (as numbered above) comprises a computation of $3^n \bmod p$, for $p = 257$ with $n = 1, \ldots 256$. The final entry illustrates a corollary to Fermat's Little Theorem. Fermat's Little Theorem is:

$$a^p = a \bmod p$$

for any positive integer a and any prime p. A simple corollary of this is that:

$$a^{p-1} = 1 \bmod p, \text{any a with } (a, p) = 1$$

as is illustrated by $3^{256} = 1 \bmod 257$.

It is curiously of interest that Francis chose $p = 257$ since it is the fourth Fermat prime. Fermat primes are those of the form $p = 2^{(2n)} + 1$, and hence rather special. Only five such primes are known, the largest being 65,537. It is not known how many more, or even if there is a finite or infinite number of them. The fact that Francis Skinner used a Fermat prime does not appear to be relevant to the rest of the work. Nevertheless, appearances can be deceptive and it is not impossible that actual family resemblances hover below the surface of our ignorance.

Page 2 starts by writing out the terms in the sum:

$$s_1 = \sum_{n=1}^{128} \left(3^{2n} \bmod 257\right)$$

Curiously, this sum is not evaluated. But it is easy to see that this sum is zero.

Indeed, if one looks at s_1 one finds that all terms pair up so that the sum can be written as a sum of terms of the form:

$$3^q + 3^{q+128} = 3^q \left(3^{128} + 1\right)$$

Evaluated mod 257, $3^{128} + 1 = 0$. So we get that $s_1 = 0$.

The other summations are all of the same general type and can be evaluated by similar calculations. Each of them is:

$$\sum 3^{an+b} \mathrm{mod}\ 257,$$

and each of these sums is made up entirely of pairs separated by 3^{128} and so when added will give 0 mod 257. On manuscript pages 2, 3 and 4 we have some exercises, like the above, for some terms called t_i, u_i, w_i, x_i, y_i and z_i with $a = 4, 8, 16, 32, 64$ and 128 respectively. They are all 0.

Page 5 notes that these are the roots of some algebraic equations. If these equations are to be taken mod 257 then they are trivially true since all entries are zero. If they are to be taken in usual arithmetic, then they are false. It is unclear what the significance of these is. It may be that some pages are missing; or that the author is presupposing something that he supposed was relevant and could introduce yet omitted to leave on record.

With reference to MS page 6: If these z_i are the ones on Page 4, then everything is zero. Presumably this is not what is meant. As noted, in the entries from z_{12} onwards there seems to be missing an S with just the subscript present. The manuscript does not present original or significant results, though its routes could be a source for illumination. As sometimes happens even in advanced mathematical research, what is left out or presupposed, or missing, could be what furnishes significance for what is present.

It is possible that the above calculations presuppose some early 1930s Cambridge context that reflects an undeveloped interest in routes that lead into cryptanalysis or number theory. There was some research seeding the growth for this area in Cambridge at the time.[34] Although the later Diffie-Hellman cryptographic system is based on modular arithmetic of this type, it was not devised in the 1930s. Its *explicit* discovery route cannot be presupposed here; *yet* already in the 1930s pathways were being devised that eventually led to the Diffie-Hellman breakthrough. (It may be that not all pages are preserved in the published manuscript, with some more important pages lost.) As we noted in the Introduction, Turing was a Cambridge student, and in contact with Wittgenstein by 1933. For some years around this date both in varying ways were in the states of mutual and different entanglement as patterns in the web of progress. There is no space here to review further ranges of evidence as to how this connects to this Chapter's concerns and its historical context. There is considerable information to show that Cambridge pure maths research, which contributed to later breakthroughs at Bletchley Park in WWII,

[34]See the end of Chap. 1 for background on this. For much later developments of this, cf. Diffie and Hellman (1976).

was well underway by the early 1930s.[35] Wittgenstein was giving some attention to number theory, and we can see signs of such a matter connecting with his interest in Euler, since Euler researched Fermat primes and the Fermat Little Theorem, as well as the latter's relation to congruence.[36] Further attention to this subject is reserved for another publication.[37]

At the side of all this, we would do well to exercise humility when we reflect on Francis Skinner's exercise reproduced above – influenced as it may be by Wittgenstein, not least also due to facets of Wittgenstein's investigations documented in the Introduction. Certainly, the material in this chapter is basic, albeit with original twists. Yet Wittgenstein's technical limitations, contrasted with the stunning expertise of research pure mathematicians, should be put alongside the respect in which he was and is held, and they ways he probed unknown areas, by some leading members of the latter specialists.[38] Fermat's Little Theorem is contiguous to intractable depths in mathematics.

Erdos famously stated, about the domain of primes, that it will be a million years before we understand them. We do well to reflect on how – launched from an unexpected direction – in mathematical research, investigating the familiar, the elementary or the seeming trivial, with originality, can adduce stunningly unexpected theorems and results. Consequently, Erdos and other mathematicians turn to ponder the 'simple' matters of primes, so as to discover what has lain unnoticed in the familiar matter of family resemblances in them. At least in this respect, Francis's probing the familiar, in the calculations in this Appendix, was following Wittgenstein's aims – attempting to explore within obvious patterns hidden ones that can lead to new games.

---

[35]See A. Gibson [forthcoming *d*]. Consider the case of Tutte (of Trinity, Cambridge), who until recently has been vastly underestimated; Turing, and to a smaller degree his teacher, Newman. There were also Watson, Bronowski, Hardy, Littlewood, Davenport, whose work or activities were unpublished, though note researches by Heilbronn (1934) and Sastry (1934). (MI6 and UK Foreign Office files have yet to be opened to discover if Wittgenstein was involved. His 1930s' activities in related realms have yet to be explored, e.g.: his various trips to Moscow, Berlin, Switzerland; his close association with Keynes; meetings with Turing and others, and with Nazi government officials in Germany, and the fact that German agents kept him under surveillance, including a team of agents on his trip to New York; cf. Waugh (2008: 131–302)).

[36]See Varadarajan (2006: 37–42).

[37]A. Gibson, 'Euler, Fermat, and Wittgenstein' [forthcoming b]; cf. Floyd (2013, 2012a, b); and Marion (2011, 1998).

[38]As sketched in the Introduction, Sec. 'Wittgenstein and Cambridge Mathematics'.

Appendices

Appendix [A]: Formulae for Sums of Powers of Roots of a Polynomial

Symmetric Functions of Roots of a Polynomial

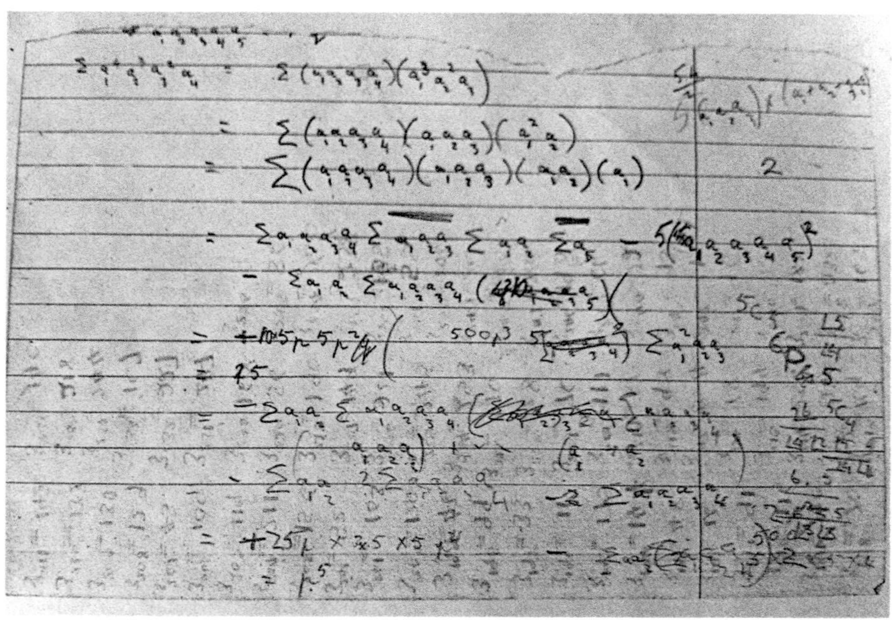

A. Gibson, N. O'Mahony, *Ludwig Wittgenstein: Dictating Philosophy*,
https://doi.org/10.1007/978-3-030-36087-0

Appendix [B]: Transcript of Letter from Alice Ambrose to Mrs G. E. Moore, Feb. 8, 1936

Editorial Note

Arthur Gibson

Some of the content in this letter is personal and domestic. These aspects also in varying ways colour Miss Ambrose's engagement with Wittgenstein and some of her contact with other Cambridge people. As such it is worth retaining such details since they modify and interact with more directly significant evidence in her context. Part of this concerns her communications with Wittgenstein and his with Mrs. Moore and G. E. Moore. McGuinness (2008) contains important letters that, given context, his judgements should be deployed to guide measurement of Miss Ambrose's interpretations in the following letter, and others. The later part of the letter directly concerns her view of Skinner and Wittgenstein.

Here is her hitherto unpublished letter:

443 S. Division St.
Apt. 2, Ann Arbor, Mich.
Feb. 8, 1936.

Dear Mrs. Moore,

I'm about to be bare-legged for lack of mended stockings, but I'm not going to let your good letter go un-answered any longer. First of all, my thanks for the photos and my reproaches for not sending one of yourself. It's like you not to. I hope you will write me again and that then you will send me a snapshot of yourself, and if you've one of Elsie Norman I'd be ever so pleased to have that, too. I don't think she'd mind. I had a nice little book about England from her at Christmas, but haven't heard from her since then. The photo of Prof. Moore sits constantly on my study desk. It is quite a good likeness, and certainly a good one of Tim. The picture of Nick was not as good a likeness, but still not bad. Your letter came in good time – it was forwarded from Ann Arbor to my aunt in Illinois (where I spent the holidays) and still came before Christmas.

It was indeed good to have the news of Cambridge. Russell's visit must have been exciting, and I was sorry to have missed it. Three years there, and then I missed hearing Russell by a few months.

I suppose the King's death has pushed important issues into the background. The papers here were full of it, and doubtless they were fuller in England. I was exceedingly pleased to have your account of Prof. Moore's new activities in connection with the election and with the Council for Civil Liberties. Sometimes I think there aren't any civil liberties here, though Lord knows those who fight for them fight bitterly enough. Once more the Scottsboro negroes have been found guilty, and Patterson has been given life imprisonment. A recent divorce case resulted in giving the father the custody of the children because the mother said if she had them she'd educate them to think for themselves (whether such thinking was congenial to the social order or not). I believe I told you that I had to swear to support the state and federal constitutions before I could hold my post here. Perhaps you read that a Harvard professor refused the oath and contested the state's requiring it. I don't know how it has ended with him, but several professors have resigned at various universities. I did nothing about it. I'm too small fry, there's no organization here, and the university is very conservative (its leaders – regents – etc. are). We're just now having an investigation of the President's arbitrary dismissal of some students for deficiency in non-academic qualifications for admission (the students were active in the National Student League, and were Jews).

As for me, my nose has been too close to the grind-stone to know what has been going on in any detail. I've just now completed a term of teaching, and am in the midst of the term examinations. These are a beast. I've to get out marks, 80 some – and in borderline cases I've no idea what to do. Whichever way I joggle my marks they average out on the borderline. And marks are such an obsession with the students that one is under pressure not to make a blunder in the direction of being too hard. Yesterday a man in my exam bit his fountain pen cap squarely in two in his agitation. He had cause to. He is a hair's breadth from failure and I don't know what to do. I like my students very much – I'm always afraid, too much. For when there's any question in these cases I give them the benefit of the doubt. And that means that the already low standard falls.

I've learned a great deal this past term both about teaching and about the subject. And I must make pretty drastic revision next term of what I've given and the way I've taught. Now that I've landed on my feet with the teaching and can see a bit better what I'm about, I've time to worry about the seminar in logic which I help to give. It's scandalous and there's nothing I can do. The man knows nothing about logic (the man with whom I collaborate) and doesn't prepare half of the time. (He doesn't prepare because he looks on students as the means whereby he earns enough to do "research" and what research. He plans, for example, to push his own bad logic book on his classes in *April*, when the book appears, merely to sell it. This means that his students do without a text until the course is over half through, and then are forced to buy one. Further, he's even asked one of the young assistants to do the same thing. He's a Russian whom Russia did well to get rid of.) The trouble is that I'm not good enough to make something good out of a mess of a discussion and a series of trivialities, and obviously I can't intrude. I'm done with collaboration. It's no good unless each person takes charge of 4–6 discussions straight anyway. I'm bad enough at this very difficult subject, but I at least prepare when I've got to take charge. The matter weighs on my mind, and haunts my sleep even.

The people here I get on with very nicely despite all this. They have been exceedingly kind to me, and one of the older men I like. The younger men – two of them – I find very seriously trying to do philosophy in a thorough way. I like them. We each hold discussion groups with the older students now and then, and try to have sober meetings. I suspect, knowing Wittgenstein, that you have already heard from him that finding people here who are very bad has made me inordinately conceited. This is just wrong, and I am very angry that he should interpret a letter, written him after the maddest discussion you could possibly imagine, as just weak on my part. As if I couldn't really be interested in a high standard, but wanted to make invidious comparisons. Further, he gives *his* impression of my letter as a statement of fact (I know he has done this with at least one person – a person whom he has now alienated from being my friend). The point is that he'd made up his mind that having this job would have a bad effect on me, and construed my letter accordingly. It is odd. I would indeed be a fool if I thought to convince him of my merit by showing how badly others were doing the job. Surely that doesn't mean that I do my job either better or worse, and my memory is not as short as to have forgotten last May. He wrote me a letter indicating that I do not yet think lowly enough of myself. If I thought any more lowly I should at once be as incapacitated for doing anything at all as I was ever since last May. I judge from other sources that he also thought I was expressing a mean contempt for these medi[cal] training – which again, I wasn't, no-one can help his training. But I've seen people with the same sort of training who were yet sober, conscientious, reasonably humble. In fact I was upset by the very arrogance of spirit of which he accuses me. The spirit of that particular discussion (in which the Russian told the mad one that he had seen matters more clearly than Leibniz, in order to have acclaim for his own absurdities) was as bad, as pompous, as anything I've ever known. And I think the *results* of the philosophizing were the results of just such a spirit. Either Wittgenstein or Moore would have made a scene. I'm not saying at all that scenes are justified. Of course I did nothing – just kept a poker face to hide my amazement. It *would* have been arrogant for *me* in those circumstances to have criticized them, and obviously tactless. They didn't want criticism, and I would need to be much better than I am to have made a sweeping one.

But Wittgenstein chose to interpret my letter as an expression of conceit, and passed off that interpretation as a fact. He seems to have indulged in a most minute criticism of me with my friend, *even* down to a discussion of my clothes. My friend, who is obviously now a disciple of his, has forgotten my warning. Wittgenstein has no business with young people who are unstable. I know when I'm beaten. I know, and knew it last summer, that my friend would become his worshipper and would forget that he was going to defend me if W[ittgenstein] urged him to the mad pursuit. But that's entirely his affair.

I wrote Wittgenstein an answer to his letter, defending myself as I have here. And then I told him what I thought of his own conceit. I'm tired of his going about laying down the moral law. I told him outright that he wasn't aware of how much of an egotist he was, that he doesn't question his right to pass judgment on people, that he took it upon himself to guide the lives of others. And that, it seems to me, does presuppose that one thinks oneself better fitted to judge what is right for that person than the person does himself. I hardly know what's right for myself, let alone others. I do *think* that it is better for a young person not to become a disciple of Wittgenstein, but after having made this point, I don't pursue it with that person. I think it is a person's privilege to go to hell even, if he wants to. I told Wittgenstein also that his behavior about Miss Stebbing's comments, which he hadn't even read and which made him so wild *in part* because they *touched him*, was shocking. And I told him that he used his power over people to extract worship. I suspect he will be wild at me, but I'm not going to let him get me back to the inertia of last summer. I worry myself ragged about what is right for me, but I'll lose perspective entirely if I must cope with minute analyses which W. sends across the Atlantic. He is so subtle that whether he's right or wrong one suffers terribly. I think you were perfectly right in telling him that he made everyone who disagreed with him feel like a fool. That's another expression of his egotism – though I didn't tell him that for fear he'd known where it came from. I think he made even Moore feel like a fool when Moore didn't see *that he was right*! And he is so strong that one is beaten at the start. He succeeds, and yet there is a very great deal in him to love. Were this not the case I should not have been so upset since the receipt of his and of my friend's letters as to stop work and lose the ground I'd gained in mastery of my own thoughts. I should be better off if I could forget his fine qualities, and be primitive enough to dislike him completely or strong enough to ignore him. I was taking real pleasure in my work, was utilizing every moment, was actually getting a discipline of my emotions just before his letter came. When the examinations are past I intend to get down to that discipline again. I want to think about the use of definitions in the system of elementary propositions, and *proofs* of equivalence *on the basis of them*. And I will. Moore once wrote a letter for me in which he said I was "eminently sane and reliable". I certainly haven't been, since the time he wrote that; at least not sane. I intend to get back to the point where the same thing could be said of me again. I'm tired of myself. I must say that the more I see of Wittgenstein the better I think of Moore. And even though I have inclinations to zealotry, I shall keep it in leash.

This is enough about me, and too much. Now about the weather. I hear you've been having very blustery and snowy times in England, and we've been having the same here, only more so. At Christmas, at my aunt's, it was 12° [C.] below zero. When we plucked the fowl for Christmas dinner, the feathers froze. Recently there has been another cold wave, and it was 24 °[C.] below zero down there. Here the lakes affect the climate so that we don't have such bitter weather. It is really very pleasant here – dry, though the people here complain of dampness, and I think it has not been colder than 10°–12° below zero. I like it, and flourish on it. And the magnificent heating of the houses makes one *almost* sell out to Mammon.

In the spring, Stokowski and the Philadelphia Philharmonic come here for a week's festival. I am looking forward to it. The Boston Symphony gave a concert here about two months back. And the final thing was the Kolisch string quartet. Myra Hess comes next month. Students here produced "Ruddigore", and I was carried back to England. Yes, I miss Cambridge, and the very different atmosphere which pervades a group of English people. I was rather shocked, and a bit dazed by meeting such a lot of people, when I first arrived. It is very different. But relationships with strangers, particularly with groups of them, are always difficult. And I find congenial souls here – or at least people who would be congenial if one had two or three years to build the relationship properly. I wish Prof Langford were here. I hear that he is a good person, and he is surely a stimulating mind. One of the young person's here whom I respect most is his student. You've probably met Prof Langford by now. The very odd professor, who talks politics so constantly, is just chattery and not really offensive. He is open, and bears no malice.

I'm wondering how you bore up during the strenuous holidays. I imagine you are glad they are past.

I enclose here stamps and matchbox lids I have been collecting. My good Austrian people have sent me a whole collection for Xmas, which they know I pass on to you. I hope you will find something nice in the lot.

By the way, I intend giving Prof. Moore something if Wittgenstein will let me. I have during this term typed off copies of the "*Brown Book*", the thing which W. did this last year. He lent it to Moore once, and it occurred to me that he might be glad to have it. As there is nothing I can do for Moore that would please him a great deal I thought I could send him a copy of these notes, with my contribution the work of typing the hundred some pages. Of course it is W[ittgenstein]'s manuscript, and if he objects, I suppose I can't do it. I've written to ask W[ittgenstein]. about it. Anyway, you'll know my intention was good. I don't know whether I'll hear from W [itgenstein]. or not. I wasn't offensive in my letter, but he may find the truth offensive. We all do. And he may just decide it was arrogant of me to tell him he is arrogant. Which it is. But it may give him pause when he starts picking me about again. This is I suspect unlikely as Sraffa has told him he's unegotistical, without effect.[1]

I was very pleased to hear that Newnham actually is getting things started for reconstructing living quarters.

Oh yes, I'm on the trail of jobs again. The people here say that if necessary, they will keep me on if they possibly can. But of course I don't want to stay on as excess baggage. Things are livening up in this country, and I've hopes of a post. Do write, Mrs. Moore. My good wishes to Prof. Moore. I miss you both.

Yours,
Alice Ambrose

[1]The expression "unegotistical, without effect" is broken up and written above and below the line, so reconstruction of it is not quite certain, though this seems to be the sense.

Appendix [C]: Letter of 20.10.1941 from Wittgenstein to British Academy

Trinity College
Cambridge
20.10.41

Dear Sir,

I very much regret to inform you that I shall be unable to deliver the Philosophical Lecture of the British Academy which I promised to deliver next March. I am leaving the University to take up some war-work which will leave me no time to <u>write</u> the lecture. I am sorry to have to trouble & to let you down in this way.

I could, if you think that this is all right, deliver a purely oral lecture & should, of course, in this case not expect to be paid for it.

Yours sincerely
L. Wittgenstein

Appendix [D]: Selection of Margaret MacDonald's Unpublished Lecture Notes of Wittgenstein's 1935 Lecture Series

Introduction

Arthur Gibson

There are unpublished notes of these lectures by MacDonald (1934–35), which to the present Editor's view are secondary in quality and precision to the manuscript being reproduced here. They are much shorter, and less technical than most of the present Chapters. For example, her Lecture VII is of this sort. Lecture VII in McDonald's notes are about 680 words long, whereas Skinner's for the same lecture are over 2200 words long, plus with diagrams that are not present in McDonald's. She also omitted the opening sentence, and commences the lecture with:

> What[2] is the criterion for a proposition being a proposition of logic.
> By some it was taken to be self-evidence[3].

Her text is thus absent of the notion of Wittgenstein's use of "chapter", and his announcement that what follows is a deviation from Lecture VI, which Skinner omits. Among the reasons Wittgenstein appears to have offered for this deviation, which MacDonald does not include, is his concern with one of the greatest difficulties of logic, that of generality.

This use of "chapter" may be an explicit indication that here Wittgenstein was attempting to compose – dictate – part of a book. If so, this evidently enables us to allow this text a role that punches above its apparent weight as dictated lecture notes. That is, it signals a stage in a composition process aimed at book composition, which perhaps indicates that Wittgenstein had moved compositionally on from the stage manifested in McDonald's notes.

On the first page of this text, there is the trace of an impress left by a sentence written on paper above it. This trace corresponds to a sentence written on a brown paper bag found with other containers, randomly placed separately at the bottom of the Archive. In Skinner's handwriting, it states:

> Notes taken by myself of Dr. Wittgenstein's lectures Lent Term 1935.

Skinner's manuscript includes a few questions from Wisdom to Wittgenstein (in the first person of each speaker).

---

[2]There is no 'is' in McDonald's notes.

[3]In McDonald's (11.2.35.) version the lecture notes begin as follows: "What [sic] the criterion for a proposition being a proposition of logic? Criterion used to be self-evidence, a word taken from psychology." (There is no reference to Russell in McDonald's version.)

The text of Chap. 8 has some detailed parallels with Margaret MacDonald's (1935) unpublished lecture notes[4], though her notes lack some of the technical elements and certain general statements that contextualise remarks in Skinner's manuscript. These are absent from Ambrose's notes, while some of these are present in McDonald's (1935) unpublished account – though sometimes the latter is in third-person reported speech. Included in the manuscript are instances of Wittgenstein's first-person dialogue with Wisdom.

The commencement of this manuscript affords some support for the proposal that the manuscript could be reproduced in a book in keeping with Wittgenstein's aim. There is a similar type of qualitative contrast between Skinner's manuscript and Ambrose's publication in length and accuracy as there is between Wittgenstein's text in Notebook 148 (see for example pp. 13–18) and Ambrose (1979: 174, 178–79) and her transposition of Wittgenstein's order in his Notebook, for example her transposition of the Pythagorean triangle with squares and the Napoleon figure, as well as differences in the attendant explanation.[5]

---

[4]These McDonald notes are held in Trinity College Library, in a collection of Rhees's papers.

[5]Ambrose (1979: ix) mentions that Rhees enabled Miss Ambrose to have copies of some of Margaret MacDonald's notes, so this confusion may also derive from either or both them, though she attended the lectures. In this regard it is worth comparing Ambrose's use of 'Napoleon' with McDonald's (1934–35) unpublished Wittgenstein lecture notes – set three, for the Summer Term 1935, LECTURE SIX 13.5.35, p.11: "I could show that by combining certain geometrical shapes I could get Napoleon. [Here there is an outline drawing of Napoleon]".

Appendix [E]: 'Yellow Book' Notes

A. Ambrose and M. Masterman

Collated by A. Ambrose

* * * * * * * *

Introduction

Arthur Gibson

The cyclostyled typescript manuscript reproduced in this Appendix [E] was found by Jonathan Smith in the Rhees Papers that are held in the Trinity College Library. I much appreciate the College's permission to publish it. As far as is known it has not been published before nor examined in a publication until now.

Hitherto the only published items from the postulated "Yellow Book" manuscript appear in Ambrose (1979), in whose Preface there is a candid statement by Miss Ambrose that she could not remember the specific circumstances of taking the notes nor how they precisely related to the occasions of the lectures, nor vouch for their accuracy. So the document below is an important basis to assist us to assess how it relates to Ambrose (1979). More particularly, it is present in this Edition to afford an opportunity to compare its contents and many differences with the *Pink Book* in the Wittgenstein-Skinner Archive. The purported copy of the Yellow Book, which has been repeatedly reported to exist, was allegedly sold at auction in New York in some haste, as a product of Miss Ambrose's Will. This copy is not currently available for inspection.

Miss Ambrose took some ten years or more to persuade Rhees to allow her to publish the short pieces in her 1979 publication. His correspondence displays some vexation with her over the matters; and no Yellow Book emerged in subsequent years. Wittgenstein himself instructed her and Miss Masterman *not* to take notes other than those that he dictated for that purpose. The alleged 'Yellow Book' contents, insofar as we fragmentarily know of them, as reported by Miss Ambrose, appear to be those that he said should not be recorded or reported. So, rather than having any authority from Wittgenstein, the Yellow Book appears to be accorded the opposite by Wittgenstein. Where there are similarities with details of what Wittgenstein intended to be copied down from his lectures, it should be borne in mind that it is precisely the absence of the refinement of detail and – to him – the less-then-acceptable crafting of his narrative by him, which were the reasons why he did not wish to allow collections to be made of expressions that appear within his permission as part of the 'Yellow Book'.[6] The narrative below shows instability by using the first-person pronoun of Wittgenstein, and – without any description accounting for the change – we find that Wittgenstein is found in the third person, for example: "The idea which Wittgenstein is struggling to explain. . .". Accordingly, we cannot be sure which expressions are those of Wittgenstein's own words, and which are those reported in the third person.

Here below is a reproduction of the manuscript, with the exact heading provided on the manuscript:

"Some Notes on 'Personal Experience' taken from the 'Yellow Book' (Companion to the Blue Book) of Undictated Notes Taken From Wittgenstein's Lectures 1933–34 by A. Ambrose and M. Masterman"

* * * * * * * * * * * * * * * * * * * *

Transcription of "Some Notes from Wittgenstein's 'Yellow Book' owned by A. Ambrose"

There is a locality of pain in what I will call 'pain-space'. It is possible to shut your eyes and yet say that we have pains (e.g. pricks) in two different places. It is even possible to say that the two pains come nearer and finally merge into each other (c/f psychologists' experiments). This does not of course mean that if we could be more exact we could have measurement in pain space; just as although in visual space we may speak of 2/3 or ¾, we are not enabled on that account to speak of .111 or square root of E. Euclidean geometry does not apply to visual space. For in visual space there is not a centagon[7] as opposed to a 101gon (because we could not distinguish between them in the way in which we distinguish between a square and a triangle). There is only a multigon.

(This is a preliminary parenthesis because you are tempted to say that although in visual space there may be a centagon you can't see accurately enough to see it. This is nonsense. In visual space there is no centagon). Take another case:

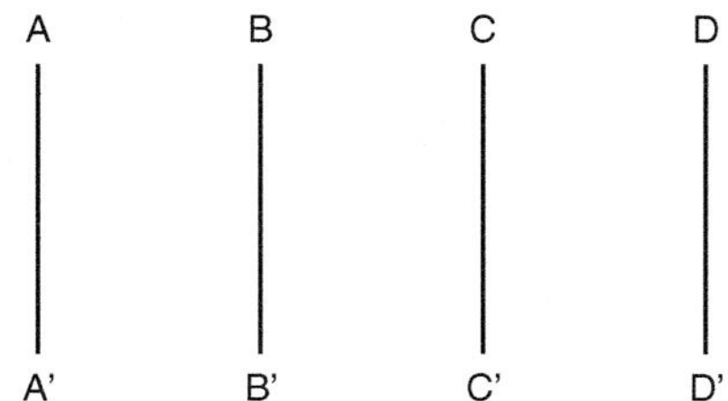

How should we express this in ordinary language? We should say that each of these dashes AA', BB', etc. is bigger than the last though it does not *seem* to be

[6]For further details on some of Miss Ambrose's unpublished items, see J. G. Slater "The Alice Ambrose Collection". http://www.rrbltd.com/item_58_essay.pdf. See also De Pellegrin, E. (Ed.) [*Forthcoming*].

[7]'centagon', of course – a rare term for a 100-sided polygon.

bigger. But I can say 'In visual space two or three may be the same as each other and the third may be bigger.'

Now I have not made a discovery. I have only translated from one kind of language to another. Now in philosophy when people talk of sense data, they are really only making this kind of translation; but they talk as though they had made some important discovery and had found some new entities. The two languages could be characterised roughly in this way, that one would talk about chairs, etc. (physical objects) and the other about 'perceptual objects'. (Philosophers often call this second language 'phenomenological language'). But these terms are misleading for they make it appear as though only a further analysis of the known facts could lead us to use such a language whereas the truth is there is nothing in the one language which cannot also be said in the other, only it will be said differently. We are talking about just the same things in each case.

Apart from what we have just said about the locality of pain there is something utterly different. Take the case of a man with a pain in his knee. I ask 'Where do you feel a pain?' The man replies 'In my leg'. I say 'Where?' and he points. I then ask, 'In what sense did you know where it was before you pointed?'

In this way, what I am saying comes to this. The words 'locality of pain' have a great many different meanings.

1. Where would you point?
2. Which part of your body, if pressed, hurts? (which is quite different).
3. Which part of your body looks discoloured?

Contrast this case with that in which I make a black dot on a piece of white paper. In this case the pointing to the dot is not the criterion for its position. If asked, Did you know where it was? I should reply, Yes, I did. If then asked 'In what sense did you know where it was?' I could reply 'If I had been asked before I pointed to it I should have been able to reply 'It is near the upper right-hand corner' (or some such thing).

In what sense did you know before where the pain was? You might have said, 'It's in my leg' and so far you did know but no further. You could also look at the spot and so far you did know but no further. What did it mean 'knowing where it was before you pointed'? The word 'where' has many meanings. Did you know e.g. that it was six inches above your knee? Have you measured it with a ruler? Will you measure where to point with your eye? No. The pointing gives you new evidence. What does it mean 'to know where pain is'? It may mean all sorts of things.

1. To make an estimate, say in inches.
2. To look at a particular spot.
3. To shut your eyes and point to the place.
4. To press to see which place hurts.

It comes to this: There are many different criteria for what we call 'the place of a pain'. And some of them have nothing to do with any sort of measurement.

The point is that straightforwardly it would seem as if the pain had only one place and that there were several ways of finding out that place. The case of the pain is more complicated than the case of a man standing a foot away from a well. In fact you might say that it has no one answer at all.

In what sense can I have toothache in another person's tooth? If you have toothache with closed eyes you will always immediately say you have toothache. Suppose I then asked you to point to it still with closed eyes. You point, and I tell you to open your eyes and look at your finger. You open them and find you are pointing to another person's tooth.

Take another case. If you sit side by side with someone wearing the same coloured sleeves. Suppose with closed eyes you say you have a pain in your hand. Then you point to the other man's hand, i.e. in the pointing sense you have a pain in the other man's hand.

Let me see if I can give you an idea where we are getting to. The case which I ought to talk about is this: If my nerves were connected to another man's so that we had teeth in common it could be asked 'Who has pains in the tooth, he or I?' Now if I were asked I could say 'I don't know where he has got a pain but I know that I have.' Or I might say, 'all that I feel is pain in this tooth. I can't say who has it.'

I want to give a few hints about the way we are going to go:

1. How do we use the word 'I'? The point about this question is that it seems at first as if this word had a meaning independent of the well-known facts about our bodies. But looking into the question further one sees that it has not.
2. But one can say that one uses it as a sign for 'coming from a certain body'. It is also one of its uses to provide an opportunity for making a noise by which the speaker may be recognised. Now suppose when people spoke the sounds [they] always came from the same loudspeaker and that the voices were all alike. Then the word 'I' could have no damn use at all. It would be absurd to say 'I have a pain'.
3. Now in addition to this suppose we only see our bodies in mirrors. Suppose then our bodies became transparent, but the mirror image remains the same. 'My throat' will now be where the mirror image is. The criterion for something being my throat is now utterly different. The use of 'I' becomes very queer.

Final remark (on a slightly different line). If one makes an investigation like this one, one is often asked, 'Do you mean to say that you alone have real toothache?' I should say 'Of course the other person has real toothache as opposed to when he is shamming'. One is also asked 'Do you think the word has a different meaning for you and for him?' Different meaning? How? We use the same word. I have toothache and he has toothache.

Take the case of the bell register of a hospital. I can co-ordinate the ringing of the bell in a room with the appearing of a light on the register. Suppose now the bell did not ring but the light flashed up. I might then say 'I did not hear it ring; I only saw it ring'. And there might be good practical reasons for this. Suppose then someone said, 'But does it really ring in the person's room' (as we might say, 'Has he really got toothache inside as well as showing its symptoms in his external behaviour? M. M.). I would say 'Yes, of course, in my sense of ring, it really rings'.

This is where my language draws its line; a rather queer line perhaps, but the line once drawn, I have now to say that the bell rings. Now suppose I am tempted to draw two different boundaries in two different places. This is what happens. The one boundary says, 'I have toothache, she has toothache, etc.' The other boundary says, 'There is toothache in one tooth' and mentions no proprietor at all. The tooth has an owner and the tooth has a place. But the toothache has no owner. For if it is logically impossible for me to say that another person has toothache, it is equally so for me to say that I have.

(p. 67 *et seq.*) Let us imagine that we see human bodies in a mirror only. (This mirror by the way is only an easy device for making a transition from one picture to another. It also only makes a difference in the case of our own bodies, not in the case of other people's.) Our own bodies then become transparent but the mirror images don't, so we use the mirror images instead.

The purpose of all this is to show not only that experience could be so arranged that I saw my own body like other people's instead of from above, but also that it is not necessary for it to have, permanently, one and the same body. Now suppose I give everybody a name. If Shaw's body is partly covered it makes sense to say 'Whose hands are these?' and a method could be found to answer it, i.e. observing the peculiarities of his hands. Of course if Shaw's body were visible we should never ask 'Are those your hands' because we should answer that they were his hands if they were attached to his body. Still, under certain circumstances, you could ask the question as I have shown. But there seems to be a totally different sort of criterion here. Imagine one mirror image in the example we imagined, touching another. Suddenly there is a shout 'Hi, this is me'. This shout won't do as a criterion because other people might shout also. But you might say, the person himself has no need of a criterion; he feels the pain. But in this mirror world as I showed last time there is no use for the word 'I'. So the person concerned would have to say 'A pain is felt'.

There are thus two kinds of criteria.

1. Bodily criteria. The criteria for which person has toothache is the behaviour of a particular body.
2. My criterion. I feel the pains.

Suppose now I call my body by the name of Wittgenstein. I can now say 'Wittgenstein has toothache' just as I can say 'Shaw has toothache'. On the other hand, I should have to say 'I feel the pain' and I might feel it when Wittgenstein had not toothache or when Shaw had. It is only a matter of fact that Wittgenstein has toothache when I feel the pain.

If I use 'I' and 'Wittgenstein' thus 'I' is no longer opposed to anything. So we could use a different kind of notation. We could talk of pain in the one case and of behaviour in the other. But does this mean the same as saying that I have real toothache and the other person has not? No, for the word 'I' has now vanished from the language. We can only say 'there is toothache', give its locality and describe its nature.

In doing this we are keeping the ordinary language and beside it I am putting another. Everything said in the one can of course be said in the other. But the two draw different boundaries, arrange the facts differently. What is queer about an ordinary notation is that it draws a boundary round a rather heterogeneous set of experiences. This fact tempts people to make another notation in which there is no such thing as the proprietor of a toothache. But without the people realising it or even realising that there are two, the two notations clash.

Put it in another way. To the person who says 'Only I can have real toothache' the reply should be 'If only you can have real toothache there is no sense in saying 'Only I can have real toothache'. Either you don't need 'I' or you don't need 'real'. Your notation makes too many specifications. You had much rather say 'There is toothache' and give the locality and the description. This is what you are trying to say and it is much clearer without too many specifications. 'Only I have real toothache' either has a common sense meaning or if it is a grammatical (philosophical) proposition it is meant to be a statement of a rule; it wishes to say 'I should like to put instead of the notation 'I have real toothache'; 'There is real toothache' or 'I have toothache'. Thus the rule does not allow 'Only I have real toothache' to be said. But the philosopher is apt to say the thing which his own rule has just forbidden him to say, by using the same words as those in which he has just stated the rule.

'I can't know whether another person has toothache' seems to indicate a barrier between me and the other person. I want to point out to you that this is a pseudo-problem. It is our language which makes it seem as though there were a barrier.

I talked before of the differences which our language stresses and the differences it hushes up. Here is a wonderful example of a difference hushed up. It is not entirely hushed up for of course all the notations must have the same multiplicity. Nothing can be said in the one which cannot be said in the others. But a notation can stress or it can minimise; and in this case it minimises.

It comes to this. Our customary language makes just the distinction which the other language without the 'I' also makes. But not only does it minimise this difference, but also by using the word *can't* blurs it entirely.

There is another thing which I can only hint at now. We are accustomed to say, 'Visual impressions presuppose an eye'. But one could perfectly well imagine a body entirely transparent so that if you looked into a mirror you would see the landscape without yourself in it. This would not prevent you from having visual impressions. But you would then never talk of moving about in a room; you would only talk of visual experience altering. It comes to this: that for the personal experience of seeing a human being is not needed. This is a grammatical statement stating the rule. We can't talk about my visual space, his visual space. *Visual space has no proprietor*. But although the human eye is not essential to visual space there is something also the *geometrical eye* which is quite different. The experience of a ring coming nearer and nearer to my eye would exist even if I had no eye. This visual space has geometrical eyes. Imagine an eye in a mirror with a ring covering it. You can imagine the ring without the eye. But that would make no difference to your saying that it covers your geometrical eye.[8]

---

[8]Added notes by the writers: "It comes [???] visual field. It starts off the visual field."

People were tempted to make a notation saying e.g. 'a red patch is to the left of a blue one'. Someone said, 'But you have left out the 'I'. They replied, 'Oh, of course *I do the seeing* but I could leave that out'. The use of this notation is that it shows us where idealism, solipsism, realism, go wrong. It is the notation which the solipsist would like to have but he does not go through with it. What the solipsist wants is not a notation in which the ego has a monopoly, but one in which the ego vanishes.

How would we distinguish the two cases 'A ring coming nearer to X's eye with no-one else in the room' and 'a ring coming nearer to X's eye and further from Y's eye'? Imagine a box with one hole and plasticine inside which you could model to describe a visual experience. Now with this clay you can do any damn thing you please, e.g. make one person and a moving ring or two people with a ring moving between them or make Miss Ambrose with the ring moving away from her and towards the hole in the box. Now this box has a language inside it, but the eye, outside the hole is not part of the language.

I might say, take the Idealist's statement, 'It's only our ideas or the ideas in our mind which exist and not the chairs, tables, etc.'. There must be some terrific temptation to say such a thing for enormously clever people such as Kant, Berkeley, Schopenhauer, have said it in different ways. If you think in a particular way it becomes almost unavoidable. It is on the face of it nonsense, for we already have a place in our language for 'the chair doesn't exist', i.e. it is a sham chair. The characteristic of the realist conception is that it agrees with common sense; but if that was all about it there could be no realism for there would be no such statement. Idealism, on the other hand, sees the possibility of a language which does not talk of the wooden chair, but of what we might call 'the visual body, the visual chair'.

I should like to talk about the beginning of our reasoning. I always postponed talking about private experience because when talking about private experience everything we have so far said about signs, manipulating signs, etc. seems suddenly so to speak to acquire a question mark. For (1) immediately the world splits up into a series of personal experiences. (2) We become uncertain whether another person's experience is real or not. We might now e.g. question the 'reality' of signs and the existence of communication, since it now seems that probably no one understands what I say, itself becomes problematical.

The answer is, 'Are we *constantly* puzzled about our personal experience'? No, any more than we are usually puzzled about the solidity of the floor because Jeans tell us it is made of electrons far apart. We can of course worry about whether the floor is strongly enough supported; just as we can worry about whether someone really has toothache or whether he is shamming, and this sort of thing is what we ordinarily mean by a problem.

As soon as we do philosophy, however, everything seems to go up in smoke, because it looks as though the philosopher is more refined (sensitive) than the ordinary person, and can see problems exactly where the ordinary man sees them roughly. That is not so. The philosopher does not see facts more closely than the ordinary man does. We can neglect philosophical difficulties in a way we can't engineers' because the philosophers' difficulties unlike the others are due to misunderstanding.

For example: it has worried many philosophers that we can never describe the only thing that is real, viz. the present, both because by the time we can describe it is over and also because we can never accurately describe any experience. This makes us feel 'what we need is a more subtle language'. This also is a misunderstanding.

We are not concerned with anything that our ordinary everyday language can't say (including in 'everyday language' the technical languages of say chemistry and physics). When in doing philosophy you get the feeling 'Gracious! Our ordinary languages are not refined enough to do it at all!' then you are making a mistake.

Put it this way: suppose someone was worried by the fleetingness of experience. We should then say to him, 'Tell me what you would like and we'll say whether we can't find a way of getting it done'.

And take the case of the unclarity (inaccuracy) of all sense experience. One might ask, 'Have we the right to complain about this?' No, because in this case there is no accuracy to compare the inaccuracy with. 'One can never see an accurate circle.' What does this mean? Show me what an accurate circle is like and perhaps I'll see one tomorrow. The man who says this is not opposing something seen to something seen but something seen to something measured. Philosophic thinking seems to complain about things which no ordinary person would think of complaining about.

To revert to what we started with: philosophy makes us complain about the troubles of sense experience; and this is a sure symptom of there being a misunderstanding. It is as though we laid our language on the facts in such a way that it can't fit upside down or wrong end on or something and then are surprised because it doesn't fit. Ordinary language described sense experiences; and then suddenly it seems as though we could no longer describe any sense experiences at all. But when we get the feeling that we can no longer describe anything it is because we are applying one normal form of language in a wrong way.

A crude way of putting this would be: we say that we have difficulties in understanding say whether everyone has special private sense experiences. But if we say this we are applying the word 'difficulty' in a wrong way. The word 'difficulty' is taken from real difficulties, and is therefore here inapplicable. Just as we made a wrong application of 'inaccurate' when we said that our description of our experiences was inaccurate so we are here making a wrong application of the word 'difficulty'.

Consider the sentence, 'I can't know if the other man has toothache; I can only conjecture it'. Now is this to mean that no-one has yet succeeded in knowing this; that people have tried to know it but failed? If one gives as a reason for not knowing whether the other person has pains that 'I am I and he is he' this points to the fact that the impossibility of which we speak – is a logical impossibility. Could we try to know? And what do we call trying in this case? If we mean that knowing in this case is logically impossible, we are just ruling out the use of the word 'to know' in this particular connection. But then this word does not point to a goal which we can't reach, but we state that in our game there is no goal. We have provided no goal, but this is so is at present difficult to see. For our usual expression stated this absence of a goal in the form that the goal can't be reached.

The sentence, 'I can't know that the other person has pain' points out something about the grammar of the sentences 'I have toothache' and 'the other person has toothache'. But it hides this behind the appearance of an assertion about pain and knowing. In the grammar of the sentences 'I have pain,' 'The other person has pain', an asymmetry strikes us. But one might ask, 'Is this difference not simply that you can know the one and not know the other?' Exactly so, but remember that it is a logical impossibility which you are talking about; and that if this statement of the difference is a statement of logic – therefore a grammatical rule – or just states an asymmetry in the grammar. For consider whether our case is analogous to this: I know that I have got a gold tooth in my mouth, but I can't know whether Mr. N. has.

But isn't it quite simple, you might say, when the other person has toothache he has what I have when I have toothache. Let us compare the grammar of the word 'having' in the expression 'having toothache' and on the other hand 'having a gold tooth'. The criterion for a person having a gold tooth, having a sum of money, etc. consists in there being some relation between the body of this person and the object owned. The relation need not be as simple as that of A holding what he owns in his hand, or carrying it in his pocket or having the gold tooth in his mouth, but let me remind you that it is the appearance of your body, or a photo of it, or the signature, or the fingerprint by which a person proves his identity.

One may ask, 'Is the person A the same thing as A's body? That the name of a person refers to a body is clear enough if only you consider how you would introduce A to someone. Also the following case is conceivable. Someone comes into the room and says 'I am your friend Smith though I don't look like it. My body has changed overnight while I slept.' Is there no evidence we would accept for his being the man we knew, although his body had changed entirely? Would we dismiss what he had said as absurd? Or, if not, what would we do to test the truth of what he had said? I think we should ask him a lot of questions about his past; we should say he was the man he claimed to be if he could tell us all the details of his life which we knew Smith could have told us. And another criterion we might apply would be that Smith's former body had actually disappeared and that his second body had come into existence in its place. (There needn't have been a continuous transition.)

But now be careful not to think that these considerations show that besides A's body there is *something else*, another object which is A. That is, we must refrain from looking for a substance which you see a substantive; but not from thoroughly examining the use of a word. We can say that these considerations show that the proper name 'A' and the expression 'A's body' do not have the same use. At least, not if we decide to use the criteria for the identity of A which we used in our example. But nothing compels us to use these criteria; I might perfectly well have said that the man who came into my room even if he satisfied my tests was not Smith, but someone else who had the same memories as Smith, or some such thing. But anyhow the proper name referred in one way or another to a human body; for Smith in his new body had to remember his old body. The proper name of a person can therefore be said to be connected with a human body in a way not essentially different from that in which the words 'King's Chapel' designates a structure of stones even though the structure changes with time.

Also consider this: the sentence 'the person A and the person B both have [']a' has a different grammar if, for 'a' we put such an expression as 'this house' or, on the other hand, 'ten pounds'. We here distinguish between A and B possessing £10 and A and B possessing 'this identical £10 note'. And we may ask, 'has it sense to say that A and B have the identical toothache as opposed to the same toothache, i.e. a toothache of the same intensity, character and place in their heads?'

Let us compare having toothache with having 'a' where 'a' stands for such an expression as 'the house I am pointing at' or 'this copy of so-and-so's works'. There one may say that if, at one time, the person A had 'a' and at another time B had 'a', what changed was roughly speaking that one human body took the place of another. I want to remind you here again that our example about Smith's two bodies doesn't *prove* anything about the existence of a personality apart from the body, etc.; it just shows the uses of words. Frege said that as the numbers were not the same things as the numerals, there was a realm of reality not perceivable by the senses. The argument here is this: the word 'three' does not signify the sign '3', nor any other material object. The conclusion drawn in this argument is utterly meaningless, and at the same time very misleading, because it seems to say that we have proved the existence of a thing in something like a realm of spirits, shades, etc. Now does the difference between my having pain and someone else's having pain consist in the fact that in the second case another body takes the place of mind? Please don't say that in the case of the pains it is the *mind* which possesses them, and not the body; and that one mind takes the place of another. To say this gets us no further, as we have not yet investigated the grammar of such a phrase as 'the mind A possesses so-and-so'. If in the experience which I described by saying that I have toothache we replaced that human body which to my friends is known as W's body by a body of different appearance, say, Mr. Smith's body, the experience described does not change into Smith's toothache, but into one which I should have to describe by saying that I had toothache in Smith's body. Let us for the sake of abbreviation use personal proper names only as the names of human bodies. Then it is conceivable that I should have toothache in W. or in Smith. And if I imagined having toothache in Smith, I would not thereby imagine what is usually called 'having toothache in Smith' (or 'having Smith's toothache'? MM.).

But can't I imagine that another person has toothache? Doesn't one imagine such a thing? If, for instance, I see him writhing with pain and pity him, am I not imagining his feeling pain? 'Isn't there in fact a particular kind of ghost of a pain which I feel in this case? This certainly does happen, and there is such an image; but let us see how we apply this image of the pain. For in saying 'can't we imagine, etc.' our fault was just to overlook a difference. Suppose I was told that Smith had had a fight and got a black eye; I then imagine him with a black eye (I could also have drawn a picture of him with a black eye from my imagination). Now when I see him I compare this picture with his actual appearance. But if I imagined that he has toothache does anything correspond, in this case, to comparing the image of his black eye with his eye? Do you make the same use of what you call the image of his toothache and of the image of his black eye?

Let's revert to the difference between Smith having pains and my having pains in one of his teeth. What I pointed out could be expressed by saying that there is a difference between telling you the place of a pain and telling you its owner. We shall have to ask what are the criteria for a pain being in such a place? This will lead us into describing fictitious cases which will make you feel less and less at home in our normal notation, not of course that this notation is shown to be wrong only we shall see that if certain facts with which we are infinitely[9] familiar were different, you would feel a strong urge to change your mode of expression. This mode of expression would be influenced by the acts in the same way in which that of a geometry is influenced by the facts of physics; facts don't make it false but they make it clumsy, cumbrous, impracticable.

(p. I06). One of the questions that comes up in the discussion of memory is this: Can I and my body be identified? If you discover that the word 'I' does not mean 'my body', i.e. that it is used differently, this does not mean that you have discovered a new entity, an ego, besides the body. All that we have discovered is that 'I' is not used in the same way as 'my body'. The argument then says, 'Since *I* isn't identical with *my body*, therefore there is something else', and it is this which Wittgenstein is trying to squash.

Wittgenstein then discussed the idea that the ego is a sort of collection of memories. What would it mean to say that two people can inhabit the same body? Someone might say 'I' isn't used to refer to my body because two persons might have the same body. Of course it is a question of terminology whether one says Dr. Jekyll and Mr. Hyde are two people or one. The idea of personality is not so primitive and rock bottom as is supposed. Ideas which to philosophers seem so simple are not actually so – words are used in many different ways. Take an example from medicine and see how the meanings of the words 'health' and 'sickness' have changed during the last 200 years. Examine 'A cure should be of the same nature as the illness', and 'The cure should be of an opposite nature to the illness'. Doctors today don't think in such terms but this does not necessarily mean that the doctors of 200 years ago were stupid. Philosophers should try to find out what steps were made from those arguments to the arguments which are used in medicine today. Once we are in a new terminology we can't go back to an old one, but it is a mistake to think that because doctors talked what clearly seems bosh 200 years ago, they don't still talk bosh, in fact they clearly do as e.g. when they talk about 'nerves'. Wittgenstein thinks that if you read medical books and read 'contemporary philosophy' you would find surprising similarity.

Nor does philosophy become commonplace when we see that the Ego breaks up and becomes a common word? Supposing someone addressed a statue as a god and then changed his religion and said 'The statue is just marble'. You cannot say this man has necessarily progressed unless you mean merely that the second attitude is later than the first. Wittgenstein wants to guard against the stupid idea that progress consists in the abolition of the marble statue from his religion.

In philosophy if one compares one's past thoughts with some of the ideas of the last century one sees that the change is enormous and one cannot revert to their

---

[9]Note by A. Gibson: Although Wittgenstein never seems to have used this phrase 'infinitely familiar', the original Ambrose MS (p. 10) employs it. If the manuscript is based on something Wittgenstein said, 'infinitely' may be the dictatee's error for 'intimately'.

terminology with a good mental conscience; but this does not necessarily mean they were stupid.

Let us examine the case of a man with a peculiar memory who on alternate days when asked, What did you do yesterday? Replied with a description of the day before yesterday's events and with a description of what in fact happened on the morrow. One might say, here are two people. But now suppose that suddenly one got the ordinary phenomenon of memory. Here one might feel tempted to say, 'Mr. X has gone'. Suppose on the other hand that I (having normal memory) didn't one day remember a thing I had done yesterday (say that I slept or had no occasion to remember). Here one would not feel inclined to say that I had died that day. Obviously the more one behaved on such days *like* the way one behaved on other days the more one would be inclined to say it was the *same* person.

Suppose that a race of people existed who looked pretty much the same as each other excepting for colour of hair, tone of voice and the fact that each had a certain number of creases on his forehead; suppose that these had each specific character with specific characteristic such as e.g. slowness and that these characters shifted about from body to body. Should we not then inclined to give the name of the person to the character other than to the body which contained it?

Let us consider further the possibility of groups of characteristics travelling from body to body. In such a world you would get over the confusion between one person giving another person's answers by asking that the person's memories were and thus distinguish between people by means of their memories.

We must of course also talk of a kind of memory where we remember a poem or tune and not some event in the past. 'To remember' here means 'To be able to reproduce it'. In such a case we don't first recall the image and then see the poem. We just start right off and what puzzles one in such cases is the lack of any transition.

Is 'willing to do a certain action' a particular state of mind? What connection has 'being willing to do A' got to do with it and what is the connection between the state of mind and the action? From the fact of being willing to sing A you must know *what* you are willing to do because you have no further test such as 'this state of mind of often followed by A', If 'being willing' is a certain state of the muscles, then *what* we are willing to do is a matter of experience. For it can be determined by experiment. To wish and what is wished are here connected empirically but in the ordinary sense they are not. If they were it would make sense to ask 'How do you know you are willing to do A'? We do not want the act of willing to be connected empirically with what happens. We want an act in which what we are going to do is already *(pre)-performed.*

Solipsism and Idealism. Wittgenstein comments on the non-contradictory appearance of solipsism. This is the due to the fact that when you state the solipsistic position you already speak the solipsistic language and so in a sense it can't be contradicted because you have made it correct. But in that sense is this true? Suppose in our language we had no word for the eyes of a stag but called them 'lights'. A foreigner will then say 'are not these two the same' meaning simply that he has both words in his language. We might then say, No they are not the same' and thus controversy would begin. This is like the controversy between the solipsist and the idealist. This [is] only a pointer as to how solipsism appears so obviously correct W [ittgenstein] will try to explain clearly what that means. Say two people are accustomed to use the word 'toothache'. They will then be tempted to think that without that word our language would be incomplete, but this is not at all the case for one could say something indirect to refer to toothache, like 'Give me something' or 'I must go to a dentist'. This language would be perfectly complete but we might be tempted to say it is indirect. But why and what would be meant by *direct*? Take another example, suppose that all ordinary verbs were used with the words standing for flower, except for a special sacred flower. If then a new philosophical school came which used these verbs with the name of the sacred flower we should have a queer linguistic situation. This situation is paralleled by the coming of the Solipsist, Idealist, and the Realist schools of philosophy. The flower example shows how with the arrival of a new school of philosophy one could make an entirely different grammar instead of just replacing one substantive by another.

How then would one get into the solipsistic mood? The solipsistic mood would begin by not using the word 'I' when describing a personal experience. Instead of saying 'I think' one might say 'It thinks' like 'It rains'. When Wittgenstein gave an example of having toothache in someone else's tooth it was to show that under certain circumstances one might be strongly tempted to do away with the simple use of 'I'. His idea is to show that our use of the word 'I' is suggested by certain invariable experiences and that if we imagine these changed, the ordinary use of 'I' breaks up and we see clearly that the 'I' notation is not the only notation which can be used.

In what cases would I say 'I can move a chair if I can move other things besides my limbs?' I might have feelings in the chair or pains correlated with different parts of the chair. This state of affairs might in fact be so clear that I might complain of pains in the chair, have the chair taken to the doctor and have it comforted in the usual way by something being done to *me*. What all this comes to is that there are various criteria of a thing being mine (and we could imagine circumstances which could destroy any one of them). Take the case where the criterion is pointing, as in 'the toothache is in this tooth'. You can imagine cases where the feeling of pain is in one tooth and you are pointing to another. Or take the case where the criterion is feeling. You could imagine a case where I had feelings in a chair and took it to a doctor; or take the case where the criterion of a person having pain is that the words come out of their mouth. Here the word 'I' refers to a speaking mouth. You can imagine a person who is possessed so that someone else took control over his mouth or a man who had two mouths. The idea which Wittgenstein is struggling to explain is that the use of the word 'I' depends on an experienced correlation between the mouth and certain limbs. In order to show that the word 'I' would have no meaning without such a correlation W. tried to describe a case where at first sight there might seem to be a use for the word 'I' but where on closer investigation there seems to be no use. But the trouble in such cases is that he removes the temptation to use the word 'I' and the possibility of using it at the same time. The confusion is that we persist in thinking that if my experience changes I must make alterations in the whole language as though in omitting something from one language we have mangled the *other* language. We have the idea that certain changes in our symbolism are really omissions. Thus if the 'I' is left out we feel that the language which is left is incomplete but this is not so. One symbolism is in fact as good as the next, so *one* symbolism is needed. We constantly judge a language from the standpoint of the language we are used to and hence we think we describe phenomena incompletely if we leave out personal pronouns. It is as if we had omitted to *point* to something, since the word 'I' seems to do this, to point to the personality. We feel we have a certain physical impossibility of using the language left us – like trying to play chess without a board. Likewise we feel that the absence of certain correlations would make something physically impossible, e.g. if I had no mouth I could not speak. But this impossibility is a logical impossibility: 'mouth' is part of the symbolism of which 'I' is also a part (here referring to the case where *my* hand is pinched and *my* hand is that person's hand whose mouth shouts when it is pinched). One is inclined e.g. to think that *2-o-clock* will keep its meaning even if there are no clocks, but it actually has meaning only in a game in which clocks also have a part.

All this leads us to the confusion about the word 'meaning'. If the essential thing about the 'meaning' of a word is its use then we can see that I can leave out such a word as 'I' and still describe the phenomenon formerly described. If I keep to the idea that there is something present when the word is uttered one can see the reason for feeling that something is omitted when 'I' is omitted. The 'I' is part of the picture. Of course if this is so then by leaving out 'I' we leave out part of the picture; but with a different picture this would not be so.

W[ittgenstein] talking about the grammar of the word 'I' said that it was essential that it should be said by a mouth (in the case of pinching my hand, etc.). He wanted merely to say that 'I' was one among symbols having a *practical* use and should be discarded if necessary. The word 'I' doesn't stand out among all the other words we use in practical life unless we start to use it as Descartes did. W. has tried to convince us of just the opposite of D's emphasis on 'I'. He would say that if he said (as he wouldn't), 'My body has toothache', this would be merely expressing something wrongly; it would be saying that there was no such thing as 'I' and replacing 'I' by 'my body'. This is like the mathematician saying (rightly) that there is no such thing as number as an entity, and then saying wrongly that numbers are scratches on paper. We think that when one answers 'I' pointing to the body that is an indirect way of pointing to the Self. It is bound up with a particular sort of counting of objects in visual space in which we understand what we oppose our bodies to. We can count bodies but how do we count selves? What do I oppose myself to? Since other names refer to bodies what is the use of names for selves? We are inclined to say that the latter refer to hypothetical entities connected with bodies but this is a mistake. Like supposing everyone has a 1/- but I don't[10] know that they have. I know only that I have one. But it is essential to this supposition that I should be able to *draw a shilling out*. But can I make a picture of what you and I both have if we each have a self? The word 'I' can't be replaced by 'this body' but has meaning only with reference to a body. Like the King of Chess. Not even a voice will do for the same voice may come from many places at once.

[10]After "don't", the remainder of this text is in handwriting at the end of the typed script.

Appendix [F]: Notes on Reuben Goodstein's Letter to Wittgenstein

Arthur Gibson

Professor Brian McGuinness has published Professor's Goodstein's letter to Wittgenstein.[11] I am very grateful to Professor McGuinness for passing me his copy of the letter before it was published, as well as offering his helpful insights.

In the following I have inserted the *italicized*, parenthesized alphabetical notation – *(a)* through to *(k)* – into it to facilitate referenced discussion below subsequent to quoting it:

Dear Dr. Wittgenstein,

Very many thanks for your letters. Three parcels of Francis' papers reached me on the 22nd. I am very grateful to you for sending them to me, as I very much wanted to have them. I have so far sorted out two of the parcels. These consist of

1. *(a)* Work done at School
2. *(b)* University lecture notes and *(c)* worked examples
3. *(d)* One volume of rough notes on your lectures taken by Francis himself *(e)* and a fair copy of these notes dated Michaelmas 1934.

(f) I need not tell you how beautifully neat, careful and thorough all the work is – a true piece of Francis himself.

I suppose *(g)* the dictated notes to which you refer are in the third parcel.

If I find anything of *(h)* Francis' work sufficiently complete for publication, I shall get in touch with you about it.

What *(i)* Francis's family perhaps don't realise is that his chief work was his *life* and now that we have lost him the most precious thing that we have left is the *memory* of that life, not something that can be dressed in words for a *(j)* philosophical article. You say that they don't seem to realise what it is they have lost, but perhaps it is that *they* lost him already years ago. If I do hear from his family, which I think very unlikely, though of course I wrote to them, I shall try to explain to them that what they are attempting is *(k)* the very last thing Francis would have wished.

I shall let you know the next time I am going to London in the hope that we shall be able to meet.

Ever yours
Louis Goodstein.

There are a number of items worth noting or raising here in addition to discussion in the Introduction and footnotes to the Archive:

[11] McGuinness (2008: 346); Letter no. 295.

(a) The school work is not included in the present book, though examples in it are occasionally referred to in this Edition.
(b) All the extant lecture notes in the Skinner Archive are of Wittgenstein's lectures. In the Archive there are more than one set of lectures notes for different lecture series.
(c) There is an extremely long worked example: a fully written out original calculations involving Fermat's Little Theorem. It is not in the standard form for such a proof in the 1930s. It appears to be Skinner's own work, probably exhibiting Wittgenstein's influence on the method.
(d) There is more than one set of notes of different lectures. So we may conclude that the other sets come from the third parcel.
(e) There is more than one set of fair copies, which perhaps were in the third parcel.
(f) There are a considerable number of additions in Wittgenstein's own handwriting, including sentences, words, phrases and punctuation. Insofar as can be assessed, these all display great respect for Skinner's own handwritten contributions. It is worth noting that while Professor Goodstein applied the expression "rough notes", he also rightly judges that "beautifully neat, careful and thorough all the work is", and it represents "a true piece of Francis himself". This is clearly a strong recommendation about their reliability.
(g) Some of the descriptions in the letter may not be particular enough for us certainly to determine their application, especially when balanced with our not knowing what was in the third parcel.
(h) "Francis's work" could be seen as an ambivalent designation. Apart from the calculations on Fermat's Little Theorem, and the individual mathematical sheets, for example in Appendices [A], [L], [P], there appears not to be original work, in terms of content, by Francis in the Archive. So either it has been removed, and we do not know about it; or, more likely, it is fully recognised by Prof. Goodstein that he is referring to what also Wittgenstein accepts: namely, that "Francis's work" is his reproduction of Wittgenstein's words. It is reasonable to suppose that a presupposition of Goodstein's use is a rubric about what he takes to be Wittgenstein's wishes concerning the publication of Wittgenstein's philosophy. This leaves us with some alternatives and problems: as to how exactly they are to be formulated; how he and Professor Goodstein understood this situation. May we provisionally present as possibilities in the following form?

    (I) Professor Goodstein is aware that all this Archive precedes and is at some compositional distance from Wittgenstein's then-current writing, which is rapidly evolving 'beyond', or in a different direction from the Archive's earlier writings.[12] So perhaps he sides with what he takes to be Wittgenstein's view, that nothing in it should be published. On this reading,

[12]According to Pichler (2013), by September 1936 Wittgenstein had decided that (the German version of) the *Brown Book* was not the vehicle he needed for his emerging *Philosophical Investigations*' project. This does not entail that he ceased to regard it as worthy of revision, marginal though the significance of this may be.

the passage marked (j) in the above letter would be a sign of diplomatic sensitivity – a caring manner of speaking to handle Francis's parents' hope for publication. Simultaneously this somewhat dissolves the grounds of that hope, not least since he would be aware of Wittgenstein's sense of obligation and guilt towards Francis.

(II) In Skinner's letter of 17 September 1935, Skinner writes of "the work which we are going to do". This designation is Skinner's, and his confident use of it in the letter reports to Wittgenstein what they appear to have agreed, or at least what Francis hopes is still the plan.[13] Ray Monk's justified argument is that the subject of this letter is the *Brown Book*. Since Skinner's complete copy of the *Brown Book* is in the Archive, and is replete with Wittgenstein's changes, revisions and additions, it is a short step from these considerations to view the prospect that Skinner has been invited by Wittgenstein to prepare the work for publication. The letter allows the interpretation that Skinner is viewing himself as joint collaborator or amanuensis. We do not know how Wittgenstein would read this, though Francis hardly seems the sort who would write in such a way as to disagree with Wittgenstein on such a matter.

(III) Possibly Wittgenstein might have signaled in his letter that he was willing to defer to Goodstein as Skinner's friend. (Maybe in the perspective of Wittgenstein feeling debarred from the decision by the guilt of neglecting Skinner or paralysis over the decision about what to do when approached by Skinner's family.)

(IV) Another option, which might assume elements of II and or III, is that Wittgenstein left Goodstein (as he did with his other literary heirs) to toss a coin, or act in accordance with such a chance. See the conclusion of Chap. 2 concerning this, and Gibson (2013).

[13]Discussion of this possibility will benefit from reading the study about what it is to be a work of Wittgenstein by Joachim Schulte (2006).

Appendix [G]: Some Genealogical and Biographical Data about Francis Skinner

Arthur Gibson and Niamh O'Mahony[14]

Almost nothing has been published about Francis Skinner. The task of researching this arena must also fall to others. This appendix is offered as a very provisional startup account for those who wish to explore further. Its merit is that it is the first such published impression of such information about Francis Skinner, which otherwise is not available to a public readership. It should be stressed that it is a first impressionistic picture, with an incomplete account of data, and connections, as the basis for further research. These are supplied in the main by very helpful members of Francis Skinner's family. It is hoped that this will facilitate more detailed research.

Part 1: European Ancestry

1. a. Francis Skinner's maternal family has German ancestry.[15] Via the maternal line Francis came from a line of German biblical scholars going back to the 17th century, via the name Michaelis; for example, J. H. Michaelis 1668–1738 (Chief Director, Francke Colligium orientale theologicum, Leipzig). Francis's maternal grandfather was Otho E. Michaelis (1843–1890), who was an engineer, and a son of Emil Michaelis. The latter emigrated from Germany to the USA. Perhaps due to anti-German feeling in the USA, the family changed its name in 1917 to Woodbridge.
2. b. Further detail on Francis's Jewish ancestry[16]:
   MICHAELIS FAMILY TREE: Mayer Michaelis's descendants "Michaelis Family Genealogy (Hildburghausen)" on the basis of miscellaneous data in the Central Archives for the History of the Jewish people in Jerusalem held in the Stanton Collection. Otho is not mentioned in this; but names of relatives (in letters to his fiancé) names of sisters (also in New York census) and mother are correct[17].

   I. **Descendants of Mayer MICHAELIS** b. 1777 d 6/5/1850 (merchant son of a "Michael" migrated from Küps, Bavaria (German Bayern) buried Hildburghausen, who had another son Joseph Michael Hirsch MICHAELIS no children who wrote Synagogue consecration text in 1811) m Betty BRÜLL b 1788 d 17/12/1849 (s/o Emanuel BRÜLL grandfather of Jacob SIMON)

[14]Much of these data have been gathered and communicated by relations of Francis Skinner, in particular Dr. Margaret Harper, which I gratefully acknowledge.

[15]It is worth placing Skinner's genealogy alongside Wittgenstein's – as finely mapped by McGuinness (2005b: Chapter 1), in view of their years together in discussion.

[16]Provided by Dan M. Barlev (Bruell) via Dr. M. Harper.

[17]I appreciate the Stanton information from Nancy Bray.

II. **Emil** (Otto Emil 1st son's name?) **MICHAELIS** b c1815 Saxony returned to New York 18 Jan 1848 as a merchant on the "Missouri" from Le Havre, France m 9/8/1844 **Clara** WOLFE (or Klara WOLFF?) b c.1818 Prussia returned to NY on 20 September 1848 on the "Panama" from Hamburg, Germany (listed as Clara McAlis aged 32) lvd NY 1861 & 1881

(a) **Otho** MICHAELIS b 3/8/1843 Saxony arrived NY 20 September 1848 from Hamburg Germany (listed as Otto McAlis aged 6) d 1/5/1890 m 29/12/1868 **Kate** Kercheval WOODBRIDGE b 2/2/1846 Detroit lvd Kennebec Maine 1890 d 1926 (obit says from family of scholars but mother from business family)

(i). Guy Ernest MICHAELIS b 14/9/1869 d 11/10/1887 (TB)

(ii). **Marion** Field MICHAELIS b 1/9/1871 m 16/4/1906 **Sidney** SKINNER b 23/8/1863 d 4/11/1944 Principal South Western (later Chelsea) Polytechnic then Kings College London

(a) Katharine Anne b 5/7/1908; died 1996? Raleigh N. Carolina m 28/7/1934 Letchworth George WOODBRIDGE d 1995 cousin from USA

(b) Priscilla Margaret SKINNER b 30/5/1910 d 22/8/1989 Ely m David Netherclift TRUSCOTT 28/8/1936 Letchworth

(c) **Sydney George Francis Guy SKINNER** b 9/6/1912 London friend of L. Wittgenstein d 11/10/1941 Cambridge (b cert Jun/1/4 Kensington 1a 200, d cert Dec1/4 Cambridge 3b 877, (*had Osteomyelitis, probably TB and died of Polio*)

(d) Marion Agnes SKINNER b 20/3/1914 d 27/3/1914 (blue baby).[18]

Otho's uncles (∗brothers of Emil's) from notes made about letters of Otho to his future wife Kate Kercheval MICHAELIS:

1. Theodore retired by 1868 (son Dennie) m Betty, Adela 19 & 5 cousins Anna 16, Ida 14, Clara, Jenny and Rosa? Leipzig,
2. Moritz businessman m Henrietta dau Rosa 3 b c 1865
3. ∗David MICHAELIS notar m Joanna 5 children including Agnes. In Hildburghausen
4. ∗Carl (or Karl or Charles) MICHAELIS m Helena (or Hanna) no issue.)

3. A senior member of Francis's family, Dr. William Truscott offered the following helpful sketch on Francis' and family origins with some pertinent history:

---

[18]**KEY** art = arrived**;** b born, bed buried, bp baptised, c circa/about, chd child, chdn children, chkd checked, cmn cremation, d died, est estate (mainly from Bob Brown), f/o father of, iss issue, m married m1 1st marriage (italics link children of marriage with relevant spouse), mother, probably separated s/o son of, sis/o sister of, sibs siblings.

"The Skinner family begins with a boy (George Edward if I remember correctly) whose father deserted his wife and two young boys in Lymington. The wife took the youngest son (who might have been a baby) back to her family (maybe on the Isle of Wight) leaving the boy to be brought up by the parish. As soon as the Parish was able to; it found the boy employment in the local Customs House, which stopped him from being a burden on the parish. The civil service had just become a meritocracy and an able person could move up the ranks by passing a sequence of exams. Harriet has, in Toft, the exercise book used as the young man worked his way up through these exams – it is a fascinating record of self improvement from simple arithmetical work to foreign languages. He was also, I understand, strongly encouraged by his parents-in-law (Banks). He rose through the system to become, if I remember correctly, the head civil servant in the Paymaster General's office.

He had two sons and several daughters. One son, Arthur Banks Skinner, became a Director of the Art Museum, an antecedent to the V&A. Francis' father, Sidney, was a researcher (University demonstrator) in the Cavendish Lab with JJ Thompson and became Director of Chelsea Polytechnic. His mother came from an intellectual family in Boston. One, Muriel, (or more) of his aunts were among the first women to attend Cambridge.

With this family background great things were foreseen for Francis, who showed considerable promise in his academic studies. The family's disappointment was therefore at least as much to do with Francis' apparent rejection of his own contribution to the intellectual world in becoming an instrument-maker rather than a Cambridge don."

Part 2: Francis Skinner's Schooling

4. St. Paul's School

Francis attended St Paul's School, and prize books awarded to him include: *The Third British Empire* by Alfred Zimmern (1927), awarded Apposition.

1929 at St Paul's School for 1st prize in Mathematical VIII (John Bell High Master) *English Wayfaring Life in the Middle Ages* by J. J. Jusserand (1929), awarded Apposition 1929 at St Paul's School for 1st prize in Mathematical VIII (John Bell High Master).

The Note Books of Samuel Butler ed. Henry F. Jones (1926), undated St Paul's School Apposition prize (John Bell High Master).

Western Civilization in its Economic Aspect ((Ancient Times) by W. Cunningham (1924), for the St Paul's School Apposition Prize (John Bell High Master).

London Past and Present by William J. Claxton, undated St Paul's Preparatory School Upper Third, Second Prize, Summer 1922.

Stories from the Faerie Queen by Lawrence H. Dawson, Lower Third, Easter 1922.

Part 3: Fragments of Family Biography and European Travel

5. The family possesses a diary written by Francis' aunt (reported that she was Mildred Skinner) on a trip with Francis, Priscilla (Francis' sister) and Margaret Banks (possibly a cousin) to Grindelwald August 1 – 26, 1929. A reason for mentioning these entries is that they document aspects of the summer trip that Francis made immediately before coming up to Cambridge for his first year as an undergraduate, which soon resulted in meeting Wittgenstein and it appears became a basis for launching and developing their social relationship.

6. The writer of the diary clearly had competent German, she writes of sheltering in a rain storm: ". . . . and I close by with a veteran Swiss who was cutting wood in his veranda. He was a very ancient old gentleman, with only two teeth. He gave me much information, rather difficult to gather. The chalet, he said, was more than 200 years old, he himself was very old, he could not tell how many years, and he cut wood for the winter. The storm would last 'ein viertel Stund'. He had no cow, no goat, no dog, but his daughter had beautiful cows and goats. He had fine milk. He had an apple tree but no pear tree."

Further on she writes: "When we returned we found Madame (the Hotel Sans Souci's patron's wife) had looked out a great volume on the Berner Oberland. A most interesting volume, compiled from documents many preserved at Bern some in the National Library in Paris, and illustrated with fascinating reproductions of old maps and woodcuts and water colours. The earliest mention of this country was 598–99 in a Latin doc: it is related that the waters of the Thuner See boiled so strongly that a great multitude of fish were cooked. There was an interesting account of the founding of the Augustinian Priory at Interlaken and of the Nunnery. A curious story of the Ritter von Stralliger who fought for England against France 1297. He became England's champion in a duel and defeated his opponent by snoring! We sat in the garden, later in the verandah, after lunch and I translated some 24 pages for Francis' benefit – He was much interested." On the next day the diarist writes: "Later I translated a good deal more of the Landbuch for Francis. We read a great deal about the town of Thun and its rise to power in the middle ages [N.B. Grindelwald and Interlaken were two of the oldest places in the Bernese Oberland.]"

"At the time of the trip my mother was already at Cambridge and Francis (the diary implies) was due to start that autumn. The Skinner group was joined by Mrs. Grigg with Elizabeth, Joseph (Billy?) and Margie Grigg. In the course of their stay they were visited by Mr. Goodchild and Maisie. Later they called on Sir Alfred Ewing[19] and his wife who knew Sidney Skinner well, recalled the diarist. He was introduced to Francis and told them that they were being joined by Professor (Lionel Robert?) Wilberforce. The diary recounts a number of combined activities of the Skinners, Griggs, Ewings and Wilberforces. At one of these Francis acts as Antony

[19]This is the scientist, Sir James Alfred Ewing (of Trinity); not Alfred Cyril Ewing (of Jesus College) – the critic of Wittgenstein.

and delivers a Greek oration over Caesar "- Very misleading for audience". A Sir Richard Glazebrook also appears and he had lately seen Sir Napier (Shaw?) The person referred to as Billy, who is probably Joseph Grigg took several photographs on this trip.

Most of the people described in this diary, who were the people Francis met on this holiday before going up to Cambridge, had connections with the Physics rather than the Mathematics world. My mother was a very able mathematician and, I understand, performed exceptionally in the Final Tripos, which caused some difficulty given the status of women (there may even be records of this). She would therefore have known several of the mathematics teachers who might well have been interested in meeting. . .her equally promising brother."

The picture of Francis given by the diary and the prize books (if he were allowed to choose these) is of a person whose reading German is poor or non-existent and who was more interested in social history than anything else. It seems that as an undergraduate he mastered German with Wittgenstein's assistance.

Part 4: Sidney Skinner
Select Bibliography of Sidney Skinner, Francis Skinner's Father[20]

It is worth availing ourselves a basis for future assessment of likely influences on Francis Skinner's education, or his family background in science. One dimension of this is his father's profession as a scientist. This could be explored in relation to some of Francis' interests as they may perhaps be obliquely echoed through samples of his work in some of the present Edition's other Appendices [A], [L], [P], and Chap 9.

1887, Skinner, Sidney: On Phosphonium Chloride. Proc. R. Soc. Lond. 42:283–289. doi:10.1098/rspl.1887.0052

1888, Skinner, Sidney and Ruhemann. S.: The Action of Phenylhydrazine on Urea and some of its Derivatives. J. Chem. Soc., Trans., 53:550–558. doi:10.1039/CT8885300550.

1889, Skinner, Sidney and Ruhemann. S.: Contributions to the knowledge of citric and aconitic acids. J. Chem. Soc., Trans., 1889, 55:235–242. DOI: 10.1039/CT8895500235,

1890, Skinner, Sidney: Apparatus for measuring the compressibility of liquids. Proc. Phys. Soc. London 11: 147. (http://iopscience.iop.org/1478-7814/11/1/320).

1891, Skinner, Sidney: Apparatus for measuring the compressibility of liquids. Philosophical Magazine Series 5, 32: [194] 79–80 DOI:10.1080/14786449108621396

1892, Glazebrook, R. T. and Skinner, S.: On the Clark Cell as a Standard of Electromotive Force. Proc. R. Soc. Lond. 51:60–67. doi:10.1098/rspl.1892.0007

1892, Skinner, Sidney: Physical properties of solutions of some metallic chlorides. J. Chem. Soc, Trans 1892 339–344. doi:10.1039/CT8926100339

[20]With the assistance of Francis Skinner's relations Margaret Harper and Ruth Lynden-Bell.

1894, Skinner, Sidney: The Clark Cell when Producing a Current. Philosophical Magazine Series 5 1894 38 [232] 271–279. DOI:10.1080/14786449408620633

1894, Skinner, S.: The Tin-Chromic Chloride Cell. **Proc. Phys. Soc. London** 13 477. doi: https://doi.org/10.1088/1478-7814/13/1/340

1895, Skinner, Sidney: Letter – The Clark Cell when Producing a Current in reply to Professor Threlfall criticisms. Philosophical Magazine Series 5, 39:375–6

1895, Skinner, Sidney: The Tin-Chromic Chloride Cell. Philosophical Magazine Series 5, 19:Issue [240] 444–47. DOI:10.1080/14786449508620740

1898, Skinner, Sidney: Affinity constants of dihydroxymaleic, dihydroxyfumaric, dihydroxytartaric, and tartronic acids. J. Chem. Soc., Trans., 73: 483–490. DOI: 10.1039/CT8987300483

1900, Skinner, Sidney: On the electro-chemical equivalent of carbon. Proc. Camb. Phil. Soc. 10 [5] 261–267.

1901, Skinner, Sidney: Observations on the minute structure of the surface ice of glaciers. Proc. Camb. Phil. Soc. 1901 [1] 33–36 + 1pl.

1901, Skinner, Sidney: Note on a comparison of the silver deposited on voltmeters containing different solvents. Proceedings British Association for the Advancement of Science Glasgow 1901 (1–3)

1903, Skinner, Sidney: On the Occurrence of Cavitation in Lubrication. Philosophical Magazine Series 5 Vol 7 pls 17–19

1903, Skinner, Sidney: Note on the slipperiness of ice. Proc. Camb. Phil. Soc. 1901 12 [2] 33–36 + 1pl.

1903 Skinner, S.: The Photographic Action of Radium Rays. Proc. Phys. Soc. London 19 82–86??

1903 Skinner, S:, The Photographic Action of Radium Rays. Philosophical Magazine Series 5

Volume 7, Issue 39, 288–292 1904 DOI:10.1080/14786440409463113.

1904/3, Skinner, Sidney: Note on the action of Radium rays on mercurous salts. Proc Camb. Phil. Soc., 12 Pt 4, 260—261?

1904, Skinner, Sidney: Note on the action of Radium rays on mercurous salts. Proc. Phys Soc. London 19 288–292

1904 Skinner, Sidney: On the occurrence of cavitation in lubrication. Philosophical magazine series 6 7 [40] 329–335 DOI: 10.1080/14786440409463123

∗1905, Skinner, Sidney: Experiment on Pressure due to Waves. Nature, Volume 71, Issue 1852, pp. 609 (1905). DOI: https://doi.org/10.1038/071609b0

∗1908, Skinner, Sidney: Chemical Analyses of Water from Dew Ponds. Nature, Vol. 78, Issue 2011, pp. 30 doi: https://doi.org/10.1038/078030a0

1912, Skinner, Sidney: The Drosometer, or Measurer of Dew, QJ Roy. Meteor. Soc., vol. 38i, 1912, p. 131–136.

1915, Skinner, Sidney: Experiments on the Flow of Heat in **Metal Sheets**. Proc. Phys. Soc. London 28 119 doi: 10.1088/1478-7814/28/1/313.

1915, Skinner, Sidney and Entwistle, F.: The Effect of Temperature on the Hissing of Water When Flowing through a Constricted Tube. Proc. R. Soc. Lond. A August 2, 1915 91:481-485; doi:10.1098/rspa.1915.0039

∗1916, Skinner, Sidney: Negative Liquid Pressure at High Temperatures. Nature, Volume 97, Issue 2437, pp. 402 (1916). doi:https://doi.org/10.1038/097402a0

1918, Skinner, Sidney and Burfitt, R. W.: Temperature Coefficient of Tensile Strength of Water. Proc. Phys. Soc. London 31, 131–136 doi: https://doi.org/10.1088/1478-7814/31/1/314

∗1918, Skinner, S.: Notes on Lubrication. **Proc. Phys. Soc. London** 31 94 doi: https://doi.org/10.1088/1478-7814/31/1/311

Appendix [H]: Kreisel's Letter Concerning Wittgenstein's *Remarks on the Foundations of Mathematics*

Analytical Note

Arthur Gibson

The acclaim of Wittgenstein by some leading mathematicians, reported in the above Introduction (Chap. 1), conflicts with George Kreisel's view. Kreisel was also notorious for his controversial remarks attacking Gödel, Frank Ramsey, and others, as can be seen below.[21] Subsequent to Wittgenstein's death, Kreisel infamously castigated Wittgenstein's book *Remarks on the Foundations of Mathematics*.[22] Given the extensive publicity that Kreisel received on this account, and the damage it did, it is appropriate to publish its less well-known aftermath: Kreisel, in his own later handwritten, hitherto unpublished, letter – seemingly largely unknown – dismissed his own lengthy condemnation. This letter removes his credibility in criticizing Wittgenstein here. For example, in this letter (the relevant page is reproduced below in this Appendix) Kreisel stated: "In a sense I might be said to have made fun of Wittgenstein in a review I wrote in the 1950s of his *Remarks on the Foundations of Mathematics*. . . I had made a mistake, which I noticed some 20 years later."[23]

This explicit admission in the letter remained unpublished during Kreisel's lifetime. His brief recantation it seems feeble, as contrasted with his fierce attack in the review. In that review Kreisel made a lengthy attempt to demolish Wittgenstein's reputation. He concluded: "It seems to me to be a surprisingly insignificant product of a sparkling mind."[24] Upon hearing this and being asked about the worth of Kreisel's own research, the distinguished Trinity College mathematician Sir Peter Swinnerton-Dyer, who knew Kreisel well, stated that, "Kreisel's life-long research contribution was small and insignificant. This output is a striking contrast to his boastful manner and destructive attitude to other mathematicians, not

---

[21] So as to formulate a broader perspective about Kreisel, it is instructive to read P. Odifreddi, ed., *Kreiseliana: about and around Georg Kreisel* (Wellesley, MA: A. K. Peters, 1996). Arthur Gibson made a number of attempts to meet Kreisel about this matter, but without any success.

[22] See the end of Appendix [H] for the portion of Kreisel's letter (Letter, p. 3).

[23] A supposed basis by Kreisel for attack can be found in his own, hitherto unknown, handwritten remarks (headed by him "UNSINN") superimposed on an unpublished Wittgenstein MS 213 – a copy of *The Big Typescript*, held in Trinity College Cambridge (this version is not noted in any of the published editions). See Gibson (2016: Secs. V–VI), and for a fuller account of the errors in Kreisel's remarks: Gibson, Leader, I, & Smith, J. [forthcoming].

[24] G. Kreisel, "Remarks on the Foundations of Mathematics," *The British Journal for the Philosophy of Science* 9, no. 34 (August 1958): 35–158.

least young ones whom he on occasion eviscerated in public, clearly attempting to ruin their careers."[25] It is worth noting that Kreisel was not a researcher in central areas of pure mathematics. He was a logician who specialised in narrow select areas of logic.

The Cambridge mathematician Professor Martin Hyland, who knew Kreisel well, states that he precisely remembers Kreisel having stated that he had just read a study by Ramsey presenting his [Ramsey's] original results on logic and intuitionism. Kreisel then roundly denounced them and stated: "Ramsey clearly did not understand them at all and did not have the faintest idea about the subject."[26] Ramsey's work in this arena outpaced Kreisel and now stands as having historic significance. Furthermore, with Ramsey's own 'Ramsey's Theorem', and his significant Ramsey Theory, which took up some strands of Ramanujan's research and advanced it, Ramsey was a mathematical giant, in contrast with Kreisel.

In his review of Wittgenstein's book mentioned above, Kreisel denounces Wittgenstein's use of Gödel. The renowned specialist on Gödel and related matters – Solomon Feferman[27], a long-time colleague of Kreisel at Stanford, was greatly troubled by Kreisel's attacks on Gödel's mathematical writings, which are in the same arena as Kreisel's criticism of Ramsey above. For example, Kreisel asserted, "Gödel used crude, hackneyed formulations." Feferman's analysis of Kreisel's remarks concludes: "This tells us more about Kreisel than about Gödel."[28] We have enough warrant with these above cases about Kreisel – and there are others – to generalise Feferman's conclusion: many of Kreisel's vehement attacks are his own inadvertent autobiographical disclosures, rather than extensionally valid critiques. Feferman believes that Kreisel's wartime work with applied mathematics "encouraged a casual disregard for mathematical fastidiousness". Feferman also pinpoints how Kreisel made mistakes and failed in his published attempt to advance research on Littlewood's 1914 theorem. This may explain why Littlewood and Hardy did not fulfil Kreisel's wish that he be considered for a Trinity fellowship.[29]

---

[25]Quoted with permission from Professor Swinnerton-Dyer, former Cambridge University V.C. and Professor of Pure Mathematics.

[26]Quoted with permission from Professor Hyland, Cambridge University.

[27]Kreisel, "Gödel's Excursions into Intuitionistic Logic," in *Gödel Remembered: Salzburg, 10–12 July 1983*, Eds. Paul Weingartner and Leopold Schmmetterer (Naples: Bibliopolis, 1987), 65–186; cf. 161. (Even these insults will seem mild compared to Kreisel's much more extreme excesses concerning his as yet unpublished writings about other people.)

[28]S. Feferman, "Lieber Herr Bernays! Lieber Herr Gödel! Gödel on Finitism, Constructivity, and Hilbert's Programme", in *Dialectica* 62 (2008): 179–203. Feferman quotes Rosser's very critical review of Kreisel's technical work in the *Journal of Symbolic Logic* 18 (1953): 78–9, where "there are too many errors to permit a complete listing." *Notre Dame Journal of Formal Logic* 31, no. 4 (1990): 602–41, doi: 10.1305/ndjfl/1093635594.

[29]The Cambridge University Library Archives communicated this information; also affirming that Kreisel was not registered as a Ph.D. student; nor did he complete or finish this PhD.

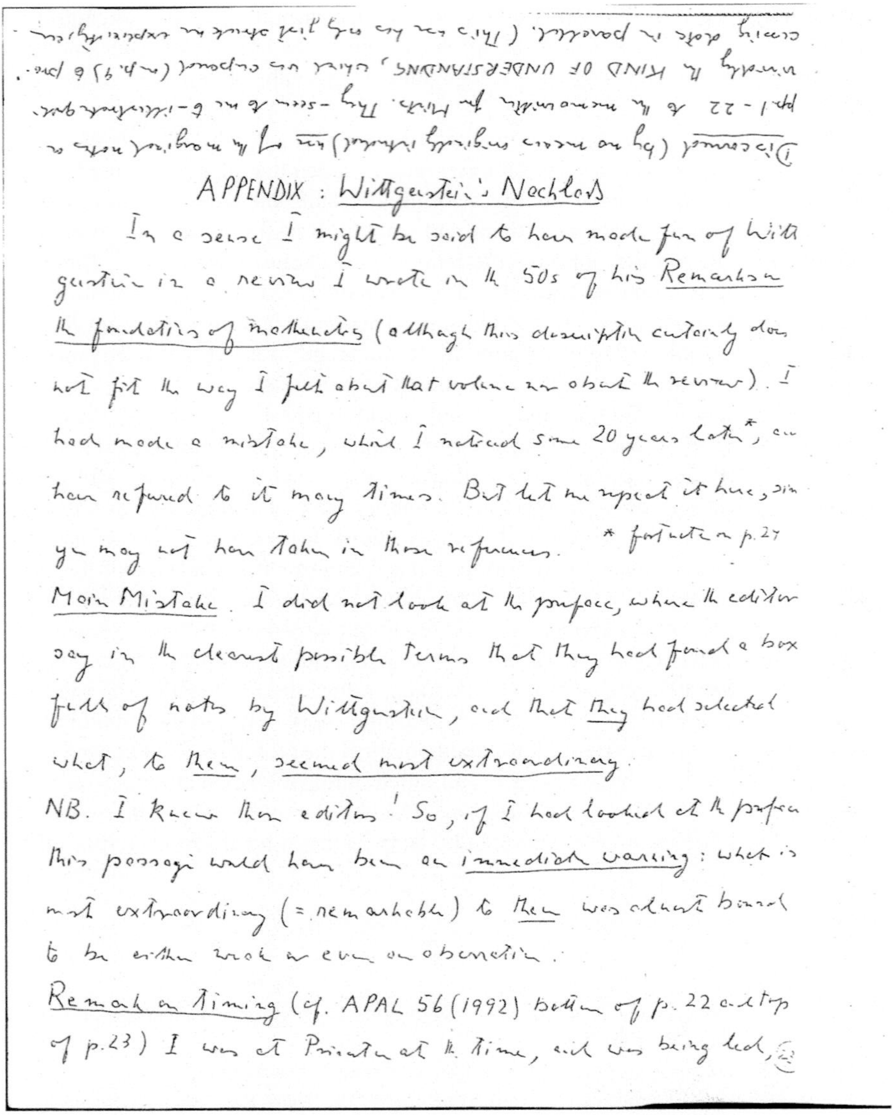

Discovered (by no means originally intended) use of the marginal notes on pp. 1–22 to the memo written for Mints. They seem to me to illustrate quite vividly the KIND OF UNDERSTANDING, which was expounded (on p. 9) to processing data in parallel. (This use has only just struck me explicitly, even.

APPENDIX: Wittgenstein's Nachlass

In a sense I might be said to have made fun of Wittgenstein in a review I wrote in the 50s of his Remarks on the foundations of mathematics (although this description certainly does not fit the way I felt about that volume nor about the review). I had made a mistake, which I noticed some 20 years later*, and have referred to it many times. But let me repeat it here, since you may not have taken in those references. * footnote on p. 24

Main Mistake. I did not look at the preface, where the editors say in the clearest possible terms that they had found a box full of notes by Wittgenstein, and that they had selected what, to them, seemed most extraordinary.

NB. I knew those editors! So, if I had looked at the preface this passage would have been an immediate warning: what is most extraordinary (= remarkable) to them was almost bound to be either weak or even an aberration.

Remark on timing (cf. APAL 56 (1992) bottom of p. 22 and top of p. 23) I was at Princeton at the time, and was being led, (2)

Plate 1 Kreisel's Letter, Page 28. (We are grateful to Mathieu Marion for furnishing this copy of Kreisel's handwritten letter). Martin Hyland verified Kreisel's handwriting for us. This letter seems to have been addressed to Gregori Mints, and requires further investigation. (Note Gregori Papers, SC1225, OAC, University of California's California Digital Library); cf. Gibson (2019), Section V)

Appendix [I]: Re. Fermat's Little Theorem

Here is the Skinner manuscript's ending of calculations that are transcribed in Chap. 9, expressing the key prime number, which ends the calculation:

$3^{188} = 76$
$3^{189} = 210$
$3^{190} = 116$
$3^{191} = 91$
$3^{192} = 16$
$3^{193} = 48$
$3^{194} = 144$
$3^{195} = 175$
$3^{196} = 11$
$3^{197} = 33$
$3^{198} = 99$
$3^{199} = 46$
$3^{200} = 120$
$3^{201} = 103$
$3^{202} = 52$
$3^{203} = 156$
$3^{204} = 211$
$3^{205} = 119$
$3^{206} = 100$
$3^{207} = 43$
$3^{208} = 129$
$3^{209} = 130$
$3^{210} = 133$
$3^{211} = 142$
$3^{212} = 169$
$3^{213} = 250$
$3^{214} = 236$
$3^{215} = 194$
$3^{216} = 68$
$3^{217} = 204$
$3^{218} = 98$
$3^{219} = 37$
$3^{220} = 111$
$3^{221} = 76$
$3^{222} = 228$
$3^{223} = 170$
$3^{224} = 253$
$3^{225} = 245$
$3^{226} = 221$
$3^{227} = 149$
$3^{228} = 190$
$3^{229} = 56$
$3^{230} = 168$
$3^{231} = 247$
$3^{232} = 227$
$3^{233} = 167$
$3^{234} = 244$
$3^{235} = 218$
$3^{236} = 140$
$3^{237} = 163$
$3^{238} = 232$
$3^{239} = 182$
$3^{240} = 32$
$3^{241} = 96$
$3^{242} = 31$
$3^{243} = 93$
$3^{244} = 22$
$3^{245} = 66$
$3^{246} = 198$
$3^{247} = 80$
$3^{248} = 240$
$3^{249} = 206$
$3^{250} = 104$
$3^{251} = 55$
$3^{252} = 165$
$3^{253} = 238$
$3^{254} = 200$
$3^{255} = 86$
$3^{256} = 1$

Appendix [J]: Five Pushkin Poems that Wittgenstein Wrote in Russian

Notebook MS 166, 1941

(For a German Translation of these Poems, See Biesenbach (2014, 413–17).)

Translated by Niamh O'Mahony

Пророк (Prophet) (1826)

Tormented by spiritual void,
I stumbled through a grim wasteland
and there, at a crossroads,
a six-winged seraph appeared before me.
He caressed my eyelids with fairy fingers
light as a dream.
My eyes were opened wide in prophecy
like a startled eaglet.
He stroked my ears
and they were filled with clamour and tumult
and I heard the heavens quake
and a celestial flight of angels
and the glide of deep-sea beasts
and vines thrusting upward
in distant valleys
and then he crouched down to my lips
and tore out my sinful tongue
fluent in deceit and vanity
and inserted with a bloodied right-hand
the venom of snakes
into my petrified mouth.
And cleft my bosom with a sword-stroke
and pulled out my trembling heart,
and thrust into the gaping wound,
a shard of living fire.
I lay in the wilderness, corpse-like,
and the voice of God called out to me:
"Arise, O Prophet, perceive and hear,
fulfill My Will.
Go forth over land and sea
and sear the hearts of men
with my Word."

Анчар (The Upas Tree) (1828)

In a charred and stunted desert
in fierce and brutal heat
the Upas Tree stands alone
like a baleful sentinel.

Begot of the parched plain
on a day of wrath.
Nature poured green venom
into its roots and boughs.

Poison oozes through its bark,
and melts in the blazing midday sun.
It hardens at the fall of night
into thick transparent tar.

And birds do not visit it.
And the tiger shuns it.
Only a whirlwind brushes by
and speeds away, tainted.

And if a stray cloudburst
wets its dense foliage,
poisoned rain from the branches
seeps into the burning sand.

When an imperious man beckoned
and sent a man off to the Upas tree,
he obeyed and returned by morning,
with the poison.

He brought the deadly resin
and a bough with leaves already withered.
Sweat on his pale brow
streamed down in cold rivulets.

He brought it – and weakened,
lay down in a vaulted tent
and thus the poor slave died
at the feet of his invincible master.

And the tsar soaked his arrows
in the very same poison
and to his neighbours in other lands
he dispatched death.

Бесы (The Demons) (1830)

Storm-clouds hurtling, storm-clouds whirling,
murky sky and murky night.
Furtive moon illuminating
flying snow with lunar light.
On and on through vast expanses
our little bell goes ding-ding-ding.
Unknown valleys, fear and terror,
Who knows what the night may bring?

Onward coachman! No, we can't.
The horses, Sir, can do no more.
I am blinded by the snowstorm
and the roads are blocked, for sure.
For the life of me, can't see the tracks
We've gone off course, we've lost our way.
It's clear to me the devil's nabbed us
and he's leading us astray.

Look, he's there! He's toying with us,
spitting, blowing, what a scene.
Now he's bent on driving my horse
to the brink of that ravine.
Now he's looming up before us
like a signpost, very queer,
Now with evil flame he flares
up, only then to disappear.

Storm-clouds hurtling, storm-clouds whirling
murky sky and murky night.
Furtive moon illuminating
flying snow with lunar light.
We cannot circle any longer.
The bell stops dead. What's happening here?
Is that a wolf, perhaps a tree-stump?
Horses halt, so gripped with fear.

Snowstorm's raging, snowstorm's howling
Horses whinnying in fright.
Now the demon gallops further
eyes still burning in the night.
Horses pull us back a little,
ding-ding-ding goes the little bell.
I look and see the spirits gathering
in the valley, straight from hell.

Countless sundry ugly demons
cavorting in the eerie light.
Like whirling leaves in late November
swirling, dancing, in their flight.
Why so many! Why so restless?
What's that awful plaintive cry?
Perhaps a witch is being married?
Or a goblin's off to die?

Storm clouds hurtling, storm-clouds whirling
murky sky and murky night,
Furtive moon iluminating
flying snow with lunar light.
Devils swarming in a frenzy,
soaring through the endless plain.
Piteous squeals and mournful howling
Rip my frightened heart in twain.

Воспоминание (Remembrance) (1828)

At the far end of a busy day
when night shadows descend on mute city squares
and after a long day's work, sleep is settling in.
That's when the torture starts –

hours dragging on and on
tormenting me
and during the sluggish night
jagged pangs of remorse rear up in me
like serpents.
My imagination goes berserk
and my head hurts from the incessant onslaught
of agonising thoughts.
Memory silently unfurls before me
its long winding scroll.
Fierce acrid tears stream down my face
Yet, the words I cannot wash away.

Элегия (Elegy) (1830)

My reckless crazy years
are spent
and weigh heavily on me,
like a bleary hangover.
Like wine, nostalgia for days gone by
matures and grows stronger with time.
My future is set,
and looks bleak.
The turbulent sea of the future forebodes
only work and grief.

Yet, I don't want to die, my friends;
I want to live, to think and suffer.
There will be some pleasures among the cares and woes.
I will be drunk with joy again.
I will shed tears over my ideas.
And, who knows?
Perhaps as I head off towards the sunset,
love may once more bestow on me
a parting smile.

Appendix [K]: Alice Ambrose (Laverowitz)

Letter to Rush Rhees, April 9, 1977

SMITH COLLEGE
NORTHAMPTON, MASSACHUSETTS 01060
DEPARTMENT OF PHILOSOPHY

Newhall Road
Conway, Mass.01341
April 9, 1977.

Dear Rush Rhees:

The notes of Margaret Macdonald's which you sent me by air have arrived, and for these, and for your good letter, I thank you very much. Also my thanks for your efforts to find other notes of that year. I also have tried without success, though I have learnt from Goodstein that he has deposited in the Leicester Univ. library Francis Skinner's notes of the lectures, "with Wittgenstein's manuscript corrections on a few pages". I am not sure, from his letter, whether these were notes from 1934-35 or notes taken during the year when _The Blue Book_ was dictated. Wittgenstein stopped dictating and gave lectures from time to time, and Francis' notes on one or two lectures are included in The Yellow Book; but they are not very full, and I remember Margaret Masterman saying at the time that they were not very adequate. There are only about 6 pages, and the notes are disjointed. Goodstein was one of the five to whom _The Blue Book_ was dictated, but according to my memory, he was not in _Cambridge in 1934-35_. He has had the terrible misfortune to have had a stroke in 1975. Fortunately he has almost totally recovered, but he is to be away from Leicester for six months, and whether he returns in the autumn will depend on his blood pressure. I wrote him immediately on receiving his letter of March 11, to inquire about the notes in the Leicester library, but I suspect he had already gone; for I have had no reply, and he is always prompt. It is shocking how many of us in those classes have fallen ill or died. As for Margaret Masterman, I wonder if Mr. Feather will have a reply from her.

I am working very steadily at the notes. I am extremely pleased to have Margaret's notes from you, though they are not as different from mine as I had hoped. I mean that they are not very much fuller. I took full notes, and Margaret also did--in fact hers may have been transcribed from short-hand, as I know she could do short-hand. But they are not appreciably fuller.

I have not been at all sure what is the best procedure in editing, and for this reason the job is proving a very big one. It makes me appreciate what a job the literary executors have faced and how well you have done it (Cora Diamond as well). As you well know, asides and returns to a topic, with considerable repetition, are characteristic of the lectures. In order not to make Wittgenstein appear confused, I think I must delete those of the notes which are confused. And in a lecture or two there is repetition of material (on tautologies) which appears in the _Tractatus_. There is really no point, in my opinion, in including it. _But if I_ make omissions, or reorder the material, I then fail to present each lecture as fully as possible. I could state, for a given lecture, the omission of material, with an indication of its subject matter; and I could gather together his remarks on a topic during that year in one lecture or section, and indicate it. If I did the latter I would no longer be faithful to the order of material as actually presented.

I have been going through my 1932-33 notes carefully with the idea of incorporating material on solipsism into The Yellow Book, indicating where this is done. The 1932-33 notes have an abundance of material on solipsism, as well as on other topics, and my notes appear quite good. (Since The Blue Book deals with solipsism, The Yellow Book in consequence does too.) To incorporate material into The Yellow Book coming from the year preceding it would mean that The Yellow Book as an entity on its own would no longer exist, and it might be best to keep the two things separate. The notes I have on 1932-33, my first year, are surprisingly full. Their ratio to the 1934-35 notes is about 5 to 7, and I should thinkthe 1932-33 notes might even come to more than The Yellow Book itself, particularly if I don't use Margaret Masterman's part.

As you can see from tis, I have been going through all the materials I have with the idea of seeing how they might best be used. I feel pretty confident that notes on "Philosophy for Mathematicians" (1932-33) should be added to the 1934-35 lecture notes, at the end, even though there are two years in between.

You must have faced just these problems in dealing with the materials you edited. Any advice you have will be very welcome. And again my sincere thanks for your letter and for Margaret's notes. I can't help feeling that if we had been in touch during these years things might have been different.

Yours,

Alice A. Lazerowitz

Appendix [L]: Skinner's Work Sheet on Equipotentials

Analytical Note

Arthur Gibson and Malcolm Perry

At first glance the illustration below appears to be a depiction of equipotentials with two equal negative charges, though the spread of numbers needs some investigations, which future publication may explore.[30] This sheet does not have any explanation on or attached to it. This type of illustration and attendant mathematics can be found in James Clerk Maxwell (1873: 148ff). Wittgenstein had considerable interest in Maxwell's work, which might have influenced Skinner here. On another interpretation, there could be some connection with harmonics.[31] Such possibilities might be cross-related, though it not clear how at this stage.

Readers may like to compare examples of equipotentials with two equal negative charges on a site that enables analysis of modelling of equipotentials, for example at:
http://vnatsci.ltu.edu/s_schneider/physlets/main/equipotentials.shtml
with Skinner's illustration below:

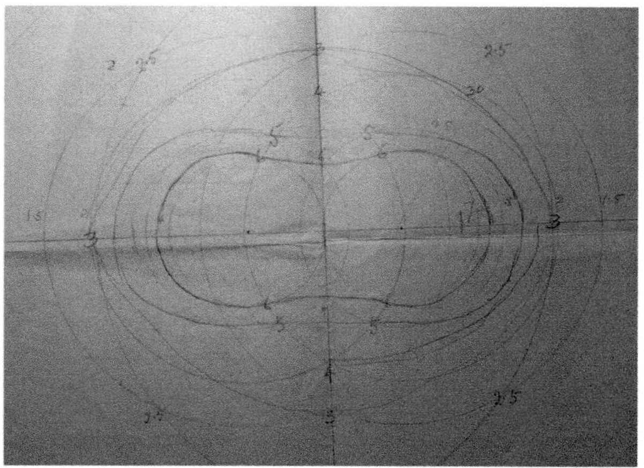

Skinner manuscript analysing Equipotentials

Below is a partial interpretation by Malcolm Perry and Arthur Gibson, which is not intended to be complete. Rather, it is to stimulate exploration of the diagram.

[30]See Gibson [forthcoming] 'Skinner on Equipotentials'.

[31]Cf. J. C. Maxwell (2010), (1873).

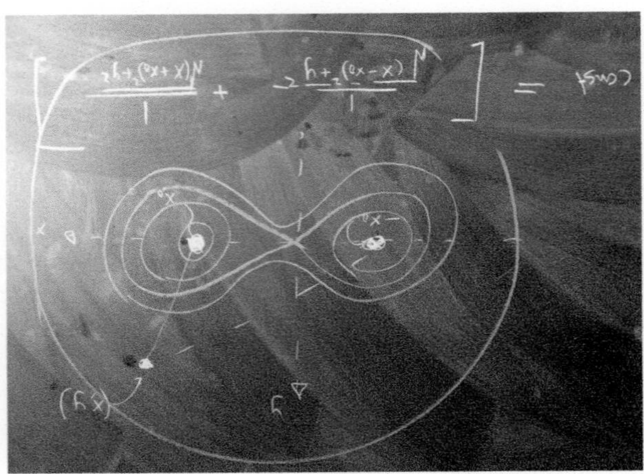

Perry's and Gibson's outline interpretation of Skinner's diagram on Equipotentials

As with many cases of a sheet without a framework, it is difficult to assign a specific context and fixed significance to Skinner's example here. Another reason for the ambiguity is that a brief treatment such as this may belong to either an unimportant exercise or be a precursor to significant insight. Further attention is needed to such examples in Skinner's Archive (see Appendices [I] and [P]), and specially to the prospects of the examples being related in some unknown way, for example with Chap. 9.

Appendix [M]: Francis Skinner's Death Certificate

HD148445

CERTIFIED COPY of an ENTRY OF DEATH
Pursuant to the Births and Deaths Registration Act 1953

Registration District Cambridgeshire

1941 Death in the Sub-district of Cambridge in the County of Cambridge

295	Eleventh October 1941 Infectious Diseases Hospital U.D.	Sydney George Francis Guy SKINNER	male	29 years	of 31 East Road Cambridge U.D. a Mechanic (Scientific Instrument Works)	(a) Acute Anterior Poliomyelitis Certified by Henry E. Nourse M.D.	L Wittgenstein Present at the death Trinity College Cambridge	Thirteenth October 1941	Arthur Leamington

Deputy

5.8.2009

WARNING: A CERTIFICATE IS NOT EVIDENCE OF IDENTITY

Appendix [N]: Samples of Wittgenstein's Handwritten Changes in the Archive's *Brown Book* Manuscript (MS Pages 1–3)

Introductory Note

Arthur Gibson

These first three pages contain hitherto unknown sentences in Wittgenstein's own handwriting in the handwritten *Brown Book* comprising part of the Wittgenstein-Skinner Archive. These are inserted into this *Brown Book* text, clearly included to be part of the narrative. These sentences are additional to and different from the published Rhees' Edition of the *Brown Book*, and appear not to have counterparts in other manuscripts.

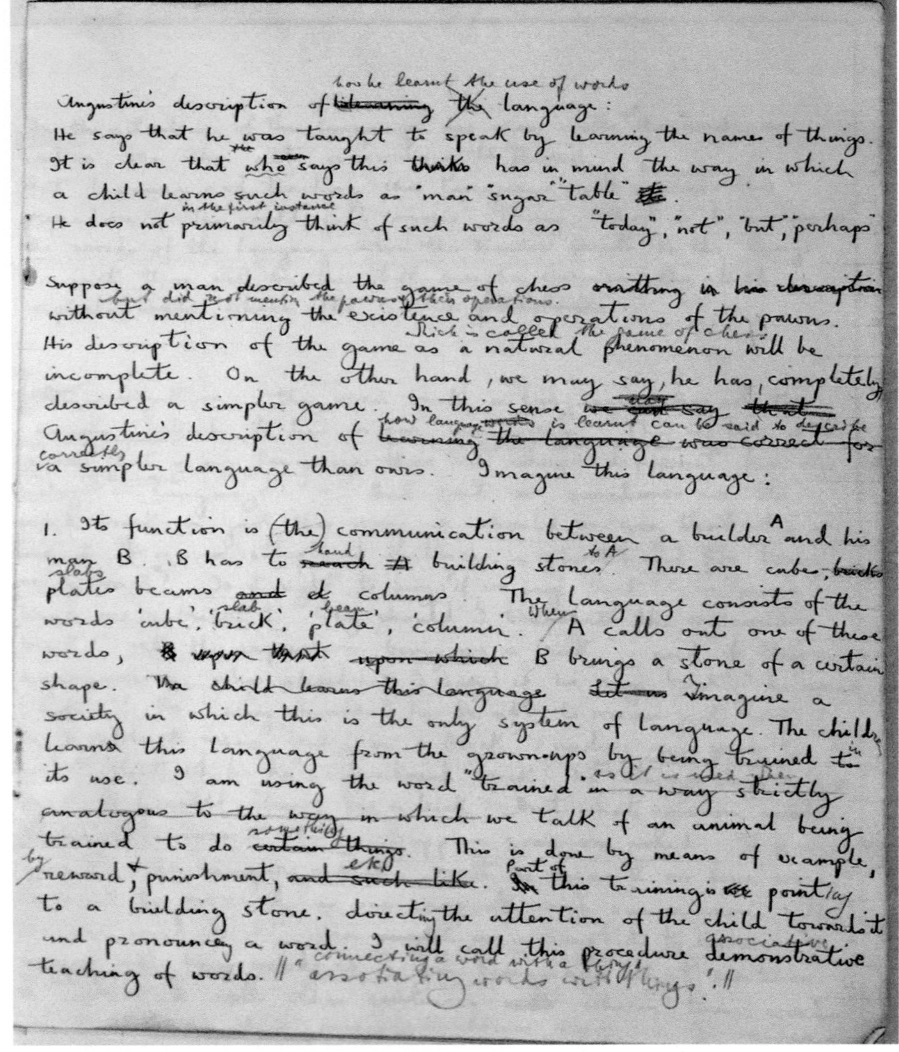

how he learnt the use of words
Augustine's description of ~~learning~~ ~~the~~ language:
He says that he was taught to speak by learning the names of things.
It is clear that the who says this ~~thinks~~ has in mind the way in which
a child learns such words as "man" "sugar" "table" [illegible].
in the first instance
He does not primarily think of such words as "today", "not", "but", "perhaps"

Suppose a man described the game of chess ~~omitting in his description~~
but did not mention the pawns & their operations.
without mentioning the existence and operations of the pawns.
which is called the game of chess
His description of the game as a natural phenomenon will be
incomplete. On the other hand, we may say, he has completely
described a simpler game. In this sense ~~we can say that~~
how language is learnt can be said to describe
Augustine's description of ~~learning the language was correct for~~
correctly
a simpler language than ours. Imagine this language:

1. Its function is (the) communication between a builder A and his
hand
man B. B has to ~~reach~~ ~~A~~ building stones to A. There are cubes, ~~bricks~~
slabs
plates beams ~~and~~ & columns. The language consists of the
slab beam
words 'cube', 'brick', 'plate', 'column'. When A calls out one of these
words, ~~B upon that upon which~~ B brings a stone of a certain
shape. ~~The child learns this language~~ ~~Let us~~ Imagine a
society in which this is the only system of language. The children
learns this language from the grown-ups by being trained in
its use. I am using the word "trained" ~~in a way strictly~~ as it is used when
~~analogous to the way in which~~ we talk of an animal being
something
trained to do ~~certain things~~. This is done by means of example,
etc.
by reward, punishment, ~~and such like~~. Part of ~~In~~ this training is ~~to~~ pointing
to a building stone, directing the attention of the child towards it
and pronouncing a word. I will call this procedure demonstrative
associative
"connecting a word with a thing"
teaching of words. // "associating words with things". //

In the actual use of this language one man calls out the words, as orders, the other acts according to them.
But learning and teaching this language will include this procedure: The child just 'names' things, that is he pronounces the words of the language when the teacher points to the things. In fact there will be the still simpler exercise: the child repeats words which the teacher says for him.

[Objection: the word brick in language 1 has not the meaning which it has in our language. — This is true if we wish to say that in our language there are ways of using the word 'brick' different from the way this word is used in language 1. — But don't we sometimes use the word brick in just this way or should we say that when we so use it it is an elliptical sentence, a shorthand for "bring me a brick"? Is it right to say: "if we say "brick!", we mean "bring me a brick"? Why should I translate the expression "brick!" into the expression "bring me a brick", and if they are synonymous, why shouldn't I say: if he says "brick!" he means "brick!"? Or: why shouldn't he be able to mean just "brick!" if he is able to mean "bring me a brick"; unless you wish to assert that while he is saying aloud "brick" he as a matter of fact always says in his mind, to himself, "bring me a brick". — But what reason have we to assert this? Suppose some-one asked: If a man gives the order "bring me a brick" must he mean it as four words or can't he mean it as one composite word synonymous with the one word "brick!" One is tempted to answer: "He means all four words, if in his language that sentence stands in contrast with other sentences in which these words
if they are not for him inseparably bound together in this sentence
if that sentence stands in contrast with other sentences

; in the end, however, we were led to an idea of meaning different from that of a state of mind

are used, such as for instance "Take these bricks away" etc. "But how is our sentence contrasted with these others? Must he have thought them simultaneously or one shortly before, or after the other, or is it sufficient that he should have at any time learnt them?" It appears when we asked ourselves these questions that it is irrelevant which of these alternatives is the case and we are inclined to say that all that is really relevant is that the possibilities of these varying combinations should exist in the system of language which he is using, and that they need not in any sense be present in his mind when he utters the sentence. Compare this statement of the case with our original question. When we asked it, we seemed to ourselves to be asking a question about the state of mind of the man who says the sentence. Whereas the idea of meaning which we arrived at in the end was not that of a state of mind. | We think of the meaning which a sign has sometimes as being a state of mind of the person who is using it, sometimes as the role which the sign plays in the usage of a system of language. The connexion between these two ideas is that the mental experiences which accompany the use of a sign are certainly conditioned by the usage of this sign in a particular system of language.

William James speaks of specific feelings accompanying the use of such words as "and" "if" "or"; and there is no doubt that at least certain gestures are often connected with such words as a collecting gesture with "and" and a dismissing gesture with "not" and there obviously are visual and muscular sensations connected with these gestures. On the other hand it is clear enough that these gestures do not accompany every use of the words "not" and "and".

Appendix [O]: Dates in Skinner's Unpublished Handwritten Copy of the *Blue Book*

Note by Arthur Gibson

There is an unpublished, incomplete written draft of the *Blue Book* in Francis Skinner's handwriting. The manuscript consists of a bound single volume hard-backed exercise book.[32] It extends to cover approximately the first half of the Blue Book. It is appears to be an early draft, which has sentences that are absent from the duplicated and published *Blue Book*.

A few of these sentences appear in the present Edition's Pink Book, as documented in the present edition.

This Skinner handwritten copy is the only draft that has dates on it, and in Skinner's hand, as if written on the dates of writing, and perhaps dictation by Wittgenstein. The first date in this draft of the *Blue Book* is November 8th 1933. Here is a list of the other dates, each along with the opening expression quoted following each date, so that the readers can locate its position in the *Blue Book*. Note that where the day name is linked to a date, this enables us to identify the years as 1933, moving into 1934 with the January dates. Accordingly, this comprises a new source for identifying some timing and features involved in the of the emergence of the *Blue Book*. Here is the list of dates in this manuscript:

Wed. Nov. 8th ("What is the meaning of a word?");
Nov. 15th ("It seems that there are...)
Fri. Nov. 17th ("It seems at first sight...")
Wed. 22nd ("Perhaps the main reason...")
Fri. Nov. ("In saying that the idea")
Wed. 29th ("We want to return to the diviners...")
Fri. Dec. 1st ("If we are taught the meaning...")
Wed. Dec. 6th ("Suppose I pointed...")
Fri. Dec. 8th ("Let us go back...")
Sat. Dec. 9th ("This craving for generality...")
Fri. Jan. 19th ("What happens if...")
Mon. Jan. 22nd ("One may characterise the meaning...")
Wed. Jan 24th ("Now we may describe...").

[32] Acknowledgements to Dr Michael Nedo for showing this MS to the editor, A. Gibson.

Appendix [P]: Skinner's Geometrical Diagram

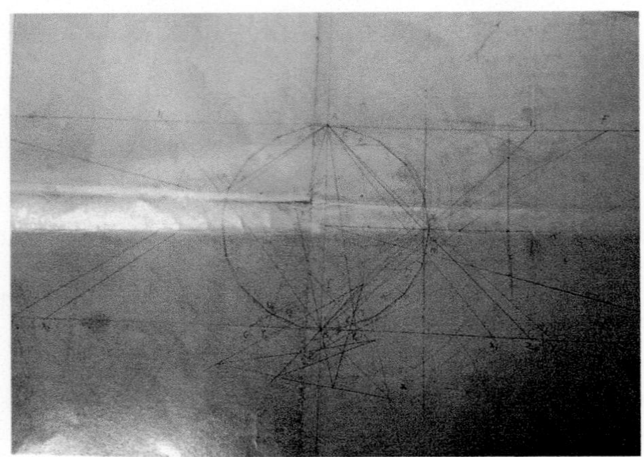

Appendix [Q]: Document on Francis' Life, By his Elder Sister *Katharine*

Editorial Note

Arthur Gibson and Niamh O'Mahony

The manuscript below is a result of much-appreciated research by Margaret Harper and Marion Eastwood in kind response to Arthur Gibson's enquiries. We acknowledge their locating and sending this copy for use in the present book.[33]

The four pages of manuscript reproduced below is in the handwriting of Katharine Skinner.[34] They present memories of her younger brother, which Marion Eastwood reports were still painful over 40 years later. These include an account of the circumstances leading up to his Francis's death, with Wittgenstein in attendance. Katharine appears in the photograph with her brother Francis, in Plate 4 in the Introduction.

The opening of the document is missing from the facsimile below. It should read, "Francis Skinner 1912 – June 9 – Born at 38 Drayton Gardens, S Kensington, London SW". There is a summarized transcription of these pages, placed after their reproduction below:

[33]A niece of Francis's Skinner's – Dr. Margaret A. Harper (of the School of Geography, Environment and Earth Sciences, Victoria University of Wellington) – has been of special value and enabled these manuscripts to be used here, as has the industry of some of her other family relations (especially Marion Eastwood (Francis' great-niece), who suspects that these notes date from the 1980s.)

[34]Katharine Skinner's married name is Woodbridge.

London SW

3rd child of Sidney Skinner (Principal of the Chelsea Polytechnic) & of Marion Field Michaelis—

"Nanny" (Edith Ashford, who died a couple of years ago) took care of Priscilla & me—

"Hilda" (Hilda Jackson, later became a trained nurse & then Mrs Jones) came as undernurse & took care of Francis. He loved her & said he would marry her when he grew up

We children were taken twice daily, morning & afternoon, to Kensington Gardens—Our chief playmates were the Grigg children. Francis played specially with his contemporary, Elizabeth Grigg—

First school for a few classes was Challoner School (Blue Froebel)

From 1914 through 1919 our family went every summer (1–August) to Home Lodge, Great Amwell, Herts. During later summers of this period, the Griggs also took a house there, & we would meet every day, as in winter

1919 Family moved to 77 Brook Green, Hammersmith

Francis transferred to Froebel Educational Institute for boys & girls in West Kensington (Green Froebel)—

1920 Summer holidays at Ventnor, Isle of Wight, where we were joined by Aunt Clare, Patsy & Micky, who stayed in a hotel nearby.

Later Francis went to Colet Court (boys' school) where he began to do well

1921 Summer holidays in boarding house at Beer in Devon. Our Mother went to America to see the family after the long wartime gap. Aunt Mildred Skinner ("Pount") came to take care of us instead—She & Aunt Joyce Skinner generally spent part of

(2) of the holidays with us, both in summer & throughout the year

1922 Summer holidays in Grindelwald, Switzerland, where we met Aunt Clare, Patsy & Micky again - Aunt Mildred was there, - but did my Mother & Francis come too? Perhaps not, perhaps he was considered too young.

Fall 1922 My father's serious illness (carbuncle leading to leg infection & general poisoning)

1923(?) Fall - Entered St Paul's School as one of their 153 foundation scholars - On Classical Side -

1924 - Second visit to Grindelwald

About 1925 Onset of his terrible illness in the spring - Osteomyelitis - Operation on leg to drain poison from marrow of bone - Later, tonsillectomy, in case tonsils had been cause of infection - In West London Hospital about 3 months, where one leg grew faster than the other. Convalescent at a children's home Birchington-on-Sea, Kent

August - Joined family at Westgate, Kent

1925-26 Spent scholastic year at home - As soon as he was well enough, he was tutored by various masters from St Paul's after school hours.

1926 or 27 Returned to school & was transferred to Mathematical Side - In Mathematical VIIIth form - Made new friends including R.L. Goodstein, who became his close working companion & rival.

1930 Won scholarship to Trinity College, Cambridge, & went up in the Fall - R.L.G also went up.

1931 spring Family moved to 17 Sollershott West, Letchworth, Herts (our Father having retired a couple of years earlier)

③ 1931 or 2 Passed 1st part of Mathematical Tripos with First Class Honours
1933 Ditto with 2nd part (with Sched. B. added)

He first met Dr Wittgenstein in the winter of 1931–32 or 1932–33, attending his lectures – He fell under his influence –

Summer 1933 – Trinity gave him a studentship for 2 or 3 years – Used this time and money to help Dr Wittgenstein, rather than pursuing his own work (When first at Trinity he had a room on Whewell's Court – Later he had a beautiful room on Great Court) My family were deeply distressed that he did not use this opportunity to do research in a branch of mathematics for which he had more gifts than for the philosophical mathematics which was Dr Wittgenstein's field – If he accomplished a successful mathematical project of his own, Trinity might have given him a research fellowship, & ultimately he might have won a place on the mathematical faculty at Cambridge or some other University – All his masters at St Paul's & his ~~other~~ teachers at Cambridge thought him extremely gifted & promising – I do not think he realized how much pain his actions gave my family – Ever since his serious boyhood illness he had periodic returns of infection. His leg would swell & his temperature rise, he would return home & my Mother would nurse him.

(4)

The Doctor would lance his leg, the poison (& sometimes a chip of bone) would drain away. As soon as he was well again, he immediately returned to work. Because of these recurring illnesses, my family were very anxious to see him professionally & financially established. In addition, the things of the mind were all important to them –

About 1936 the money from Trinity came to an end – Francis took a job in Cambridge as a skilled workman with the {Pye Instrument Co / or Cambridge Scientific Instrument Co}

One of his principal motives was to remain close to Dr Wittgenstein, whose 7-years Fellowship at Trinity was also coming to an end. They took lodgings together over a shop at 81 East Road, Cambridge, & lived very austere lives like "intellectual monks", as someone said.

[In 1934 I married & came to this country. Was preoccupied with my husband & babies & even less in touch with things, in spite of periodic visits to England]

In 1939 War began & in the fall of 1941 there was a severe air raid at Cambridge & many casualties were brought to Addenbrooke's Hospital – At the same time Francis had one of his recurrent infections (this time it was polio) & Dr Wittgenstein took him to the hospital but did not notify my family – The personnel was overwhelmed & it was not till morning that the Head Nurse came roun[d]

to my brother & realized how ill he was. She immediately telephoned my family who came at once. But it was too late; Francis died while they were on the way – They never forgave Dr Wittgenstein for not telling them

Transcription of Katharine's Document

Note by Editors

This transcription adheres exactly to the style of Katharine's typed presentation.

{The earlier part lists schools and holidays pre-1922}.

1923(?) Fall {Francis} entered St Paul's school as one of their 153 foundation scholars on the Classical side.

1924 {summer} 2nd visit to Grindelwald.

About 1925 Onset of his terrible illness in the Spring.

Osteomyelitis – operation on leg to drain poison from the marrow of bone – Later tonsillectomy, in case tonsils had been cause of infection. In West London Hospital about 3 months, where one leg grew faster than the other. Convalescent at a children's home Birchington-on-Sea, Kent.

August joined family at Westgate, Kent

1925–26 spent scholastic year at home – as soon as he was well enough, he was tutored by various masters from St Paul's school after school hours.

1926 or 27 returned to school & was transferred to the Mathematical side – In Mathematical VIII form – made new friends including R. L. Goodstein, who became his close working companion and rival.

1930 won scholarship to Trinity College Cambridge & went up in the Fall – R. L. G. also went up.

1932 Family moved to 17 Sollershott West, Letchworth, Herts (father having retired a couple of years earlier)

1931 or 2 passed 1st part Mathematical Tripos with 1st Class Honours

1933 Ditto with 2nd part (Schedule B added)

He 1st met Dr. Wittgenstein in the winter of 1931–32 or 1932–33 attending his lectures – he fell under his influence.

Summer 1933 – Trinity gave him a studentship for 2–3 years. Used this time and money to help Dr. W, rather than pursuing his own work. (When 1st at Trinity he had a room on Whewell's Court- Later he had a beautiful room on Great Court). My family was deeply distressed that he did not use this opportunity to do research in a branch of Mathematics for which {they thought} he had more gifts than for Philosophical mathematics. If he accomplished a successful mathematical project of his own Trinity might have given him a research fellowship & ultimately, he might have won a place on the mathematical faculty at Cambridge or some other University. All his masters at St Paul's & his teachers at Cambridge thought him extremely gifted & promising. I do not think he realised how much pain his actions gave my family – Ever since his serious boyhood illness her had periodic returns of infection & his leg would swell & his temperature rise, he would return home & my mother would nurse him. The Doctor would lance his leg, the poison (& sometimes a chip of bone) would drain away. As soon as he was well again, he immediately returned to work. Because of these recurring illnesses, my family were anxious to see him professionally and financially established. In addition, things of the mind were important to them.

About 1936 the money from Trinity came to an end. Francis took a job in Cambridge as a skilled workman with the Pye Instrument Co. or Cambridge Scientific Instrument Co.

One of his principal motives was to remain close to Dr. W.; whose seven-year fellowship at Trinity was also coming to an end. They took lodgings together over a shop at 81 East Road Cambridge & lived very austere lives like "intellectual monks", as someone said.

[In 1934 I married & came to U.S. preoccupied with husband and babies]

In 1939 war began & in the fall of 1941 there was a severe air raid at {near} Cambridge & many casualties were brought to Addenbrooke's Hospital. At the same time Francis had one of his recurrent infections (this time it was polio). Dr. W. took him to hospital but did not notify my family – the personnel were overwhelmed and it was not until morning that the Head Nurse came round to my brother & realized how ill he was. She immediately telephoned my family who came at <u>once</u>. But it was too late; Francis died while they were on the way & they never forgave Dr. W. for not telling them.

Appendix [R]: Letter from Pat Sloan (Moscow 24.9.1935) to Maurice Dobb

Concerning Wittgenstein in Moscow

Editorial Note

Niamh O'Mahony and Arthur Gibson

This sketch expands the introduction of Sloan's letter at the end of Chap. 2, Sect. 3.

The spies Anthony Blunt[35] and Michael Straight, while travelling separately from Wittgenstein (all from Trinity College Cambridge), also went to Russia in 1935 – as did Wittgenstein.[36] Although this manifested concurrence with a fashionable trend among a select stream of clientele in England (which was at its height in 1935),[37] nonetheless their combined Cambridge origins and asymmetric cross-connections there with mutual acquaintances are worthy of further investigation, which is not practical here. One of Wittgenstein's reasons for the visit, aimed at employment as a labourer, was motivated by Tolstoy's views. He had special further motive for his visit, which according to Rhees was to go one of the newly colonized "primitive" areas.[38] An obvious candidate for this is the Jewish Autonomous Republic of Birobidjan, which was established in 1934 – the year before Wittgenstein's trip to Moscow.[39] The Moscow correspondent for the *Daily Worker*, a former Cambridge Economics student Pat Sloan, a friend of Sraffa and Maurice Dobb, was a Moscow point of contact for Wittgenstein.

Sloan's letter affords us an independent view of Wittgenstein in Moscow. Sloan wrote from Moscow to Dobb in Cambridge, a few days after Francis Skinner had written to Wittgenstein in Moscow. Sloan's letter is reproduced below as Plates 2.

---

[35]See Gibson (2007: 101–10).

[36]See O'Mahony (2013: 166).

[37]See O'Mahony (2013: 170).

[38]According to Rhees (1984: 228).

[39]O'Mahony (2013: 171); cf. Marley's (1935) report.

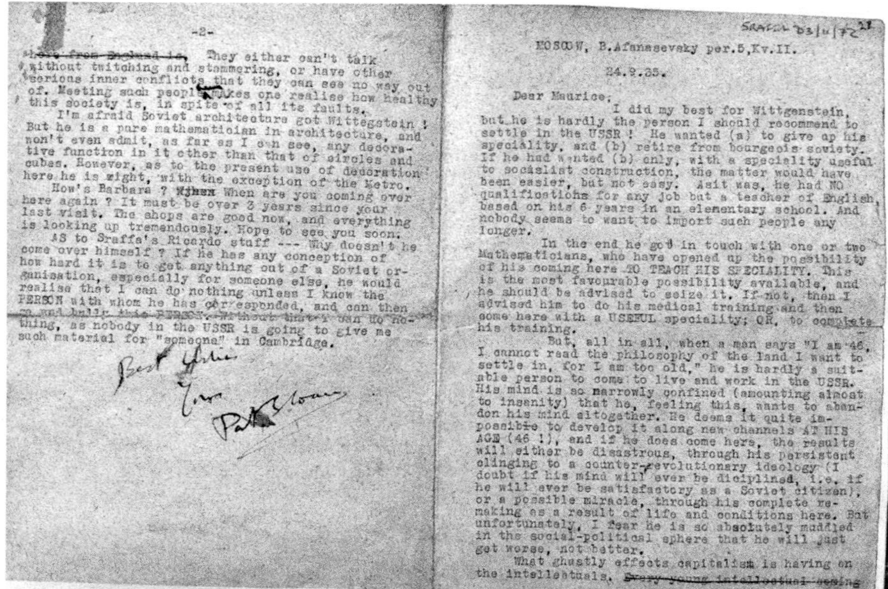

-2-

They either can't talk without twitching and stammering, or have other serious inner conflicts that they can see no way out of. Meeting such people makes one realise how healthy this society is, in spite of all its faults.

I'm afraid Soviet architecture got Wittegstein ! But he is a pure mathematician in architecture, and won't even admit, as far as I can see, any decorative function in it other than that of circles and cubes. However, as to the present use of decoration here he is right, with the exception of the Metro.

How's Barbara ? When are you coming over here again ? It must be over 3 years since your last visit. The shops are good now, and everything is looking up tremendously. Hope to see you soon.

AS to Sraffa's Ricardo stuff --- Why doesn't he come over himself ? If he has any conception of how hard it is to get anything out of a Soviet organisation, especially for someone else, he would realise that I can do nothing unless I know the PERSON with whom he has corresponded, and can then ... thing, as nobody in the USSR is going to give me such material for "someone" in Cambridge.

Best wishes
Yours
Pat Sloan

SRAFFA D3/11/72/21

MOSCOW, B.Afanasevsky per.5,Kv.II.

24.9.35.

Dear Maurice,

I did my best for Wittgenstein, but he is hardly the person I should recommend to settle in the USSR ! He wanted (a) to give up his speciality, and (b) retire from bourgeois soviety. If he had wanted (b) only, with a speciality useful to socialist construction, the matter would have been easier, but not easy. Asit was, he had NO qualifications for any job but a teacher of English, based on his 6 years in an elementary school. And nobody seems to want to import such people any longer.

In the end he got in touch with one or two Mathematicians, who have opened up the possibility of his coming here TO TEACH HIS SPECIALITY. This is the most favourable possibility available, and he should be advised to seize it. If not, then I advised him to do his medical training and then come here with a USEFUL speciality; OR, to complete his training.

But, all in all, when a man says "I am 46, I cannot read the philosophy of the land I want to settle in, for I am too old," he is hardly a suitable person to come to live and work in the USSR. His mind is so narrowly confined (amounting almost to insanity) that he, feeling this, wants to abandon his mind altogether. He deems it quite impossible to develop it along new channels AT HIS AGE (46 !), and if he does come here, the results will either be disastrous, through his persistent clinging to a counter-revolutionary ideology (I doubt if his mind will ever be diciplined, i.e. if he will ever be satisfactory as a Soviet citizen), or a possible miracle, through his complete remaking as a result of life and conditions here. But unfortunately, I fear he is so absolutely muddled in the social-political sphere that he will just get worse, not better.

What ghastly effects capitalism is having on the intellectuals.

Plate 2 Letter from Pat Sloan (Moscow 24.9.1935) to Maurice Dobb (Cambridge) concerning Wittgenstein. (This letter is owned by Trinity College, Cambridge, and is held in its Sraffa manuscript collection. Trinity Wren Library designation: Add.MS.a.522(1): Sraffa D3/11/72/21, or D3_11_72_21. The address "MOSCOW, B. Afanasevsky per. 5, Kv. II." is translated as Flat 2, 5 Bol'shoy Afanas'yevskiy Lane, Moscow, where Sloan was living while working as a journalist in Russia.)

Bibliography

A

Adams, L. S. (1996). *Methodologies of art: An introduction*. New York: Westview Press.
Ahonen, H. (2005). Wittgenstein and the conditions of musical communication. *Philosophy, 80*(4), 513–530.
Ahonen, H. (2006). Wittgenstein and formalism. *Acta Philosophica Fennica, 80*, 223–239.
Allen, R., & Turvey, M. (2001). *Wittgenstein, theory and the arts*. London: Routledge.
Ambrose, A. (Ed.). (1979). *Wittgenstein's lectures: Cambridge 1932–35*. Totowa/Oxford: Rowman & Littlefield/Blackwell.
Ammereller, E., & Fisher, E. (2004). *Wittgenstein at work: Method in the philosophical investigations*. London: Routledge.
Andrew, C. M. (2018). *Secret World: A History of Intelligence*. New Haven: Yale University Press.
Andrews, H. C. (1935). *Sidelights on Brasses in Hertfordshire Churches*. EHAT.
Andrews, D. R. (1996). Nothing is hidden: A Wittgensteinian interpretation of Sraffa. *Cambridge Journal of Economics, 20*(6), 763–778.
Anscombe, G. E. M. (1957). *Intention*. Oxford: Basil Blackwell.
Anscombe, G. E. M. (1996). *An introduction to Wittgenstein's Tractatus*. Bristol: Thoemmes Press.
Appel, A. W. (2012). The birth of computer sciences at Princeton in the 1930s. In A. W. Appel (Ed.), *Alan Turing's system of logic: The Princeton Thesis* (pp. 1–12). Princeton: Princeton University Press.
Augustine. (1991/2009). *Confessions*. (H. Chadwick, Ed., Trans.). Oxford: Oxford University Press.
Augustine. (1995). *Confessions. Books I-IV*. (G. Clark Ed.). Cambridge: Cambridge University Press.

B

Baaz, M., et al. (Eds.). (2011). *Kurt Gödel and the foundations of mathematics: Horizons of truth*. Cambridge: Cambridge University Press.
Baez, J. C. (2011, June 6). *Renyi entropy and free energy*. arXiv: 1102.2098v3 [quant-phi].
Baker, H. F. (1937). On the proof of a lemma enunciated by Severi. *Proceedings of the Cambridge Mathematical Society, 32*, 253–259.

A. Gibson, N. O'Mahony, *Ludwig Wittgenstein: Dictating Philosophy*,
https://doi.org/10.1007/978-3-030-36087-0

Baker, G. (2002). Wittgenstein on metaphysical/everyday use. *Philosophical Quarterly, 52*(208), 289–302.

Baker, G. P. (2003). *The voices of Wittgenstein, The Vienna circle: Ludwig Wittgenstein and Friedrich Waismann*. London: Routledge.

Baker, G. P. (2004). *Wittgenstein's method, neglected aspects: Essays on Wittgenstein*. (K. J. Morris. Ed.). Oxford: Blackwell.

Bar-Elli, G. (2006). Wittgenstein on the experience of meaning and the meaning of music. *Philosophical Investigations, 29*(3), 217–249.

Barrett, C. (1984). Ethics and aesthetics are one. *Ästhetik: Akten des, 8*, 17–23.

Barrett, C. (1991). *Wittgenstein on ethics and religious belief*. Oxford: Blackwell.

Beardsley, M. C. (1958). *Aesthetics: Problems in the philosophy of criticism*. New York: Harcourt, Brace.

Beck, S. (2006). These Bizarre fictions: Thought-experiments, our psychology and our selves. *Philosophical Papers, 35*(1), 29–54.

Beck, J. (2008). *Combinatorial games: Tic-tac-toe theory*. Cambridge: University Press.

Bermúdez, J. L. (2003). *Thinking without words*. Oxford: Oxford University Press.

Bermúdez, J. L., & Millar, A. (2002). *Reason and nature: Essays in the theory of rationality*. Oxford: Clarendon Press.

Bernstein, J., & Gelbart, S. (2004). *An Introduction to the Langlands Program*. Boston: Birkhäuser.

Bibiloni, L., Viader, P., & Paradís, J. (2004). On a series of Goldbach and Euler. *Annals of The Mathematical Association of America, 113*, 206–220.

Biesenbach, H. (2011). Anspielungen und Zitate im Werk Ludwig Wittgensteins. In A. Pichler (Ed.), *Publications from the Wittgenstein Archives of the University of Bergen* (Vol. 22). Bergen: University of Bergen.

Biesenbach, H. (2014). *Anspielungen und Zitate im Werk Ludwig Wittgensteins. Erweiterte Neuausgabe* (2nd ed.). Sofia: Sofia University Press.

Binkley, T. (1973). *Wittgenstein's language*. The Hague: Nijhoff.

Birk, A. (2004). The later Wittgenstein and his readers. In A. Coliva & E. Picardi (Eds.), *Wittgenstein today* (pp. 465–475). Padova: Il Poligrafo casa editrice.

Birkar, C. (2016). *Singularities of linear systems and boundedness of Fano varieties*. ArXiv eprint: 1609.05543, 09/2016 (Revised 25 Jan 2018 (version, v3).

Black, M. (1962). *Models and metaphors: Studies in language and philosophy*. Ithaca: Cornell University Press.

Blunt, A. F. (2009). *Autobiographical confession held in reserve collection* (unpublished manuscript). London: British Library.

Boghossian, P. (2001). How are objective epistemic reasons possible? *Philosophical Studies, 106* (1–2), 1–40.

Bohm, D., & Nichol, L. (1998). *On creativity*. London: Routledge.

Bombieri, E. (2011). The Mathematical Infinity. In M. Heller & W. Woodin (Eds.). *New Research Frontiers* (pp. 55–75). Cambridge: Cambridge University Press.

Bourbon, B. (2005). Wittgenstein's preface. *Philosophy and Literature, 29*(2), 428–443.

Bouvresse, J. (1995). *Wittgenstein reads Freud: The myth of the unconscious* (C. Cosman, Trans.). Princeton: Princeton University.

Bouwsma, O. K. (1961, March 16). On a Series of Goldbach and Euler in *The Blue Book. Journal of Philosophy, LVIII*, 141–162.

Bouwsma, O. K. (1986). *Wittgenstein: Conversations, 1949–1951* (J. L. Craft & R. E. Hustwit. Eds.). Indianapolis: Hackett Pub. Co.

Bowyer, M. J. F. (1990). *Action stations. 1, Military airfields of East Anglia, 1939–1945* (2nd ed.). Wellingborough: Stephens.

Brill, S. B. (1995). *Wittgenstein and critical theory: Beyond postmodern criticism and toward descriptive investigations*. Athens: Ohio University Press.

Brock, S., & Pichler, A. (2015). Contributions of a conceptual ontology for Wittgenstein. *Wittgenstein-Studien, 6*, 257–275. de Gruyter.

Burnyeat, M. R. (1987). Wittgenstein and Augustine De Magistro. *Proceedings of the Aristotelian Society: Supplementary Volume, 61*(1987), 1–24.

C

Camp, E. (2006). Contextualism, metaphor, and what is said. *Mind and Language, 21*(3), 280–309.

Carruthers, P. (1990). *The metaphysics of the tractatus*. Cambridge: Cambridge University Press.

Carruthers, P. (2004). On being simple minded. *American Philosophical Quarterly, 41*(3), 205–220.

Carter, Z. A. (2006). Wittgenstein, Rush Rhees, and the Measure of Language. *New Blackfriars, 87* (1009), 288–301.

Caruana, L. (2004). Wittgenstein and the status of contradictions. In A. Coliva & E. Picardi (Eds.), *Wittgenstein today* (pp. 223–232). Padova: Il Poligrafo.

Cater, J., & Lemco, I. (2009). Wittgenstein's combustion chamber. *Records of the Royal Society, 63*, 95–104.. Published online 6 January 2009 (pp. 94–104). http://rsnr.royalsocietypublishing.org/. https://doi.org/10.1098/rsnr.2008.0031.

Cavell, S. (1962). The availability of Wittgenstein's later philosophy. *Philosophical Review, 71*, 67–93.

Cavell, S. (1976). *Must we mean what we say?* Cambridge: Cambridge University Press.

Cavell, S., Conant, J., Diamond, C., Dilman, I., Hacker, P. M. S., McGuinness, B. F., et al. (2001). On Wittgenstein. *Philosophical Investigations, 24*(2), 89–184.

Chadwick, H. (1991). *Augustine*. (1991/2009). Confessions. (H. Chadwick, Ed., Trans. Oxford: Oxford University Press.

Champernowne, D. (2000, December 4). David Champernowne (1912–2000). *Journal of the International Computer Chess Association, 23*.

Cheney, C. R. (2000). *A handbook of dates*. In M. Jones (Ed.), *Royal historical society guides and handbooks* (Vol. 4). Cambridge: Cambridge University Press.

Cheung, L. K. (2006). The unity of language and logic in Wittgenstein's Tractatus. *Philosophical Investigations, 29*(1), 22–50.

Chinca, M. (2020). *Meditating death in medieval and early modern devotional writing: From Bonaventure to Luther. Oxford studies in medieval literature & culture*. Oxford: Oxford University Press.

Cioffi, F. (1998). *Freud and the question of pseudoscience*. Chicago: Open Court.

Citron, G. (2013). Moore's notes on Wittgenstein's lectures, Cambridge 1930-1933: Text, context, and content. In Stern, D. G., Citron, G., & Rogers, B.. Nordic Wittgenstein Review, 2 (1), 161–179.

Clark, G. (1995). *Augustine: Confessions books I-IV*. Cambridge: Cambridge University Press.

Cohen, L. J. (1992). *An essay on belief and acceptance*. Oxford: Clarendon Press.

Colgough, S., & Vincent, D. (2009). Reading. In D. McKitterick (Ed.), *The Cambridge history of the book* (Vol. VI, pp. 281–323). Cambridge: Cambridge University Press.

Coliva, A., & Picardi, E. (Eds.). (2000). *Wittgenstein today* (pp. 477–488). Padova: Il Poligrafo casa editrice.

Conway, J. H. (1976). *On numbers and games*. London/New York/San Francisco: Academic Press.

Copi, I. M., & Beard, R. W. (1966). *Essays on Wittgenstein's Tractatus*. London: Routledge.

Currie, G., & Ravenscroft, I. (2002). *Recreative minds: Imagination in philosophy and psychology*. Oxford: Clarendon Press.

D

Danto, A. C. (1986). *The philosophical disenfranchisement of art*. New York: Columbia University Press.

Deacon, R. (1985). *The Cambridge Apostles*. London: Robert Royce.

De Hamel, C. (2016). *Meetings with remarkable manuscripts*. London: Allen Lane, Penguin, Random House.

De Iaco, M. (2019). Wittgenstein to Sraffa: Two newly discovered letters from February and March 1934. *Nordic Wittgenstein Review* (A. Pichler, Ed.), *8*(1), 93–106.

De Pellegrin, E. (2019). The *Brown Book* of Alice Ambrose: remarks on Ludwig Wittgenstein's dictated notes of 1934 – 35. *Wittgenstein-Studien, 10*(1), 1–36.

De Pellegrin, E. (Ed.) (Forthcoming). *Wittgenstein: The Yellow Book*. A compilation of lecture notes, Cambridge 1932–34.

Deluty, E. W. (2005). Wittgenstein's Paradox: Philosophical investigations, Paragraph 242. *International Philosophical Quarterly, 45*(177), 87–102.

Diamond, C. (1991). *The realistic spirit: Wittgenstein, philosophy, and the mind*. Cambridge, Mass.: MIT Press.

Diamond, C. (2005). Logical syntax in Wittgenstein's Tractatus. *Philosophical Quarterly, 55*(218), 78–89.

Dickson, L. E. (1919). *History of the theory of numbers* (Vol. 3). New York: Chelsea Publishing Company.

Diffie, W., & Hellman, M. (1976). New directions in cryptography. *IEEE Transactions on Information Theory, 22*(6): 644.

Dilman, I. (1998). The philosopher and the fly bottle. *Ratio, 11*(2), 102–124.

Donatelli, P. (2004). Wittgenstein, ethics and religion: Earlier and later. In A. Coliva & E. Picardi (Eds.), *Wittgenstein today* (pp. 447–464). Padova: Il Poligrafo casa editrice.

Dromm, K. (2006). Wittgenstein on language-learning. *History of Philosophy Quarterly, 23*(1), 79–94.

Drury, M. O'. C. (1984). Conversations with Wittgenstein. In R. Rhees (Ed.), *Recollections of Wittgenstein* (pp. 97–171). Oxford: Oxford University Press.

Drury, M. O'. C. (2017). *The selected writings of Maurice O'Connor Drury: On Wittgenstein, philosophy, religion and psychiatry* (J. Hayes, Ed.). London: Bloomsbury Academic. (e-book, 2017; paperback November 2018).

Dummett, M. A. E. (1981). *Frege: Philosophy of language* (2nd ed.). London: Duckworth.

Dumoncel, J.-C. (2011). Suivre une règle. Wittgenstein entre Aristote et Lejeune-Dirichlet. *Revue philosophique de la France et de l'étranger, 136*(2011), 187–209.

E

Elliott, R. K., & Crowther, P. (2006). *Aesthetics, imagination and the unity of experience*. Aldershot: Ashgate.

Engelmann, M. L. (2013). *Wittgenstein's philosophical development*. New York: Palgrave Macmillan.

Engelmann, P., & Wittgenstein, L. (1967). *Letters from Ludwig Wittgenstein with a Memoir*. Oxford: Basil Blackwell.

Erden, Y. J. (2009a). *The limits of language*. Ph.D. thesis, Roehampton University. www.academia.edu/1645942/Wittgenstein_on_simile_as_the_best_thing_in_philosophy

Erden, Y. J. (2009b, December). Book review: 'Wittgenstein's house: language, space, and architecture. *Social Semiotics*, 19, 4:515–521.

Erden, Y. J. (2010). Could a created being ever be creative? Some philosophical remarks on creativity and AI development. *Minds and Machines, 20*(3), 349–362.

Erden, Y. J. (2011). Wittgenstein on Simile as the "best thing". *Philosophical Investigations, 35*(2), 127–137.

Erden, Y. J., & Magill, K. (2012). Autonomy and desire in machines and cognitive agent systems. *Cognitive Computation, 4*(3), 354–364.

Euclid. (1920, 2014). Euclid in Greek. Ed. and Notes by Sir T.L. heath. Cambridge: Cambridge University Press.

Ewald, W. B. (1996). *From Kant to Hilbert: A source book in the foundations of mathematics*. Oxford: Clarendon Press.

F

Feferman, S. (1988). Turing in the land of O(z). In R. Herken (Ed.), *The Universal Turing machine*. Oxford: Oxford University Press.

Feferman, S. (2011). *Lieber Herr Bernays*! *Lieber Herr Gödel*! Gödel on finitism, constructivity, and Hilbert's programme. In M. Baaz et al. (Eds.), *The foundations of mathematics* (pp. 111–133). Cambridge: Cambridge University Press.

Feferman, S. (2012). Turing's thesis. In A. M. Turing (Ed.), *Alan Turing's system of logic: The Princeton thesis* (pp. 13–28). Princeton: Princeton University Press.

Floyd, J. (1995). On saying what you really want to say: Wittgenstein, Gödel, and the Trisection of the angle. In J. Hintikka (Ed.), *From Dedekind to Gödel: Essays on the development of the foundations of mathematics* (pp. 373–425). Dordrecht: Springer.

Floyd, J. (1998). The Uncaptive Eye: Solipsism in Wittgenstein's Tractatus. *Boston University Studies in Philosophy and Religion, 19*, 79–108.

Floyd, J. (2004). Wittgenstein on philosophy of logic and mathematics. *Graduate Faculty Philosophy Journal, 25*, 227–268.

Floyd, J. (2007). Wittgenstein and the Inexpressible. In A. Crary & C. Diamond (Eds.), *Wittgenstein and the moral life: Essays in honour of Cora Diamond* (pp. 177–234). Cambridge, Mass.: MIT Press.

Floyd, J. (2012a). Wittgenstein's diagonal argument: A variation on Cantor and Turing. In P. Dybjer, S. Lindström, E. Palmgren, & G. Sundholm (Eds.), *Epistemology versus ontology: Essays on the philosophy and foundations of mathematics in honour of Pe Martin-Löf* (Vol. 27, pp. 25–44). Dordrecht: Springer.

Floyd, J. (2012b). Wittgenstein, Carnap, and Turing: Contrasting notions of analysis. In P. Wagner (Ed.), *Carnap's ideal of explication and naturalism* (pp. 34–46). Basingstoke: Palgrave Macmillan.

Floyd, J. (2013). Turing, Wittgenstein and types: Philosophical aspects of Turing's "The reform of mathematical notation and phraseology". In B. S. Cooper & J. van Leeuwen (Eds.), *Alan Turing – His work and impact*. Amsterdam: Elsevier Science.

Floyd, J. (2016a). Turing, Wittgenstein and Emergence. In J. Floyd, J. E. Katz, & J. E (Eds.), *Philosophy of emerging media: Understanding, appreciation, application* (pp. 219–242). Oxford: Oxford University Press.

Floyd, J. (2016b). Chains of life: Turing, Lebensform, and the emergence of Wittgenstein's later style. *Nordic Wittgenstein Review, 5*(2), 7–82.

Floyd, J. (2017a). Turing on "Common Sense": Cambridge resonances. In J. Floyd & A. Bokulich (Eds.), *Philosophical explorations of the legacy of Alan Turing; Turing 100* (Boston Studies in the Philosophy of Science Series). New York/Berlin: Springer.

Floyd, J. (2017b). Parikh and Wittgenstein. In C. Baskent, L. Moss, & R. Ramanujan (Eds.), *Rohit Parikh on logic, Language and Society. Studia Logica outstanding contributions* (pp. 1–35). Berlin: Springer.

Floyd, J. (2018). Aspects of aspects. In H. Sluga & D. Stern (Eds.), *The Cambridge companion to Wittgenstein* (2nd ed., pp. 361–388). Cambridge: Cambridge University Press.

Floyd, J. (*forthcoming*). *Lebensform versus culture*.

Floyd, J., & Kanamori, A. (2015). *Gödel vis `a vis Russell: Logic and set theory to philosophy*. In G. Crocco & E.-M. Engelen (Eds.), *Essays in honor of Kurt Gödel: Philosopher-Scientist*. Aix-en-Provence: Universitaires de Provence.

Floyd, J., & Katz, J. E. (Eds.). (2016). *Philosophy of emerging media: Understanding, appreciation, application*. Oxford: Oxford University Press.

Floyd, J., & Putnam, H. (2006). Bays, Steiner, and Wittgenstein's "Notorious" paragraph about the Gödel Theorem. *Journal of Philosophy, 103*(2), 101–110.

Fogelin, R. J. (1992). *Philosophical interpretations*. Oxford: Oxford University Press.

Fogelin, R. J. (1995). *Wittgenstein*. London: Routledge.

Forster, T. (2003). *Logic, induction and sets*. London: London Mathematical Society.

Forster, M. N. (2004). *Wittgenstein on the arbitrariness of grammar*. Princeton: Princeton University Press.

Frascolla, P. (1994). *Wittgenstein's philosophy of mathematics*. London: Routledge.

Frascolla, P. (2001). Philosophy of mathematics. In H.-J. Glock (Ed.), *Wittgenstein: A critical reader* (pp. 268–288). Oxford: Blackwell.

Frascolla, P. (2004). Wittgenstein on mathematical proof. In A. Coliva & E. Picardi (Eds.), *Wittgenstein today* (pp. 167–184). Padova: Il Poligrafo casa editrice.

Friedlander, E. (2001). *Signs of sense: Reading Wittgenstein's Tractatus*. Cambridge, MA: Harvard University Press.

G

Ganesalingam, M. (2013). *The language of mathematics*. Berlin/Heidelberg: Springer.

Garver, N. (1994). *This complicated form of life: Essays on Wittgenstein*. Chicago: Open Court.

Garver, N. (2006). *Wittgenstein and approaches to clarity*. Amherst: Humanity Books.

Gaskin, R. (2001). *Grammar in early twentieth-century philosophy*. London: Routledge.

Gasking, D. A. T., & Jackson, A. C. (1967). Wittgenstein as teacher. In K. T. Fann (Ed.), *Wittgenstein: The man and his philosophy* (pp. 49–55). New York: Dell Publishing Co.

Gauthier, Y. (2015). *Towards an arithmetical logic* (Series: Studies in Universal Logic). Basel: Springer, Birkhauser.

Geach, P. T. (1976). Saying and showing in Frege and Wittgenstein. *Acta Philosophica Fennica, 28*, 54–70.

Genova, J. (1995). *Wittgenstein: A new way of seeing*. London: Routledge.

Gerrard, S. (1991). Wittgenstein's philosophies of mathematics. *Synthese, 87*(1), 125–142.

Gerrard, S. (2002). One Wittgenstein? In E. H. Reck (Ed.), *From Frege to Wittgenstein: Perspectives on early analytic philosophy* (pp. 52–74). Oxford: Oxford University Press.

Geuss, R. (2017). *Changing the subject: Philosophy from Socrates to Adorno*. Cambridge, MA: Harvard University Press.

Giaquinto, M. (2007). *Visual thinking in mathematics*. Oxford: Oxford University Press.

Gibbons, L. (2018, June). The other side of the sky. *Dublin Review of Books, 101*. http://www.drb.ie/essays/the-other-side-of-the-sky

Gibson, A. (2000). *God and the universe*. London: Routledge.

Gibson, A. (2003). *Metaphysics and transcendence*. London: Routledge.

Gibson, A. (2007). *What is literature?* Oxford/Bern/Berlin: Lang.

Gibson, A. (2010a). Francis Skinner's original Wittgenstein Brown Book. In V. Munz et al. (Eds.), *Language and world. Part one* (pp. 351–366). Frankfurt/Paris: Ontos Verlage.

Gibson, A. (2010b). The Wittgenstein archive of Francis Skinner. In N. Venturinha (Ed.), *Wittgenstein after his nachlass* (pp. 64–77). New York: PalgraveMacmillan.

Gibson, A. (2013). Wittgenstein and the future of the mathematical sciences. In J. Padilla-Galvez & M. Gaffal (Eds.), *Doubtful Certainties* (pp. 213–224). Heusenstamm: Ontos.

Gibson, A. (2015). Peter Geach: a Few Personal Remarks. *Philosophical Investigations, 38*(1/2), 25–33.

Gibson, A. (2019). Anscombe, Cambridge, and the challenge of Wittgenstein. In J. Haldane (Ed.), *The life and philosophy of Elizabeth Anscombe* (St Andrews studies in philosophy and public affairs, pp. 23–41). Imprint-Academic: Exeter; and American Catholic Philosophical Quarterly, 90(2) (Spring 2016): 191–206.

Gibson, A. (2020). Intuition, Counter-Intuition & the Absence of Ontology for Dark Matter. In M. Gaffal (Ed.), *Language, Truth and Democracy* (pp. 135–156). Berlin: De Gruyter.

Gibson, A. (forthcoming-a). Counter-Intuition.

Gibson, A. (forthcoming-b). *Euler, Fermat, and Wittgenstein.*

Gibson, A. (forthcoming-c). *Wittgenstein and mathematics.*

Gibson, A. (forthcoming-d). *Wittgenstein, security, war, spies: 1931–1946.*

Gibson, A., Leader, I., & Smith, J. (forthcoming). *A Wittgenstein Manuscript 'corrected' by Kreisel.*

Gibson, J., & Huemer, H. (2004). *The literary Wittgenstein.* London: Routledge.

Gill, J. H. (1996). *Wittgenstein and metaphor.* Atlantic Highlands: Humanities Press.

Glock, H.-J. (2001). *Wittgenstein: A critical reader.* Oxford: Blackwell.

Glock, H.-J. (2004). Wittgenstein on Truth. *Schriftenreihe-Wittgenstein Gesellschaft, 33*, 13–31.

Glock, H.-J. (2006). Thought, language, and animals. *Grazer philosophische Studien, 71*, 139–160.

Glucksberg, S., & Haught, C. (2006). On the relation between metaphor and simile: When comparison fails. *Mind and Language, 21*(3), 360–378.

Gödel, K. (1944). Russell's mathematical logic. In P. A. Schilpp (Ed.), *The Philosophy of Bertrand Russell.* Tudor Publishing Company: New York. cf. also 1951 3rd ed.: 123–154.

Goldfarb, W. (1997). Metaphysics and nonsense: On Cora Diamond's the realistic spirit. *Journal of Philosophical Research, 22*, 57–74.

Goldstein, L. (1999). *Clear and Queer thinking: Wittgenstein's development and his relevance to modern thought.* New York: Rowman & Littlefield.

Goldstein, L. (2007). Why the substitution of co-referential expressions in a statement may result in change of truth-value (Concluding part). *The Reasoner, 1*(2), 6–7.

Gowers, T., Barrow-Green, J., & Leader, I. M. (Eds.) (2008). *The princeton companion to mathematics.* Princeton/Oxford: Oxford University Press.

Grève, S. S. (2018). Logic and philosophy of logic in Wittgenstein. *Australasian Journal of Philosophy, 96*(1), 168–182. For online reference: https://doi.org/10.1080/00048402.2017.1326512

Grève, S. S., & Kienzler, W. (2016). Wittgenstein on Gödelian 'Incompleteness', proofs and mathematical practice: Reading remarks on the foundations of mathematics, Part I, Appendix III. In S. Sunday Grève & J. Mácha (Eds.), *Wittgenstein and the creativity of language* (pp. 76–116). Basingstoke: Palgrave Macmillan.

Grève, S. S., & Mácha, J. (Eds.). *Wittgenstein and the creativity of language.* Basingstoke: Palgrave Macmillan.

Griffiths, A. P. (1991). *Wittgenstein centenary essays.* Cambridge: Cambridge University Press.

Ground, I., & Flowers, F. A. (Eds.). (2015). *Portraits of Wittgenstein.* London: Bloomsbury.

Guetti, J. L. (1993). *Wittgenstein and the grammar of literary experience.* Athens: University of Georgia Press.

Gustafsson, M. (2006). Nonsense and philosophical method. *Acta Philosophica Fennica, 80*, 11–34.

Guter, E. (2005). Critical study: An inadvertent Nemesis-Wittgenstein and contemporary aesthetics. *British Journal of Aesthetics, 45*(3), 296–306.

Guttenplan, S. (2006). The transparency of metaphor. *Mind and Language, 21*(3), 333–359.

H

Hacker, P. M. S. (1996). *Wittgenstein's place in twentieth-century analytic philosophy*. Oxford: Blackwell.

Hacker, P. M. S. (2000). Was he trying to whistle it? In A. M. Crary & R. J. Read (Eds.), *The new Wittgenstein* (pp. 353–388). London: Routledge.

Hacker, P. M. S. (2001). *Wittgenstein: Connections and controversies*. Oxford: Clarendon.

Hacker, P. M. S. (2003). Wittgenstein, Carnap and the New American Wittgensteinians. *Philosophical Quarterly, 53*(210), 1–23.

Hacking, I. (2000). What mathematics has done to some and only some philosophers. *Proceedings of the British Academy, 103*, 83–138.

Hacking, I. (2011). Wittgenstein, necessity, and the application of mathematics. *South African Journal of Philosophy, 30*(2), 61–73.

Hacking, I. (2014). *Why is there a philosophy of mathematics at all?* Cambridge: Cambridge University Press.

Haco, S., Hawking, S. W., Perry, M. J., & Strominger, A. (2017). *The Conformal BMS Group*. *arXiv*: 1701.08110.

Haco, S., Hawking, S. W., Perry, M. J., & Strominger, A. (2018). *Black hole entropy and soft hair*. *arXiv*: 1810.01847v4.

Hadersbeck, M., Pichler, A., Bruder, D., & Schweter, S. (2016). New (re)search possibilities for Wittgenstein's Nachlass II: Advanced search, navigation and feedback with the FinderApp WiTTFind. In S. Majetschak & A. Weiberg (Eds.), *Aesthetics today* (pp. 90–93). Kirchberg am Wechsel: Contributions of the Austrian Ludwig Wittgenstein Society.

Hagberg, G. L. (1995). *Art as language: Wittgenstein, meaning, and aesthetic theory*. Ithaca: Cornell University Press.

Hagberg, G. L. (2004). Wittgenstein underground. *Philosophy and Literature, 28*(2), 379–392.

Hagberg, G. L. (2008). *Describing ourselves*. Oxford: Clarendon Press.

Haller, R. (1988). *Questions on Wittgenstein*. London: Routledge.

Hamilton, K. (2001). Wittgenstein and the mind's eye. In J. C. Klagge (Ed.), *Wittgenstein: Biography and philosophy* (pp. 53–97). Cambridge: Cambridge University Press.

Hanfling, O. (2001). Wittgenstein on language, art and humanity. In R. Allen & M. Turvey (Eds.), *Wittgenstein, theory and the arts* (pp. 75–91). London: Routledge.

Hanfling, O. (2003). *Wittgenstein and the human form of life*. London: Routledge; *Proceedings, London Mathematical Society*, 22(2), 46–56.

Hardy, G. H. (1929). Mathematical proof. *Mind, 38*(149), 1–25.

Hardy, G. H. (1943). *A course in pure mathematics* (7th ed.). Cambridge: CUP.

Hardy, G. H. (1949). *Divergent series*. Oxford: Clarendon Press.

Hardy, G. H. (1967). *A mathematician's apology*. Cambridge: Cambridge University Press.

Hardy, G. H., & Littlewood, J. E. (1923). Some problems of 'Partitio numerorum'; III: On the expression of a number as a sum of primes. *Acta Mathematica, 44*, 1923): 1–1923):70.

Hardy, G. H., & Littlewood, J. E. (1925). Researches on Waring's problem. *Math Zeitschrift, 23*(4), 1–37.

Hardy, G. H., & Ramanujan, S. (1917). The normal number of prime factors of a number *n*. *The Quarterly Journal of Pure and Applied Mathematics, 48*(1917), 76–92.

Hardy, G. H., Littlewood, J. E., & Polya, B. (1934). *Inequalities*. Cambridge: CUP.

Heal, J. (1995). Wittgenstein and dialogue. In T. J. Smiley (Ed.), *Philosophical dialogues: Plato, Hume, Wittgenstein* (Vol. 85, pp. 63–83). Oxford: Oxford University Press for the British Academy.

Heal, J. (2003). *Mind, reason, and imagination: Selected essays in philosophy of mind and language*. Cambridge: Cambridge University Press.

Heilbronn, H. (1934). On the class number in imaginary quadratic fields. *Quarterly Journal of Mathematics, 5*, 150–160.

Heller, M., & Woodin, H. (2011). *Infinity: New research frontiers*. Cambridge: Cambridge University Press.

Hertzberg, L. (2006). On excluding contradictions from our language. *Acta Philosophica Fennica, 80*, 169–184.

Hilmy, S. S. (1987). *The later Wittgenstein*. Oxford: Blackwell.

Hintikka, J. (1976). *Essays on Wittgenstein in Honour of G.H. von Wright*. Amsterdam: North-Holland Pub.

Hintikka, J. (2005). Did Wittgenstein follow the rules? (or was he guided by them?). *Schriftenreihe-Wittgenstein Gesellschaft, 34*, 49–62.

Hintikka, M. B., & Hintikka, J. (1986). *Investigating Wittgenstein*. Oxford: Blackwell.

Hodges, M. P. (1990). *Transcendence and Wittgenstein's Tractatus*. Philadelphia: Temple University Press.

Hodges, A. (2008a). Alan Turing and the Turing test. In R. Epstein, G. Roberts, & G. Beber (Eds.), *Parsing the Turing test: Philosophical and methodological issues in the quest for the thinking computer*. Dordrecht: Springer. See also: http://www.turing.org.uk/publications/testbook.html.

Hodges, A. (2008b). Alan Turing. In T. Gowers, J. Barrow-Green, & I. Leader (Eds.), *The Princeton companion to mathematics* (pp. 821–822). Princeton/Oxford: Princeton University Press.

Hoerl, C., & McCormack, T. (2001). *Time and memory: Issues in philosophy and psychology*. Oxford: Clarendon Press.

Hookway, C. (2000). *Truth, rationality, and pragmatism: Themes from Peirce*. Oxford: Clarendon Press.

Horwich, P. (2012). *The metaphilosophy of Wittgenstein*. Oxford: Oxford University Press.

Howes, B. (2007). Rethinking' the preface of the Tractatus. *Philosophical Investigations, 30*(1), 3–24.

Hoyt, C. (2007). Wittgenstein and Religious Dogma. *International Journal for Philosophy of Religion, 61*(1), 39–49.

Hutto, D. D. (2003). *Wittgenstein and the end of philosophy: Neither theory nor therapy*. Basingstoke: Palgrave Macmillan.

Hyland, J. M. E. (2015). Classical lambda calculus in modern dress. *Mathematical Structures in Computer Science, 27*(05), 1–20. https://doi.org/10.1017/S0960129515000377.

Hyland, J. M. E., & Ong, C.-H. L. (2000, December). On full abstraction for CFP. *Journal of Information and Computation, 163*(2), 285–408.

Hyman, J. (2001a). The Gospel according to Wittgenstein. In R. Arrington & M. Addis (Eds.), *Wittgenstein and philosophy of religion: Self and personal identity in the eighteenth century* (pp. 1–11). London: Routledge.

Hyman, J. (2001b). The Urn and the Chamber Pot. In R. Allen & M. Turvey (Eds.), *Wittgenstein, theory and the arts* (pp. 137–153). London: Routledge.

I

Ingham, A. E. (1932). The distribution of prime numbers. *Cambridge Tracts in Mathematics, 30*, 8.

Ivić, A. (2013). *The theory of Hardy's Z-Function*. Cambridge: Cambridge University Press.

J

James, W. (1890). *The principles of psychology*. Cambridge: Harvard University Press.

Janik, A., & Toulmin, S. (1996). *Wittgenstein's Vienna*. Chicago: Elephant Paperbacks.

Jankélévitch, V., & Abbate, C. (2003). *Music and the ineffable*. Princeton: Princeton University Press.

Jeshion, R. (2016). Frege on sense identity, basic law V, and analysis. *Philosophia Mathematica, 24* (1), 9–29.

Johnson, J. (2004). Music and the ineffable. *Music and Letters, 85*(4), 643–647.

K

Kaufman, D. A. (2005). Between reason and common sense. *Philosophical Investigations, 28*(2), 134–158.

Keefe, R. (1995). Contingent identity and vague identity. *Analysis, 55*(3), 171.

Khalfa, J. (2003). Pascal's theory of knowledge. In N. Hammond (Ed.), *The Cambridge companion to Pascal* (pp. 122–143). Cambridge: Cambridge University Press.

Kieran, M., & Lopes, D. (2003). *Imagination, philosophy, and the arts*. London: Routledge.

King, J. (1984). *Recollections of Wittgenstein*. In R. Rhees (Ed.), *Recollections of Wittgenstein*. Oxford: Oxford University Press.

Kivy, P. (2007). *Music, language, and cognition: And other essays in the aesthetics of music*. Oxford: Clarendon Press.

Klagge, J. C. (2001). *Wittgenstein: Biography and philosophy*. Cambridge: Cambridge University Press.

Klagge, J. C. (2011). *Wittgenstein in exile*. Boston: MIT Press.

Klagge, J., & Nordmann, A. (Eds.). (1993a). *Philosophical occasions: 1912–1951*. Indianapolis & Cambridge: Hackett.

Klagge, J. C., & Nordmann, A. (Eds.). (1993b). *Ludwig Wittgenstein: Public and private occasions*. Lanham: Rowman and Littlefield.

Knuth, D. E. (1974). *Surreal numbers*. Reading: Addison-Wesley.

Kober, M. (2006). *Deepening our understanding of Wittgenstein*. Amsterdam/New York: Rodopi.

Koethe, J. (1996). *The continuity of Wittgenstein's thought*. Ithaca: Cornell Univ. Press.

Koethe, J. (2003). On the "Resolute" reading of the Tractatus. *Philosophical Investigations, 26*(3), 187–204.

Kölbel, M., & Weiss, B. (2004). *Wittgenstein's lasting significance*. London: Routledge.

Kolmogorov, A. N. (2003). Колмогоров. Юбилейное издание в 3-х книгах. Кн. 3, Звуков сердца тихое эхо...Из дневников [Soft echo of the heartbeat... From the Diaries]. Moscow: Fizmatlit.

Kreiger, H. (2017). *The creativity and structure of pure mathematics: An interview with Holly Krieger by Interview with K. O. Matthews*, in https://medium.com/q-e-d/the-creativity-and-structure-of-pure-mathematics-an-interview-with-holly-krieger-39bc4b203a9c

Kreisel, G. (1987). Gödel's excursions into intuitionistic logic. In P. Weingartner & L. Schmetterer (Eds.), *Gödel remembered: Salzburg, 10–12 July 1983* (pp. 65–186). Naples: Bibliopolis.

Kremer, M. (1997). Contextualism and Holism in the Early Wittgenstein: From ProtoTractatus to Tractatus. *Philosophical Topics, 25*(2), 87–120.

Kremer, M. (2001). The purpose of Tractarian nonsense. *Nous, 35*(1), 39–73.

Kripke, S. A. (1982). *Wittgenstein on rules and private language: An elementary exposition*. Cambridge, MA: Harvard University Press.

Kukla, A. (2005). *Ineffability and philosophy*. London: Routledge.

Kurz, H.-D., & Salvadori, N. (2003). Sraffa and the mathematicians: Frank Ramsey and Alister Watson. In H.-D. Kurz & N. Salvadori (Eds.), *Classical economics and modern theory: studies in long-period analysis* (pp. 187–217). London: Routledge.

Kuusela, O. (2005). From metaphysics and philosophical theses to grammar: Wittgenstein's Turn. *Philosophical Investigations, 28*(2), 95–133.

Kuusela, O. (2006a). Do the concepts of grammar and use in Wittgenstein articulate a theory of language or meaning? *Philosophical Investigations, 29*(4), 309–341.

Kuusela, O. (2006b). 'Nonsense and clarification in the Tractatus – Resolute and ineffability readings and the *Tractatus*' failure. *Acta Philosophica Fennica, 80*, 35–66.

Kuusela, O., & McGinn, M. (Eds.). (2011). *The Oxford handbook of Wittgenstein*. Oxford: Oxford University Press.

L

Lafforgue, V. (2018, March). *Shtukas for reductive groups and Langlands correspondence for function fields*. In eprint arXiv:1803.0379.

Langlands, R. P. (2007). *Reflexions on receiving the Shaw Prize*. Hong Kong: Shaw Foundation.

Langlands, R. P. (2016). *The work of Robert Langlands*. Eds. (Founder) Bill Casselman, et al. Princeton: Princeton Institute for Advanced Study on line collection at: http://publications.ias.edu/rpl/

LaRocca, D. (2007). Changing the Subject: The Auto/biographical as the philosophical in Wittgenstein. *Epoche, 12*, 169–184.

Letourneur, Jérôme (2010 [2011–13]). Keim and Bereitschaft: Readiness and embodiment. In E. Nemeth, R. Heinrich, A. W. Pichler (Eds.), *ALWS archives: A selection of papers from the International Wittgenstein Symposia in Kirchberg am Wechsel* (pp. 179–210). Bergen.

Levinson, J. (2006). *Contemplating art: Essays in aesthetics*. Oxford: Clarendon Press.

Lewis, P. (2004). *Wittgenstein, aesthetics, and philosophy*. Aldershot: Ashgate.

Littlewood, J. E. (1926). *Notes on functions* (2nd ed.). Cambridge: Heffers.

Littlewood, J. E. (1944). *Lectures on the theory of functions*. Oxford: Oxford University Press. [In Littlewood's own copy presented to the Trinity College Library, a note on the cover mentions that the first chapter was printed by C.F. Hodgson in 1931. This latter relates to lectures given in that year and so 1931 is referred to in the above footnote.]

Littlewood, J. E. (1986). *Miscellany*. (2nd ed., B. Bollobás, Ed.). Cambridge: Cambridge University Press.

Livadas, S. (2020). Why is cantor's absolute inherently inaccessible? *Axiomathes*. https://doi.org/10.1007/s10516-020-09474-y

Loner, J. D. (2018a). Alice Ambrose and life unfettered by philosophy in Wittgenstein's Cambridge, 1930–1936. https://jhiblog.org/2016/02/01/alice-ambrose-and-life-unfettered-by-philosophy-in-wittgensteins-cambridge/

Loner, J.D. (2018b). *Whewell's Court revisited: Ludwig Wittgenstein, Francis Skinner and Elite Male Space in Interwar Cambridge* [*Forthcoming*].

Luckhardt, C. G. (1979). *Wittgenstein, sources and perspectives*. Ithaca: Cornell University Press.

Lugg, A. (2003). Wittgenstein's Tractatus: True thoughts and nonsensical propositions. *Philosophical Investigations, 26*(4), 332–347.

Luntley, M. (2003). *Wittgenstein: Meaning and judgement*. Oxford: Blackwell.

Luntley, M. (2015). *Wittgenstein: Opening investigations* (pp. 148–150). Oxford: Wiley Blackwell.

M

MacDonald, M. (1934–35). *MacDonald's Lectures Notes of Wittgenstein's 1934–35 lectures on logic* [Unpublished manuscript, Rush Rhees' Papers, Trinity College Cambridge.]

MacIver, A. (circa 1930–1940). *Arthur MacIver's diary Cambridge (October 1929–March 1930)*. Unpublished manuscript, Trinity College Cambridge [Forthcoming, B. McGuinness (Ed.)].

Mackay, D. R. H. (2012). *Memories from the Cambridge firm of D. Mackay*. (M. Page, Ed.) Cambridge: Mackays.

Magee, B. (2005). Philosophy's neglect of the arts. *Philosophy, 313*, 413–422.

Maisky, I. (2015). The Maisky Diaries: 1932–1943. (Ed G. Gorodetsky; trans. T. Sorokina & O. Ready (New Haven: Yale).

Malcolm, N. (1958). *Ludwig Wittgenstein, a memoir*. London: Oxford University Press.

Malcolm, N. (1986). *Nothing is hidden: Wittgenstein's criticism of his early thought*. Oxford: Blackwell.

Mamminen, J. (2011). Waismann's testimony of Wittgenstein's fresh starts in 1931–35. *Vienna Circle Institute Yearbook, 15*, 243–265.

Marconi, D. (1997). Book reviews – Beyond the limits of thought. *The Philosophical Review, 106* (4), 620.

Margolis, J. (1995). *Interpretation radical but not unruly: The new puzzle of the arts and history*. Berkeley: University of California Press.

Marion, M. (1995). Wittgenstein and finitism. *Synthese, 105*(2), 141–176. Oxford: Clarendon Press.

Marion, M. (1998). *Wittgenstein, finitism, and the foundations of mathematics*. Oxford: Clarendon Press.

Marion, M. (2003). Wittgenstein and Brouwer. *Synthese, 137*(1/2), 103–127.

Marion, M. (2005). Sraffa and Wittgenstein: physicalism and constructivism. *Review of Political Economy, 17*, 381–406.

Marion, M. (2006). Hintikka on Wittgenstein: From language-games to game semantics. *Acta Philosophica Fennica, 78*, 255–274.

Marion, M. (2011). Wittgenstein on surveyability of proofs. In O. Kuusela & M. McGinn (Eds.), *The Oxford handbook of Wittgenstein* (pp. 138–161). Oxford: Oxford University Press.

Marion, M & Okada, M. (2018). Wittgenstein, Goodstein and the origin of the uniqueness role for primitive recursive arithmetics. In Stern, D. (Ed.), *Wittgenstein in the 1930s: Between the tractatus and the investigations* (pp. 253–271). Cambridge: Cambridge University Press.

Marley, D. L. A. (1935). *Birobidjan as I saw it*. New York: ICOR.

Maxwell, J. C. (1873). *A treatise on electricity and magnetism* (Vol. 1 & 2, 2nd ed.). Cambridge: Cambridge University Press. (2011).

Maxwell, J. C. (2010). Spherical harmonics. In *A treatise on electricity and magnetism* (pp. 157–180). (Cambridge Library Collection - Physical Sciences)). Cambridge: Cambridge University Press.

McFee, G. (1999). Wittgenstein on art and aspects. *Philosophical Investigations, 22*(3), 262–284.

McGilchrist, I. (2009). *The master and his emissary: The divided brain and the making of the Western World*. New Haven/London: Yale University Press.

McGinn, M. (2002). *Wittgenstein and the Philosophical Investigations*. London: Routledge.

McGuinness, B. F. (1988). *Wittgenstein: A life: Young Ludwig, 1889–1921* (Vol. 1). London: Duckworth. reprint: Oxford: Clarendon Press, 2005.

McGuinness, B. F. (2002). *Approaches to Wittgenstein: Collected papers*. London: Routledge.

McGuinness, B. F. (2004). What was Wittgenstein's intellectual project? In A. Coliva & E. Picardi (Eds.), *Wittgenstein today*.

McGuinness, B. F. (2005a). G. H. von Wright as Heir to Wittgenstein. *Acta Philosophica Fennica, 77*, 79–88.

McGuinness, B. F. (2005b). *Young Ludwig: Wittgenstein's life, 1889–1921*. Oxford: Clarendon Press.

McGuinness, B. F. (2006). Wittgenstein: Philosophy and literature. *Publications- Austrian Ludwig Wittgenstein Society New Series, 2*, 367–381.

McGuinness, B. F. (2007a, September). In praise of nonsense. In *Paper presented at the Reading Wittgenstein: Conference in honour of Hidé Ishiguro*.

McGuinness, B. F. (2007b). What Wittgenstein got from Sraffa. In G. Chiodi & L. Ditta (Eds.), *Sraffa or an alternative economics*. Basingstoke: Palgrave Macmillan.

McGuinness, B. F. (Ed.). (2008). *Wittgenstein in Cambridge: Letters and documents 1911–1951*. Oxford: Blackwell.

McGuinness, B. F. (2012). Two cheers for the 'new' Wittgenstein? In J. L. Zalabardo (Ed.), *Wittgenstein's early philosophy* (pp. 260–272). Oxford: Oxford University Press.

McGuinness, B. (2016). Arthur MacIver's Diary: Cambridge (October 1929 – March 1930). *Wittgenstein-Studien, 7*. https://doi.org/10.1515/witt-2016-0113.

McGuinness, B. F., Ascher, M.C., & Pfersmann, O. (1996). *Wittgenstein Familienbriefe*: *Schriftenreihe-Wittgenstein Gesellschaft, 23*, ALL.

McHale, M.E. (1966). *Ludwig Wittgenstein: A survey of source material for a philosophical biography*. Washington, D.C. MA thesis.

McHale, M.E. (1979). *The notion of the mystical in Wittgenstein's Tractatus*. Catholic University of America. Ph.D thesis.

McManus, D. (2004). *Wittgenstein and scepticism*. London: Routledge.

Medina, J. (2002). *The unity of Wittgenstein's philosophy: Necessity, intelligibility, and normativity*. Albany: State University of New York Press.

Medina, J. (2003a). On Being '"Other-Minded": Wittgenstein, Davidson, and logical aliens'. *International Philosophical Quarterly, 43*(172), 463–476.

Medina, J. (2003b). Wittgenstein and Nonsense: Psychologism, Kantianism, and the Habitus. *International Journal of Philosophical Studies, 11*(3), 293–318.

Menger, K. (Ed.). (1931–37). *Ergebnisse eines mathematischen Kolloquiums*. Berlin: B.G. Teubner.

Menger, K. (Ed.) (1939–48). *Reports of a Mathematical Colloquium*, 1-8. Stuttgart: B. G. Teubner.

Menger, K. (1994). *Reminiscences of the Vienna Circle and the Mathematical Colloquium* (L. Golland, B. McGuinness & A. Sklar, Eds.). Dordrecht: Kluwer.

Miell, D., MacDonald, R. A. R., & Hargreaves, D. J. (2005). *Musical communication*. Oxford: Oxford University Press.

Miller, A. I. (2000). *Insights of genius: Imagery and creativity in science and art*. Cambridge, Mass.: MIT Press.

Monk, R. (1990). *Ludwig Wittgenstein: The duty of genius*. London: Vintage.

Monk, R. (1995). 'Full-blooded Bolshevism: Wittgenstein's philosophy of mathematics. *Wittgenstein-Studien, 2*(1).

Monk, R. (1997). *Bertrand Russell: The spirit of solitude*. London: Cape.

Monk, R. (2000). *Bertrand Russell 1921–70: The ghost of madness*. London: Cape.

Monk, R. (2007). Bourgeois, Bolshevist or Anarchist? The reception of Wittgenstein's philosophy of mathematics. In G. Kahane et al. (Eds.), *Wittgenstein and his interpreters: Essays in memory of Gordon Baker* (pp. 269–294). Oxford: Wiley-Blackwell.

Moore, G. E. (1959). *Philosophical papers*. London: Allen and Unwin.

Moore, A. W. (1997). *Points of view*. Oxford: Clarendon Press.

Moran, J. (1972). Wittgenstein and Russia. New Left Review 1.73 May-June 1972.

Morineau, T. (2012). Hypercube algebra: A diagrammatic and sentential notation to support inferences in logic. In *European conference on cognitive ergonomics*. Edinburgh, 28–31 August 2012.

Mounce, H. O. (1981). *Wittgenstein's Tractatus: An introduction*. Chicago: University of Chicago Press.

Moyal-Sharrock, D. (2004). *The third Wittgenstein: The post-investigations works*. Aldershot: Ashgate.

Mühlholzer, F. (2005). "A mathematical proof must be surveyable": What Wittgenstein meant by this and what it implies. *Grazer Philosophische Studien, 71*, 57–86.

Mühlhölzer, F. (2008). Wittgenstein und der Formalismus. In M. Kroß (Ed.), *»Ein Netz von Normen«: Wittgenstein und die Mathematik* (pp. 107–148). Berlin: Parerga.

Mühlhölzer, F. (2010). *Bemerkungen über die Grundlagen der Mathematik Ein Kommentar des Teils III von Wittgensteins Bemerkungen über die Grundlagen der Mathematik.* Vittorio Klostermann: Frankfurt am Main.

Mulhall, S. (2007). *Wittgenstein's private language: Grammar, nonsense, and imagination in philosophical investigations, sections 243–315*. Oxford: Clarendon Press.

Munz, V. A. (2013). *Ludwig Wittgenstein and Yorick Smithies. A hitherto unknown relationship*. See online: http://wab.uib.no/agora/tools/alws/collection-1-issue-1-article-48.annotate

N

Newman, M. H. A. (1923). *The foundations of mathematics from the standpoint of physics*. Unpublished Fellowship Dissertation (August 1923), St John's College, Cambridge; held in St John's College Library.

Newman, M. H. A. (1926). *Foundations of mathematics from the standpoint of physics* (Fellowship dissertation submission). Cambridge: St John's College.

Newton, J. J. M. (2007). *The Langlands correspondence for* GL_2. Cambridge, Ph.D. thesis.

O

O'Mahony, N. (2013). Russian matters for Wittgenstein. In J. Padilla Galvez and M. Gaffal (Eds.), *Doubtful certainties: Language-games, forms of life, relativism* (pp. 149–180). Heusenstamm: Ontos, & Walter de Gruyter, Series: Aporia 7.

Oakes, M., & Pichler, A. (2013). Computational stylometry of Wittgenstein's "Diktat für Schlick". *Bergen Language and Linguistics Studies, 3*(1), 221–239.

Ostrow, M. B. (2002). Wittgenstein and the Liberating Word. In E. H. Reck (Ed.), *From Frege to Wittgenstein: Perspectives on early analytic philosophy* (pp. 353–373). Oxford: Oxford University Press.

Overgaard, S. (2004). Exposing the conjuring trick: Wittgenstein on subjectivity. *Phenomenology and Cognitive Sciences', 3*(3), 263–286.

P

Padilla-Galvez, J., & Gaffal, M. (Eds.). (2013). *Doubtful certainties*. Heusenstamm: Ontos.

Paris, J. B., & Pathmanathan, N. (2006). A note on Priest's finite inconsistent arithmetics. *Journal of Philosophical Logic, 35*(5), 529–537.

Pascal, B. (1995). *Pensées (A. Krailsheimer, Trans. & Introduction)*. London: Penguin.

Pascal, B. (2004). Provinciales, Pensées et opuscules divers. (Eds. P. Sellier & G. Ferryrolles.) Paris: La Pochothèque.

Pascal, B. (2006). Pensées. 2006; Introduction by T. S. Eliot. (Original Ed.: New York: Dutton, 1958; 2006 Edition: John Hagerson, L. N. Yaddanapudi, J. Sutherland et al. Gutenberg EBook #18269: https://www.gutenberg.org/files/18269/18269-h/18269-h.htm).

Pascal, F. (1984). Wittgenstein: A personal memoir. In R. Rhees (Ed.) (1984), *Recollections of Wittgenstein*. Oxford: Oxford University Press.

Paul, D. (2007). *Wittgenstein's Progress 1929–1951*. Bergen: Wittgenstein Archives at the University of Bergen.

Paul, G. A., Smith, H. M., & Murray, A. R. M. (1936). What can philosophy determine? In *Proceedings of the Aristotelian society, Supplementary Volume* 15. Symposium: Is there a problem about sense-data? (pp. 61–102).

Pears, D. F. (1971). *Ludwig Wittgenstein*. Glasgow: Collins.

Pears, D. F. (1987). *The false prison: A study of the development of Wittgenstein's philosophy, Volume 1*. Oxford: Clarendon Press.

Pears, D. F. (1988). *The false prison: A study of the development of Wittgenstein's philosophy, Volume 2*. Oxford: Clarendon Press.

Pears, D. F. (1989, August). Wittgenstein's account of rule-following. In *Paper presented at the Wittgenstein conference in Tallahassee*, Florida.

Pears, D. F. (2006). *Paradox and platitude in Wittgenstein's philosophy*. Oxford: Clarendon Press.

Perloff, M. (1996). *Wittgenstein's Ladder: Poetic language and the strangeness of the ordinary*. Chicago: University of Chicago Press.

Perloff, M. (2004). "But isn't the same at least the same?" Wittgenstein and the question of poetic translatability. In J. Gibson & W. Huemer (Eds.), *The literary Wittgenstein* (pp. 34–54). London: Routledge.

Pevsner, N. (1999). *Hertfordshire. Series: The buildings of England* (B. Cherry, Ed.). Harmondsworth: Penguin.

Phillips, D. Z. (1991). Religion in Wittgenstein's mirror. In A. Phillips Griffiths (Ed.), *Wittgenstein centenary essays*. Cambridge: Cambridge University Press.

Phillips, D. Z. (1993). *Wittgenstein and religion*. New York: St. Martin's Press.

Pichler, A. (1997). *Wittgensteins Philosophische Untersuchungen*. In Working Papers from the University of Bergen (Vol. 14).

Pichler, A. (2010). Towards the New Bergen electronic edition. In N. Venturinho (Ed.), *Wittgenstein after his Nachlass* (History of Analytic Philosophy Series) (pp. 157–172). Houndmills: Palgrave Macmillan.

Pichler, A. (2013). The philosophical investigations and syncretistic writing. In N. Venturinha (Ed.), *The textual genesis of Wittgenstein's philosophical investigations* (pp. 65–80). New York/London: Routledge.

Pichler, A. (2016). Review: Engelmann (2013), *Wittgenstein's philosophical development*. *Journal for the History of Analytical Philosophy, 4*(3), 1–9.

Pichler, A. (2018a). Ludwig Wittgenstein: A report of two dreams from October 1942 (Ms-126, 21–26). *Nordic Wittgenstein Review* [S.l.], *7*(1), 101–107. https://www.nordicwittgensteinreview.com/article/view/3499

Pichler, A. (2018b). Wittgenstein on understanding: Language, calculus, and practice. In D. Stern (Ed.), *Wittgenstein in the 1930s: Between the Tractatus and the investigations* (pp. 45–60). Cambridge: Cambridge University Press.

Pichler, A. [forthcoming]. *Two voices in the middle Wittgenstein*.

Pichler, A., & Säätelä, S. (2006). *Wittgenstein: The philosopher and his works*. Frankfurt: Wittgenstein Archives at the University of Bergen.

Picon, F. J. (2016). *Bakhtin and Nabokov: The dialogue that never was*. PhD. Dissertation, Columbia University, New York. https://doi.org/10.7916/D8SN08J6

Pierce, R. B. (2003). Defining "Poetry". *Philosophy and Literature, 27*(1), 151–163.

Pihlstrom, S. (2006). Shared language, transcendental listeners, and the problem of limits. *Acta Philosophica Fennica, 80*, 185–222.

Pinch, T., & Swedberg, R. (2012). Wittgenstein's visit to Ithaca in 1949: on the importance of details. *Distinktion: Scandinavian Journal of Social Theory*, 1–28.

Pitcher, G. (1964). *The philosophy of Wittgenstein*. Englewood Cliffs: Prentice-Hall.

Poincaré, H. (1902). *La science et l'hypothèse*. Paris: Flammarion.

Poincaré, H. (1905). *Science and hypothesis* (G. B. Halsted, Trans; Preface by Poincaré, H). New York: Science Press.

Pólya, G. (1940). Sur les types des propositions composées. *Journal of Symbolic Logic, 5*, 98–103.

Potter, M. (2011). Wittgenstein on mathematics. In O. Kuusela & M. McGinn (Eds.), *The Oxford handbook of Wittgenstein* (pp. 122–137). Oxford: Oxford University Press.

Price, M., & Walmsley, M. (2013). R.L. Goodstein and the mathematical association. *Mathematical Gazette, 97*, 398–408.

Priest, G. (2002). *Beyond the limits of thought*. Oxford: Clarendon Press.

Priest, G. (2006). *In contradiction: A study of the transconsistent*. Oxford: Clarendon Press.

Proessel, D. (2005). Wittgenstein on scepticism and nonsense. *Philosophical Investigations, 28*(4), 324–345.

Pushkin, A. S. (1954). *А.С.Пушкин: Сочинение в трех томах. Том. I.* Moscow: ГИХЛ.

Q

Qi, X. (2006). Comprehending the incomprehensibility: The crossing of art and philosophy. *International Studies in Philosophy, 38*(1), 119–140.

R

Ramsey, F. P. (1990). *Ramsey: Philosophical papers* (D. H. Mellor, Eds.). Cambridge: Cambridge University Press.

Reck, E. H. (2002). *From Frege to Wittgenstein: Perspectives on early analytic philosophy*. Oxford: Oxford University Press.

Redpath, R. T. H. (1990). *Ludwig Wittgenstein: A student's memoir*. London: Duckworth.

Rée, J., Ayers, M., & Westoby, A. (1978). *Philosophy and its past*. Atlantic Highlands: Humanities Press.

Reid, L. (1998). Wittgenstein's Ladder: The Tractatus and nonsense. *Philosophical Investigations, 21*(2), 97–151.

Reid, L. (2008). Wittgenstein: The philosopher and his works. *Philosophical Investigations, 31*(2), 182–190. (A. Pichler and S. Säätelä, Eds.).

Rhees, R. (1965). Some developments in Wittgenstein's view of ethics. *Philosophical Review, 74*, 17–26.

Rhees, R. (1968). Wittgenstein's notes for lectures on "Private Experience" and "Sense Data". *The Philosophical Review, 77*(I, II & 3,4), 271–320.

Rhees, R. (Ed.). (1970). *Eine Philosophische Betrachtung. L. Wittgenstein, L. Schriften* (p. 5). Suhrkamp: Frankfurt.

Rhees, R. (1984). *Recollections of Wittgenstein: Hermine Wittgenstein* (F. Pascal, F. R. Leavis, J. King, & M. O'C. Drury, Eds.). Oxford: Oxford University Press.

Richter, D. (2004). *Wittgenstein at his word*. London: Continuum.

Ricoeur, P. (1981). *The Rule of Metaphor: Multi-disciplinary studies of the creation of meaning in language*. Toronto: University of Toronto Press.

Ridley, A. (2003). Against musical ontology. *Journal of Philosophy, 100*(4), 203–220.

Riemann, B. (1868). Über die Hypothesen, welche der Geometrie zu Grunde liegen [Riemann's *Habilitationsschrift*, 1854]. In *Abhandlungen der Königlichen Gesellschaft der Wissenschaften zu Göttingen, 13*, 254–269.

Roberts, S. (2007). *King of infinite space: Donald Coxeter: The man who saved geometry*. London: Profile.

Roberts, S. (2015). *Genius at play: The curious mind of John Horton Conway*. London/New York: Bloomsbury.

Roberts, S., Weiss, S., & Ivic, A. (2009). Obituary: Harold Scott Macdonald Coxeter. *Bulletin of the London Mathematical Society, 41*(5), 943–960.

Rodych, V. (2018). Wittgenstein's Philosophy of Mathematics. In E. N. Zalta (Ed.), *The Stanford encyclopedia of philosophy*. https://plato.stanford.edu/archives/spr2018/entries/wittgenstein-mathematics/

Rothhaupt, J., & Vossenkuhl, W. (Eds.). (2013). *Kulturen und Werte: Wittgensteins "Kringel-Buch" als Initialtext*. Berlin: De Gruyter.

Rudd, A. (2004). Logic and ethics as the limits of the world. In B. Stocker (Ed.), *Post-Analytic Tractatus* (pp. 47–58). Aldershot: Ashgate.

Russell, B. (1903). *The principles of mathematics*. Cambridge: Cambridge University Press.

Russell, B. (1912). *Problems of philosophy*. London: Henry Holt & Company.

Russell, B. (1956). *Logic and knowledge: Essays, 1901–1950*. London: G. Allen & Unwin.

S

Saccheri, G. G. (1733). *Euclides ab omni naevo vindicatus sive conatus geometricus quo stabiliuntur prima ipsa universae Geometriae principia*. Mediolani: P.A. Morgan.

Saccheri, G. G. (1986[1733]). *Euclides vindicatus*. 1st ed; Ed. and trans. from Latin by G. B. Halsted. Chicago: Open Court Publishing Company, 1920; 2nd ed.; trans. from German by F. Steinhardt. New York: Chelsea Publishing Co.

Sachs, R. K. (1962). Asymptotic symmetries in gravitational theory. *Physical Review, 128*, 2851–2864.

Salvadori, N., & Gerhke, C. (Eds.). (2011). *Keynes, Sraffa and the criticism of neoclassical theory*. London: Routledge.

Sass, L. A. (2001). Wittgenstein, Freud and the nature of psychoanalytic explanation. In R. Allen & M. Turvey (Eds.), *Wittgenstein, theory and the arts* (pp. 254–296). London: Routledge.

Sastry, S. (1934, October). On sums of powers. *Journal of the London Mathematical Society, 9*(4), 242–246.

Scanlon, T. (2003). Thickness and theory. *Journal of Philosophy C, 6*, 275–287.

Schalkwyk, D. (2004). Wittgenstein's 'Imperfect Garden': The ladders and labyrinths of philosophy as Dichtung. In J. Gibson & W. Huemer (Eds.), *The literary Wittgenstein* (pp. 55–74). London: Routledge.

Schroeder, S. (2001). The coded message model of literature. In R. Allen & M. Turvey (Eds.), *Wittgenstein, theory and the arts* (pp. 211–229). London: Routledge.

Schroeder, S. (2012). Schopenhauer's influence in Wittgenstein. In B. Vandenabeele (Ed.), *A companion to Schopenhauer*. WileyBlackwell: Maldon, Oxford.

Schroeder, S. (2020). *Wittgenstein on mathematics* (Series: *Wittgenstein's Thought & Legacy*). London: Routledge.

Schulte, J. (1993). *Experience and expression: Wittgenstein's philosophy of psychology*. Oxford: Clarendon Press.

Schulte, J. (2000). Wittgenstein's Quietism. *Schriftenreihe- Wittgenstein Gesellschaft, 28*, 37–50.

Schulte, J. (2001, August). *Wittgenstein's 'Method'*. In Paper presented at the international Wittgenstein symposium, Kirchberg am Wechsel, Austria.

Schulte, J. (2004). 'The life of the sign': Wittgenstein on reading a poem. In J. Gibson & W. Huemer (Eds.), *The literary Wittgenstein* (pp. 146–164). London: Routledge.

Schulte, J. (2005a). The Pneumatic conception of thought. *Grazer Philosophische Studien, 71*, 39–56.

Schulte, J. (2005b). What is a work by Wittgenstein. In A. Pichler & S. Säätelä (Eds.), *Wittgenstein: The philosopher and his works*. Bergen: The University of Bergen.

Schulte, J. (2005c). Wittgenstein – Our untimely contemporary. *Acta Philosophica Fennica, 77*, 59–78.

Schulte, J. (2007). Ways of reading Wittgenstein. In G. Kuhane, E. Kanterian, & O. Kuusela (Eds.), *Wittgenstein and his interpreters* (pp. 145–168). Oxford: Blackwell.

Schulte, J. (2011a). Waismann as spokesman for Wittgenstein. In B. F. McGuinness (Ed.), *Friedrich Waismann- Causality and logical positivism* (Vienna Circle Institute Yearbook) (Vol. 15, pp. 225–241).

Schulte, J. (2011b). Privacy. In O. Kuusela & M. McGinn (Eds.), *The Oxford handbook of Wittgenstein* (pp. 429–450). Oxford: Oxford University Press.

Schulte, J. (2013). The role of the big typescript in Wittgenstein's later philosophy. In N. Venturinha (Ed.), *The textual genesis of Wittgenstein's philosophical investigations* (pp. 81–89). New York/Abingdon: Routledge.

Schulte, J., & Sundholm, G. (1992). *Criss-crossing a philosophical landscape: Essays on Wittgensteinian themes dedicated to Brian McGuinness*. Amsterdam: Rodopi.

Scruton, R. (2004). Wittgenstein and the understanding of music. *British Journal of Aesthetics, 44* (1), 1–9.

Shanker, S. G. (1987, October). Wittgenstein versus Turing on the nature of Church's thesis. *Notre Dame Journal of Formal Logic, 28*(4), 615–650.

Shanker, S. (1998). *Wittgenstein's remarks on the foundations of AI*. London: Routledge.

Shaw, J. T. (1993). *Pushkin's poetics of the unexpected: The Nonrhymed lines in the lined poetry and the rhymed lines in the Nonrhymed Poetry*. Columbus: Slavica Publishers.

Shiryaev, A. N. (Ed.). (2003). *Kolmogorov Book 3, Звуков сердца тихое эхо...Из дневников (Soft echo of the heartbeat... From the Diaries)*. Moscow: Fizmatlit.

Shoesmith, D. J., & Smiley, T. J. (1980 [corrected version 2009]). *Multiple conclusion logic*. Cambridge: Cambridge University Press.

Sinha, A. (2006). A comment on Sen's 'Sraffa, Wittgenstein, and Gramsci. *Journal of Economic Behavior and Organization, 61*(3), 504–512.

Sloan, P. (1935). *Letter from Pat Sloan to Maurice Dobb*. Unpublished. Cambridge: Trinity College Library. Sraffa Collection. Ref. D3|**11**| 72.

Sluga, H. (2010). Our grammar lacks surveyability. In V. Munz, K. Puhl, & J. Wang (Eds.), *Language and world: Essays on the philosophy of Wittgenstein* (pp. 185–204). Frankfurt: Ontos Verlag.

Sluga, H. D., & Stern, D. G. (1996, 2017). *The Cambridge companion to Wittgenstein* (2nd ed.). Cambridge: Cambridge University Press.

Smith, J. (1998). An archivist's apology: The papers of Piero Sraffa at Trinity College Cambridge. *Pensiero Economico Itlaniano, 6*, 36–54.

Smith, J. (2011). Sraffa and Trinity. In N. Salvadori & C. Gerhke (Eds.), *Keynes, Sraffa and the criticism of neoclassical theory: Essays in honour of Heinz Kurz*. London: Routledge.

Smith, J. (2012). Circuitous processes, jigsaw puzzles, and indisputable results: Making the best use of the manuscripts of Sraffa's Production of Commodities by means of Commodities. *Cambridge Journal of Economics, 36*(6), 1291–1301.

Smith, J. (2013). Wittgenstein's Blue Book: Reading between the lines. In N. Venturhino (Ed.), *The textual genesis of Wittgenstein's philosophical investigations*. New York/London: Routledge.

Sorensen, R. A. (2001). *Vagueness and contradiction*. Oxford: Clarendon Press.

Standish, P. (1992). *Beyond the self: Wittgenstein, Heidegger and the limits of language*. Aldershot: Avebury.

Steiner, M. (2009). Empirical regularities in Wittgenstein's philosophy of mathematics. *Philosophia Mathematica, 17*(1), 1–34.

Stern, D. G. (1993). Notes for lectures on "Private Experience" and "Sense Data". In C. Klagge & A. Nordmann (Eds.), *Philosophical occasions* (pp. 200–288). Indianopolis/Cambridge: Hackett.

Stern, D. (Ed.). (2018). *Wittgenstein in the 1930s: Between the Tractatus and the Investigations*. Cambridge: Cambridge University Press.

Stern, D. G., Citron, G., & Rogers, B. (2013). Moore's notes on Wittgenstein's lectures, Cambridge 1930–1933: Text, context, and content. *Nordic Wittgenstein Review, 2*(1), 161–179.

Stern, D. G., Citron, G., & Rogers, B. (2016). *Wittgenstein: Lectures, Cambridge 1930–1933, From the notes of G. E. Moore*. Cambridge: Cambridge University Press.

Sterrett, S. G. (2006). *Wittgenstein flies a kite: A story of models of wings and models of the world*. New York: Pi Press.

Stiers, P. (2000). Meaning and the limit of the world in Wittgenstein's early and later philosophy. *Philosophical Investigations, 23*(3), 193–217.

Stocker, B. (2004). *Post-analytic Tractatus*. Aldershot: Ashgate.

Stokhof, M. J. B. (2002). *World and life as one: Ethics and ontology in Wittgenstein's early thought*. Stanford: Stanford University Press.

Stoutland, F. (2006). Analytic Philosophy and Metaphysics. *Acta Philosophica Fennica, 80*, 67–96.

Stroll, A. (2004). Wittgenstein's foundational metaphors. In D. Moyal Sharrock (Ed.), *The third Wittgenstein* (pp. 13–25). Hampshire: Ashgate.

Strominger, A. (2018). *Lectures on the infrared structure of gravity and gauge theory*. Princeton: Princeton University Press. [Cf. the redacted transcript of the lecture course gave rise to the 2018 book: https://arxiv.org/pdf/1703.05448.pdf. The lectures themselves are available: https://www.youtube.com/playlist?list=PLwLjkVy3evOazQ3FoRH-Sz8Eoxx2oriXL&disable_polymer=true].

Sullivan, P. M. (2004). 'The general propositional form is a variable' (Tractatus 4.53). *Mind, 113* (449), 43–56.

Summerfield, D. M. (1996). Fitting versus tracking. In H. Sluga & D. M. Stern (Eds.), *The Cambridge companion to Wittgenstein*. Cambridge: Cambridge University Press.

T

Tao, T. (2008). Countable and uncountable sets. In Gowers et al. (170–172). *The Princeton companion to mathematics*. Princeton/Oxford: OUP.

Tao, T. (2016). *254A, Lecture 5: Other topological recurrence results*. https://terrytao.wordpress.com/2008/01/21/254a-lecture-5-other-topological-recurrence-results/#more-243

Taylor, R. (2007). *Reciprocity laws and density theorems*. See: http://www.math.ias.edu/~rtaylor/shaw.pdf

Thorne, J. A. (2016). 'Elliptic curves over Q_∞ are modular'. [forthcoming] *Journal of the European Mathematical Society*.

Thornton, T. (1998). *Wittgenstein on thought and language: The philosophy of content*. Edinburgh: Edinburgh University Press.

Tilghman, B. R. (1991). *Wittgenstein, ethics and aesthetics: The view from eternity*. Albany: State University of New York Press.

Totaro, B. (2008). Algebraic topology. In T. Gowers, J. Barrow-Green, & I. M. Leader (Eds.) *The Princeton companion to mathematics* (pp. 383–396). Princeton/Oxford: Princeton University Press.

Turing, A. M. (1936). On computable numbers, with an application to the *Entscheidungsproblem. Proceedings of the London Mathematical Society,* Series 2. *42*, 230–265.

Turing, A. M. [1938] (2012). *Systems of logic based on Ordinals*. Dissertation presented to the faculty of Princeton University in candidacy for the degree of Doctor of Philosophy. Electronic version of Turing's Princeton PhD. Thesis, transcribed by A. B. Matos [Artificial Intelligence and Computer Science Laboratory Universidade do Porto, Portugal] (September 18, 2014): http://www.dcc.fc.up.pt/~acm/turing-phd.pdf. Cf. A. W. Appel (Ed.), *Alan Turing's system of logic: The princeton Thesis* (pp. 31–140). Princeton: Princeton University Press.

Turing, A. M. [1944/45] (2013). Reform of mathematical notation and phraseology. In B. S. Cooper, & J. van Leeuven (Eds.), *Alan Turing: His work and impact*. Amsterdam/Oxford/Waltham: Elsevier.

U

Uschanov, T. P. (2006). On ladder withdrawal symptoms and one way of dealing with them. *Acta Philosophica Fennica, 80*, 11–168.

V

Vandenabeele, B. (Ed.). (2011). *A companion to Schopenhauer* (pp. 21–22). Malden: WileyBlackwell.

Varadarajan, V. S. (2006). *Euler through time: A new look at old themes*. Providence: American Mathematical Society.

Vasalou, S. (2015). *Schopenhauer and the aesthetic standpoint*. Cambridge: Cambridge University Press.

Venturinha, N. (2004). Wittgenstein's way of working and the nature of experience. In C. M. Marek & M. E. Reicher (Eds.), *Experience and Analysis: Contributions of the Austrian Ludwig Wittgenstein Society* (Vol. XII, pp. 398–400).

Venturinha, N, (2012). Sraffa's notes on Wittgenstein's "Blue Book". In A. Pichler, S. Saatela, & Y. Neuman (Eds.), *Nordic Wittgenstein review* (Vol. 1).

Venturinha, N. (Ed.). (2013). *The textual genesis of Wittgenstein's Philosophical Investigations*. New York/Abingdon: Routledge.

von Wright, G. H. (1969). The Wittgenstein papers. *Philosophical Review, 78*, 483–503.

von Wright, G. H. (1982). *Wittgenstein* (pp. 35–62). Oxford: Basil Blackwell.

von Wright, G. H. (1992). The troubled history of part II of the Investigations. In J. Schulte & G. Sundholm (Eds.), *Criss-crossing a philosophical landscape: Essays on Wittgensteinian themes dedicated to Brian McGuinness* (pp. 181–192). Amsterdam: Rodopi.

von Wright, G. H. (1996). *Six essays in philosophical logic*. Philosophical Society of Finland: Helsinki.

W

Waismann, F. (1965). Notes on talks with Wittgenstein. *The Philosophical Review, 74*(1), 12.

Waismann, F. (1979). *Wittgenstein and the Vienna Circle: Conversations recorded by Friedrich Waismann* (B. McGuinness, Ed., T. J. Schulte & B. McGuiness, Trans.). New York: Barnes & Noble Books.

Wallgren, T. (2006). Overcoming overcoming: Wittgenstein, metaphysics, and progress. *Acta Philosophica Fennica, 80*, 97–130.

Waugh, A. (2008). *The house of Wittgenstein: A family at war*. London: Bloomsbury.

Wearing, C. (2006). Metaphor and what is said. *Mind and Language, 21*(3), 310–332.

Weber, D. (2019). A pedagogic reading of Wittgenstein's Life and later works. In P. Standish, & A. Skilbeck (Eds.), *Wittgenstein an education*. Wiley Publishing. https://onlinelibrary.wiley.com/doi/abs/10.1111/1467-9752.12390

Weber, D. (forthcoming). *Reading Wittgenstein in politics: Normativity, judgment, and political pedagogy*

Weiss, J. (2001). Illusions of sense in the Tractatus: Wittgenstein and imaginative understanding. *Philosophical Investigations, 24*(3), 228–245.

Welch, P. D. (2017). Obtaining Woodin's cardinals. In A. E. Caicedo et al. (Eds.) *W. H. Woodin: Foundations of mathematics. Logic at Harvard. Contemporary mathematics* (Vol. 690, pp. 161–176).

Weyl, H. (2012). In P. Pesic (Ed.) *Levels of infinity/selected writings on mathematics and philosophy*. New York: Dover.

Weyl, Hermann. (1925–7). *The current epistemological situation in mathematics* [trans. of "Die heutige Erkenntnislage in der Mathematik"]. In P. Mancosu (Ed.), *From Brouwer to Hilbert* (pp. 123–142). New York: Oxford University Press, 1988.

White, R. M. (1996). *The structure of metaphor: The way the language of metaphor works*. Oxford: Blackwell.

Williams, B. A. O. (1985). *Ethics and the limits of philosophy*. London: Fontana.

Williams, B. A. O. (2002). *Truth & truthfulness: An essay in genealogy*. Princeton: Princeton University Press.

Williams, M. (2004). Nonsense and cosmic exile: The Austere reading of the Tractatus. In M. Kölbel & B. Weiss (Eds.), *Wittgenstein's lasting significance* (pp. 6–31). London: Routledge.

Williams, J. N. (2006). 'Wittgenstein, Moorean absurdity and its disappearance from speech. *Synthese, 149*(1), 225–254.

Williams, R. (2014). *The edge of words*. London: Bloomsbury.

Williamson, T. (2004). Computational limits and epistemic logic. *Schriftenreihe-Wittgenstein Gesellschaft, 33*, 126–140.

Wilson, J. (1960). *Language and the pursuit of truth*. Cambridge: Cambridge University Press.

Winch, P. (2001). On Wittgenstein. *Philosophical Investigations, 24*(2), 89–184.

Winchester, I. (2000). Beyond the bounds of thought: Speculative philosophy and the last proposition of the Tractatus. *Interchange, 31*(2/3), 292–300.

Wisdom, J. (1933–1934). Logical constructions IV & V. *Mind, XLII* (165), *43–66* (166), 186–202.

Wisdom, J. (1934). *Problems of mind and matter*. Cambridge: Cambridge University Press.

Witten, E. (2010). Geometric Langlands duality and the equations of Nahm and Bogomolny. *Proceedings of the Royal Society of Edinburgh, 140A*, 857–895.

Witten, E. (2015). More on gauge theory and geometric Langlands. arXiv:1506.04293v1 [hep-th].

Wittgenstein, L. (*circa* 1933-4). *The Blue Book* [unpublished incomplete draft in the hand of Francis Skinner; held by M. Nedo].

Wittgenstein, L. (1935) [circa]. Bemerkungen zur philosophischen Grammatik (ÖNB, Cod.Ser. n.22.021). Bergen Nachlass Edition, metadata for MS 115 (www.wittgensteinsource.org).

Wittgenstein, L. (1935a). Large C-Series Notebook, MS 148.

Wittgenstein, L. (1935b). Large C-Series Notebook, MS 151.

Wittgenstein, L. (1935c). Large C-Series Notebook, MS 149.

Wittgenstein, L. (1936). Eine Philosophische Betrachtung [= Bergen Nachlass Edition, metadata for MS 115ii].

Wittgenstein, L. (1937). Ms.-119: XV. [Ed. B. Giertsen A/S Bok- & Papirhanel. Smaastransdaten – Bergen]. Bergen Nachlass Edition, metadata for Ms-119: XV (www.wittgensteinsource.org).

Wittgenstein, L. (1938a). MS. 120 4.1.1938.

Wittgenstein, L. (1938b–39 [1940?]). Taschennotizbuch MS 162a.

Wittgenstein, L. (1958). *The Blue and Brown Books: Preliminary studies for the 'Philosophical Investigations'* (R. Rhees, Ed.). Oxford: Blackwell.

Wittgenstein, L. (1965). A lecture on ethics. *The Philosophical Review, 74*(1), 3–12.

Wittgenstein, L. (1966). *Lectures and conversations on aesthetics, psychology, and religious belief* (C. Barrett, Ed.), Oxford: Blackwell.

Wittgenstein, L. (1969). *Briefe an Ludwig von Ficker* (G. H. von Wright & W. Methlagl, Trans.), Salzburg: O. Müller.

Wittgenstein, L. (1970). *Eine Philosophische Betrachtung*. In R. Rhees (Ed.), *Schriften 5* (pp. 117–237). Frankfurt: Suhrkamp.

Wittgenstein, L. (1971). *ProtoTractatus: An early version of Tractatus Logico-Philosophicus*. Ithaca: Cornell University Press.

Wittgenstein, L. (1973). *Letters to C. K. Ogden with comments on the English translation of the Tractatus Logico-Philosophicus* (C. K. Ogden, G. H. von Wright & F. P. Ramsey, Eds.). Oxford: Basil Blackwell.

Wittgenstein, L. (1974). *Tractatus Logico-Philosophicus* (D. Pears & B. McGuinness, Trans.). London: Routledge.

Wittgenstein, L. (1975a). *On Certainty* (G. E. M. Anscombe & G. H. von Wright, Eds., G. E. M. Anscombe, Trans.). Malden/Oxford: Blackwell.

Wittgenstein, L. (1975b). *Philosophical remarks* (R. Rhees, Ed., R. Hargreaves & R. White, Trans.). Oxford: Blackwell.

Wittgenstein, L. (1976). *Wittgenstein's lectures on the foundations of mathematics, Cambridge, 1939: From the notes of R.G. Bosanquet, Norman Malcolm, Rush Rhees and Yorick Smythies* (R. G. Bosanquet, Ed.). Hassocks: Harvester Press.

Wittgenstein, L. (1977a). *Remarks on colour* (G. E. M. Anscombe, Ed., L. L. McAlister & M. Schättle, Trans.). Oxford: Blackwell.

Wittgenstein, L. (1977b). *Vermischte Bemerkungen*. Frankfurt: Suhrkamp.

Wittgenstein, L. (1978a). *Philosophical grammar* (R. Rhees, Ed., A. Kenny, Trans.). Oxford: Blackwell.

Wittgenstein, L. (1978b). *Remarks on the foundations of mathematics* (G. H. von Wright, R. Rhees & G. E. M. Anscombe, Eds. & Trans.). Oxford: Blackwell.

Wittgenstein, L. (1980a). *Culture and value* (G. H. von Wright & H. Nyman, Eds., P. Winch, Trans.). Oxford: Blackwell.

Wittgenstein, L. (1980b). *Remarks on the philosophy of psychology: Volume I* (G. E. M. Anscombe & G. H. von Wright, Trans.). Oxford: Blackwell.

Wittgenstein, L. (1980c). *Remarks on the philosophy of psychology: Volume II* (G. E. M. Anscombe & G. H. von Wright, Trans.). Oxford: Blackwell.

Wittgenstein, L. (1980d). *Wittgenstein's lectures, Cambridge, 1930–1932: From the notes of John King and Desmond Lee* (J. King & H. D. P. Lee, Eds.). Oxford: Blackwell.

Wittgenstein, L. (1981). *Zettel* (G. E. M. Anscombe & G. H. von Wright, Trans.). Oxford: Blackwell.

Wittgenstein, L. (1982). *Last writings on the philosophy of psychology: Volume I: Preliminary studies for part II of philosophical investigations* (G. H. von Wright & H. Nyman, Eds., C. G. Luckhardt & M. A. E. Aue Trans.). Oxford: Blackwell.

Wittgenstein, L. (1988). *Wittgenstein's lectures on philosophical psychology 1946–47* (P. T. Geach, Trans.). Chicago: University of Chicago Press.

Wittgenstein, L. (1992). *Last writings on the philosophy of psychology: Volume II: The inner and the outer* (G. H. von Wright & H. Nyman, Trans.). Oxford: Blackwell.

Wittgenstein, L. (1993). *Remarks on Frazer's Golden bough, in Wittgenstein (1993a) L., Philosophical occasions: 1912–1951* (pp. 118–155). J. C. Klagge & A. Nordmann A.

Wittgenstein, L. (1993a). *Philosophical occasions, 1912–1951* (J. C. Klagge & A. Nordmann, Eds.). Indianapolis: Hackett Pub. Co.

Wittgenstein, L. (1994a). *Wiener Ausgabe: Band 1 – Philosophische Bemerkungen* (M. Nedo, Ed.). Wien/New York: Springer.

Wittgenstein, L. (1994b). *Wiener Ausgabe: Band 2 – Philosophische Betrachtungen, Philosophische Bemerkungen* (M. Nedo, Ed.). Wien/New York: Springer-Verlag. 3.

Wittgenstein, L. (1994c). *Wiener Ausgabe: Band 3 – Bemerkungen, Philosophische Bemerkungen* (M. Nedo, Ed.). Wien/New York: Springer.

Wittgenstein, L. (1995a). *Wiener Ausgabe: Band 4– Bemerkungen zur Philosophie, Bemerkungen zur philosophischen Grammatik* (M. Nedo, Ed.). Wien/New York: Springer.

Wittgenstein, L. (1995b). *Ludwig Wittgenstein: Cambridge letters: Correspondence with Russell, Keynes, Moore, Ramsey and Sraffa* (B. McGuinness, Ed.). Oxford: Blackwell.

Wittgenstein, L. (1997). *Wiener Ausgabe: Konkordanz zu den Bänden 1–5* (M. Nedo, Ed.). Wien/New York: Springer.

Wittgenstein, L. (1998a). *Wiener Ausgabe: Register zu den Bänden 1–5* (M. Nedo, Ed.). Vienna: Springer.

Wittgenstein, L. (1998b). *Notebooks, 1914–1916* (G. H. von Wright & G. E. M. Anscombe, Trans.). Oxford: Blackwell.

Wittgenstein, L. (1998c). *'Notes on logic', Notebooks 1914–1916*. Oxford: Blackwell.

Wittgenstein, L. (1998d). *Wittgenstein's Nachlass: The Bergen electronic edition (BEE)*. Director: A. Pichler. Bergen: University of Bergen.

Wittgenstein, L. (2000a). 'Wiener Aufgabe Band 11': *'The Big Typescript'* (M. Nedo, Ed.). Vienna: Springer.

Wittgenstein, L. (2000b).*Wittgenstein's Nachlass: The Bergen electronic edition* (Ed. Wittgenstein Archives at the University of Bergen). Oxford: OUP.

Wittgenstein, L. (2000c). Wiener Ausgabe: Band 8.1 – Synopse der Manuskriptbände V bis X (M. Nedo, Ed.). Vienna: Springer.

Wittgenstein, L. (2003). *Ludwig Wittgenstein: Public and private occasions* (J. C. Klagge & A. Nordmann, Eds.). Lanham: Rowman & Littlefield Publishers.

Wittgenstein, L. (2005). *The Big Typescript, TS. 213* (C. G. Luckhardt & M. Aue, Eds., Trans.). Oxford: Blackwell.

Wittgenstein, L. (2008). *Wittgenstein in Cambridge: Letters and Documents, 1911–1951* (B. McGuinness, Ed.), Oxford: Blackwell.

Wittgenstein, L. (2009). *Philosophische Untersuchungen/Philosophical Investigations* (G. E. M. Anscombe, P. M. S. Hacker & J. Schulte, Trans.). (Revised 4th ed., P.M.S. Hacker and J. Schulte (Eds.). Chichester/Malden: Wiley-Blackwell.

Wittgenstein, L. (2015a). *Wittgenstein source Bergen Nachlass edition*. (Ed. The Wittgenstein Archives at the University of Bergen under the direction of Alois Pichler. Bergen: Wittgenstein Source (2009-) https://wittgensteinsource.org.

Wittgenstein, L., Rhees, R., & Citron, G. (Eds.). (2015b, January). Wittgenstein's philosophical conversations with Rush Rhees (1939–50): From the notes of Rush Rhees. *Mind, 124* (493), 1–72.

Wittgenstein, L. (2016). *Interactive Dynamic Presentation (IDP) of Ludwig Wittgenstein's philosophical Nachlass* (https://wittgensteinonline.no/). (Ed. The Wittgenstein Archives at the University of Bergen under the direction of Alois Pichler). Bergen.

Wolniewicz, B. (1979). A Wittgensteinian semantics for propositions. In C. Diamond, & J. Teichman (Eds.), *Intention and intentionality: Essays in honour of Professor G E M Anscombe* (pp. 165–78). Ithaca: Cornell University Press.

Woodin, W. H. (2011). The realm of the infinite. In M. Heller & H. Woodin (Eds.), *Infinity: New research frontiers* (pp. 89–118). Cambridge: Cambridge University Press.

Woodin, W. H. (2017), In A. E. Caicedo, et al. (Eds.), *W. H. Woodin: Foundations of mathematics. Logic at Harvard. Essays in honor of Hugh Woodin's 60th birthday*. Contemporary Mathematics. 690. American Mathematical Society, Cambridge, MA.

Woods, J. (2003). *Paradox and paraconsistency: Conflict resolution in the abstract sciences*. Cambridge: Cambridge University Press.

Wright, C. (1980). *Wittgenstein on the foundations of mathematics*. London: Duckworth.

Z

Zalabardo, J. L. (2015). *Representation and reality in Wittgenstein's Tractatus*. Oxford: Oxford University Press.

Author Index

A. Gibson, N. O'Mahony, *Ludwig Wittgenstein: Dictating Philosophy*,
https://doi.org/10.1007/978-3-030-36087-0

Subject Index

A. Gibson, N. O'Mahony, *Ludwig Wittgenstein: Dictating Philosophy*,
https://doi.org/10.1007/978-3-030-36087-0

O

P